White Noise
Distribution
Theory

White Noise Distribution Theory

Hui-Hsiung Kuo

CRC Press

Boca Raton   New York   London   Tokyo

Library of Congress Cataloging-in-Publication Data

Catalog regcord is available from the Library of Congress

Preface

White noise is generally regarded as the time derivative of a Brownian motion. It does not exist in the ordinary sense since almost all Brownian sample paths are nowhere differentiable. Informally it can be regarded as a stochastic process which is independent at different times and is identically distributed with zero mean and infinite variance. Due to this distinct feature, white noise is often used as an idealization of a random noise which is independent at different times and has large fluctuation. This leads to integrals with respect to white noise. Such integrals can not be defined as Riemann-Stieltjes integrals. In order to overcome this difficulty, K. Itô developed stochastic integration with respect to Brownian motion in 1944. The Itô theory of stochastic integration turns out to be one of the most fruitful branches of mathematics with far-reaching applications.

In the Itô theory of stochastic integration, the integrand needs to be nonanticipating. It is natural to ask whether it is possible to extend the stochastic integral wherein the integrand can be anticipating. In the 1970s such integrals were defined by M. Hitsuda, K. Itô, and A. V. Skorokhod. Since then there has been a growing interest in studying such integrals and anticipating stochastic integral equations. There are several approaches to define these integrals; for instance, the enlargement of filtrations, Malliavin calculus, and white noise analysis. An intuitive approach is to define such an integral directly with white noise as part of the integrand. This would require the building of a rigorous mathematical theory of white noise.

Motivated by P. Lévy's work on functional analysis, T. Hida introduced the theory of white noise in 1975. The idea is to realize nonlinear functions on a Hilbert space as functions of white noise. There are several advantages for this realization. First of all, white noise can be thought of as a coordinate system since it is an analogue of independent identically distributed random variables. This enables us to generalize finite dimensional results in an intuitive manner. Secondly, we can use the Wiener-Itô theorem to study functions of white noise. This allows us to define generalized functions of white noise in a very natural way. Thirdly, since Brownian motion is an integral of white noise, the use of white noise provides an intrinsic method to define stochastic integrals without the nonanticipating assumption.

During the last two decades the theory of white noise has evolved into an infinite dimensional distribution theory. Its space $(\mathcal{E})$ of test functions is the infinite dimensional analogue of the Schwartz space $\mathcal{S}(\mathbb{R}^r)$ on the finite dimensional space $\mathbb{R}^r$. The dual space $(\mathcal{E})^*$ of $(\mathcal{E})$ is the infinite dimensional analogue of the space $\mathcal{S}'(\mathbb{R}^r)$ of tempered distributions on $\mathbb{R}^r$. It contains generalized functions such as the white noise $\dot{B}(t)$ for each fixed t. Thus $\dot{B}(t)$ is a rigorous mathematical object and can be used for stochastic integration. White noise distribution theory has now been applied to stochastic integration, stochastic partial differential equations, stochastic variational equations, infinite dimensional harmonic analysis, Dirichlet forms, quantum field theory, Feynman integrals, infinite dimensional rotation groups, and quantum probability.

There have been two books on white noise distribution theory:

1. T. Hida, H.-H. Kuo, J. Potthoff, and L. Streit: *White Noise: An Infinite Dimensional Calculus.* Kluwer Academic Publishers, 1993.

2. N. Obata: *White Noise Calculus and Fock Space.* Lecture Notes in Math., Vol. 1577, Springer-Verlag, 1994.

The space $(\mathcal{E})^*$ of generalized functions used in these books is somewhat limited. By the Potthoff-Streit characterization theorem, the S-transform $S\Phi$ of a generalized function Φ in $(\mathcal{E})^*$ must satisfy the growth condition

$$|S\Phi(\xi)| \leq K \exp\left[a|\xi|_p^2\right].$$

Thus the function $F(\xi) = \exp\left[\langle\xi,\xi\rangle^2\right]$ is not the S-transform of a generalized function in $(\mathcal{E})^*$. However this function is within the scope of Lévy's functional analysis. In fact, $F(\xi) = \exp\left[t\langle\xi,\xi\rangle^2\right]$ is the S-transform of a solution of the heat equation associated with the operator $(\Delta_G^*)^2$. Here Δ_G^* is the adjoint of the Gross Laplacian Δ_G. This shows that there is a real need to find a larger space of generalized functions than $(\mathcal{E})^*$.

Recently, Yu. G. Kondratiev and L. Streit have introduced an increasing family of spaces $(\mathcal{E})_\beta^*$, $0 \leq \beta < 1$, of generalized functions with $(\mathcal{E})_0^* = (\mathcal{E})^*$. In the characterization theorem for Φ in $(\mathcal{E})_\beta^*$, the growth condition is

$$|S\Phi(\xi)| \leq K \exp\left[a|\xi|_p^{\frac{2}{1-\beta}}\right].$$

For example, the above function $F(\xi) = \exp\left[\langle\xi,\xi\rangle^2\right]$ is the S-transform of a generalized function in $(\mathcal{E})_{1/2}^*$.

There are three objectives for writing this book. The first one is to carry out white noise distribution theory for the space $(\mathcal{E})_\beta^*$ of generalized functions and the space $(\mathcal{E})_\beta$ of test functions for each $0 \leq \beta < 1$. As pointed out above, the space $(\mathcal{E})^*$ is not large enough for the study of heat equation associated with the operator $(\Delta_G^*)^2$. In general, $F(\xi) = \exp\left[t\langle\xi,\xi\rangle^k\right]$ is the S-transform of a solution of the heat equation associated with the operator $(\Delta_G^*)^k$. The solution is a generalized function in the space

$(\mathcal{E})^*_{(k-1)/k}$. There are other motivations for the spaces $(\mathcal{E})^*_\beta$, $0 \le \beta < 1$; for instance, a generalization of Donsker's delta function in Example 7.7 and the white noise integral equation in Example 13.46. An important example of generalized functions in the space $(\mathcal{E})^*_\beta$ is given by the grey noise measure ν_λ (see Example 8.5) which has the characteristic function

$$\int_{\mathcal{E}'} e^{i\langle x,\xi\rangle}\, d\nu_\lambda(x) = L_\lambda(|\xi|_0^2),$$

where L_λ, $0 < \lambda \le 1$, is the Mittag-Leffler function given by

$$L_\lambda(t) = \sum_{n=0}^{\infty} \frac{(-t)^n}{\Gamma(1 + \lambda n)}.$$

Here Γ is the Gamma function. The grey noise measure induces a generalized function in the space $(\mathcal{E})^*_{1-\lambda}$.

The second objective is to discuss some of the recent progress on Fourier transform, Laplacian operators, and white noise integration. The Fourier-Gauss transforms acting on the space $(\mathcal{E})_\beta$ of test functions are studied in detail. The probabilistic interpretation of the Lévy Laplacian is discussed. The Hitsuda-Skorokhod integral $\int_0^t \partial_s^* \varphi(s)\,ds$ is defined as a random variable, not as a generalized function. The S-transform is used to solve some anticipating stochastic integral equations. For instance, consider the anticipating stochastic integral equation in Example 13.30

$$X(t) = \mathrm{sgn}\,(B(1)) + \int_0^t \partial_s^* X(s)\, ds,$$

where the anticipation comes from the initial condition. The S-transform can be used to derive the solution

$$X(t) = \mathrm{sgn}\,(B(1) - t)\, e^{B(t) - \frac{1}{2}t}.$$

Note that $X(t)$ is not continuous in t. Consider another interesting anticipating stochastic integral equation (see Example 13.35)

$$X(t) = 1 + \int_0^t \partial_s^* \big(B(1)X(s)\big)\, ds,$$

where the anticipation comes from the integrand. We prove an anticipating Itô's formula in Theorem 13.21 which can be used to find the solution

$$X(t) = \exp\left[B(1) \int_0^t e^{-(t-s)}\, dB(s) - \frac{1}{4}B(1)^2\big(1 - e^{-2t}\big) - t \right].$$

On the other hand, we define white noise integrals as Pettis or Bochner integrals taking values in the space $(\mathcal{E})_\beta^*$ of generalized functions. For example, Donsker's delta function can be represented as a white noise integral (see Example 13.9). This representation is applied to Feynman integrals in Chapter 14. An existence and uniqueness theorem for white noise integral equations is given in Theorem 13.43.

The third objective is to give an easy presentation of white noise distribution theory. We have provided motivations and numerous examples. More importantly, many new ideas and techniques are introduced. For example, the Wick tensor $:x^{\otimes n}:$ is usually defined by induction. However, being motivated by Hermite polynomials with a parameter, we find it much easier to define $:x^{\otimes n}:$ directly by

$$:x^{\otimes n}: = \sum_{k=0}^{[n/2]} \binom{n}{2k} (2k-1)!!\, (-1)^k x^{\otimes(n-2k)} \widehat{\otimes} \tau^{\otimes k},$$

where τ is the trace operator. This definition immediately gives the relationship between Wick tensors and Hermite polynomials with a parameter. The Kubo-Yokoi theorem says that every test function has a continuous version. A beginner in white noise distribution theory usually finds the proof of this theorem very hard to understand. We give a rather transparent proof of this theorem by showing that a test function φ can be represented as in Equation (6.11) by

$$\varphi(x) = \langle\!\langle :e^{\langle\cdot,x\rangle}:, \Theta\varphi \rangle\!\rangle,$$

where the renormalization $:e^{\langle\cdot,x\rangle}:$ is a generalized function in $(\mathcal{E})^*$ for each x and Θ is a continuous linear operator from $(\mathcal{E})_\beta$ into itself for any β. In fact, this representation also yields the analytic extension of φ and produces various norm estimates. We have also introduced new ideas and techniques in the proof of Kondratiev-Streit characterization theorem for generalized functions in $(\mathcal{E})_\beta^*$.

This book is accessible to anyone with a first year graduate course in real analysis and some knowledge of Hilbert spaces. Prior familiarity with nuclear spaces and Gel'fand triples is not required. The reader can easily pick up the necessary background from Chapter 2. Anyone interested in learning the mathematical theory of white noise can quickly do so from Chapters 3 to 9. Applications of white noise in integral kernel operators, Fourier transforms, Laplacian operators, white noise integration, and Feynman integrals are given in Chapters 10 to 14. Although there are some relationships among them, these five chapters can be read independently after Chapter 9. The material in Chapter 15 can be read right after Chapter 8. We have put it after Chapter 13 since white noise integral equations provide a natural motivation for studying positive generalized functions. In Appendix A we give bibliographical notes and comment on several open problems relating to other fields. Appendix B contains a collection of miscellaneous formulas which are needed for this book. Some of them are actually proved as lemmas or derived in examples. They are put together for the convenience of the reader.

Since this book is intended to be an introduction to white noise distribution theory, we can not cover every aspect of its applications. For example, we have omitted topics such as quantum probability and Yang-Mills equations, infinite dimensional rotation groups, stochastic partial differential equations, stochastic variational equations, Dirichlet forms, and quantum field theory. We only make a few comments on these topics in Appendix A. On the other hand, we believe that the material in Chapters 3 to 9 will provide adequate background for applications to many problems which use white noise. In Appendix A we mention several open problems for further research. It is our sincere hope that the reader will find white noise distribution theory interesting. If the reader can apply this theory to other fields, it will be our greatest pleasure.

This book is based on a series of lectures which I gave at the Daewoo Workshop and at several universities in Korea (Seoul National University, Sogang University, Sook Myung Women's University, Korean Advanced Institute of Science and Technology, Chonnam National University, and Yonsei University) during the summer of 1994. Toward the end of my visit Professor D. M. Chung suggested that I should write up my lecture notes, which eventually became the manuscript for this book. Were it not for his suggestion, this book would not have materialized. I certainly was not thinking about writing a book before my visit to Korea.

My visit to Korea was arranged by Professor D. M. Chung with financial support from Daewoo Foundation, Global Analysis Research Center at Seoul National University, Sogang University, and Sook Myung Women's University. I would like to express my gratitude to him and these institutions. I want to express my appreciation for the warm hospitality of the following persons: K. S. Chang, S. J. Chang, D. P. Chi, S. M. Cho, B. D. Choi, T. S. Chung, U. C. Ji, S. H. Kang, S. J. Kang, D. H. Kim, S. K. Kim, K. Sim Lee, Y. M. Park, Y. S. Shim, I. S. Wee, and M. H. Woo.

I want to take this opportunity to thank each of the following persons and their institutions for inviting me to give a series of lectures: B. D. Choi (Korean Advanced Institute of Science and Technology), P. L. Chow (Wayne State University), T. Hida (Meijo University and Nagoya University), C. R. Hwang (Academia Sinica), Y.-J. Lee (Cheng Kung University), T. F. Lin (Soochow University), Y. Okabe (Hokkaido University), R. Rebolledo (Pontificia Universidad Católica de Chile), H. Sato (Kyushu University), N. R. Shieh (Taiwan University), L. Streit (Universität Bielefeld), and Y. Zhang (Fudan University). The lecture notes resulting from these visits have influenced the writing of this book.

I am glad to thank D. Betounes, D. M. Chung, W. G. Cochran, J. A. Hildebrant, U. C. Ji, R. J. Koch, I. Kubo, S. K. Ngobi, N. Obata, M. Redfern, K. Saitô, A. Sengupta, P. Sundar, and J. J. Whitaker for reading portions of the manuscript and making numerous corrections and valuable suggestions. The preparation of this book is partially supported by a grant from the U. S. Army Research Office. I want to thank CRC Press editor R. Stern for his assistance in publishing this book. I also want to thank C. N. Delzell for helping me with highly nontrivial TEXnical questions.

Most of all, with the warmest affection, I would like to express my deepest gratitude to my teachers Professors L. Gross and T. Hida for their constant encouragement. They have always been a source of new ideas for me.

Hui-Hsiung Kuo

Baton Rouge, Louisiana
January 1996

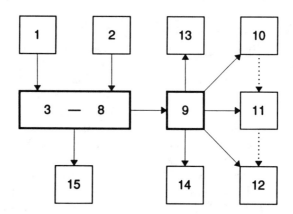

Note: Chapters 1 and 2 can be skipped. Chapters 3 to 9 constitute the mathematical theory of white noise. Applications are given in Chapters 10 to 14, each of which can be read independently after Chapter 9. The dotted arrows indicate some dependencies. From Chapter 8 one may go directly to Chapter 15.

Contents

Contents

1

Introduction to White Noise

1.1 What is white noise?

A musician thinks of white noise as a sound with equal intensity at all frequencies within a broad band. Some examples are the sound of thunder, the roar of a jet engine, and the noise at the stock exchange.

Engineers use white noise as a noise that disturbs an input so that the output is a function of the input and the noise. They regard white noise as a random process $\dot{z}(t)$ with mean 0 and variance ∞ such that $\dot{z}(t)$ and $\dot{z}(s)$ are independent for $t \neq s$ and

$$E\left(\int f(t)\dot{z}(t)\, dt \right)^2 = \int f(t)^2\, dt.$$

Thus $\dot{z}(t)$ can be thought of as a continuous analogue of an independent and identically distributed random sequence. At each time t, $\dot{z}(t)$ has infinite fluctuation.

A mathematician often thinks of white noise as the time derivative $\dot{B}(t)$ of a Brownian motion $B(t)$. A Brownian motion starting at 0 is a stochastic process $B(t)$ satisfying the following conditions:

(1) $P\{B(0) = 0\} = 1$.
(2) $B(t)$ has independent increments, i.e., for any $0 \le t_0 < t_1 < t_2 < \cdots < t_n$, the random variables $B(t_1) - B(t_0), B(t_2) - B(t_1), \ldots, B(t_n) - B(t_{n-1})$ are independent.
(3) For any $s < t$, the random variable $B(t) - B(s)$ is normally distributed with mean 0 and variance $t - s$.
(4) $P\{\omega; B(\cdot, \omega) \text{ is continuous}\} = 1$.

By the third condition, the random variable $\epsilon^{-1}(B(t+\epsilon) - B(t))$ has mean 0 and variance ϵ^{-1}. Thus $\dot{B}(t)$ does not exist. In fact, $B(t)$ is nowhere differentiable.

White noise is also often thought of as a "generalized" Gaussian process. A stationary Gaussian process X_t is determined by

(1) $EX_t = m$ (independent of t).

(2) $E(X_{t+u} - m)(X_u - m) = \theta(t)$ (independent of u).

The function $\theta(t)$ is positive definite and $\theta(0) = E(X_u - m)^2 \equiv \sigma^2$. Thus if we assume that $\theta(t)$ is continuous, then by the Bochner theorem

$$\theta(t) = \int e^{it\lambda} f(\lambda) \, d\lambda.$$

Here $f(\lambda)$ is called the spectral density of X_t. Note that

$$\int f(\lambda) \, d\lambda = \theta(0) = \sigma^2 \neq 0.$$

Informally, white noise is regarded as a stationary Gaussian process such that the spectral density function is constant. Thus the variance $\sigma^2 = \infty$. Such a process does not exist as above since the function $\theta(t)$ is given by the Dirac function $\theta(t) = \delta_0(t)$ at the origin.

So, what is white noise in rigorous mathematical terms? It can be defined as a generalized stochastic process X_ξ with index ξ in a space of test functions such that

(1) $X_{a\xi + b\eta} = aX_\xi + bX_\eta$.

(2) X_ξ is normally distributed with mean 0 and variance $\int |\xi(t)|^2 \, dt$.

For instance, $X_\xi = \int \xi(t) \, dB(t)$ (a Wiener integral, see Kuo [98]) is a white noise. This definition of white noise is quite acceptable. However, with this definition, it is impossible to study functions of white noise such as $\dot{B}(t)^2$ and $e^{\dot{B}(t)}$.

The best way to understand white noise is to regard it as *a generalized function on an infinite dimensional space*. This is an infinite dimensional distribution theory introduced by T. Hida in 1975 [45]. During the last two decades there has been an extensive study on this subject. For instance, see the book by Hida et al. [58], the survey articles by Kuo [110] and [112], or the monograph by Obata [142]. In this book we will give a comprehensive treatment of white noise distribution theory and study recent progress.

1.2 A simple example

Before we define the white noise $\dot{B}(t)$, let us see how it can be used as a noise in a simple equation. The solution of the equation

$$\frac{du}{dt} = -cu, \quad u(0) = u_0, \quad (c > 0)$$

is given by $u(t) = u_0 e^{-ct}$. Now, consider the following equation with "noise" $f(t)$:

$$\frac{du}{dt} = -cu + f(t), \quad u(0) = u_0, \quad (c > 0).$$

The solution u (output) now depends on both the initial condition u_0 (input) and the function f (noise). It is given by

$$u(t) = u_0 e^{-ct} + \int_0^t e^{-c(t-s)} f(s) \, ds. \tag{1.1}$$

The meaning of the integral depends on what the noise function f is. We discuss this problem in several cases.

Case 1: f is deterministic.

Assume that f is locally integrable (e.g., bounded and measurable). The integral in equation (1.1) is a Lebesgue integral. The solution u is a deterministic function.

Example 1.1. Suppose $f \equiv 1$. Then we have

$$u(t) = u_0 e^{-ct} + \frac{1}{c}\left(1 - e^{-ct}\right) \to \frac{1}{c} \quad \text{as } t \to \infty.$$

Example 1.2. Suppose $f \equiv -1$. Then we have

$$u(t) = u_0 e^{-ct} - \frac{1}{c}\left(1 - e^{-ct}\right) \to -\frac{1}{c} \quad \text{as } t \to \infty.$$

Example 1.3. Suppose $f = \sin t$. Then we have

$$u(t) = u_0 e^{-ct} + \frac{1}{1+c^2}\left(c\sin t - \cos t + e^{-ct}\right)$$

$$\approx \frac{1}{1+c^2}(c\sin t - \cos t) \quad \text{for large } t.$$

Case 2: f is stochastic.

In this case, $f(t)$ is a random variable for each t. Suppose almost all sample paths $f(\cdot)$ are locally integrable. The integral in equation (1.1) is defined as a Lebesgue integral with probability 1. The solution $u(t)$ is a random variable for each t. The randomness of u is a result of the noise f.

Example 1.4. Suppose $f(t) = X$ for all t (X is a random variable). Then

$$u(t) = u_0 e^{-ct} + \frac{X}{c}\left(1 - e^{-ct}\right) \approx \frac{X}{c} \quad \text{for large } t.$$

Example 1.5. Suppose $f(t) = B(t)$, a Brownian motion. The solution $u(t)$ has a normal distribution with mean $u_0 e^{-ct}$ and variance

$$\text{var}(u(t)) = \frac{t}{c^2} - \frac{3}{2c^3} + \frac{2}{c^3} e^{-ct} - \frac{1}{2c^3} e^{-2ct}.$$

This shows that the solution $u(t)$ is asymptotically normally distributed with mean 0 and variance $\frac{t}{c^2} - \frac{3}{2c^3}$.

Case 3: f is super-stochastic.

In this case, $f(t)$ is a generalized function for each t. The integral in equation (1.1) can be a Wiener integral, an Itô integral, or a white noise integral. The solution $u(t)$ can be a random variable or a generalized function.

Example 1.6. Suppose $f(t) = \dot{B}(t)$. The integral in equation (1.1) is interpreted as a Wiener integral, i.e.,

$$\int_0^t e^{-c(t-s)} \dot{B}(s) \, ds = \int_0^t e^{-c(t-s)} \, dB(s).$$

Hence the solution $u(t)$ is given by

$$u(t) = u_0 e^{-ct} + \int_0^t e^{-c(t-s)} \, dB(s),$$

where the Wiener integral is normally distributed with mean 0 and variance

$$E\left(\int_0^t e^{-c(t-s)} \, dB(s) \right)^2 = \int_0^t e^{-2c(t-s)} \, ds$$

$$= \frac{1}{2c}(1 - e^{-2ct}) \to \frac{1}{2c} \quad \text{as } t \to \infty.$$

Example 1.7. Suppose $f(t) = B(t)\dot{B}(t)$. The integral in equation (1.1) is interpreted as an Itô integral, i.e.,

$$\int_0^t e^{-c(t-s)} B(s)\dot{B}(s) \, ds = \int_0^t e^{-c(t-s)} B(s) \, dB(s).$$

The solution $u(t)$ is given by

$$u(t) = u_0 e^{-ct} + \int_0^t e^{-c(t-s)} B(s) \, dB(s).$$

It follows from Itô's formula that

$$\int_0^t e^{cs} B(s) \, dB(s) = \frac{1}{2} e^{ct} B(t)^2 - \frac{1}{2c}(e^{ct} - 1) - \frac{c}{2} \int_0^t e^{cs} B(s)^2 \, ds.$$

Thus the solution $u(t)$ can be expressed as:

$$u(t) = u_0 e^{-ct} + \frac{1}{2}B(t)^2 - \frac{1}{2c}(1 - e^{-ct}) - \frac{c}{2}\int_0^t e^{-c(t-s)}B(s)^2\, ds.$$

Hence $u(t)$ is a function of the Brownian motion. For large t,

$$u(t) \approx \frac{1}{2}B(t)^2 - \frac{1}{2c} - \frac{c}{2}\int_0^t e^{-c(t-s)}B(s)^2\, ds.$$

Example 1.8. We may try the noise $\dot{B}(t)^2$. But this noise is not a generalized function and we need to take the renormalization $f(t) =\, :\dot{B}(t)^2:$, which is a generalized function (see §3.4). In this case, the integral in equation (1.1) is a white noise integral and

$$u(t) = u_0 e^{-ct} + \int_0^t e^{-c(t-s)} :\dot{B}(s)^2:\, ds.$$

The solution $u(t)$ is not a random variable. Instead, it is a generalized function. We will study the renormalization in Chapter 5 and white noise integrals in Chapter 13.

2

Background

In this chapter we describe briefly the background concerning the concepts of abstract Wiener space, countably-Hilbert space, nuclear space, and Gel'fand triple. For details, see Gross [41] or Kuo [98] (for abstract Wiener spaces) and Gel'fand and Vilenkin [38] (for countably-Hilbert spaces and nuclear spaces).

2.1 Abstract Wiener spaces

The standard Gaussian measure on $\mathbb{R}^r$ is the unique probability measure μ_r such that

$$\int_{\mathbb{R}^r} e^{i(x,y)} \, d\mu_r(x) = e^{-\frac{1}{2}|y|^2}, \quad y \in \mathbb{R}^r.$$

Let H be a real separable Hilbert space with norm $|\cdot|$. An obvious extension of the standard Gaussian measure to H is the unique probability μ on H such that

$$\int_H e^{i(x,h)} \, d\mu(x) = e^{-\frac{1}{2}|h|^2}, \quad h \in H.$$

However, such a measure μ does not exist if H is infinite dimensional. To see this, suppose such a measure exists and let $\{h_n; n = 1, 2, \ldots, \}$ be an orthonormal basis for H. Then

$$\int_H e^{i(x,h_n)} \, d\mu(x) = e^{-\frac{1}{2}}.$$

Note that for every x in H, $(x, h_n) \to 0$ as $n \to \infty$. Hence by the Lebesgue dominated convergence theorem we get a contradiction $1 = e^{-\frac{1}{2}}$.

A norm $\|\cdot\|$ on H is called *measurable* if for any $\epsilon > 0$, there exists a finite dimensional orthogonal projection P_ϵ such that for any finite dimensional orthogonal projection P with $PP_\epsilon = 0$, we have

$$\mu_P\{x \in PH; \|x\| > \epsilon\} < \epsilon,$$

where μ_P is the standard Gaussian measure on PH. It can be shown that a measurable norm $\|\cdot\|$ is weaker than $|\cdot|$. Moreover, an inner product norm $\|\cdot\|$ is measurable if and only if $\|x\| = |Tx|$ for an injective Hilbert-Schmidt operator T of H.

Let B be the completion of H with respect to a measurable norm $\|\cdot\|$. The pair (H, B) is called an *abstract Wiener space*. An important fact is that the non-σ-additive Gauss measure on H is σ-additive on the larger space B. More precisely, let μ be the cylindrical measure on H defined by

$$\mu\{x \in H; Px \in D\} = \mu_P\{x \in PH; x \in D\}.$$

Then μ is not σ-additive by the above contradiction.

Fact 2.1. (Gross theorem) *Let (H, B) be an abstract Wiener space and i the inclusion map from H into B. Then the cylindrical measure $\mu \circ i^{-1}$ on B is σ-additive.*

Let μ denote also the σ-additive extension to the Borel σ-field of B. It is called the standard *Gaussian measure* on the abstract Wiener space (H, B). By using the Riesz representation theorem to identify H^* with H, we have the continuous inclusion maps $B^* \subset H \subset B$ and $\langle \xi, h \rangle = (\xi, h)$ for all $\xi \in B^*$ and $h \in H$. Here $\langle \cdot, \cdot \rangle$ is the natural bilinear pairing of B^* and B. If $\xi \in B^*$, then $\langle \xi, \cdot \rangle$ is defined everywhere on (B, μ) and has normal distribution with mean 0 and variance $|\xi|^2$. If $h \in H$, take a sequence $\{\xi_n\}$ from B^* such that $|\xi_n - h| \to 0$ and define $\langle h, \cdot \rangle$ to be the $L^2(\mu)$-limit of $\langle \xi_n, \cdot \rangle$. Then $\langle h, \cdot \rangle$ is well-defined and has normal distribution with mean 0 and variance $|h|^2$. Moreover,

$$\int_B e^{i\langle x, h \rangle} \, d\mu(x) = e^{-\frac{1}{2}|h|^2}, \quad h \in H.$$

Let μ_x be the translation of μ by x, i.e., $\mu_x(\cdot) = \mu(\cdot - x)$. The measure μ_x is equivalent to μ if and only if $x \in H$. Moreover, for $h \in H$, the Radon-Nikodym derivative $d\mu_h/d\mu$ is given by

$$\frac{d\mu_h}{d\mu} = e^{\langle \cdot, h \rangle - \frac{1}{2}|h|^2}.$$

Theorem 2.2. *Let $U(x, a) \equiv \{y \in B; \|y - x\| < a\}$. Then $\mu\big(U(x, a)\big) > 0$ for any $x \in B$ and $a > 0$.*

Proof. Choose $h \in H$ such that $\|h - x\| < \frac{a}{2}$. Then $U(h, \frac{a}{2}) \subset U(x, a)$. Thus it suffices to prove the case when $x \in H$. But μ_{-x} is equivalent to μ if $x \in H$. Hence we need only to prove the case $x = 0$. Let $U = \{y \in B; \|y\| < a\}$. Choose a countable set $\{h_n\} \subset H$ such that $\{h_n\}$ is dense in B. Let $U_n = h_n + U$. If $\mu(U) = 0$, then $\mu_{-h_n}(U) = 0$ for all n. But $\mu_{-h_n}(U) = \mu(U_n)$. Thus $\mu(U_n) = 0$ for all n. Hence $\mu(B) = 0$ since $B = \cup_{n=1}^{\infty} U_n$. This contradiction shows that we must have $\mu(U) > 0$. $\square$

Fact 2.3. (Fernique theorem) *There exists a constant $c > 0$ such that $\int_B e^{c\|x\|^2} d\mu(x) < \infty$.*

2.2 Countably-Hilbert spaces

Let V be a topological vector space over $\mathbb{C}$ with the topology given by a family $\{|\cdot|_n; n = 1, 2, \ldots\}$ of inner product norms. For $u, v \in V$, define

$$d(u, v) = \sum_{n=1}^{\infty} 2^{-n} \frac{|u - v|_n}{1 + |u - v|_n}.$$

Then d is a metric on V and a net is Cauchy in this metric if and only if it is Cauchy in each norm $|\cdot|_n$. In particular, this metric generates the same topology on V.

A topological vector space V with a family $\{|\cdot|_n; n \geq 1\}$ of inner product norms is called a *countably-Hilbert space* if it is complete with respect to its topology or, equivalently, with respect to the corresponding metric.

By considering the new norms $\|v\|_n = \left(\sum_{k=1}^{n} |v|_k^2\right)^{1/2}$, if necessary, we may assume that the family $\{|\cdot|_n; n \geq 1\}$ of norms is increasing, i.e.,

$$|v|_1 \leq |v|_2 \leq \cdots \leq |v|_n \leq \cdots, \quad \forall v \in V.$$

Let V_n denote the completion of V in the norm $|\cdot|_n$. Then V_n is a Hilbert space and the following inclusions are continuous

$$V \subset \cdots \subset V_{n+1} \subset V_n \subset \cdots \subset V_1.$$

It can be shown that V is complete if and only if $V = \cap_{n=1}^{\infty} V_n$.

Conversely, suppose $\{(U_n, |\cdot|_n); n \geq 1\}$ is a sequence of Hilbert spaces such that U_{n+1} is continuously imbedded in U_n for each n. Let $U \equiv \cap_{n=1}^{\infty} U_n$ and endow U with the *projective limit topology*, i.e., the coarsest topology such that for each n the inclusion from U into U_n is continuous. This topological vector space U is called the *projective limit* of $\{U_n; n \geq 1\}$. Obviously U is a countably-Hilbert space with the norms $\{|\cdot|_n; n \geq 1\}$.

Next we consider the dual spaces. Let V be a countably-Hilbert space associated with an increasing sequence $\{|\cdot|_n\}$ of norms. Let V_n be the completion of V with respect to the norm $|\cdot|_n$. Then the dual space V' of V is given by $V' = \cup_{n=1}^{\infty} V_n'$ and we have the inclusions

$$V_1' \subset \cdots \subset V_n' \subset V_{n+1}' \subset \cdots \subset V'. \qquad (2.1)$$

Let $\langle \cdot, \cdot \rangle$ denote the natural bilinear pairing of V' and V. There are several topologies on the dual space V':

A. Weak topology

The *weak topology* on V' is defined to be the coarsest topology on V' such that the functional $\langle \cdot, v \rangle$ is continuous for any $v \in V$. Equivalently, this topology has a base of neighborhoods of zero given by sets of the form:

$$N(v_1, \ldots, v_k; \epsilon) = \{x \in V'; |\langle x, v_j \rangle| < \epsilon, 1 \le j \le k\},$$

where $v_1, \ldots, v_k \in V$ and $\epsilon > 0$. Obviously the inclusion map from V_n' into V' is continuous and V_n' is dense in V' relative to the weak topology of V'.

B. Strong topology

A subset D of V is said to be *bounded* if for any neighborhood N of zero in V there is a positive number c such that $D \subset cN$. It is easy to check that D is bounded if and only if $\sup_{v \in D} |v|_n < \infty$ for all n. The *strong topology* of V' is defined to be the topology with a base of neighborhoods of zero given by

$$N(D; \epsilon) = \{x \in V'; \ \sup_{v \in D} |\langle x, v \rangle| < \epsilon\},$$

where D is any bounded subset of V and $\epsilon > 0$. Obviously the strong topology is finer than the weak topology. Moreover, the inclusion map from V_n' into V' is continuous and V_n' is dense in V' relative to the strong topology of V'.

It can be shown that a sequence $\{x_n\}$ in V' converges strongly if and only if it converges weakly. Moreover, V' is sequentially complete with respect to the weak and strong topologies.

Similar to the above discussion we can define strongly bounded subsets of V'. Then the strong topology can be introduced in V''. With this strong topology the space V'' is isomorphic to V. Thus any countably-Hilbert space V is reflexive.

C. Inductive limit topology

Let $\{(W_n, |\cdot|_n); n \ge 1\}$ be a sequence of Hilbert spaces such that W_n is continuously imbedded in W_{n+1} for each n. Let $W \equiv \cup_{n=1}^{\infty} W_n$ and endow W with the finest locally convex topology such that for each n the inclusion from W_n into W is continuous. This topological vector space W is called the *inductive limit* of the sequence $\{W_n; n \geq 1\}$. A base of neighborhoods of zero in this inductive limit topology is given by the collection of balanced convex hulls of sets $\{x \in W_n; |x|_n < \epsilon_n\}$, $n \geq 1$. A sequence $\{x_k\}$ converges to x in W in this topology if and only if there exists some n such that $x_k \in W_n$ for all k and $|x_k - x|_n \to 0$, i.e., x_k converges to x in W_n.

In particular, let V be a countably-Hilbert space. Its dual space V', in view of (2.1), can be regarded as the inductive limit of $\{V'_n; n \geq 1\}$. Thus V' has the inductive limit topology. It turns out that this inductive limit topology is the same as the strong topology of V'.

2.3 Nuclear spaces

Let V be a countably-Hilbert space associated with an increasing sequence $\{|\cdot|_n\}$ of norms. Let V_n be the completion of V with respect to the norm $|\cdot|_n$. The countably-Hilbert space V is called a *nuclear space* if for any n, there exists $m \geq n$ such that the inclusion map from V_m into V_n is a Hilbert-Schmidt operator, i.e., there is an orthonormal basis $\{v_k\}$ for V_m such that $\sum_{k=1}^{\infty} |v_k|_n^2 < \infty$.

Note that a trace class operator is also a Hilbert-Schmidt operator and that the product of two Hilbert-Schmidt operators is a trace class operator. Therefore V is a nuclear space if and only if for any n, there exists $m \geq n$ such that the inclusion map from V_m into V_n is a trace class operator.

Nuclear spaces have many properties similar to those of the finite dimensional space $\mathbb{R}^r$. For instance a subset of a nuclear space is compact if and only if it is closed and bounded. This implies that an infinite dimensional Banach space is not nuclear.

If V is a nuclear space, then the σ-fields on V' generated by the three topologies (weak, strong, and inductive limit) are all the same. This σ-field is regarded as the Borel field of V'.

Fact 2.4. (Minlos theorem) *Let V be a real nuclear space. A complex-valued function φ on V is the characteristic function of a unique probability measure ν on V', i.e.,*

$$\varphi(v) = \int_{V'} e^{i\langle x, v \rangle} \, d\nu(x), \quad v \in V,$$

if and only if (1) $\varphi(0) = 1$, (2) φ is continuous, and (3) φ is positive definite, i.e., for any $z_1, \ldots, z_n \in \mathbb{C}$ and $v_1, \ldots, v_n \in V$,

$$\sum_{j,k=1}^{n} z_j \overline{z_k} \, \varphi(v_j - v_k) \geq 0.$$

For the proof of the above fact, see Gel'fand and Vilenkin [38] or Hida [44]. Another important fact about nuclear spaces is the abstract kernel theorem. Suppose A is a continuous linear operator from V into V' with the weak topology for V'. Then the bilinear functional $F(u,v) = \langle Au, v \rangle$ on $V \times V$ is continuous. The abstract kernel theorem says that the converse is also true.

Fact 2.5. (Abstract kernel theorem) *Let V be a nuclear space associated with an increasing sequence $\{|\cdot|_n\}$ of norms and let V_n be the completion of V with respect to $|\cdot|_n$. Suppose $F : V \times V \to \mathbb{C}$ is a separately continuous bilinear functional. Then there exist $n, p \geq 1$, and a Hilbert-Schmidt operator A from V_n into V_p' such that*

$$F(u,v) = \langle Au, v \rangle, \quad u, v \in V,$$

where $\langle \cdot, \cdot \rangle$ is the natural bilinear pairing of V' and V.

Note that for any n, there exists $m \geq n$ (by the nuclearity of V) such that the inclusion map from V_m into V_n is a trace class operator. Hence the assertion "Hilbert-Schmidt operator" in the abstract kernel theorem can be replaced by "trace class operator."

We remark that the abstract kernel theorem implies the following fact. Suppose A is a continuous linear operator from V into V' with the weak topology for V'. Then there exist n and p such that A is a Hilbert-Schmidt operator from V_n into V_p'.

2.4 Gel'fand triples

It is often useful and desirable to imbed a nuclear space V in its dual space V'. Suppose V is imbedded in some complex Hilbert space E and that V is dense in E relative to the norm of E. By using the Riesz representation theorem to identify E with its dual space E', we get the triple

$$V \subset E \subset V'.$$

Such a triple is called a *Gel'fand triple*. Note that E is dense in V' with the weak topology of V'. Here the Riesz representation theorem is used to identify E with its dual space E' in the sense as follows. Each $h \in E$ is identified as the element φ_h in E' defined by

$$\varphi_h(x) = (x, Jh), \quad x \in E,$$

where $(\cdot, \cdot)$ is the inner product in E and J is a conjugation of E. Thus the natural bilinear pairing of V' and V is related to the inner product of E by

$$\langle h, v \rangle = (h, Jv), \quad \forall h \in E, v \in V.$$

Let V be a nuclear space associated with a sequence $\{|\cdot|_n\}$ of norms and let V_n be the completion of V with respect to $|\cdot|_n$. By using $E = V_1$, we get a Gel'fand triple $V \subset V_1 \subset V'$ and the continuous inclusions

$$V \subset \cdots \subset V_{n+1} \subset V_n \subset \cdots \subset V_1 \subset \cdots \subset V_n' \subset V_{n+1}' \subset \cdots \subset V'.$$

On the other hand, suppose $\{U_n\}$ is a sequence of Hilbert spaces such that U_{n+1} is imbedded in U_n for each n and the inclusion map is a Hilbert-Schmidt operator. Let U be the projective limit of $\{U_n; n \geq 1\}$. Then $U \subset U_1 \subset U'$ is a Gel'fand triple and U' is the inductive limit of $\{U_n'; n \geq 1\}$.

3

White Noise as an Infinite Dimensional Calculus

3.1 White noise space

As mentioned at the end of §1.1, the white noise $\dot{B}(t)$ is a generalized function on an infinite dimensional space. First we shall specify such a space. A common choice is the Schwartz space $\mathcal{S}(\mathbb{R})$ of real-valued rapidly decreasing functions on $\mathbb{R}$. A function ξ on $\mathbb{R}$ is called rapidly decreasing if it is smooth and for all nonnegative integers n and k,

$$|x^n \xi^{(k)}(x)| \to 0 \quad \text{as } |x| \to \infty.$$

For any $n, k \geq 0$, define a norm $|\cdot|_{n,k}$ on $\mathcal{S}(\mathbb{R})$ by

$$|\xi|_{n,k} = \left(\int_{\mathbb{R}} |x^n \xi^{(k)}(x)|^2 \, dx \right)^{1/2}.$$

This family $\{|\cdot|_{n,k}; n, k \geq 0\}$ of norms generates a topology on $\mathcal{S}(\mathbb{R})$. Thus $\mathcal{S}(\mathbb{R})$ is a topological vector space. In fact, it is a nuclear space (see the next section). Let $\mathcal{S}'(\mathbb{R})$ denote the dual space of $\mathcal{S}(\mathbb{R})$. Then we have the following Gel'fand triple

$$\mathcal{S}(\mathbb{R}) \subset L^2(\mathbb{R}) \subset \mathcal{S}'(\mathbb{R}),$$

where $L^2(\mathbb{R})$ is identified with its dual space by the Riesz representation theorem.

Consider the following function defined on $\mathcal{S}(\mathbb{R})$:

$$C(\xi) = e^{-\frac{1}{2}|\xi|_0^2}, \quad \xi \in \mathcal{S}(\mathbb{R}),$$

where $|\cdot|_0$ denotes the $L^2(\mathbb{R})$-norm.

Lemma 3.1. *The function C is positive definite, i.e., for any $z_j \in \mathbb{C}$ and $\xi_j \in S(\mathbb{R}), j = 1, 2, \ldots, n$,*

$$\sum_{j,k=1}^{n} z_j C(\xi_j - \xi_k)\overline{z_k} \geq 0.$$

Proof. Let V be the subspace of $S(\mathbb{R})$ spanned by $\xi_1, \xi_2, \ldots, \xi_n$ with norm $|\cdot|_0$. Let μ_V be the standard Gaussian measure on V. Then for any $\xi \in V$

$$\int_V e^{i(\xi,y)} \, d\mu_V(y) = e^{-\frac{1}{2}|\xi|_0^2},$$

where $(\cdot, \cdot)$ is the inner product of $L^2(\mathbb{R})$. Therefore

$$\sum_{j,k=1}^{n} z_j C(\xi_j - \xi_k)\overline{z_k} = \sum_{j,k=1}^{n} \int_V z_j e^{i(\xi_j - \xi_k, y)}\overline{z_k} \, d\mu_V(y)$$

$$= \int_V \left| \sum_{j=1}^{n} z_j e^{i(\xi_j, y)} \right|^2 d\mu_V(y)$$

$$\geq 0. \qquad \square$$

The function C is obviously continuous on $S(\mathbb{R})$ and $C(0) = 1$. Hence by the Minlos theorem (see Fact 2.4), there exists a unique probability measure μ on the dual space $S'(\mathbb{R})$ of $S(\mathbb{R})$ such that

$$\int_{S'(\mathbb{R})} e^{i\langle x, \xi \rangle} \, d\mu(x) = e^{-\frac{1}{2}|\xi|_0^2}, \quad \xi \in S(\mathbb{R}),$$

where $\langle \cdot, \cdot \rangle$ denotes the bilinear pairing of $S'(\mathbb{R})$ and $S(\mathbb{R})$.

Definition 3.2. *The probability space $(S'(\mathbb{R}), \mu)$ is called a white noise space. The measure μ is called the standard Gaussian measure on $S'(\mathbb{R})$.*

The reason for calling $(S'(\mathbb{R}), \mu)$ a white noise space is as follows. Note that for each $\xi \in S(\mathbb{R})$, the random variable $\langle \cdot, \xi \rangle$ is defined everywhere on $S'(\mathbb{R})$. It is normally distributed with mean 0 and variance $|\xi|_0^2$. Now, suppose $f \in L^2(\mathbb{R})$ (real-valued). Take a sequence $\{\xi_n\}$ in $S(\mathbb{R})$ such that $\xi_n \to f$ in $L^2(\mathbb{R})$. Then the sequence $\{\langle \cdot, \xi_n \rangle\}$ of random variables is Cauchy in $L^2(S'(\mathbb{R}), \mu)$. Define

$$\langle \cdot, f \rangle = \lim_{n \to \infty} \langle \cdot, \xi_n \rangle \text{ in } L^2(S'(\mathbb{R}), \mu).$$

The limit is independent of the sequence. Moreover, the random variable $\langle \cdot, f \rangle$ is normally distributed with mean 0 and variance $|f|_0^2$. Define

$$B(t)(x) = \begin{cases} \langle x, 1_{[0,t]} \rangle, & \text{if } t \geq 0, x \in S'(\mathbb{R}); \\ -\langle x, 1_{[t,0]} \rangle, & \text{if } t < 0, x \in S'(\mathbb{R}). \end{cases} \tag{3.1}$$

It is easy to check that $B(t)$ is a Brownian motion. Moreover, by taking the time derivative informally, we get $\dot{B}(t) = x(t)$. Thus the elements of $\mathcal{S}'(\mathbb{R})$ can be regarded as the sample paths of white noise. However, we have not yet answered the following

Question: For each t, what is $\dot{B}(t)$?

3.2 A reconstruction of the Schwartz space

In this section we will reconstruct the Schwartz space $\mathcal{S}(\mathbb{R})$ from the real Hilbert space $L^2(\mathbb{R})$ and the operator $A = -d^2/dx^2 + x^2 + 1$. Then the same procedure will be used to introduce a space of test functions on the white noise space $(\mathcal{S}'(\mathbb{R}), \mu)$.

The operator A is densely defined on $L^2(\mathbb{R})$ and has the following properties:

(1) Let $H_n(x) = (-1)^n e^{x^2}(d/dx)^n e^{-x^2}$ be the Hermite polynomial of degree n and let

$$e_n(x) = \frac{1}{\sqrt{\sqrt{\pi} 2^n n!}} H_n(x) e^{-x^2/2}$$

be the corresponding Hermite function. Then the set $\{e_n; n \geq 0\}$ is an orthonormal basis for $L^2(\mathbb{R})$. Moreover, these Hermite functions are eigenfunctions of A:

$$A e_n = (2n+2) e_n, \quad n = 0, 1, 2, \ldots.$$

(2) A^{-1} is a bounded operator of $L^2(\mathbb{R})$ with $\|A^{-1}\| = \frac{1}{2}$.
(3) For any $p > \frac{1}{2}$, A^{-p} is a Hilbert-Schmidt operator of $L^2(\mathbb{R})$. In fact,

$$\|A^{-p}\|_{HS}^2 = \sum_{n=0}^{\infty} (2n+2)^{-2p}.$$

Now, for each $p \geq 0$, define

$$|f|_p = |A^p f|_0,$$

where $|\cdot|_0$ is the $L^2(\mathbb{R})$-norm. Equivalently, $|f|_p$ is given by

$$|f|_p = \left(\sum_{n=0}^{\infty} (2n+2)^{2p} (f, e_n)^2 \right)^{1/2},$$

where $(\cdot, \cdot)$ is the inner product of $L^2(\mathbb{R})$. Let

$$\mathcal{S}_p(\mathbb{R}) \equiv \{ f \in L^2(\mathbb{R}); \ |f|_p < \infty \}.$$

Remark. Here and from now on, we use the following convention about the "norm" $|f|_p$, i.e., if f is not specified, then $|f|_p = \infty$ is allowed. Of course, $|\cdot|_p$ is a norm on the space $\{f; |f|_p < \infty\}$. Similar convention is used for the "norm" $\|\varphi\|_p$ in §3.3 below and other "norms" to be defined in later chapters.

Note that $\mathcal{S}_p(\mathbb{R})$ is a Hilbert space with norm $|\cdot|_p$. Moreover, we have the following facts:

(1) $\mathcal{S}(\mathbb{R}) = \cap_{p \geq 0} \mathcal{S}_p(\mathbb{R})$.
(2) The families $\{|\cdot|_{n,k}; n, k \geq 0\}$ and $\{|\cdot|_p; p \geq 0\}$ are equivalent, i.e., they generate the same topology on $\mathcal{S}(\mathbb{R})$.
(3) $\mathcal{S}(\mathbb{R})$ is a nuclear space. This is a consequence of the next lemma.

Lemma 3.3. *For any $p \geq 0$, the inclusion map $\mathcal{S}_{p+1}(\mathbb{R}) \hookrightarrow \mathcal{S}_p(\mathbb{R})$ is a Hilbert-Schmidt operator.*

Proof. Let $\xi_n = (2n+2)^{-(p+1)}e_n$. Then $\{\xi_n; n = 0, 1, 2, \ldots\}$ is an orthonormal basis for $\mathcal{S}_{p+1}$ and

$$\sum_{n=0}^{\infty} |\xi_n|_p^2 = \sum_{n=0}^{\infty} \frac{1}{(2n+2)^2} < \infty.$$

Hence the inclusion map from $\mathcal{S}_{p+1}(\mathbb{R})$ into $\mathcal{S}_p(\mathbb{R})$ is a Hilbert-Schmidt operator. $\qquad\square$

Thus $\mathcal{S}(\mathbb{R})$ has been reconstructed as the projective limit of $\{\mathcal{S}_p(\mathbb{R}); p \geq 0\}$ (by facts (1)-(2)) and $\mathcal{S}(\mathbb{R}) \subset L^2(\mathbb{R}) \subset \mathcal{S}'(\mathbb{R})$ is a Gel'fand triple (by fact (3)). Moreover, we have the following continuous inclusion maps:

$$\mathcal{S}(\mathbb{R}) \subset \mathcal{S}_p(\mathbb{R}) \subset L^2(\mathbb{R}) \subset \mathcal{S}'_p(\mathbb{R}) \subset \mathcal{S}'(\mathbb{R}).$$

The norm of the dual space $\mathcal{S}'_p(\mathbb{R})$ of $\mathcal{S}_p(\mathbb{R})$ can be shown to be given by

$$|f|_{-p} = |A^{-p}f|_0, \quad p > 0.$$

That is, the space $\mathcal{S}'_p(\mathbb{R})$ is the completion of $L^2(\mathbb{R})$ with respect to the norm $|\cdot|_{-p}$.

3.3 The spaces of test and generalized functions

In finite dimensional distribution theory, one takes the space $\mathbb{R}^r$ with the Lebesgue measure and considers the Gel'fand triple

$$\mathcal{S}(\mathbb{R}^r) \subset L^2(\mathbb{R}^r) \subset \mathcal{S}'(\mathbb{R}^r).$$

It is well-known that the Lebesgue measure does not exist on an infinite dimensional space (e.g., see Kuo [98, p.1]). However, the Gaussian measure does exist on an infinite dimensional space and is often used for infinite dimensional analysis.

In white noise distribution theory, the white noise space $(\mathcal{S}'(\mathbb{R}), \mu)$ is taken as an infinite dimensional analogue of $(\mathbb{R}^r, dx)$. For simplicity, we will use (L^2) to denote the space $L^2(\mathcal{S}'(\mathbb{R}), \mu)$. We need to construct a space X of test functions and its dual space X^* of generalized functions such that the following is a Gel'fand triple

$$X \subset (L^2) \subset X^*.$$

We will use the Wiener-Itô theorem and the second quantization operator to construct such a space X.

A. *Wiener-Itô theorem*

Any function $\varphi \in (L^2)$ can be decomposed uniquely as

$$\varphi = \sum_{n=0}^{\infty} I_n(f_n), \quad f_n \in \widehat{L}_c^2(\mathbb{R}^n),$$

where I_n is the multiple Wiener integral of order n with respect to the Brownian motion in (3.1) (for the multiple Wiener integral, cf. §4.2) and $\widehat{L}_c^2(\mathbb{R}^n)$ denotes the space of symmetric complex-valued L^2-functions on $\mathbb{R}^n$.

Notation: The subscript c will always denote the complexification of a real vector space. For instance, V_c denotes the complexification of V.

The (L^2)-norm $\|\varphi\|_0$ of φ is given by

$$\|\varphi\|_0 = \left(\sum_{n=0}^{\infty} n! |f_n|_0^2 \right)^{1/2},$$

where $|\cdot|_0$ denotes the $L_c^2(\mathbb{R}^n)$-norm for any n.

B. *Second quantization operator*

Let $A = -d^2/dx^2 + x^2 + 1$ as before. For $\varphi = \sum_{n=0}^{\infty} I_n(f_n)$ satisfying the condition

$$\sum_{n=0}^{\infty} n! |A^{\otimes n} f_n|_0^2 < \infty,$$

we define $\Gamma(A)\varphi \in (L^2)$ by

$$\Gamma(A)\varphi = \sum_{n=0}^{\infty} I_n(A^{\otimes n} f_n).$$

The operator $\Gamma(A)$ is called the *second quantization operator* of A. It is densely defined on (L^2) and has properties similar to those of A given in §3.2 (see the proof of Theorem 1.5.1 in Kuo [110] or see §4.2 for the general case):

(1) $\Gamma(A)$ has a set of eigenfunctions which forms an orthonormal basis for (L^2).
(2) $\Gamma(A)^{-1}$ is a bounded operator of (L^2) with $\|\Gamma(A)^{-1}\| = 1$.
(3) For any $p > 1$, the operator $\Gamma(A)^{-p}$ is a Hilbert-Schmidt operator of (L^2).

We now apply the same procedure described in §3.2 to introduce a space of test functions on the space $\mathcal{S}'(\mathbb{R})$. For each $p \geq 0$, define

$$\|\varphi\|_p = \|\Gamma(A)^p\varphi\|_0,$$

where $\|\cdot\|_0$ is the (L^2)-norm. Let

$$(\mathcal{S}_p) \equiv \{\varphi \in (L^2); \|\varphi\|_p < \infty\}.$$

Then $(\mathcal{S}_p)$ is a Hilbert space with norm $\|\cdot\|_p$. Define

$$(\mathcal{S}) \equiv \text{projective limit of } \{(\mathcal{S}_p); p \geq 0\}.$$

Then $(\mathcal{S})$ is a nuclear space by the above property (3) of $\Gamma(A)$. We will call $(\mathcal{S})$ a space of *test functions*. The dual space $(\mathcal{S})^*$ of $(\mathcal{S})$ is called a space of *generalized functions* (or *Hida distributions*). Thus we have a Gel'fand triple:

$$(\mathcal{S}) \subset (L^2) \subset (\mathcal{S})^*.$$

The bilinear pairing of $(\mathcal{S})^*$ and $(\mathcal{S})$ will be denoted by $\langle\!\langle \cdot, \cdot \rangle\!\rangle$. It is related to the inner product of (L^2) by

$$\langle\!\langle \varphi, \psi \rangle\!\rangle = (\varphi, \overline{\psi})_{(L^2)}, \quad \varphi \in (L^2), \; \psi \in (\mathcal{S}).$$

The same convention is also used for the inner product of $(\mathcal{S}_p)$ in the following continuous inclusions:

$$(\mathcal{S}) \subset (\mathcal{S}_p) \subset (L^2) \subset (\mathcal{S}_p)^* \subset (\mathcal{S})^*.$$

However, we will rarely use the inner products on (L^2) and on $(\mathcal{S}_p)$. It can be shown that $(\mathcal{S})^* = \cup_{p \geq 0}(\mathcal{S}_p)^*$ and the norm on the dual space $(\mathcal{S}_p)^*$ of $(\mathcal{S}_p)$ is given by

$$\|\Phi\|_{-p} = \|\Gamma(A)^{-p}\Phi\|_0, \quad p > 0.$$

That is, the space $(\mathcal{S}_p)^*$ is the completion of (L^2) with respect to the norm $\|\cdot\|_{-p}$. An element Φ in $(\mathcal{S}_p)^*$ can be expressed as $\Phi = \sum_{n=0}^{\infty} I_n(F_n)$, $F_n \in \widehat{\mathcal{S}'_c}(\mathbb{R}^n)$, such that

$$\|\Phi\|_{-p}^2 = \sum_{n=0}^{\infty} n! |(A^{-p})^{\otimes n} F_n|_0^2 < \infty.$$

3.4 Some examples of test and generalized functions

We first answer the question at the end of §3.1: For each $t \in \mathbb{R}$, $\dot{B}(t)$ is a generalized function, i.e., $\dot{B}(t) \in (\mathcal{S})^*$. In our setup, $\dot{B}(t) = \langle \cdot, \delta_t \rangle = I_1(\delta_t)$ and

$$\|\dot{B}(t)\|_{-p} = |A^{-p}\delta_t|_0 = |\delta_t|_{-p}.$$

By using the expansion $\delta_t = \sum_{n=0}^{\infty} e_n(t)e_n$ in terms of Hermite functions, we get

$$|\delta_t|_{-p} = \left(\sum_{n=0}^{\infty} (2n+2)^{-2p} e_n(t)^2 \right)^{1/2}.$$

But it is well-known that $\sup_{t \in \mathbb{R}} |e_n(t)| = O(n^{-1/12})$. Hence for any $p > \frac{5}{12}$, we have $|\delta_t|_{-p} < \infty$ and so $\dot{B}(t) \in (\mathcal{S}_p)^*$. Thus $\dot{B}(t)$ is a generalized function. It can be shown that all time derivatives of Brownian motion are elements in $(\mathcal{S})^*$. In fact, the k-th derivative $B^{(k)}(t)$ belongs to $(\mathcal{S}_p)^*$ for any $p > \frac{5}{12} + \frac{k-1}{2}$.

We extend the bilinear pairing $\langle \cdot, \cdot \rangle$ of $\mathcal{S}'(\mathbb{R})$ and $\mathcal{S}(\mathbb{R})$ to $\mathcal{S}'_c(\mathbb{R})$ and $\mathcal{S}_c(\mathbb{R})$. Then the function $\langle \cdot, f \rangle$ is everywhere defined on $\mathcal{S}'(\mathbb{R})$ if $f \in \mathcal{S}_c(\mathbb{R})$, almost everywhere defined on $\mathcal{S}'(\mathbb{R})$ if $f \in L^2_c(\mathbb{R})$. But it is a generalized function if $f \in \mathcal{S}'_c(\mathbb{R})$.

Now, let $f \in L^2_c(\mathbb{R})$. We have the following Wiener-Itô decomposition for the function $\langle \cdot, f \rangle^2$,

$$\begin{aligned} \langle \cdot, f \rangle^2 &= (\langle \cdot, f \rangle^2 - \langle f, f \rangle) + \langle f, f \rangle \\ &= I_2(f^{\otimes 2}) + \langle f, f \rangle. \end{aligned} \tag{3.2}$$

Note that $\langle \cdot, \xi \rangle^2 \in (\mathcal{S})$ if $\xi \in \mathcal{S}_c(\mathbb{R})$ and $\langle \cdot, f \rangle^2 \in (L^2)$ if $f \in L^2_c(\mathbb{R})$. But $\langle \cdot, y \rangle^2$ has no meaning if $y \in \mathcal{S}'_c(\mathbb{R})$. We need to consider its renormalization in order to get a generalized function. The renormalization is motivated by equation (3.2), i.e., define the *renormalization* $:\langle \cdot, y \rangle^2:$ of $\langle \cdot, y \rangle^2$ by

$$:\langle \cdot, y \rangle^2: \equiv I_2(y^{\otimes 2}).$$

Then we have

$$\|:\langle \cdot, y \rangle^2:\|_{-p} = \sqrt{2} \left| (A^{-p})^{\otimes 2} y^{\otimes 2} \right|_0 = \sqrt{2} \, |y|^2_{-p}.$$

This shows that $:\langle \cdot, y \rangle^2: \in (\mathcal{S})^*$ since $|y|_{-p} < \infty$ for some $p \geq 0$. In particular, for $y = \delta_t$, we have the generalized function $:\dot{B}(t)^2:$, which we mentioned in Example 1.8.

In general, for $y \in \mathcal{S}'_c(\mathbb{R})$, we define the renormalization $:\langle \cdot, y \rangle^n:$ of $\langle \cdot, y \rangle^n$ by

$$:\langle \cdot, y \rangle^n: \equiv I_n(y^{\otimes n}).$$

Thus $:\langle\cdot,y\rangle^n:\ \in(\mathcal{S})^*$ for any $y\in\mathcal{S}'_c(\mathbb{R})$. In particular, let $y=\delta_t$ and we have a generalized function $:\dot{B}(t)^n:$ for any real number t.

We mention some exponential functions of white noise. Let $f\in L^2(\mathbb{R})$. The function $e^{\langle\cdot,f\rangle}$ can be shown to have the following Wiener-Itô decomposition

$$e^{\langle\cdot,f\rangle}=e^{\frac{1}{2}|f|_0^2}\sum_{n=0}^{\infty}\frac{1}{n!}I_n(f^{\otimes n}).\tag{3.3}$$

It is easy to check that $e^{\langle\cdot,\xi\rangle}\in(\mathcal{S})$ if $\xi\in\mathcal{S}(\mathbb{R})$ and $e^{\langle\cdot,f\rangle}\in(L^2)$ if $f\in L^2(\mathbb{R})$. However, $e^{\langle\cdot,y\rangle}$ has no meaning if $y\in\mathcal{S}'(\mathbb{R})$. Again, we need a renormalization. This is suggested by equation (3.3), i.e., for $y\in\mathcal{S}'_c(\mathbb{R})$, define the *renormalization* $:e^{\langle\cdot,y\rangle}:$ of $e^{\langle\cdot,y\rangle}$ by

$$:e^{\langle\cdot,y\rangle}:\ \equiv\sum_{n=0}^{\infty}\frac{1}{n!}I_n(y^{\otimes n}).$$

It is easy to check that

$$\|:e^{\langle\cdot,y\rangle}:\|_{-p}=e^{\frac{1}{2}|y|^2_{-p}}.$$

Since $|y|_{-p}<\infty$ for some $p\geq0$, we see that $:e^{\langle\cdot,y\rangle}:$ is a generalized function. Note that for f in $L^2_c(\mathbb{R})$, we have

$$:e^{\langle\cdot,f\rangle}:\ =e^{\langle\cdot,f\rangle-\frac{1}{2}\langle f,f\rangle}.$$

Thus for $\xi\in\mathcal{S}_c(\mathbb{R})$, both $e^{\langle\cdot,\xi\rangle}$ and $:e^{\langle\cdot,\xi\rangle}:$ are functions in $(\mathcal{S})$.

4

Constructions of Test and Generalized Functions

4.1 General ideas for several constructions

Let (L^2) denote the Hilbert space of complex-valued square integrable functions on the dual space $\mathcal{N}'$ of a nuclear space $\mathcal{N}$ with respect to the standard Gaussian measure on $\mathcal{N}'$. In white noise analysis there have been several constructions of a space X of test functions and the corresponding space X^* of generalized functions with the following continuous inclusions:

$$X \subset (L^2) \subset X^*.$$

Here the bilinear pairing $\langle\!\langle \cdot, \cdot \rangle\!\rangle$ of X^* and X and the inner product $(\cdot, \cdot)_{(L^2)}$ on (L^2) are related by (as mentioned in §3.3)

$$\langle\!\langle \varphi, \psi \rangle\!\rangle = (\varphi, \overline{\psi})_{(L^2)}, \quad \varphi \in (L^2), \ \psi \in X.$$

This convention will be assumed throughout the book although we will rarely use the inner product on (L^2) in the calculation.

In the construction of X and X^* one usually uses the Wiener-Itô theorem, i.e., each function φ in (L^2) can be decomposed uniquely as a sum of multiple Wiener integrals (as mentioned in §3.3)

$$\varphi = \sum_{n=0}^{\infty} I_n(f_n). \tag{4.1}$$

We first describe the general ideas for these constructions and then discuss some of them in detail in later sections.

A. *Original construction of Hida*

The theory of white noise was initiated by Hida in his 1975 Carlton lecture notes [45]. One of his motivations was to introduce a space of generalized functions that would contain white noise functions such as $\dot{B}(t)$, $:\dot{B}(t)^n:$, and $:e^{\dot{B}(t)}:$.

Take the nuclear space $\mathcal{N}$ to be $\mathcal{S}(\mathbb{R})$. Let $(L^2)^+$ be the set of all φ's in equation (4.1) with f_n belonging to the Sobolev space of order $\frac{n+1}{2}$. Let $(L^2)^-$ be the dual space of $(L^2)^+$ so that we have the following continuous inclusions

$$(L^2)^+ \subset (L^2) \subset (L^2)^-.$$

The dual space $(L^2)^-$ consists of all φ's in equation (4.1) with f_n belonging to the Sobolev space of order $-\frac{n+1}{2}$ (the integrals are generalized multiple Wiener integrals introduced in Hida [46]).

It turns out that the space $(L^2)^+$ is not easy to use. For instance, it is not closed under multiplication. Moreover, in order to study functions involving $\ddot{B}(t)$, it is necessary to change the Sobolev orders $\frac{n+1}{2}$ and $-\frac{n+1}{2}$ to $\frac{n+3}{2}$ and $-\frac{n+3}{2}$, respectively.

B. *General construction of Kubo and Takenaka*

Let E be a real separable Hilbert space with a certain family of norms. Take $\mathcal{N}$ to be the nuclear space $\mathcal{E}$ arising from this family of norms. By using the Fock space representation of (L^2), Kubo and Takenaka ([92], [93], [94], and [95]) constructed and used the following Gel'fand triple

$$(\mathcal{E}) \subset (L^2) \subset (\mathcal{E})^*.$$

This space $(\mathcal{E})$ of test functions is closed under multiplication. Moreover, it has many properties similar to those of the Schwartz space $\mathcal{S}(\mathbb{R})$.

We will give a modification of this construction in §4.3. The Wiener-Itô decomposition in equation (4.1), rather than the Fock space representation of (L^2), will be used for the modified construction.

C. *Special cases of general construction*

Several special cases of Kubo-Takenaka's construction are often used. The special case with $E = L^2(\mathbb{R})$ and the Sobolev norms is used in Kuo [108] and Kuo et al. [113].

An important special case is given by $E = L^2(\mathbb{R})$ and the norms arising from the operator $A = -d^2/du^2 + u^2 + 1$. The resulting Gel'fand triple is $(\mathcal{S}) \subset (L^2) \subset (\mathcal{S})^*$ which we constructed in Chapter 3. This Gel'fand triple was introduced by Potthoff and Streit in [150].

Another important special case is when $E = L^2(T, \nu)$ and the norms given by a certain operator A. The resulting Gel'fand triple was introduced by Hida et al. [60] (when T is an interval of $\mathbb{R}$ with a measure ν) and Hida et al. [57] (when T is a Riemannian manifold with volume element ν).

Yet another special case of Kubo-Takenaka is when E is a real separable Hilbert space and the norms are given by an operator A satisfying certain conditions. The associated Gel'fand triple was introduced by Kubo and Kuo [90]. In the next section §4.2, we will construct this Gel'fand triple.

D. *Construction of Kondratiev and Streit*

In [84], Kondratiev and Streit introduced a family $(\mathcal{S})^*_\beta$, $0 \leq \beta < 1$, of generalized functions. Their novel idea is to replace $n!$ in the definition of various norms by $(n!)^{1+\beta}$. The resulting Gel'fand triple is $(\mathcal{S})_\beta \subset (L^2) \subset (\mathcal{S})^*_\beta$ and the following relationship holds:

$$(\mathcal{S})_\beta \subset (\mathcal{S}) \subset (L^2) \subset (\mathcal{S})^* \subset (\mathcal{S})^*_\beta.$$

We will discuss the construction of this Gel'fand triple in §4.4.

E. *Construction of Lee*

In [122], Y.-J. Lee introduced a Gel'fand triple $\mathcal{A} \subset (L^2) \subset \mathcal{A}^*$. In Lee's construction, the Wiener-Itô theorem is not used and the elements of $\mathcal{A}$ have analytic property. This construction will be described in §15.2.

4.2 Construction from a Hilbert space and an operator

Let E be a real separable Hilbert space with norm $|\cdot|_0$. Let A be an operator on E such that there exists an orthonormal basis $\{\zeta_j; j = 1, 2, \ldots\}$ for E satisfying the conditions:

(1) $A\zeta_j = \lambda_j \zeta_j$, $j = 1, 2, \ldots$.
(2) $1 < \lambda_1 \leq \lambda_2 \leq \cdots \leq \lambda_n \leq \cdots$.
(3) $\sum_{j=1}^{\infty} \lambda_j^{-\alpha} < \infty$ for some positive constant α.

Note that by condition (2) the operator A^{-1} is bounded with the operator norm $\|A^{-1}\| = \frac{1}{\lambda_1}$. Moreover, by condition (3), the operator $(A^{-1})^{\alpha/2}$ is a Hilbert-Schmidt operator of E with

$$\|(A^{-1})^{\alpha/2}\|^2_{HS} = \sum_{j=1}^{\infty} \lambda_j^{-\alpha}.$$

Now, with the given Hilbert space E and the operator A, we can use the following standard procedure to construct a Gel'fand triple $(\mathcal{E}) \subset (L^2) \subset (\mathcal{E})^*$.

Step 1 *White noise space associated with E and A*

For each $p \geq 0$, define $|\xi|_p = |A^p \xi|_0$ and let

$$\mathcal{E}_p = \{\xi \in E; \ |\xi|_p < \infty\}.$$

It is easy to see that $\mathcal{E}_p \subset \mathcal{E}_q$ for any $p \geq q \geq 0$ and the inclusion map $\mathcal{E}_{p+\frac{\alpha}{2}} \hookrightarrow \mathcal{E}_p$ is a Hilbert-Schmidt operator for any $p \geq 0$. Let

$$\mathcal{E} = \text{ the projective limit of } \{\mathcal{E}_p; \ p \geq 0\},$$
$$\mathcal{E}' = \text{ the dual space of } \mathcal{E}.$$

Then $\mathcal{E}$ is a nuclear space and we get a Gel'fand triple $\mathcal{E} \subset E \subset \mathcal{E}'$ with the following continuous inclusions:

$$\mathcal{E} \subset \mathcal{E}_p \subset E \subset \mathcal{E}'_p \subset \mathcal{E}', \quad p \geq 0,$$

where the norm on $\mathcal{E}'_p$ can be checked to be given by

$$|f|_{-p} = |A^{-p}f|_0.$$

Step 2 *Spaces of test and generalized functions*

By the Minlos theorem there is a unique probability measure μ on $\mathcal{E}'$ such that

$$\int_{\mathcal{E}'} e^{i\langle x, \xi \rangle} \, d\mu(x) = e^{-\frac{1}{2}|\xi|_0^2}, \quad \xi \in \mathcal{E}.$$

Let (L^2) denote the complex Hilbert space $L^2(\mathcal{E}', \mu)$. By the Wiener-Itô theorem each function φ in (L^2) can be expressed uniquely as

$$\varphi = \sum_{n=0}^{\infty} I_n(f_n), \quad f_n \in E_c^{\widehat{\otimes} n},$$

where $E_c^{\widehat{\otimes} n}$ is the complexification of the symmetric tensor product $E^{\widehat{\otimes} n}$. Moreover, for the (L^2)-norm $\|\varphi\|_0$ of φ, we have

$$\|\varphi\|_0^2 = \sum_{n=0}^{\infty} n! |f_n|_0^2,$$

where $|\cdot|_0$ denotes the $E_c^{\widehat{\otimes} n}$-norm. We mention that the multiple Wiener integral I_n is defined as the linear functional on $E_c^{\widehat{\otimes} n}$ such that for any $n_1 + n_2 + \cdots = n$,

$$I_n\left(\zeta_1^{\otimes n_1} \widehat{\otimes} \zeta_2^{\otimes n_2} \widehat{\otimes} \cdots \right) = \mathcal{H}_{n_1}(\langle \cdot, \zeta_1 \rangle) \mathcal{H}_{n_2}(\langle \cdot, \zeta_2 \rangle) \cdots, \qquad (4.2)$$

where $\mathcal{H}_n(x) = (-1)^n e^{x^2/2} D_x^n e^{-x^2/2}$.

The second quantization operator $\Gamma(A)$ of A is a densely defined operator on (L^2) given by: For $\varphi = \sum_{n=0}^{\infty} I_n(f_n)$,

$$\Gamma(A)\varphi = \sum_{n=0}^{\infty} I_n(A^{\otimes n} f_n).$$

The operator $\Gamma(A)$ inherits the same properties as the operator A. To see this, let

$$\varphi_{n_1, n_2, \ldots} \equiv \frac{1}{\sqrt{n_1! n_2! \cdots}} I_n\left(\zeta_1^{\otimes n_1} \widehat{\otimes} \zeta_2^{\otimes n_2} \widehat{\otimes} \cdots \right), \quad n_1 + n_2 + \cdots = n.$$

Then the family $\{\varphi_{n_1,n_2,\dots};\ n_1 + n_2 + \cdots = n,\ n = 0,1,2,\dots\}$ forms an orthonormal basis for (L^2) and

$$\Gamma(A)\varphi_{n_1,n_2,\dots} = (\lambda_1^{n_1}\lambda_2^{n_2}\cdots)\varphi_{n_1,n_2,\dots}.$$

Thus the conditions (2) and (3) for A can be easily checked to hold for $\Gamma(A)$. Moreover, it is easy to see that $(\Gamma(A)^{-1})^{\alpha/2}$ is a Hilbert-Schmidt operator of (L^2) and

$$\|(\Gamma(A)^{-1})^{\alpha/2}\|_{HS}^2 = \left(\prod_{j=1}^{\infty}\left(1 - \lambda_j^{-\alpha}\right)\right)^{-1}.$$

Now we can use the same method as in step 1 to construct a Gel'fand triple, i.e., for each $p \geq 0$, define

$$\|\varphi\|_p = \|\Gamma(A)^p\varphi\|_0$$

and let

$$(\mathcal{E}_p) = \{\varphi \in (L^2);\ \|\varphi\|_p < \infty\}.$$

Obviously $(\mathcal{E}_p) \subset (\mathcal{E}_q)$ for any $p \geq q \geq 0$ and the inclusion map $(\mathcal{E}_{p+\frac{\alpha}{2}}) \hookrightarrow (\mathcal{E}_p)$ is a Hilbert-Schmidt operator. Let

$$(\mathcal{E}) = \text{ the projective limit of } \{(\mathcal{E}_p);\ p \geq 0\},$$
$$(\mathcal{E})^* = \text{ the dual space of } (\mathcal{E}).$$

Then $(\mathcal{E})$ is a nuclear space and we have a Gel'fand triple $(\mathcal{E}) \subset (L^2) \subset (\mathcal{E})^*$ with the following continuous inclusions:

$$(\mathcal{E}) \subset (\mathcal{E}_p) \subset (L^2) \subset (\mathcal{E}_p)^* \subset (\mathcal{E})^*, \quad p \geq 0,$$

where the norm on $(\mathcal{E}_p)^*$ can be checked to be given by

$$\|\varphi\|_{-p} = \|\Gamma(A)^{-p}\varphi\|_0.$$

That is, $(\mathcal{E}_p)^*$ is the completion of (L^2) with respect to the norm $\|\cdot\|_{-p}$.

4.3 General construction of Kubo and Takenaka

In this section we will give a modified version of the construction due to Kubo and Takenaka in [92] and [93]. The modification is the use of the Wiener-Itô theorem rather than the Fock space representation. Let E be a real separable Hilbert space with norm $|\cdot|_0$. Suppose $\{|\cdot|_p;\ p = 1,2,\dots\}$

is a sequence of densely defined inner product norms on E satisfying the following conditions:

(1) There is a number $0 < \rho < 1$ such that $|\cdot|_0 \leq \rho |\cdot|_1 \leq \cdots \leq \rho^p |\cdot|_p \leq \cdots$.
(2) Let $\mathcal{E}_p \equiv \{\xi \in E; \ |\xi|_p < \infty\}$. For any $p \geq 0$, there exists $q > p$ such that the inclusion map $i_{q,p} : \mathcal{E}_q \hookrightarrow \mathcal{E}_p$ is a Hilbert-Schmidt operator and $\|i_{q,p}\|_{HS} < 1$.

Note that $\mathcal{E}_p$ is a Hilbert space with norm $|\cdot|_p$ and $\mathcal{E}_p \subset \mathcal{E}_q$ for any $p \geq q$.
Let

$$\mathcal{E} = \text{ the projective limit of } \{\mathcal{E}_p; \ p \geq 0\},$$

$$\mathcal{E}' = \text{ the dual space of } \mathcal{E}.$$

Then $\mathcal{E} = \cap_{p \geq 0} \mathcal{E}_p$ with the topology generated by the family $\{|\cdot|_p; \ p \geq 0\}$ of norms. It is a nuclear space so that $\mathcal{E} \subset E \subset \mathcal{E}'$ is a Gel'fand triple and we have the following continuous inclusions:

$$\mathcal{E} \subset \mathcal{E}_p \subset E \subset \mathcal{E}'_p \subset \mathcal{E}', \quad p \geq 0.$$

Let μ be the standard Gaussian measure on $\mathcal{E}'$, i.e., its characteristic functional is given by

$$\int_{\mathcal{E}'} e^{i\langle x, \xi \rangle} \, d\mu(x) = e^{-\frac{1}{2}|\xi|_0^2}, \quad \xi \in \mathcal{E}.$$

Again by the Wiener-Itô theorem each element φ in $(L^2) \equiv L^2(\mathcal{E}', \mu)$ can be expressed uniquely as

$$\varphi = \sum_{n=0}^{\infty} I_n(f_n), \quad f_n \in E_c^{\widehat{\otimes} n},$$

and for the (L^2)-norm $\|\varphi\|_0$ of φ, we have

$$\|\varphi\|_0^2 = \sum_{n=0}^{\infty} n! |f_n|_0^2,$$

where $|\cdot|_0$ denotes the $E_c^{\widehat{\otimes} n}$-norm induced from the norm $|\cdot|_0$ on E.
Now, for each positive integer p, define

$$\|\varphi\|_p = \left(\sum_{n=0}^{\infty} n! |f_n|_p^2 \right)^{1/2},$$

where $|\cdot|_p$ is the $\mathcal{E}_{p,c}^{\widehat{\otimes} n}$-norm induced from the norm $|\cdot|_p$ on $\mathcal{E}_p$. Let

$$(\mathcal{E}_p) = \{\varphi \in (L^2); \ \|\varphi\|_p < \infty\}.$$

Obviously $(\mathcal{E}_p) \subset (\mathcal{E}_q)$ for any $p \geq q$. Moreover, for any $p \geq 0$, there exists $q > p$ such that the inclusion map $I_{q,p} : (\mathcal{E}_q) \hookrightarrow (\mathcal{E}_p)$ is a Hilbert-Schmidt operator and

$$\|I_{q,p}\|_{HS}^2 \leq \left(1 - \|i_{q,p}\|_{HS}^2\right)^{-1},$$

where $i_{q,p}$ is the inclusion map from $\mathcal{E}_q$ into $\mathcal{E}_p$, which is a Hilbert-Schmidt operator by assumption (2).

Finally, define

$$(\mathcal{E}) = \text{ the projective limit of } \{(\mathcal{E}_p); \, p \geq 0\},$$
$$(\mathcal{E})^* = \text{ the dual space of } (\mathcal{E}).$$

Then $(\mathcal{E}) = \cap_{p \geq 0}(\mathcal{E}_p)$ with the topology generated by the family $\{\|\cdot\|_p; \, p \geq 0\}$ of norms. It is a nuclear space so that $(\mathcal{E}) \subset (L^2) \subset (\mathcal{E})^*$ is a Gel'fand triple and we have the following continuous inclusions:

$$(\mathcal{E}) \subset (\mathcal{E}_p) \subset (L^2) \subset (\mathcal{E}_p)^* \subset (\mathcal{E})^*, \quad p \geq 0.$$

4.4 Construction of Kondratiev and Streit

The original construction of Kondratiev-Streit uses the white noise space $(\mathcal{S}'(\mathbb{R}), \mu)$ and the operator $A = -d^2/du^2 + u^2 + 1$. We can easily apply their ideas to the general construction in §4.3. But, for the sake of simplicity, we will use the white noise space $(\mathcal{E}', \mu)$ associated with an operator A satisfying the conditions in §4.2.

Recall that each $\varphi \in (L^2)$ can be represented by

$$\varphi = \sum_{n=0}^{\infty} I_n(f_n), \quad f_n \in E_c^{\widehat{\otimes}n}$$

and its (L^2)-norm is given by

$$\|\varphi\|_0 = \left(\sum_{n=0}^{\infty} n!|f_n|_0^2\right)^{1/2}.$$

Now, let $0 \leq \beta < 1$ be a fixed number. For each $p \geq 0$, define

$$\|\varphi\|_{p,\beta} = \left(\sum_{n=0}^{\infty} (n!)^{1+\beta}\left|(A^p)^{\otimes n}f_n\right|_0^2\right)^{1/2}$$

and let
$$(\mathcal{E}_p)_\beta = \{\varphi \in (L^2); \|\varphi\|_{p,\beta} < \infty\}.$$

Note that $(\mathcal{E}_0)_\beta \neq (L^2)$ unless $\beta = 0$. As in §4.2, $(\mathcal{E}_p)_\beta \subset (\mathcal{E}_q)_\beta$ for any $p \geq q \geq 0$ and the inclusion map $(\mathcal{E}_{p+\frac{a}{2}})_\beta \hookrightarrow (\mathcal{E}_p)_\beta$ is a Hilbert-Schmidt operator. Let

$$\begin{aligned}
(\mathcal{E})_\beta &= \text{ the projective limit of } \{(\mathcal{E}_p)_\beta; \, p \geq 0\}, \\
(\mathcal{E})_\beta^* &= \text{ the dual space of } (\mathcal{E})_\beta.
\end{aligned}$$

Then $(\mathcal{E})_\beta$ is a nuclear space and we have a Gel'fand triple $(\mathcal{E})_\beta \subset (L^2) \subset (\mathcal{E})_\beta^*$ with the following continuous inclusions:

$$(\mathcal{E})_\beta \subset (\mathcal{E}_p)_\beta \subset (L^2) \subset (\mathcal{E}_p)_\beta^* \subset (\mathcal{E})_\beta^*, \quad p \geq 0,$$

where the norm on $(\mathcal{E}_p)_\beta^*$ can be checked to be given by

$$\|\varphi\|_{-p,-\beta} = \left(\sum_{n=0}^{\infty} (n!)^{1-\beta} |(A^{-p})^{\otimes n} f_n|_0^2 \right)^{1/2}.$$

That is, $(\mathcal{E}_p)_\beta^*$ is the completion of (L^2) with respect to the norm $\|\cdot\|_{-p,-\beta}$.

The relationship between this new Gel'fand triple and the one discussed in §4.2 is given by

$$(\mathcal{E})_\beta \subset (\mathcal{E}) \subset (L^2) \subset (\mathcal{E})^* \subset (\mathcal{E})_\beta^*, \quad 0 \leq \beta < 1.$$

Note that $(\mathcal{E})_0 = (\mathcal{E})$ and $(\mathcal{E})_{\beta_1} \subset (\mathcal{E})_{\beta_2}$ for any $1 > \beta_1 \geq \beta_2 \geq 0$. Moreover, for any $p \geq 0$,

$$(\mathcal{E}_p)_\beta \subset (\mathcal{E}_p) \subset (L^2) \subset (\mathcal{E}_p)^* \subset (\mathcal{E}_p)_\beta^*.$$

5

The S-transform

From now on we will use the following spaces of test and generalized functions and adopt the same notations.

(A) A Hilbert space E and an operator A

This case was discussed in §4.2. For a given real Hilbert space E and an operator A of E satisfying the conditions (1)–(3) in §4.2, we have the sequence associated with E

$$\mathcal{E} \subset \mathcal{E}_p \subset E \subset \mathcal{E}'_p \subset \mathcal{E}'$$

and the sequence associated with (L^2)

$$(\mathcal{E}) \subset (\mathcal{E}_p) \subset (L^2) \subset (\mathcal{E}_p)^* \subset (\mathcal{E})^*.$$

(B) Special case with $E = L^2(T, \nu)$ and an operator A

Let T be a topological space and ν a σ-finite Borel measure on T. For the special case $E = L^2(T, \nu)$ (real-valued functions) and A an operator satisfying the conditions (1)–(3) in §4.2, the corresponding sequences will be denoted by

$$\mathcal{S} \subset \mathcal{S}_p \subset L^2(T, \nu) \subset \mathcal{S}'_p \subset \mathcal{S}'$$

$$(\mathcal{S}) \subset (\mathcal{S}_p) \subset (L^2) \subset (\mathcal{S}_p)^* \subset (\mathcal{S})^*.$$

For this special case we assume three additional conditions: (1) each $\xi \in \mathcal{S}$ has a unique continuous version $\widetilde{\xi}$, i.e., $\widetilde{\xi}$ is continuous and $\widetilde{\xi} = \xi$, ν-a.e., (2) $\delta_t \in \mathcal{S}'$ for all $t \in T$, and (3) the mapping $t \mapsto \delta_t$ is continuous with the strong topology for $\mathcal{S}'$. We need these conditions for differential operators and integral kernel operators. Thus elements in $\mathcal{S}$ will be understood to be continuous and $\widetilde{\xi}$ will be simply denoted by ξ.

(C) Spaces with index β

This is the case from §4.4. For the case (a) above, we have the following sequence from §4.4

$$(\mathcal{E})_\beta \subset (\mathcal{E}_p)_\beta \subset (L^2) \subset (\mathcal{E}_p)_\beta^* \subset (\mathcal{E})_\beta^*.$$

For the special case (b) above with $E = L^2(T,\nu)$, the corresponding sequence will be denoted by

$$(\mathcal{S})_\beta \subset (\mathcal{S}_p)_\beta \subset (L^2) \subset (\mathcal{S}_p)_\beta^* \subset (\mathcal{S})_\beta^*.$$

Remark. We have chosen the above Gel'fand triples for the convenience of calculations. Most of the theorems for these Gel'fand triples remain valid for the general Gel'fand triples in §4.3 with suitable modifications in the statements.

5.1 Wick tensors and multiple Wiener integrals

A very useful operator in white noise analysis is the *trace operator* τ. It is an element in $(\mathcal{E}_c')^{\widehat{\otimes}2}$ defined by

$$\langle \tau, \xi \otimes \eta \rangle = \langle \xi, \eta \rangle, \quad \xi, \eta \in \mathcal{E}_c.$$

Obviously the trace operator τ can be represented as

$$\tau = \sum_{j=1}^{\infty} \zeta_j \otimes \zeta_j,$$

where ζ_js are the eigenvectors of A in §4.2. Hence for $p > 0$,

$$|\tau|_{-p}^2 = \sum_{j=1}^{\infty} \left| (A^{-p})^{\otimes 2} (\zeta_j \otimes \zeta_j) \right|_0^2$$

$$= \sum_{j=1}^{\infty} \left| A^{-p} \zeta_j \right|_0^4$$

$$= \sum_{j=1}^{\infty} \lambda_j^{-4p}.$$

For any $p \geq \frac{\alpha}{4}$, we have $\lambda_j^{-4p} = \lambda_j^{-4p+\alpha} \lambda_j^{-\alpha} \leq \lambda_1^{-4p+\alpha} \lambda_j^{-\alpha}$. Hence

$$|\tau|_{-p} \leq \lambda_1^{-2p+\frac{\alpha}{2}} \left(\sum_{j=1}^{\infty} \lambda_j^{-\alpha} \right)^{1/2}. \tag{5.1}$$

Thus $\tau \in (\mathcal{E}'_{p,c})^{\widehat{\otimes}2}$ for any $p \geq \frac{\alpha}{4}$.

The *Hermite polynomial* of degree n with parameter σ^2 is defined by

$$: x^n :_{\sigma^2} = (-\sigma^2)^n e^{\frac{x^2}{2\sigma^2}} D_x^n e^{-\frac{x^2}{2\sigma^2}}.$$

The generating function for these Hermite polynomials is given by

$$\sum_{n=0}^{\infty} \frac{t^n}{n!} : x^n :_{\sigma^2} = e^{tx - \frac{1}{2}\sigma^2 t^2}. \tag{5.2}$$

Moreover, we have the following formulas:

$$: x^n :_{\sigma^2} = \sum_{k=0}^{[n/2]} \binom{n}{2k} (2k-1)!! \, (-\sigma^2)^k x^{n-2k}, \tag{5.3}$$

$$x^n = \sum_{k=0}^{[n/2]} \binom{n}{2k} (2k-1)!! \, \sigma^{2k} : x^{n-2k} :_{\sigma^2}, \tag{5.4}$$

where $(2k-1)!! = (2k-1)(2k-3)\cdots 3\cdot 1$ and by convention $(-1)!! = 1$.

Being motivated by equation (5.3), we define Wick tensors of elements in $\mathcal{E}'$ by a similar formula as follows.

Definition 5.1. The *Wick tensor* $: x^{\otimes n} :$ of an element x in $\mathcal{E}'$ is defined to be

$$: x^{\otimes n} := \sum_{k=0}^{[n/2]} \binom{n}{2k} (2k-1)!! \, (-1)^k x^{\otimes(n-2k)} \widehat{\otimes} \tau^{\otimes k},$$

where τ is the trace operator.

We have a formula similar to equation (5.4) for the Wick tensor, i.e.,

$$x^{\otimes n} = \sum_{k=0}^{[n/2]} \binom{n}{2k} (2k-1)!! : x^{\otimes(n-2k)} : \widehat{\otimes} \tau^{\otimes k}.$$

Lemma 5.2. *For any* $x \in \mathcal{E}'$ *and* $\xi \in \mathcal{E}$, *the following equalities hold:*

$$\langle : x^{\otimes n} :, \xi^{\otimes n} \rangle = : \langle x, \xi \rangle^n :_{|\xi|_0^2},$$

$$\| \langle : \cdot^{\otimes n} :, \xi^{\otimes n} \rangle \|_0 = \sqrt{n!} \, |\xi|_0^n.$$

Proof. From the definition of the Wick tensor for $: x^{\otimes n} :$, we get

$$\langle : x^{\otimes n} :, \xi^{\otimes n} \rangle = \sum_{k=0}^{[n/2]} \binom{n}{2k} (2k-1)!! \, (-|\xi|_0^2)^k \langle x, \xi \rangle^{n-2k}.$$

Thus in view of the formula (5.3) we get the first conclusion. The second conclusion follows from the following formula which can be easily checked from the generating function in equation (5.2)

$$\int_{\mathbb{R}} :x^n:_{\sigma^2}^2 \, d\mu_{\sigma^2}(x) = n!\sigma^{2n},$$

where μ_{σ^2} is the one-dimensional Gaussian measure with mean zero and variance σ^2. □

Corollary 5.3. *Let* $\xi_1, \xi_2, \ldots \in \mathcal{E}$ *be orthogonal in* E. *Then for all* $x \in \mathcal{E}'$,

$$\langle :x^{\otimes n}:, \xi_1^{\otimes n_1} \widehat{\otimes} \xi_2^{\otimes n_2} \widehat{\otimes} \cdots \rangle = :\langle x, \xi_1 \rangle^{n_1}:_{|\xi_1|_0^2} :\langle x, \xi_2 \rangle^{n_2}:_{|\xi_2|_0^2} \cdots,$$

where $n_1 + n_2 + \cdots = n$. *Moreover,*

$$\|\langle :\cdot^{\otimes n}:, \xi_1^{\otimes n_1} \widehat{\otimes} \xi_2^{\otimes n_2} \widehat{\otimes} \cdots \rangle\|_0 = \sqrt{n_1! n_2! \cdots} \, |\xi_1|_0^{n_1} |\xi_2|_0^{n_2} \cdots.$$

We now define $\langle :x^{\otimes n}:, f \rangle$ for $f \in E_c^{\widehat{\otimes} n}$. First note that if $h \in \mathcal{E}^{\widehat{\otimes} n}$, then the bilinear pairing $\langle :x^{\otimes n}:, h \rangle$ is defined for all $x \in \mathcal{E}'$. It follows from Corollary 5.3 that

$$\|\langle :\cdot^{\otimes n}:, h \rangle\|_0 = \sqrt{n!} \, |h|_0.$$

Next, suppose $f \in E^{\widehat{\otimes} n}$. Take a sequence $\{h_j\}$ in $\mathcal{E}^{\widehat{\otimes} n}$ such that $h_j \to f$ in $E^{\widehat{\otimes} n}$. Then the sequence $\{\langle :\cdot^{\otimes n}:, h_j \rangle\}$ is Cauchy in (L^2). Define

$$\langle :\cdot^{\otimes n}:, f \rangle = \text{ limit in } (L^2) \text{ of } \langle :\cdot^{\otimes n}:, h_j \rangle \text{ as } j \to \infty.$$

Obviously the limit is independent of the sequence $\{h_j\}$ and so $\langle :\cdot^{\otimes n}:, f \rangle$ is well-defined. Thus for $f \in E^{\widehat{\otimes} n}$, $\langle :x^{\otimes n}:, f \rangle$ is defined for almost all $x \in \mathcal{E}'$ and we have

$$\|\langle :\cdot^{\otimes n}:, f \rangle\|_0 = \sqrt{n!} \, |f|_0. \tag{5.5}$$

Finally, for $f = f_1 + if_2 \in E_c^{\widehat{\otimes} n}$ with $f_1, f_2 \in E^{\widehat{\otimes} n}$, we define

$$\langle :x^{\otimes n}:, f \rangle = \langle :x^{\otimes n}:, f_1 \rangle + i\langle :x^{\otimes n}:, f_2 \rangle.$$

Then $\langle :x^{\otimes n}:, f \rangle$ is defined for almost all $x \in \mathcal{E}'$ and it is easy to check that equation (5.5) still holds for $f \in E_c^{\widehat{\otimes} n}$.

The next theorem shows that multiple Wiener integrals can be expressed in terms of Wick tensors.

Theorem 5.4. (1) Let $h_1, h_2, \ldots \in E$ be orthogonal and $n_1 + n_2 + \cdots = n$. Then for almost all x in $\mathcal{E}'$,

$$\langle :x^{\otimes n}:, h_1^{\otimes n_1} \widehat{\otimes} h_2^{\otimes n_2} \widehat{\otimes} \cdots \rangle = :\langle x, h_1 \rangle^{n_1}:_{|h_1|_0^2} :\langle x, h_2 \rangle^{n_2}:_{|h_2|_0^2} \cdots .$$

(2) Let $f \in E_c^{\widehat{\otimes} n}$. Then for almost all x in $\mathcal{E}'$,

$$I_n(f)(x) = \langle :x^{\otimes n}:, f \rangle.$$

Proof. For the first assertion it suffices to show that for any $h \in E$, the following equality holds for almost all $x \in \mathcal{E}'$

$$\langle :x^{\otimes n}:, h^{\otimes n} \rangle = :\langle x, h \rangle^n:_{|h|_0^2} .$$

Choose a sequence $\{\xi_j\} \subset \mathcal{E}$ such that $\xi_j \to h$ in E. By considering $|h|_0 |\xi_j|_0^{-1} \xi_j$, we may assume that $|\xi_j|_0 = |h|_0$ for all j. Then by Lemma 5.2

$$\begin{aligned}
\langle :x^{\otimes n}:, h^{\otimes n} \rangle &= \lim_{j \to \infty} \langle :x^{\otimes n}:, \xi_j^{\otimes n} \rangle \\
&= \lim_{j \to \infty} :\langle x, \xi_j \rangle^n:_{|h|_0^2} \\
&= :\langle x, h \rangle^n:_{|h|_0^2} .
\end{aligned}$$

To prove the second assertion, note that for the function $\mathcal{H}_n(u)$ in equation (4.2) we have $\mathcal{H}_n(u) = :u^n:_1$. Hence by equation (4.2) and Corollary 5.3, the second assertion holds for $f = \zeta_1^{\otimes n_1} \widehat{\otimes} \zeta_2^{\otimes n_2} \widehat{\otimes} \cdots$. This implies that the assertion holds for any $f \in \mathcal{E}_c^{\widehat{\otimes} n}$. Since $\mathcal{E}_c^{\widehat{\otimes} n}$ is dense in $E_c^{\widehat{\otimes} n}$, the second assertion follows immediately. □

Thus by the above theorem an element φ in (L^2) can be expressed in terms of Wick tensors by

$$\varphi(x) = \sum_{n=0}^{\infty} \langle :x^{\otimes n}:, f_n \rangle, \quad \mu-\text{a.e. for } x \in \mathcal{E}',$$

where $f_n \in E_c^{\widehat{\otimes} n}$. For any $p \geq 0$ and $0 \leq \beta < 1$, we have

$$\|\varphi\|_{p,\beta}^2 = \sum_{n=0}^{\infty} (n!)^{1+\beta} |(A^p)^{\otimes n} f_n|_0^2. \tag{5.6}$$

This implies, in particular, that if $\varphi \in (\mathcal{E})_\beta$, then $f_n \in \mathcal{E}_c^{\widehat{\otimes} n}$ for all n.

We will use the same notation to express an element $\Phi \in (\mathcal{E}_p)_\beta^*$ as follows:

$$\Phi(x) = \sum_{n=0}^{\infty} \langle :x^{\otimes n}:, F_n \rangle,$$

where $F_n \in (\mathcal{E}'_{p,c})^{\widehat{\otimes}n}$. For the $(\mathcal{E}_p)^*_\beta$-norm $\|\Phi\|_{-p,-\beta}$ of Φ, we have

$$\|\Phi\|^2_{-p,-\beta} = \sum_{n=0}^{\infty} (n!)^{1-\beta} |(A^{-p})^{\otimes n} F_n|^2_0. \tag{5.7}$$

Moreover, if $\varphi = \sum_{n=0}^{\infty} \langle :x^{\otimes n}:, f_n \rangle \in (\mathcal{E})_\beta$, then

$$\langle\!\langle \Phi, \varphi \rangle\!\rangle = \sum_{n=0}^{\infty} n! \langle F_n, f_n \rangle. \tag{5.8}$$

5.2   Definition of the S-transform

Let $\varphi \in (L^2)$ and consider its μ-convolution

$$\mu\varphi(h) = \int_{\mathcal{E}'} \varphi(x + h)\, d\mu(x), \quad h \in E.$$

This μ-convolution has a smoothing property similar to the finite dimensional Gaussian measure (see Gross [42] or Kuo [98] [102]). By using the translation formula $d\mu_h/d\mu$ in §2.1, we can rewrite $\mu\varphi$ as

$$\mu\varphi(h) = \int_{\mathcal{E}'} \varphi(x) e^{\langle x,h \rangle - \frac{1}{2}|h|^2_0}\, d\mu(x)$$

$$= e^{-\frac{1}{2}|h|^2_0} \int_{\mathcal{E}'} \varphi(x) e^{\langle x,h \rangle}\, d\mu(x), \quad h \in E. \tag{5.9}$$

Question: Can we extend this μ-convolution to generalized functions?

In order to do so, we need to consider the convolution $\mu\varphi$ as a function on the space $\mathcal{E}$ (in fact, on $\mathcal{E}_c$) and to require that $e^{\langle \cdot, \xi \rangle}$ be a test function in $(\mathcal{E})_\beta$ for any $\xi \in \mathcal{E}$ (in fact, $\mathcal{E}_c$). Then the integral in equation (5.9) can be replaced by the bilinear pairing $\langle\!\langle \cdot, \cdot \rangle\!\rangle$ of $(\mathcal{E})^*_\beta$ and $(\mathcal{E})_\beta$. First we define $:e^{\langle \cdot, h \rangle}:$ for $h \in E$ by

$$:e^{\langle \cdot, h \rangle}: \equiv e^{\langle \cdot, h \rangle - \frac{1}{2}|h|^2_0}.$$

Lemma 5.5. *Let $h \in E$. Then the Wiener-Itô decomposition of $:e^{\langle \cdot, h \rangle}:$ is given by*

$$:e^{\langle x,h \rangle}: = \sum_{n=0}^{\infty} \frac{1}{n!} \langle :x^{\otimes n}:, h^{\otimes n} \rangle.$$

Proof. By the generating function in equation (5.2) with $t = 1$ and $\sigma^2 = |h|_0^2$, we have

$$:e^{\langle x,h\rangle}: = e^{\langle x,h\rangle - \frac{1}{2}|h|_0^2}$$

$$= \sum_{n=0}^{\infty} \frac{1}{n!} :\langle x,h\rangle^n:_{|h|_0^2} .$$

This proves the lemma since $:\langle x,h\rangle^n:_{|h|_0^2} = \langle :x^{\otimes n}:, h^{\otimes n}\rangle$ by Theorem 5.4 (1). $\qquad\qquad\square$

In view of Lemma 5.5 we make the following definition.

Definition 5.6. For $h \in E_c$, the *renormalization* $:e^{\langle \cdot,h\rangle}:$ of $e^{\langle \cdot,h\rangle}$ is defined by

$$:e^{\langle \cdot,h\rangle}: = \sum_{n=0}^{\infty} \frac{1}{n!} \langle :\cdot^{\otimes n}:, h^{\otimes n}\rangle.$$

Remark. It is easy to check that for any $h \in E_c$,

$$:e^{\langle \cdot,h\rangle}: = e^{\langle \cdot,h\rangle - \frac{1}{2}\langle h,h\rangle}.$$

Theorem 5.7. *If $\xi \in \mathcal{E}_c$, then $:e^{\langle \cdot,\xi\rangle}: \in (\mathcal{E})_\beta$ for all $0 \leq \beta < 1$. In fact, for any $p \geq 0$,*

$$\||:e^{\langle \cdot,\xi\rangle}:\|_{p,\beta} \leq 2^{\beta/2} \exp\left[(1-\beta)2^{\frac{2\beta-1}{1-\beta}}|\xi|_p^{\frac{2}{1-\beta}}\right],$$

where the equality holds when $\beta = 0$.

Remark. Note that by the previous remark we have

$$e^{\langle \cdot,\xi\rangle} = e^{\frac{1}{2}\langle \xi,\xi\rangle}:e^{\langle \cdot,\xi\rangle}: .$$

Hence we also have $e^{\langle \cdot,\xi\rangle} \in (\mathcal{E})_\beta$. Moreover, the proof below shows that for any ξ with $|\xi|_p \geq 1$, we have

$$\||:e^{\langle \cdot,\xi\rangle}:\|_{p,\beta} = \infty \quad \text{when } \beta = 1.$$

This is the reason why $(\mathcal{E})_\beta$ is defined only for $0 \leq \beta < 1$ since we want to include all functions $:e^{\langle \cdot,\xi\rangle}:$ in the space $(\mathcal{E})_\beta$ for $\xi \in \mathcal{E}_c$.

Proof. By Definition 5.6

$$:e^{\langle x,\xi\rangle}: = \sum_{n=0}^{\infty} \frac{1}{n!} \langle :x^{\otimes n}:, \xi^{\otimes n}\rangle.$$

First consider the case $\beta = 0$. We have

$$\| : e^{\langle \cdot, \xi \rangle} : \|_{p,0}^2 = \sum_{n=0}^{\infty} n! \frac{1}{(n!)^2} |(A^p)^{\otimes n} \xi^{\otimes n}|_0^2$$

$$= \sum_{n=0}^{\infty} \frac{1}{n!} |A^p \xi|_0^{2n}$$

$$= \sum_{n=0}^{\infty} \frac{1}{n!} |\xi|_p^{2n}$$

$$= e^{|\xi|_p^2}.$$

Hence $\| : e^{\langle \cdot, \xi \rangle} : \|_{p,0} = e^{\frac{1}{2}|\xi|_p^2}$. Next, suppose $0 < \beta < 1$. By equation (5.6)

$$\| : e^{\langle \cdot, \xi \rangle} : \|_{p,\beta}^2 = \sum_{n=0}^{\infty} (n!)^{1+\beta} \frac{1}{(n!)^2} |(A^p)^{\otimes n} \xi^{\otimes n}|_0^2$$

$$= \sum_{n=0}^{\infty} \frac{1}{(n!)^{1-\beta}} |A^p \xi|_0^{2n}$$

$$= \sum_{n=0}^{\infty} \frac{1}{(n!)^{1-\beta}} |\xi|_p^{2n}.$$

Now, write the n-th term of the above series as

$$\left(\frac{1}{2^{n\beta}} \right) \left(\frac{2^{n\beta}}{(n!)^{1-\beta}} |\xi|_p^{2n} \right)$$

and apply the Hölder inequality with the pair $\{ \frac{1}{\beta}, \frac{1}{1-\beta} \}$ to get

$$\| : e^{\langle \cdot, \xi \rangle} : \|_{p,\beta}^2 \leq \left(\sum_{n=0}^{\infty} \frac{1}{2^n} \right)^{\beta} \left(\sum_{n=0}^{\infty} \frac{1}{n!} 2^{\frac{n\beta}{1-\beta}} |\xi|_p^{\frac{2n}{1-\beta}} \right)^{1-\beta}$$

$$= 2^{\beta} \left(\exp \left[2^{\frac{\beta}{1-\beta}} |\xi|_p^{\frac{2}{1-\beta}} \right] \right)^{1-\beta}.$$

This yields immediately the inequality for the case $0 < \beta < 1$.

Observe that $|\xi|_p < \infty$ for all $p \geq 0$ since $\xi \in \mathcal{E}_c$. Hence $\| : e^{\langle \cdot, \xi \rangle} : \|_{p,\beta} < \infty$ for all $p \geq 0$ and so $: e^{\langle \cdot, \xi \rangle} : \in (\mathcal{E})_\beta$ for any $0 \leq \beta < 1$.   □

Now, in view of Theorem 5.7 and the remark preceding it, we can extend the convolution in equation (5.9) to generalized functions if we let h be in $\mathcal{E}_c$ and interpret the integral in equation (5.9) as the bilinear pairing of $(\mathcal{E})_\beta^*$ and $(\mathcal{E})_\beta$.

Definition 5.8. The *S-transform* $S\Phi$ of a generalized function $\Phi \in (\mathcal{E})^*_\beta$ is defined to be the function

$$S\Phi(\xi) = \langle\!\langle \Phi, :e^{\langle \cdot, \xi \rangle}: \rangle\!\rangle, \quad \xi \in \mathcal{E}_c,$$

or equivalently

$$S\Phi(\xi) = e^{-\frac{1}{2}\langle \xi, \xi \rangle} \langle\!\langle \Phi, e^{\langle \cdot, \xi \rangle} \rangle\!\rangle, \quad \xi \in \mathcal{E}_c.$$

The next proposition gives the S-transform from the Wiener-Itô decomposition. This is a useful result.

Proposition 5.9. If $\Phi(x) = \sum_{n=0}^{\infty} \langle :x^{\otimes n}:, F_n \rangle \in (\mathcal{E})^*_\beta$, then its S-transform is given by

$$S\Phi(\xi) = \sum_{n=0}^{\infty} \langle F_n, \xi^{\otimes n} \rangle, \quad \xi \in \mathcal{E}_c.$$

Proof. Simply apply Definition 5.6 and equation (5.8) □

The S-transform is a fundamental tool in white noise distribution theory. Often we will specify a generalized function by its S-transform. This is justified by the fact that the S-transform is injective as stated in the next proposition.

Proposition 5.10. Let $\Phi, \Psi \in (\mathcal{E})^*_\beta$. If $S\Phi = S\Psi$, then $\Phi = \Psi$.

Remark. It follows from this proposition that the linear span of the set $\{e^{\langle \cdot, \xi \rangle}; \xi \in \mathcal{E}_c\}$ is dense in $(\mathcal{E})_\beta$. Note that the linear span of the set $\{e^{\langle \cdot, \xi \rangle}; \xi \in \mathcal{E}\}$ is also dense in $(\mathcal{E})_\beta$.

Proof. It suffices to show that $S\Phi = 0$ implies $\Phi = 0$. Suppose $\Phi(x) = \sum_{n=0}^{\infty} \langle :x^{\otimes n}:, F_n \rangle$. Then by Proposition 5.9

$$S\Phi(\xi) = \sum_{n=0}^{\infty} \langle F_n, \xi^{\otimes n} \rangle, \quad \xi \in \mathcal{E}.$$

Thus for any $u \in \mathbb{R}$

$$S\Phi(u\xi) = \sum_{n=0}^{\infty} u^n \langle F_n, \xi^{\otimes n} \rangle = 0.$$

Obviously this implies that $F_0 = 0$ and for $n \geq 1$, $\langle F_n, \xi^{\otimes n} \rangle = 0$ for all $\xi \in \mathcal{E}$. But we have the polarization identity for $F \in (\mathcal{E}')^{\otimes n}$:

$$\langle F, \xi_1 \widehat{\otimes} \cdots \widehat{\otimes} \xi_n \rangle = \frac{1}{n!} \sum_{k=1}^{n} (-1)^{n-k} \sum_{j_1 < \cdots < j_k} \langle F, (\xi_{j_1} + \cdots + \xi_{j_k})^{\otimes n} \rangle.$$

Therefore $F_n = 0$ for any $n \geq 1$. Hence we have shown that $F_n = 0$ for all $n \geq 0$ and so $\Phi = 0$. $\square$

5.3 Examples of generalized functions

In this section we will give several classes of generalized functions. The subscript β below is understood to be a number in the interval $[0, 1)$. More examples of generalized functions will be given in later chapters.

A. *Renormalized polynomials*

Let $f \in E$. Then the function $\langle \cdot, f \rangle^n$ belongs to (L^2). In fact, it is easy to check from equation (5.4) and Theorem 5.4 that $\langle \cdot, f \rangle^n$ has the following decomposition

$$\langle \cdot, f \rangle^n = \sum_{k=0}^{[n/2]} \left\langle : \cdot^{\otimes(n-2k)} :, \binom{n}{2k}(2k-1)!! \, |f|_0^{2k} \, f^{\otimes(n-2k)} \right\rangle.$$

This equality shows that if $\eta \in \mathcal{E}$, then $\langle \cdot, \eta \rangle^n \in (\mathcal{E})_\beta$ for any β. It also shows that if $y \in \mathcal{E}' \backslash E$, then $\langle \cdot, y \rangle^n$ has no meaning. In this case, we need a renormalization.

Definition 5.11. Let $y \in \mathcal{E}'_c$. The *renormalization* $: \langle \cdot, y \rangle^n :$ of $\langle \cdot, y \rangle^n$ is defined by

$$: \langle \cdot, y \rangle^n : = \langle : \cdot^{\otimes n} :, y^{\otimes n} \rangle.$$

Observe that if $y = f \in E$, then $\langle : \cdot^{\otimes n} :, f^{\otimes n} \rangle = : \langle \cdot, f \rangle^n :_{|f|_0^2}$ by Theorem 5.4. Thus we have $: \langle \cdot, f \rangle^n : = : \langle \cdot, f \rangle^n :_{|f|_0^2}$. This shows that for $f \in E$ the renormalization $: \langle \cdot, f \rangle^n :$ of $\langle \cdot, f \rangle^n$ in Definition 5.11 is nothing but $: \langle \cdot, f \rangle^n :$ $= : \langle \cdot, f \rangle^n :_{|f|_0^2}$ (the n-th Hermite polynomial of $\langle \cdot, f \rangle$ with parameter $|f|_0^2$). By Proposition 5.9

$$(S : \langle \cdot, y \rangle^n :)(\xi) = \langle y, \xi \rangle^n, \quad \xi \in \mathcal{E}_c.$$

In particular, when $E = L^2(\mathbb{R})$ and $A = -d^2/du^2 + u^2 + 1$, we have the k-th derivative $B^{(k)}(t)$ of Brownian motion given by

$$B^{(k)}(t) = (-1)^{k-1} \langle \cdot, \delta_t^{(k-1)} \rangle, \tag{5.10}$$

where $\delta_t^{(k-1)}$ denotes the $(k-1)$-st distribution derivative of the delta function δ_t. Hence we get the the renormalization $: B^{(k)}(t)^n :$ with S-transform

$$(S : B^{(k)}(t)^n :)(\xi) = \left(\xi^{(k-1)}(t) \right)^n.$$

B. *Renormalized exponential functions*

Let $f \in E_c$. Then $e^{\langle \cdot, f \rangle}$ and its renormalization $: e^{\langle \cdot, f \rangle} :$ (see Definition 5.6) are both in (L^2) and we have

$$: e^{\langle \cdot, f \rangle} : = \sum_{n=0}^{\infty} \frac{1}{n!} \langle : \cdot^{\otimes n} :, f^{\otimes n} \rangle.$$

If $\eta \in \mathcal{E}_c$, then by Theorem 5.7 the functions $e^{\langle \cdot, \eta \rangle}$ and $: e^{\langle \cdot, \eta \rangle} :$ are both in $(\mathcal{E})_\beta$.

Now, if $y \in \mathcal{E}'_c \backslash E_c$, then $e^{\langle \cdot, y \rangle}$ has no meaning and we need a renormalization.

Definition 5.12. Let $y \in \mathcal{E}'_c$. The *renormalization* $: e^{\langle \cdot, y \rangle} :$ of $e^{\langle \cdot, y \rangle}$ is defined by

$$: e^{\langle \cdot, y \rangle} : = \sum_{n=0}^{\infty} \frac{1}{n!} \langle : \cdot^{\otimes n} :, y^{\otimes n} \rangle.$$

Note that this definition is consistent with Definition 5.6 if $y \in E_c$. The next theorem shows that this renormalization is indeed a generalized function.

Theorem 5.13. Let $y \in \mathcal{E}'_c$. Then $: e^{\langle \cdot, y \rangle} : \in (\mathcal{E})^*$ and its S-transform is given by

$$(S : e^{\langle \cdot, y \rangle} :)(\xi) = e^{\langle y, \xi \rangle}, \quad \xi \in \mathcal{E}_c.$$

In fact, if $y \in \mathcal{E}'_{p,c}$, then $: e^{\langle \cdot, y \rangle} : \in (\mathcal{E}_p)^*$ and $\| : e^{\langle \cdot, y \rangle} : \|_{-p} = e^{\frac{1}{2}|y|^2_{-p}}$.

Proof. Since $\mathcal{E}'_c = \cup_{p \geq 0} \mathcal{E}'_{p,c}$, there is some $p \geq 0$ such that $|y|_{-p} < \infty$. For this p,

$$\| : e^{\langle \cdot, y \rangle} : \|^2_{-p} = \sum_{n=0}^{\infty} n! \frac{1}{(n!)^2} |A^{-p} y|^{2n}$$

$$= \sum_{n=0}^{\infty} \frac{1}{n!} |y|^{2n}_{-p}$$

$$= e^{|y|^2_{-p}}.$$

Hence $\| : e^{\langle \cdot, y \rangle} : \|_{-p} < \infty$ and so $: e^{\langle \cdot, y \rangle} : \in (\mathcal{E}_p)^*$. To find the S-transform, we use equation (5.8) and Definition 5.12

$$(S : e^{\langle \cdot, y \rangle} :)(\xi) = \sum_{n=0}^{\infty} n! \langle \frac{1}{n!} y^{\otimes n}, \frac{1}{n!} \xi^{\otimes n} \rangle$$

$$= \sum_{n=0}^{\infty} \frac{1}{n!} \langle y, \xi \rangle^n$$

$$= e^{\langle y, \xi \rangle}. \qquad \square$$

When $E = L^2(\mathbb{R})$ and $A = -d^2/du^2 + u^2 + 1$, we can use the k-th derivative $B^{(k)}(t)$ of the Brownian motion $B(t)$ as given by equation (5.10) to get the renormalized exponential function $:e^{B^{(k)}(t)}: \in (S)^*$ with the S-transform

$$\left(S : e^{B^{(k)}(t)} : \right)(\xi) = e^{\xi^{(k-1)}(t)}.$$

In particular, when $k = 1$, we have

$$\left(S : e^{\dot{B}(t)} : \right)(\xi) = e^{\xi(t)}. \tag{5.11}$$

C. *Other exponential functions*

Let $f \in E$. Consider the exponential function $e^{\langle \cdot, f \rangle^2}$. First note that this function may not be in (L^2). In fact, $e^{\langle \cdot, f \rangle^2} \in (L^2)$ if and only if $|f|_0 < \frac{1}{2}$. Nevertheless, we can study the additive renormalization of this function.

Definition 5.14. Let $y \in \mathcal{E}'_c$ and $z \in \mathbb{C}$. The *additive renormalization* $:e^{z\langle \cdot, y \rangle^2}:$ of $e^{z\langle \cdot, y \rangle^2}$ is defined by

$$:e^{z\langle \cdot, y \rangle^2}: = \sum_{n=0}^{\infty} \frac{z^n}{n!} \langle :\cdot^{\otimes 2n}:, y^{\otimes 2n} \rangle.$$

Theorem 5.15. Let $y \in \mathcal{E}'_c$ and $z \in \mathbb{C}$. Then $:e^{z\langle \cdot, y \rangle^2}: \in (\mathcal{E})^*$ with S-transform given by

$$\left(S : e^{z\langle \cdot, y \rangle^2} : \right)(\xi) = e^{z\langle y, \xi \rangle^2}, \quad \xi \in \mathcal{E}_c.$$

Proof. Since $\mathcal{E}'_c = \cup_{p \geq 0} \mathcal{E}'_{p,c}$, there is some $p \geq 0$ such that $|y|_{-p} < \infty$. Then for any $r \geq 0$,

$$\begin{aligned}
|y|_{-(p+r)} &= |A^{-(p+r)}y|_0 \\
&= |A^{-r}A^{-p}y|_0 \\
&\leq \lambda_1^{-r}|A^{-p}y|_0 \\
&= \lambda_1^{-r}|y|_{-p},
\end{aligned}$$

where λ_1 is the smallest eigenvalue of A. Therefore

$$\begin{aligned}
\| :e^{z\langle \cdot, y \rangle^2}: \|^2_{-(p+r)} &= \sum_{n=0}^{\infty} (2n)! \frac{|z|^{2n}}{(n!)^2} |y|^{4n}_{-(p+r)} \\
&\leq \sum_{n=0}^{\infty} (2n)! \frac{|z|^{2n}}{(n!)^2} \lambda_1^{-4rn} |y|^{4n}_{-p}.
\end{aligned}$$

Note that by the Stirling formula we have

$$\frac{(2n)!}{(n!)^2} \approx \frac{2^{2n}}{\sqrt{\pi n}}.$$

Choose $r > 0$ so large that $2\lambda_1^{-2r}|z|\,|y|^2_{-p} < 1$. Then the last series above is convergent. Hence $:e^{z\langle\cdot,y\rangle^2}: \in (\mathcal{E}_{p+r})^*$. By using equation (5.8), we get the S-transform

$$(S:e^{z\langle\cdot,y\rangle^2}:)(\xi) = \sum_{n=0}^{\infty}(2n)!\langle\frac{z^n}{n!}y^{\otimes 2n}, \frac{1}{(2n)!}\xi^{\otimes 2n}\rangle$$

$$= \sum_{n=0}^{\infty}\frac{z^n}{n!}\langle y,\xi\rangle^{2n}$$

$$= e^{z\langle y,\xi\rangle^2}. \qquad\qquad \square$$

We can use the same idea to renormalize exponential functions with higher order exponents. But the conclusion is a little bit different.

Definition 5.16. Let $y \in \mathcal{E}'_c$ and $z \in \mathbb{C}$. The *additive renormalization* $:e^{z\langle\cdot,y\rangle^3}:$ of $e^{z\langle\cdot,y\rangle^3}$ is defined by

$$:e^{z\langle\cdot,y\rangle^3}: = \sum_{n=0}^{\infty}\frac{z^n}{n!}\langle:\cdot^{\otimes 3n}:, y^{\otimes 3n}\rangle.$$

Theorem 5.17. Let $y \in \mathcal{E}'_c$ and $z \in \mathbb{C}$. Then $:e^{z\langle\cdot,y\rangle^3}: \in (\mathcal{E})^*_{1/3}$ with S-transform given by

$$(S:e^{z\langle\cdot,y\rangle^3}:)(\xi) = e^{z\langle y,\xi\rangle^3}, \quad \xi \in \mathcal{E}_c.$$

Remark. In fact, the proof below also shows that $:e^{z\langle\cdot,y\rangle^3}: \notin (\mathcal{E})^*$ if $z \neq 0$ and $y \neq 0$.

Proof. As in the proof of Theorem 5.15, for any $p, r \geq 0$,

$$\||:e^{z\langle\cdot,y\rangle^3}:\|^2_{-(p+r),-\beta} = \sum_{n=0}^{\infty}((3n)!)^{1-\beta}\frac{|z|^{2n}}{(n!)^2}|y|^{6n}_{-(p+r)}. \qquad (5.12)$$

By the Stirling formula, we have

$$\frac{((3n)!)^{1-\beta}}{(n!)^2} \approx \frac{(6\pi)^{\frac{1-\beta}{2}}3^{3(1-\beta)n}n^{(1-3\beta)n}}{2\pi n^{\frac{1+\beta}{2}}e^{(1-3\beta)n}}.$$

By using this estimate, it is easy to see that when $\beta = 0$ the series in equation (5.12) is divergent for any $p, r \geq 0$ if $z \neq 0$ and $y \neq 0$. In that case, $:e^{z\langle\cdot,y\rangle^3}: \notin (\mathcal{E})^*$. On the other hand, as in the proof of Theorem 5.15, we have

$$\||:e^{z\langle\cdot,y\rangle^3}:\|^2_{-(p+r),-\beta} \leq \sum_{n=0}^{\infty}((3n)!)^{1-\beta}\frac{|z|^{2n}}{(n!)^2}\lambda_1^{-6rn}|y|^{6n}_{-p}.$$

Thus when $\beta = \frac{1}{3}$, we can choose $r \geq 0$ so large that $3\lambda_1^{-3r}|z|\,|y|_{-p}^3 < 1$. Then the last series is convergent and so $:e^{z\langle\cdot,y\rangle^3}: \in (\mathcal{E}_{p+r})_{1/3}^*$. The S-transform can be derived by the same argument as in the proof of Theorem 5.15. □

In general, we can define the *additive renormalization* $:e^{z\langle\cdot,y\rangle^k}:$ of $e^{z\langle\cdot,y\rangle^k}$ by

$$:e^{z\langle\cdot,y\rangle^k}: \, = \sum_{n=0}^{\infty} \frac{z^n}{n!}\langle:\cdot^{\otimes kn}:, y^{\otimes kn}\rangle.$$

It is straightforward to verify that $:e^{z\langle\cdot,y\rangle^k}: \in (\mathcal{E})_{\frac{k-2}{k}}^*$ with the S-transform

$$(S:e^{z\langle\cdot,y\rangle^k}:)(\xi) = e^{z\langle y,\xi\rangle^k}, \quad \xi \in \mathcal{E}_c.$$

6

Continuous Versions and Analytic Extensions

6.1 Continuous versions of test functions

In this section we will prove a basic result in white noise distribution theory, i.e., every test function has a continuous version. This result is due to Kubo and Yokoi [96]. We will follow essentially their proof.

Note that *a priori* a test function φ in $(\mathcal{E})$ is defined only μ–a.e. Thus it makes no sense to say that φ is continuous. However, since $\varphi \in (L^2)$, it can be represented by

$$\varphi(x) = \sum_{n=0}^{\infty} \langle :x^{\otimes n}:, f_n \rangle, \quad \mu-\text{a.e.}$$

On the other hand, since $\varphi \in (\mathcal{E})$, we have $\|\varphi\|_p^2 = \sum_{n=0}^{\infty} n! |f_n|_p^2 < \infty$ for all $p \geq 0$. This shows that $|f_n|_p < \infty$ for all $p \geq 0$ and so $f_n \in \mathcal{E}_c^{\widehat{\otimes} n}$ for each n. Hence $\langle :x^{\otimes n}:, f_n \rangle$ is defined everywhere on $\mathcal{E}'$. Let

$$\widetilde{\varphi}(x) \equiv \sum_{n=0}^{\infty} \langle :x^{\otimes n}:, f_n \rangle. \tag{6.1}$$

We will see below that this series converges absolutely for each $x \in \mathcal{E}'$. Moreover, if we use Definition 5.1 and change the order of summation informally, then we get

$$\widetilde{\varphi}(x) = \sum_{n=0}^{\infty} \langle x^{\otimes n}, g_n \rangle, \tag{6.2}$$

where g_n is given by

$$g_n = \sum_{k=0}^{\infty} \binom{n+2k}{2k} (2k-1)!! (-1)^k \langle \tau^{\otimes k}, f_{n+2k} \rangle.$$

Note: After the bilinear pairing, $\langle \tau^{\otimes k}, f_{n+2k} \rangle$ is a symmetric function of n variables.

The next proposition implies that the series in equation (6.1) converges absolutely for each $x \in \mathcal{E}'$ and that we can change the order of summation to obtain the representation in equation (6.2).

Proposition 6.1. *Let* $\varphi(x) = \sum_{n=0}^{\infty} \langle : x^{\otimes n} :, f_n \rangle$, μ*-a.e., be in* $(\mathcal{E})$. *Then for any* $x \in \mathcal{E}'$,

$$\sum_{n=0}^{\infty} \sum_{k=0}^{\infty} \left| \binom{n+2k}{2k} (2k-1)!! \langle x^{\otimes n} \widehat{\otimes} \tau^{\otimes k}, f_{n+2k} \rangle \right| < \infty.$$

Proof. Let

$$C_{n,p} \equiv \sum_{k=0}^{\infty} \binom{n+2k}{2k} (2k-1)!! |\tau|_{-p}^{k} |f_{n+2k}|_p.$$

Note that $|f_m|_p \leq \lambda_1^{-mq} |f_m|_{p+q}$. Hence

$$C_{n,p} \leq \sum_{k=0}^{\infty} \frac{(n+2k)!}{n! 2^k k!} |\tau|_{-p}^{k} \lambda_1^{-(n+2k)q} |f_{n+2k}|_{p+q}$$

$$= \frac{1}{n!} \sum_{k=0}^{\infty} \left[\frac{1}{2^k k!} \sqrt{(n+2k)!} \, |\tau|_{-p}^{k} \lambda_1^{-(n+2k)q} \right] \left[\sqrt{(n+2k)!} \, |f_{n+2k}|_{p+q} \right].$$

But $\sum_{k=0}^{\infty} (n+2k)! |f_{n+2k}|_{p+q}^{2} \leq \|\varphi\|_{p+q}^{2}$ for any n. Thus by the Schwarz inequality,

$$C_{n,p} \leq \|\varphi\|_{p+q} \frac{1}{n!} \left(\sum_{k=0}^{\infty} \frac{1}{(2^k k!)^2} (n+2k)! |\tau|_{-p}^{2k} \lambda_1^{-2(n+2k)q} \right)^{1/2}.$$

By using the inequalities $(n+m)! \leq 2^{n+m} n! m!$ and $(2k)! \leq (2^k k!)^2$, we get

$$C_{n,p} \leq \|\varphi\|_{p+q} (n!)^{-1/2} 2^{n/2} \lambda_1^{-nq} \left(\sum_{k=0}^{\infty} 2^{2k} |\tau|_{-p}^{2k} \lambda_1^{-4kq} \right)^{1/2}.$$

Therefore, if we choose q large enough such that $\lambda_1^{2q} > 2|\tau|_{-p}$, then

$$C_{n,p} \leq \|\varphi\|_{p+q} (n!)^{-1/2} 2^{n/2} \lambda_1^{-nq} \left(1 - 4|\tau|_{-p}^{2} \lambda_1^{-4q} \right)^{-1/2}.$$

Finally, for any $x \in \mathcal{E}'$, there exists some $p \geq 0$ such that $|x|_{-p} < \infty$. With this p, choose a positive number q such that $\lambda_1^{2q} > 2|\tau|_{-p}$ as above. Then we have

$$\sum_{n=0}^{\infty} |x|_{-p}^n C_{n,p}$$

$$\leq \|\varphi\|_{p+q}\left(1 - 4|\tau|_{-p}^2 \lambda_1^{-4q}\right)^{-1/2} \sum_{n=0}^{\infty} (n!)^{-1/2} |x|_{-p}^n 2^{n/2} \lambda_1^{-nq}$$

$$\leq \|\varphi\|_{p+q}\left(1 - 4|\tau|_{-p}^2 \lambda_1^{-4q}\right)^{-1/2} \left(\sum_{n=0}^{\infty} \lambda_1^{-2nq}\right)^{1/2} \left(\sum_{n=0}^{\infty} \frac{1}{n!} |x|_{-p}^{2n} 2^n\right)^{1/2}$$

$$= \|\varphi\|_{p+q}\left(1 - 4|\tau|_{-p}^2 \lambda_1^{-4q}\right)^{-1/2} \left(1 - \lambda_1^{-2q}\right)^{-1/2} e^{|x|_{-p}^2}.$$

This gives the assertion in the proposition immediately. □

Next we define a mapping Θ from $(\mathcal{E})_\beta$ into itself. Let $\varphi \in (\mathcal{E})_\beta$ be represented by

$$\varphi = \sum_{n=0}^{\infty} \langle :\cdot^{\otimes n}:, f_n \rangle.$$

Define $\Theta\varphi$ by

$$\Theta\varphi = \sum_{n=0}^{\infty} \langle :\cdot^{\otimes n}:, g_n \rangle, \tag{6.3}$$

where g_n is given by

$$g_n = \sum_{k=0}^{\infty} \binom{n+2k}{2k} (2k-1)!!(-1)^k \langle \tau^{\otimes k}, f_{n+2k} \rangle.$$

Theorem 6.2. *The mapping Θ is a continuous linear operator from $(\mathcal{E})_\beta$ into itself for any $0 \leq \beta < 1$. In fact, for any $p \geq 0$ and $q > p$ satisfying $\lambda_1^{2(q-p)} > \max\{2^{1-\beta}|\tau|_{-p}, 2^{1-\beta}\}$, the following inequality holds for all $\varphi \in (\mathcal{E})_\beta$*

$$\|\Theta\varphi\|_{p,\beta} \leq C_{p,q,\beta}\|\varphi\|_{q,\beta},$$

where $C_{p,q,\beta} = \left[1 - \left(2^{1-\beta}\lambda_1^{-2(q-p)}|\tau|_{-p}\right)^2\right]^{-1/2}\left(1 - 2^{1-\beta}\lambda_1^{-2(q-p)}\right)^{-1/2}$.

Remark. For the proof of continuous versions of test functions, we need only the fact that $\Theta\varphi \in (\mathcal{E})$ for any $\varphi \in (\mathcal{E})$. We prove this general result for later use. For instance, the inequality will be used in §6.2 and in Chapter 11 on Fourier transforms.

Proof. For any $p \geq 0$, we have

$$|g_n|_p \leq \sum_{k=0}^{\infty} \frac{(n+2k)!}{n!2^k k!} |\tau|_{-p}^k |f_{n+2k}|_p.$$

But $|f_{n+2k}|_p \leq \lambda_1^{-(n+2k)(q-p)}|f_{n+2k}|_q$ for $q > p$. Hence

$$|g_n|_p \leq \sum_{k=0}^{\infty} \frac{(n+2k)!}{n!2^k k!}|\tau|_{-p}^k \lambda_1^{-(n+2k)(q-p)}|f_{n+2k}|_q$$

$$= \frac{1}{n!}\lambda_1^{-(q-p)n} \sum_{k=0}^{\infty} \frac{(n+2k)!}{2^k k!}|\tau|_{-p}^k \lambda_1^{-2k(q-p)}|f_{n+2k}|_q. \qquad (6.4)$$

Note that for any $n \geq 0$

$$\sum_{k=0}^{\infty}((n+2k)!)^{1+\beta}|f_{n+2k}|_q^2 \leq \|\varphi\|_{q,\beta}^2.$$

This suggests to us to write

$$\frac{(n+2k)!}{2^k k!}|\tau|_{-p}^k \lambda_1^{-2k(q-p)}|f_{n+2k}|_q$$

as the following product

$$\left[\frac{((n+2k)!)^{(1-\beta)/2}}{2^k k!}\left(\lambda_1^{-2(q-p)}|\tau|_{-p}\right)^k\right]\left[((n+2k)!)^{(1+\beta)/2}|f_{n+2k}|_q\right]$$

and then apply the Schwarz inequality to (6.4) to get

$$|g_n|_p^2 \leq \|\varphi\|_{q,\beta}^2 \frac{1}{(n!)^2}\lambda_1^{-2(q-p)n} \sum_{k=0}^{\infty} \frac{((n+2k)!)^{1-\beta}}{(2^k k!)^2}\left(\lambda_1^{-2(q-p)}|\tau|_{-p}\right)^{2k}$$

$$= \|\varphi\|_{q,\beta}^2 \lambda_1^{-2(q-p)n} K_{p,q,n},$$

where $K_{p,q,n}$ is defined to be $(n!)^{-2}$ times the last summation. By using the inequalities $(n+m)! \leq 2^{n+m}n!m!$ and $(2k)! \leq (2^k k!)^2$, we get

$$K_{p,q,n} \leq \frac{1}{(n!)^{1+\beta}}2^{(1-\beta)n} \sum_{k=0}^{\infty} \frac{2^{2(1-\beta)k}((2k)!)^{1-\beta}}{(2^k k!)^2}\left(\lambda_1^{-2(q-p)}|\tau|_{-p}\right)^{2k}$$

$$\leq \frac{1}{(n!)^{1+\beta}}2^{(1-\beta)n} \sum_{k=0}^{\infty} \frac{2^{2(1-\beta)k}}{(2^k k!)^{2\beta}}\left(\lambda_1^{-2(q-p)}|\tau|_{-p}\right)^{2k}$$

$$\leq \frac{1}{(n!)^{1+\beta}}2^{(1-\beta)n} \sum_{k=0}^{\infty} 2^{2(1-\beta)k}\left(\lambda_1^{-2(q-p)}|\tau|_{-p}\right)^{2k}$$

$$= \frac{1}{(n!)^{1+\beta}}2^{(1-\beta)n}\left[1 - \left(2^{1-\beta}\lambda_1^{-2(q-p)}|\tau|_{-p}\right)^2\right]^{-1},$$

where we have used the condition that $2^{1-\beta}\lambda_1^{-2(q-p)}|\tau|_{-p} < 1$. But q also satisfies the condition that $2^{1-\beta}\lambda_1^{-2(q-p)} < 1$. Hence

$$\|\Theta\varphi\|_{p,\beta}^2 = \sum_{n=0}^{\infty}(n!)^{1+\beta}|g_n|_p^2$$

$$\leq \|\varphi\|_{q,\beta}^2\Big[1 - \big(2^{1-\beta}\lambda_1^{-2(q-p)}|\tau|_{-p}\big)^2\Big]^{-1}\sum_{n=0}^{\infty}\lambda_1^{-2(q-p)n}2^{(1-\beta)n}$$

$$= \|\varphi\|_{q,\beta}^2\Big[1 - \big(2^{1-\beta}\lambda_1^{-2(q-p)}|\tau|_{-p}\big)^2\Big]^{-1}\big(1 - 2^{1-\beta}\lambda_1^{-2(q-p)}\big)^{-1}. \quad (6.5)$$

This shows that $\Theta\varphi \in (\mathcal{E})_\beta$ for any $\varphi \in (\mathcal{E})_\beta$. Obviously Θ is linear. Thus by (6.5), Θ is a continuous linear operator from $(\mathcal{E})_\beta$ into itself. $\quad\square$

The next thing we need for the proof of continuous versions of test functions is the renormalized exponential function $:e^{\langle\cdot,x\rangle}:$ for $x \in \mathcal{E}'$ as given in Definition 5.12, i.e.,

$$:e^{\langle\cdot,x\rangle}: = \sum_{n=0}^{\infty}\frac{1}{n!}\langle:\cdot^{\otimes n}:,x^{\otimes n}\rangle.$$

If $x \in \mathcal{E}_p'$, then by Theorem 5.13

$$\|:e^{\langle\cdot,x\rangle}:\|_{-p} = e^{\frac{1}{2}|x|_{-p}^2}.$$

Hence the mapping $x \mapsto :e^{\langle\cdot,x\rangle}:$ is a function from $\mathcal{E}'$ into $(\mathcal{E})^*$. We will show that this function is continuous with the inductive limit topology for both $\mathcal{E}'$ and $(\mathcal{E})^*$. Recall from §2.2 that a base of neighborhoods of zero for the inductive limit topology of $\mathcal{E}'$ is given by the collection of balanced convex hulls of sets $\{w \in \mathcal{E}_p'; |w|_{-p} < a_p\}, p \geq 0$. We have a similar base for the inductive limit topology of $(\mathcal{E})^*$.

For convenience and without loss of generality, we will assume that the indices in the spaces $\mathcal{E}_p'$ and $(\mathcal{E}_p)^*$ to be nonnegative integers for the rest of this section.

Lemma 6.3. *Let $x \in \mathcal{E}_q'$ be fixed and let W be a neighborhood of zero given by the balanced convex hull of the sets $\{w \in \mathcal{E}_p'; |w|_{-p} < \delta_p\}, p \geq q, 0 < \delta_p \leq 1$. Then for any $y \in x + W$, there exist positive numbers $\alpha_p, q \leq p \leq N$, with $\sum_{p=q}^{N}\alpha_p = 1$ such that for any n,*

$$y^{\otimes n} = x^{\otimes n} + \sum_{p=q}^{N}\alpha_p v_{n,p},$$

where $v_{n,p} \in (\mathcal{E}_p')^{\widehat{\otimes}n}$ and $|v_{n,p}|_{-p} \leq n\big(1 + |x|_{-p}\big)^{n-1}\delta_p$.

Proof. Since $y \in x + W$, there exist $\alpha_p > 0$ and $w_p \in \mathcal{E}'_p$, $q \leq p \leq N$, such that $\sum_{p=q}^{N} \alpha_p = 1$, $|w_p|_{-p} < \delta_p$, and

$$y = x + \sum_{p=q}^{N} \alpha_p w_p.$$

Then for any $n \geq 1$, we have

$$y^{\otimes n} = \left(x + \sum_p \alpha_p w_p \right)^{\otimes n}$$

$$= x^{\otimes n} + \sum_{k=1}^{n} \binom{n}{k} x^{\otimes(n-k)} \widehat{\otimes} \left(\sum_p \alpha_p w_p \right)^{\otimes k}.$$

Therefore

$$y^{\otimes n} = x^{\otimes n} + \sum_{k=1}^{n} \binom{n}{k} x^{\otimes(n-k)} \widehat{\otimes} \left[\sum_{p_i} \alpha_{p_1} \cdots \alpha_{p_k} w_{p_1} \widehat{\otimes} \cdots \widehat{\otimes} w_{p_k} \right]$$

$$= x^{\otimes n} + \sum_{p=q}^{N} \left(\sum_{k=1}^{n} \binom{n}{k} \sum_{p_1 \vee \cdots \vee p_k = p} \alpha_{p_1} \cdots \alpha_{p_k} x^{\otimes(n-k)} \widehat{\otimes} w_{p_1} \widehat{\otimes} \cdots \widehat{\otimes} w_{p_k} \right).$$

Let

$$v_{n,p} \equiv \alpha_p^{-1} \sum_{k=1}^{n} \binom{n}{k} \sum_{p_1 \vee \cdots \vee p_k = p} \alpha_{p_1} \cdots \alpha_{p_k} x^{\otimes(n-k)} \widehat{\otimes} w_{p_1} \widehat{\otimes} \cdots \widehat{\otimes} w_{p_k}.$$

Then $y^{\otimes n} = x^{\otimes n} + \sum_{p=q}^{N} \alpha_p v_{n,p}$. Note that $v_{n,p} \in (\mathcal{E}'_p)^{\widehat{\otimes} n}$ and

$$|w_{p_j}|_{-p} \leq \lambda_1^{-(p-p_j)} |w_{p_j}|_{-p_j} \leq |w_{p_j}|_{-p_j} < \delta_{p_j}.$$

Therefore

$$|v_{n,p}|_{-p} \leq \alpha_p^{-1} \sum_{k=1}^{n} \binom{n}{k} \sum_{p_1 \vee \cdots \vee p_k = p} \alpha_{p_1} \cdots \alpha_{p_k} |x|_{-p}^{n-k} \delta_{p_1} \cdots \delta_{p_k}. \qquad (6.6)$$

It is easy to see that

$$\sum_{p_1 \vee \cdots \vee p_k = p} \alpha_{p_1} \cdots \alpha_{p_k} \delta_{p_1} \cdots \delta_{p_k} = \left(\sum_{j=q}^{p} \alpha_j \delta_j \right)^k - \left(\sum_{j=q}^{p-1} \alpha_j \delta_j \right)^k.$$

But $|u^k - v^k| \leq k|u - v|$ for any $0 \leq u, v \leq 1$. Hence we obtain

$$\sum_{p_1 \vee \cdots \vee p_k = p} \alpha_{p_1} \cdots \alpha_{p_k} \delta_{p_1} \cdots \delta_{p_k} \leq k \alpha_p \delta_p. \qquad (6.7)$$

Hence by inequalities (6.6) and (6.7),

$$|v_{n,p}|_{-p} \leq \sum_{k=1}^{n} \binom{n}{k} k |x|_{-p}^{n-k} \delta_p. \tag{6.8}$$

But it is easy to check that

$$\sum_{k=1}^{n} \binom{n}{k} k |x|_{-p}^{n-k} = n\left(1 + |x|_{-p}\right)^{n-1}. \tag{6.9}$$

By equations (6.8) and (6.9) we have $|v_{n,p}|_{-p} \leq n\left(1 + |x|_{-p}\right)^{n-1} \delta_p.$ $\qquad \square$

Theorem 6.4. *The function $x \mapsto\, :e^{\langle \cdot, x \rangle}:$ is continuous from $\mathcal{E}'$ into $(\mathcal{E})^*$ with the inductive limit topology (equivalently, the strong topology) for both $\mathcal{E}'$ and $(\mathcal{E})^*$.*

Remarks. (1) For the proof for continuous versions of test functions, we need only the continuity of the function $x \mapsto\, :e^{\langle \cdot, x \rangle}:$ from $\mathcal{E}'$ with the inductive limit topology into $(\mathcal{E})^*$ with the weak topology.

(2) Consider the renormalization $:e^{\langle \cdot, x \rangle^k}:$ given at the end of Chapter 5

$$:e^{\langle \cdot, x \rangle^k}: = \sum_{n=0}^{\infty} \frac{1}{n!} \langle :\cdot^{\otimes kn}:, x^{\otimes kn} \rangle, \quad x \in \mathcal{E}'.$$

The proof below can be modified to show that the function $x \mapsto\, :e^{\langle \cdot, x \rangle^k}:$ is continuous from $\mathcal{E}'$ into $(\mathcal{E})^*_{\frac{k-2}{k}}$.

Proof. For convenience in the proof, we will use Exp to denote the function $x \mapsto\, :e^{\langle \cdot, x \rangle}:$, namely,

$$\mathrm{Exp}(x) \equiv\, :e^{\langle \cdot, x \rangle}:, \quad x \in \mathcal{E}'.$$

We prove the continuity of Exp at each $x \in \mathcal{E}'$. Since $\mathcal{E}' = \cup_{p \geq 0} \mathcal{E}'_p$, there is some $q \geq 0$ such that $x \in \mathcal{E}'_q$. Let V be a balanced convex neighborhood of 0 in $(\mathcal{E})^*$ of the form

$$V = \mathrm{conv}\left(\bigcup_{p \geq q} \left\{ \Phi \in (\mathcal{E}_p)^*; \|\Phi\|_{-p} < \epsilon_p \right\} \right).$$

Define a neighborhood W of 0 in $\mathcal{E}'$ by

$$W = \text{balanced convex hull of the sets } \left\{ w \in \mathcal{E}'_p;\ |w|_{-p} < \delta_p \right\}, p \geq q,$$

where $\delta_p = \min\left\{ 1, \epsilon_p\left(1 + (1 + |x|_{-p})^2\right)^{-1/2} e^{-\frac{1}{2}(1 + |x|_{-p})^2} \right\}.$

Now, for any $y \in x + W$, let $0 < \alpha_p \le 1$ and $v_{n,p} \in (\mathcal{E}_p')^{\widehat{\otimes} n}$ be associated with y as given in Lemma 6.3. Then for any $n \ge 1$

$$y^{\otimes n} = x^{\otimes n} + \sum_{p=q}^{N} \alpha_p v_{n,p}.$$

Thus we have

$$\mathrm{Exp}(y) - \mathrm{Exp}(x) = \sum_{n=1}^{\infty} \Big\langle :\cdot^{\otimes n}:, \frac{1}{n!} \sum_{p=q}^{N} \alpha_p v_{n,p} \Big\rangle$$

$$= \sum_{p=q}^{N} \alpha_p \Phi_p, \tag{6.10}$$

where Φ_p is given by

$$\Phi_p = \sum_{n=1}^{\infty} \Big\langle :\cdot^{\otimes n}:, \frac{1}{n!} v_{n,p} \Big\rangle.$$

Then from Lemma 6.3 we get

$$\|\Phi_p\|_{-p}^2 = \sum_{n=1}^{\infty} n! \frac{1}{(n!)^2} |v_{n,p}|_{-p}^2$$

$$\le \sum_{n=1}^{\infty} \frac{1}{n!} n^2 \left(1 + |x|_{-p}\right)^{2n-2} \delta_p^2$$

$$= \delta_p^2 \sum_{n=1}^{\infty} \frac{n}{(n-1)!} \left(1 + |x|_{-p}\right)^{2n-2}$$

$$= \delta_p^2 \left(1 + (1 + |x|_{-p})^2\right) e^{(1+|x|_{-p})^2}.$$

Therefore, by the choice of δ_p, we have $\|\Phi_p\|_{-p} < \epsilon_p$ for any $q \le p \le N$. In view of equation (6.10), this means that $\mathrm{Exp}(y) - \mathrm{Exp}(x) \in V$, i.e., $\mathrm{Exp}(y) \in \mathrm{Exp}(x) + V$ for any $y \in x + W$. Hence Exp is continuous at x. $\square$

Now we are ready to use Theorems 6.2 and 6.4 to prove the Kubo-Yokoi theorem for the continuous versions of test functions.

Theorem 6.5. *Every test function φ in $(\mathcal{E})$ has a unique continuous version. In fact, the continuous version $\widetilde{\varphi}$ of φ is given by*

$$\widetilde{\varphi} = \sum_{n=0}^{\infty} \langle :\cdot^{\otimes n}:, f_n \rangle, \quad f_n \in \mathcal{E}_c^{\widehat{\otimes} n}.$$

Proof. Note that $(E, \mathcal{E}'_p)$ is an abstract Wiener space for any $p \geq \frac{\alpha}{2}$. Hence any open ball of positive radius in $\mathcal{E}'_p$ has positive μ measure (see Theorem 2.2). This implies the uniqueness of a continuous version for φ. On the other hand, as pointed out in the beginning of this section, $\varphi \in (\mathcal{E})$ can be represented by

$$\varphi(x) = \sum_{n=0}^{\infty} \langle :x^{\otimes n}:, f_n \rangle, \quad x \in \mathcal{E}', \quad \mu-\text{a.e.},$$

where $f_n \in \mathcal{E}_c^{\widehat{\otimes} n}$. Define

$$\widetilde{\varphi}(x) = \sum_{n=0}^{\infty} \langle :x^{\otimes n}:, f_n \rangle, \quad x \in \mathcal{E}'.$$

Obviously $\widetilde{\varphi}$ is a version of φ, i.e., $\widetilde{\varphi} = \varphi$, μ–a.e. Moreover, Proposition 6.1 implies that the last series converges absolutely for every $x \in \mathcal{E}'$ and so $\widetilde{\varphi}$ is defined everywhere on $\mathcal{E}'$. Thus to finish the proof we need only to show the continuity of $\widetilde{\varphi}$. Recall from equation (6.2) that $\widetilde{\varphi}$ can be rewritten as

$$\widetilde{\varphi}(x) = \sum_{n=0}^{\infty} \langle x^{\otimes n}, g_n \rangle,$$

where g_n is given by

$$g_n = \sum_{k=0}^{\infty} \binom{n + 2k}{2k} (2k - 1)!!(-1)^k \langle \tau^{\otimes k}, f_{n+2k} \rangle.$$

Now, note that

$$\Theta \widetilde{\varphi} = \sum_{n=0}^{\infty} \langle :\cdot^{\otimes n}:, g_n \rangle,$$

and

$$:e^{\langle \cdot, x \rangle}: = \sum_{n=0}^{\infty} \frac{1}{n!} \langle :\cdot^{\otimes n}:, x^{\otimes n} \rangle.$$

By Theorem 6.2, $\Theta \widetilde{\varphi} \in (\mathcal{E})$. Hence by equation (5.8), we get

$$\langle\!\langle :e^{\langle \cdot, x \rangle}:, \Theta \widetilde{\varphi} \rangle\!\rangle = \sum_{n=0}^{\infty} \langle x^{\otimes n}, g_n \rangle.$$

Therefore

$$\widetilde{\varphi}(x) = \langle\!\langle :e^{\langle \cdot, x \rangle}:, \Theta \widetilde{\varphi} \rangle\!\rangle, \quad x \in \mathcal{E}'.$$

Thus the continuity of $\widetilde{\varphi}$ follows from Theorems 6.2 and 6.4. $\qquad\square$

Note: From now on, we will assume that the continuous versions of test functions are already taken and $\tilde{\varphi}$ will be simply denoted by φ. That is, a test function $\varphi \in (\mathcal{E})$ is expressed as

$$\varphi(x) = \sum_{n=0}^{\infty} \langle :x^{\otimes n}:, f_n \rangle,$$

where $f_n \in \mathcal{E}_c^{\widehat{\otimes} n}$ and any $p \geq 0$

$$\|\varphi\|_p^2 = \sum_{n=0}^{\infty} n! |f_n|_p^2 < \infty.$$

6.2 Growth condition and norm estimates

We have shown in the previous section that for any test function φ in $(\mathcal{E})$

$$\varphi(x) = \langle\langle :e^{\langle \cdot, x \rangle}:, \Theta\varphi \rangle\rangle, \quad x \in \mathcal{E}', \tag{6.11}$$

where Θ is the linear operator defined by equation (6.3). By Theorem 5.13, $\| :e^{\langle \cdot, x \rangle}: \|_{-p} = e^{\frac{1}{2}|x|^2_{-p}}$. Thus φ satisfies the following growth condition

$$|\varphi(x)| \leq \|\Theta\varphi\|_p \, e^{\frac{1}{2}|x|^2_{-p}}.$$

In general we can obtain the growth condition for test functions in the space $(\mathcal{E})_\beta$. To do so, we need an inequality in the next lemma.

Lemma 6.6. *For any real number $\gamma \geq 1$, the following inequality holds*

$$\sum_{n=0}^{\infty} \left(\frac{u^n}{n!}\right)^{\gamma} \leq e^{\gamma u}, \quad u \geq 0.$$

Proof. The conclusion is obvious when $\gamma \geq 1$ is an integer (when $\gamma = 1$, it is an equality). This implies that we need only to prove the lemma for $1 < \gamma < 2$. By writing

$$\left(\frac{u^n}{n!}\right)^{\gamma} = \left(\frac{u^n}{n!}\right)^{2-\gamma} \left(\left(\frac{u^n}{n!}\right)^2\right)^{\gamma-1},$$

we can apply Hölder's inequality with $p = \frac{1}{2-\gamma}$ and $q = \frac{1}{\gamma-1}$ to get

$$\sum_{n=0}^{\infty} \left(\frac{u^n}{n!}\right)^{\gamma} \leq \left(\sum_{n=0}^{\infty} \frac{u^n}{n!}\right)^{2-\gamma} \left(\sum_{n=0}^{\infty} \left(\frac{u^n}{n!}\right)^2\right)^{\gamma-1}$$

$$\leq \left(e^u\right)^{2-\gamma} \left(e^{2u}\right)^{\gamma-1}$$

$$= e^{\gamma u}. \qquad \square$$

The next theorem is a counterpart of Theorem 5.7 for generalized functions.

Lemma 6.7. *If* $x \in \mathcal{E}'_{p,c}$, *then* $:e^{\langle \cdot, x \rangle}: \in (\mathcal{E}_p)^*_\beta$ *and*

$$\| :e^{\langle \cdot, x \rangle}: \|_{-p,-\beta} \le \exp\left[\frac{1}{2}(1+\beta)|x|^{\frac{2}{1+\beta}}_{-p}\right].$$

Proof. Note that

$$:e^{\langle \cdot, x \rangle}: = \sum_{n=0}^{\infty} \frac{1}{n!}\langle :\cdot^{\otimes n}:, x^{\otimes n}\rangle.$$

Therefore

$$\| :e^{\langle \cdot, x \rangle}: \|^2_{-p,-\beta} = \sum_{n=0}^{\infty}(n!)^{1-\beta}\frac{1}{(n!)^2}|x|^{2n}_{-p}$$

$$= \sum_{n=0}^{\infty}\frac{1}{(n!)^{1+\beta}}|x|^{2n}_{-p}.$$

Now, apply Lemma 6.6 with $\gamma = 1 + \beta$ and $u = |x|^{\frac{2}{1+\beta}}_{-p}$ to get

$$\sum_{n=0}^{\infty}\frac{1}{(n!)^{1+\beta}}|x|^{2n}_{-p} \le \exp\left[(1+\beta)|x|^{\frac{2}{1+\beta}}_{-p}\right].$$

This gives the conclusion immediately. □

The growth condition for test functions in the space $(\mathcal{E})_\beta$ is given in the next theorem.

Theorem 6.8. *Let* $\varphi \in (\mathcal{E})_\beta$, $0 \le \beta < 1$. *Then* φ *satisfies the growth condition for any* $p \ge 0$,

$$|\varphi(x)| \le \|\Theta\varphi\|_{p,\beta}\exp\left[\frac{1}{2}(1+\beta)|x|^{\frac{2}{1+\beta}}_{-p}\right], \quad x \in \mathcal{E}'_p,$$

where Θ *is the continuous linear operator from* $(\mathcal{E})_\beta$ *into itself defined by equation (6.3).*

Proof. Simply use equation (6.11) and Lemma 6.7. □

Now we regard $\varphi(x)$ as a function of x in $\mathcal{E}'_p$ for a fixed p and estimate various norms related to φ.

Lemma 6.9. *Let* $x, y \in \mathcal{E}'_p$. *Then for any* $n \ge 1$,

$$|y^{\otimes n} - x^{\otimes n}|_{-p} \le n|y - x|_{-p}\left(|x|_{-p} + |y - x|_{-p}\right)^{n-1}.$$

Proof. Let $\tilde{y} = y - x$. Then

$$|y^{\otimes n} - x^{\otimes n}|_{-p} \leq \sum_{k=1}^{n} \binom{n}{k} |x^{\otimes(n-k)} \widehat{\otimes} \tilde{y}^{\otimes k}|_{-p}$$

$$\leq \sum_{k=1}^{n} \binom{n}{k} |x|_{-p}^{n-k} |\tilde{y}|_{-p}^{k}. \tag{6.12}$$

But it is easy to check that for any $a, b \geq 0$,

$$\sum_{k=1}^{n} \binom{n}{k} a^{n-k} b^k \leq nb(a+b)^{n-1}. \tag{6.13}$$

The lemma follows from the inequalities in (6.12) and (6.13). $\square$

Lemma 6.10. *For any $x, y \in \mathcal{E}_p'$, the following inequality holds*

$$\| :e^{\langle \cdot, y \rangle}: - :e^{\langle \cdot, x \rangle}: \|_{-p,-\beta} \leq K_{p,\beta,x,y} |y - x|_{-p},$$

where $K_{p,\beta,x,y} = \left(1 + |x|_{-p} + |y-x|_{-p}\right) \exp\left[\frac{1}{2}(1+\beta)\left(|x|_{-p} + |y-x|_{-p}\right)^{\frac{2}{1+\beta}}\right]$.

Proof. Let $\tilde{y} = y - x$. By Lemma 6.9,

$$\| :e^{\langle \cdot, y \rangle}: - :e^{\langle \cdot, x \rangle}: \|_{-p,-\beta}^2$$

$$= \sum_{n=1}^{\infty} (n!)^{1-\beta} \frac{1}{(n!)^2} |y^{\otimes n} - x^{\otimes n}|_{-p}^2$$

$$\leq \sum_{n=1}^{\infty} \frac{1}{(n!)^{1+\beta}} n^2 |\tilde{y}|_{-p}^2 \left(|x|_{-p} + |\tilde{y}|_{-p}\right)^{2n-2}$$

$$\leq |\tilde{y}|_{-p}^2 \sum_{k=0}^{\infty} \frac{k+1}{(k!)^{1+\beta}} \left(|x|_{-p} + |\tilde{y}|_{-p}\right)^{2k}. \tag{6.14}$$

Use the fact that $k \leq k^{1+\beta}$ for any $k \geq 1$ to get

$$\sum_{k=0}^{\infty} \frac{k+1}{(k!)^{1+\beta}} \left(|x|_{-p} + |\tilde{y}|_{-p}\right)^{2k}$$

$$\leq \sum_{k=1}^{\infty} \frac{1}{((k-1)!)^{1+\beta}} \left(|x|_{-p} + |\tilde{y}|_{-p}\right)^{2k} + \sum_{k=0}^{\infty} \frac{1}{(k!)^{1+\beta}} \left(|x|_{-p} + |\tilde{y}|_{-p}\right)^{2k}$$

$$= \left(1 + (|x|_{-p} + |\tilde{y}|_{-p})^2\right) \sum_{k=0}^{\infty} \frac{1}{(k!)^{1+\beta}} \left(|x|_{-p} + |\tilde{y}|_{-p}\right)^{2k}.$$

Then we apply Lemma 6.6 to derive

$$\sum_{k=0}^{\infty} \frac{k+1}{(k!)^{1+\beta}} \left(|x|_{-p} + |\tilde{y}|_{-p} \right)^{2k}$$

$$\leq \left(1 + \left(|x|_{-p} + |\tilde{y}|_{-p} \right)^2 \right) \exp\left[(1+\beta) \left(|x|_{-p} + |\tilde{y}|_{-p} \right)^{\frac{2}{1+\beta}} \right]$$

$$\leq \left(1 + |x|_{-p} + |\tilde{y}|_{-p} \right)^2 \exp\left[(1+\beta) \left(|x|_{-p} + |\tilde{y}|_{-p} \right)^{\frac{2}{1+\beta}} \right]. \qquad (6.15)$$

Obviously the inequalities in (6.14) and (6.15) give the inequality in the lemma. $\qquad \square$

Theorem 6.11. *Let $\varphi \in (\mathcal{E})_\beta, 0 \leq \beta < 1$. Then for any $p \geq 0$ and $x, y \in \mathcal{E}_p'$*

$$|\varphi(y) - \varphi(x)| \leq \|\Theta\varphi\|_{p,\beta} K_{p,\beta,x,y} |y - x|_{-p},$$

where Θ is defined by equation (6.3) and $K_{p,\beta,x,y}$ is given in Lemma 6.10.

Proof. The conclusion follows from equation (6.11) and Lemma 6.10. $\qquad \square$

We can apply the estimate in Theorem 6.2 to Theorems 6.8 and 6.11. Suppose $p \geq 0$ and $q \geq 0$ satisfy the condition $\lambda_1^{2q} > \max\{2|\tau|_{-p}, 2^{1-\beta}\}$. Then for any $\varphi \in (\mathcal{E})_\beta$ and any $x, y \in \mathcal{E}_p'$, we have

$$|\varphi(x)| \leq C_{p,q,\beta} \|\varphi\|_{p+q,\beta} \exp\left[\frac{1}{2}(1+\beta)|x|_{-p}^{\frac{2}{1+\beta}} \right], \qquad (6.16)$$

$$|\varphi(y) - \varphi(x)| \leq C_{p,q,\beta} K_{p,\beta,x,y} \|\varphi\|_{p+q,\beta} |y - x|_{-p}, \qquad (6.17)$$

where $C_{p,q,\beta}$ and $K_{p,\beta,x,y}$ are given by

$$C_{p,q,\beta} = \left(1 - \left(2\lambda_1^{-2q}|\tau|_{-p} \right)^2 \right)^{-1/2} \left(1 - 2^{1-\beta}\lambda_1^{-2q} \right)^{-1/2},$$

$$K_{p,\beta,x,y} = \left(1 + |x|_{-p} + |y - x|_{-p} \right) \exp\left[\frac{1}{2}(1+\beta) \left(|x|_{-p} + |y - x|_{-p} \right)^{\frac{2}{1+\beta}} \right].$$

6.3 Analytic extensions of test functions

It has been shown in Y.-J. Lee [122] that every test function has an analytic extension (see also the book by Hida et al. [58]). We prove this property for test functions in the space $(\mathcal{E})$ and give estimates for norms related to this analytic extension. The key idea is to use the representation for $\varphi \in (\mathcal{E})$ in equation (6.11), i.e.,

$$\varphi(x) = \langle\!\langle :e^{\langle \cdot, x \rangle}:, \Theta\varphi \rangle\!\rangle, \quad x \in \mathcal{E}',$$

where Θ is the linear operator defined by equation (6.3). Recall that Θ is continuous from $(\mathcal{E})_\beta$ into itself for any $0 \leq \beta < 1$ and that

$$:e^{\langle \cdot, x \rangle}: = \sum_{n=0}^\infty \frac{1}{n!} \langle : \cdot^{\otimes n} :, x^{\otimes n} \rangle, \quad x \in \mathcal{E}'.$$

Note that the renormalized exponential function $: e^{\langle \cdot, w \rangle}:$ is a generalized function in $(\mathcal{E})^*$ for any $w \in \mathcal{E}'_c$ and

$$:e^{\langle \cdot, w \rangle}: = \sum_{n=0}^\infty \frac{1}{n!} \langle : \cdot^{\otimes n} :, w^{\otimes n} \rangle.$$

Hence the function $w \mapsto : e^{\langle \cdot, w \rangle}:$ from $\mathcal{E}'_c$ into $(\mathcal{E})^*$ is an extension of $:e^{\langle \cdot, x \rangle}:$, $x \in \mathcal{E}'$. Moreover, $:e^{\langle \cdot, w \rangle}: \in (\mathcal{E}_p)^*$ if $w \in \mathcal{E}'_{p,c}$. In fact, we have

$$\| :e^{\langle \cdot, w \rangle}: \|_{-p} = e^{\frac{1}{2}|w|^2_{-p}}.$$

This shows that $\varphi(x), x \in \mathcal{E}'$, has an extension to $\varphi(w), w \in \mathcal{E}'_c$, and the analytic property of this extension follows from the analyticity of the renormalized exponential function.

Lemma 6.12. *The function* $w \mapsto :e^{\langle \cdot, w \rangle}:$ *is Fréchet differentiable from* $\mathcal{E}'_{p,c}$ *into* $(\mathcal{E}_p)^*$ *for any* $p \geq 0$. *Moreover, for any* $0 \leq \beta < 1$,

$$\| :e^{\langle \cdot, w \rangle}: \|_{-p,-\beta} \leq \exp\left[\frac{1}{2}(1+\beta)|w|^{\frac{2}{1+\beta}}_{-p} \right], \quad w \in \mathcal{E}'_{p,c}.$$

Proof. Fix any $w \in \mathcal{E}'_{p,c}$. We show that the function

$$f(x) \equiv :e^{\langle \cdot, x \rangle}:, \quad x \in \mathcal{E}'_{p,c},$$

is Fréchet differentiable at w. Define a linear operator T from $\mathcal{E}'_{p,c}$ into $(\mathcal{E}_p)^*$ by

$$Tu = \sum_{n=1}^\infty \frac{1}{(n-1)!} \langle : \cdot^{\otimes n} :, w^{\otimes(n-1)} \widehat{\otimes} u \rangle.$$

The operator T is continuous since

$$\|Tu\|^2_{-p} \leq \sum_{n=1}^\infty n! \frac{1}{((n-1)!)^2} |w|^{2(n-1)}_{-p} |u|^2_{-p}$$

$$= |u|^2_{-p} \sum_{k=0}^\infty \frac{k+1}{k!} |w|^{2k}_{-p}$$

$$= |u|^2_{-p}(1 + |w|^2_{-p})e^{|w|^2_{-p}}.$$

Let $A_n \equiv (w + u)^{\otimes n} - w^{\otimes n} - n w^{\otimes (n-1)} \widehat{\otimes} u$. Then

$$|A_n|_{-p} = \left| \sum_{k=2}^{n} \binom{n}{k} w^{\otimes(n-k)} \widehat{\otimes} u^{\otimes k} \right|_{-p}$$

$$\leq \sum_{k=2}^{n} \binom{n}{k} |w|_{-p}^{n-k} |u|_{-p}^{k}$$

$$= |u|_{-p}^{2} \sum_{m=0}^{n-2} \frac{n(n-1)}{(m+2)(m+1)} \binom{n-2}{m} |w|_{-p}^{n-2-m} |u|_{-p}^{m}.$$

But $(m+2)(m+1) \geq 2$ for any $m \geq 0$. Hence

$$|A_n|_{-p} \leq \frac{n(n-1)}{2} |u|_{-p}^{2} \sum_{m=0}^{n-2} \binom{n-2}{m} |w|_{-p}^{n-2-m} |u|_{-p}^{m}$$

$$= \frac{n(n-1)}{2} |u|_{-p}^{2} (|w|_{-p} + |u|_{-p})^{n-2}.$$

Let $B \equiv f(w + u) - f(w) - Tu$. Then

$$B = \sum_{n=2}^{\infty} \frac{1}{n!} \langle : \cdot^{\otimes n} :, A_n \rangle.$$

By using the above estimate for $|A_n|_{-p}$, we get

$$\|B\|_{-p}^{2} = \sum_{n=2}^{\infty} n! \frac{1}{(n!)^2} |A_n|_{-p}^{2}$$

$$\leq \frac{1}{4} |u|_{-p}^{4} \sum_{k=0}^{\infty} \frac{(k+2)(k+1)}{k!} (|w|_{-p} + |u|_{-p})^{2k}.$$

It is easy to derive the sum of the last series to show that

$$\|B\|_{-p}^{2}$$
$$\leq \frac{1}{4} |u|_{-p}^{4} \left[(|w|_{-p} + |u|_{-p})^{4} + 4(|w|_{-p} + |u|_{-p})^{2} + 2 \right] e^{(|w|_{-p} + |u|_{-p})^{2}}.$$

This implies that

$$\lim_{|u|_{-p} \to 0} \frac{\|f(w + u) - f(w) - Tu\|_{-p}}{|u|_{-p}} = 0.$$

That is, the function $f(w) =: e^{\langle \cdot, w \rangle} :$ is Fréchet differentiable at w. The norm estimate for $\| : e^{\langle \cdot, w \rangle} : \|_{-p,-\beta}$ can be obtained by the same argument as in the proof of Lemma 6.7. $\qquad \square$

A function f from a complex Hilbert space H into $\mathbb{C}$ is said to be *analytic* if it is single-valued, locally bounded, and Fréchet differentiable (see Hille and Phillips [63, Definition 3.17.2 and Theorem 3.17.1]). The next theorem shows that every test function in $(\mathcal{E})_\beta$ has an analytic extension.

Theorem 6.13. *Every test function φ in $(\mathcal{E})_\beta$ has a unique extension $\varphi(w)$, $w \in \mathcal{E}'_c$, such that φ is analytic on $\mathcal{E}'_{p,c}$ for any $p \geq 0$ and*

$$|\varphi(w)| \leq C_{p,q,\beta} \|\varphi\|_{p+q,\beta} \exp\left[\frac{1}{2}(1+\beta)|w|_{-p}^{\frac{2}{1+\beta}}\right], \quad w \in \mathcal{E}'_{p,c},$$

where $C_{p,q,\beta} = \left(1 - \left(2\lambda_1^{-2q}|\tau|_{-p}\right)^2\right)^{-1/2}\left(1 - 2^{1-\beta}\lambda_1^{-2q}\right)^{-1/2}$.

Proof. The uniqueness of an analytic extension is obvious. The other assertion in the theorem follows from Lemma 6.12 and the fact that

$$\varphi(w) = \langle\!\langle : e^{\langle \cdot, w \rangle} :, \Theta\varphi \rangle\!\rangle, \quad w \in \mathcal{E}'_c,$$

where Θ is the linear operator defined by equation (6.3). $\qquad \square$

Finally we mention that the analytic extension $: e^{\langle \cdot, w \rangle} :$, $w \in \mathcal{E}'_c$, has a local Lip-1 estimate as in Lemma 6.10. Similarly, the analytic extension $\varphi(w)$, $w \in \mathcal{E}'_c$, for $\varphi \in (\mathcal{E})$ has a local Lip-1 estimate as in Theorem 6.11 or equation (6.17).

7

Delta Functions

7.1 Donsker's delta function

Let δ_a be the Dirac delta function at a and $B(t)$ a Brownian motion. It was shown in Kuo [104] that $\delta_a(B(t))$ is a generalized function in white noise distribution theory. More generally, it was shown in Kubo [88] that the composition $F(\langle \cdot, f \rangle)$ is a generalized function for any $F \in \mathcal{S}'(\mathbb{R})$ and nonzero $f \in L^2(\mathbb{R})$. Recently, generalized functions of this form have been characterized in Kubo and Kuo [90]. The Donsker delta function has also been studied in Lascheck et al. [118], H. Watanabe [172], and S. Watanabe [173] [174] [175].

In this section we will give a simple derivation of Kubo's theorem and obtain new generalized functions in the class $(\mathcal{S})^*_\beta$. Our method is based on the following

Fact. The set $\left\{ \xi_n(u) \equiv \frac{1}{\sqrt{n! \, \sigma^n}} : u^n :_{\sigma^2}; n = 0, 1, 2, \dots \right\}$ forms an orthonormal basis for the Hilbert space $L^2 \left(\mathbb{R}, \frac{1}{\sqrt{2\pi} \, \sigma} e^{-\frac{u^2}{2\sigma^2}} \, du \right)$.

Now, let $F \in \mathcal{S}'_c(\mathbb{R})$ (in fact, F can be in a larger space than $\mathcal{S}'_c(\mathbb{R})$; see Kubo and Kuo [90]). Then by the above fact, we can express F as

$$F(x) = \sum_{n=0}^{\infty} (F, \xi_n) \, \xi_n(x),$$

where $(\cdot, \cdot)$ is the bilinear pairing with respect to the measure $\frac{1}{\sqrt{2\pi} \, \sigma} e^{-\frac{y^2}{2\sigma^2}} \, dy$ and the series converges to F in $\mathcal{S}'_c(\mathbb{R})$. The above equation can be rewritten as

$$F(x) = \frac{1}{\sqrt{2\pi} \, \sigma} \sum_{n=0}^{\infty} \frac{1}{n! \sigma^{2n}} : x^n :_{\sigma^2} \int_{\mathbb{R}} F(y) : y^n :_{\sigma^2} e^{-\frac{y^2}{2\sigma^2}} \, dy, \qquad (7.1)$$

where the integral is understood to be the bilinear pairing of $\mathcal{S}'_c(\mathbb{R})$ and $\mathcal{S}_c(\mathbb{R})$ with respect to the Lebesgue measure.

Let f be a nonzero function in $L^2(\mathbb{R})$. From Theorem 5.4, $:\langle\cdot,f\rangle^n:_{|f|_0^2} = \langle:\cdot^{\otimes n}:,f^{\otimes n}\rangle$. Hence, by letting $x = \langle\cdot,f\rangle$ and $\sigma = |f|_0$ in equation (7.1), we get

$$F(\langle\cdot,f\rangle) = \frac{1}{\sqrt{2\pi}\,|f|_0} \sum_{n=0}^{\infty} \frac{1}{n!|f|_0^{2n}}$$

$$\times \left(\int_{\mathbb{R}} F(y):y^n:_{|f|_0^2}\, e^{-\frac{y^2}{2|f|_0^2}}\, dy\right)\langle:\cdot^{\otimes n}:,f^{\otimes n}\rangle. \qquad (7.2)$$

Lemma 7.1. Let $F \in S'_c(\mathbb{R})$ and $f \in L^2(\mathbb{R})$, $f \neq 0$. Then

$$\int_{\mathbb{R}} F(y):y^n:_{|f|_0^2}\, e^{-\frac{y^2}{2|f|_0^2}}\, dy$$

$$= \sqrt{2}\sqrt[4]{\pi}\sqrt{n!}\,|f|_0^{n+1} \int_{\mathbb{R}} F_{\sqrt{2}|f|_0}(y)e^{-\frac{y^2}{2}}e_n(y)\,dy,$$

where $F_\lambda(y) = F(\lambda y)$ and e_n's are Hermite functions given in §3.2.

Proof. By making a change of variables $y = \sqrt{2}|f|_0\,u$, we get easily the equality

$$\int_{\mathbb{R}} F(y):y^n:_{|f|_0^2}\, e^{-\frac{y^2}{2|f|_0^2}}\, dy$$

$$= \sqrt{2}|f|_0 \int_{\mathbb{R}} F_{\sqrt{2}|f|_0}(u):(\sqrt{2}|f|_0 u)^n:_{|f|_0^2}\, e^{-u^2}\, du. \qquad (7.3)$$

Note that Hermite polynomials with a parameter (§5.1) have the property:

$$:\left(\frac{\sigma u}{c}\right)^n:_{\sigma^2} = \frac{\sigma^n}{c^n}:u^n:_{c^2}.$$

By using this property, we can derive easily from equation (7.3)

$$\int_{\mathbb{R}} F(y):y^n:_{|f|_0^2}\, e^{-\frac{y^2}{2|f|_0^2}}\, dy$$

$$= (\sqrt{2}|f|_0)^{n+1} \int_{\mathbb{R}} F_{\sqrt{2}|f|_0}(u):u^n:_{\frac{1}{2}}\, e^{-u^2}\, du. \qquad (7.4)$$

Now, observe that

$$:u^n:_{\frac{1}{2}} = \frac{1}{2^n}H_n(u) = \frac{\sqrt[4]{\pi}\sqrt{n!}}{\sqrt{2^n}}\, e_n(u)e^{u^2/2}, \qquad (7.5)$$

where H_n is the Hermite polynomial given in §3.2. The equality in the lemma follows immediately from equations (7.4) and (7.5). $\qquad\square$

Lemma 7.2. *Let* $F \in S'_c(\mathbb{R})$ *and* $f \in L^2(\mathbb{R})$, $f \neq 0$. *Then*

$$\left| \int_{\mathbb{R}} F(y) : y^n :_{|f|_0^2} e^{-\frac{y^2}{2|f|_0^2}} \, dy \right| \le K_p \sqrt{n!} \, |f|_0^{n+1} (n+1)^p,$$

where $K_p = 2^{p+\frac{1}{2}} \sqrt[4]{\pi} \, |\theta|_{-p}$ *with* $\theta(y) = F_{\sqrt{2}|f|_0}(y) e^{-\frac{y^2}{2}}$ *and* $|\theta|_{-p} < \infty$ *for some* $p \ge 0$.

Proof. Note that $\theta \in S'_c(\mathbb{R})$. Hence $\theta \in S'_{p,c}(\mathbb{R})$ for some $p \ge 0$. Thus the lemma follows from Lemma 7.1 and the following estimate

$$\begin{aligned}
\left| \int_{\mathbb{R}} F(y) : y^n :_{|f|_0^2} e^{-\frac{y^2}{2|f|_0^2}} \, dy \right| &= \sqrt{2} \sqrt[4]{\pi} \sqrt{n!} \, |f|_0^{n+1} |\langle \theta, e_n \rangle| \\
&= \sqrt{2} \sqrt[4]{\pi} \sqrt{n!} \, |f|_0^{n+1} |\langle A^{-p}\theta, A^p e_n \rangle| \\
&\le \sqrt{2} \sqrt[4]{\pi} \sqrt{n!} \, |f|_0^{n+1} |A^{-p}\theta|_0 |A^p e_n|_0 \\
&= \sqrt{2} \sqrt[4]{\pi} \sqrt{n!} \, |f|_0^{n+1} |\theta|_{-p} (2n+2)^p. \qquad\square
\end{aligned}$$

Theorem 7.3. *Let* $F \in S'_c(\mathbb{R})$ *and* $f \in L^2(\mathbb{R})$, $f \neq 0$. *Then* $F(\langle \cdot, f \rangle)$ *is a generalized function in* $(S)^*$ *with the Wiener-Itô decomposition as given in equation (7.2) and for any* $\xi \in S_c(\mathbb{R})$

$$(SF(\langle \cdot, f \rangle))(\xi) = \frac{1}{\sqrt{2\pi} |f|_0} \int_{\mathbb{R}} F(y) \exp \left[-\frac{1}{2|f|_0^2} (y - \langle f, \xi \rangle)^2 \right] dy,$$

where the integral is understood to be the bilinear pairing of $S'_c(\mathbb{R})$ *and* $S_c(\mathbb{R})$.

Proof. From equation (7.2) and Lemma 7.2 we get

$$\begin{aligned}
&\|F(\langle \cdot, f \rangle)\|^2_{-p} \\
&\le \frac{1}{2\pi |f|_0^2} \sum_{n=0}^{\infty} n! \frac{1}{(n!)^2 |f|_0^{4n}} K_p^2 \, n! |f|_0^{2n+2} (n+1)^{2p} |f^{\otimes n}|^2_{-p}.
\end{aligned}$$

But

$$|f^{\otimes n}|_{-p} = |f|^n_{-p} \le 2^{-np} |f|^n_0.$$

Therefore

$$\|F(\langle \cdot, f \rangle)\|^2_{-p} \le \frac{1}{2\pi} K_p^2 \sum_{n=0}^{\infty} (n+1)^{2p} 2^{-2np}.$$

We can choose $p > 0$ such that the function θ (see Lemma 7.2) belongs to $\mathcal{S}'_{p,c}(\mathbb{R})$. Then the last series is convergent. This shows that $F(\langle \cdot, f \rangle)$ is in $(\mathcal{S})^*_p$. Hence $F(\langle \cdot, f \rangle) \in (\mathcal{S})^*$. To obtain the S-transform of $F(\langle \cdot, f \rangle)$, note that by Proposition 5.9

$$(SF(\langle \cdot, f \rangle))(\xi) = \frac{1}{\sqrt{2\pi}\,|f|_0} \sum_{n=0}^{\infty} \frac{1}{n!|f|_0^{2n}}$$

$$\times \left(\int_{\mathbb{R}} F(y) : y^n :_{|f|_0^2} e^{-\frac{y^2}{2|f|_0^2}} \, dy \right) \langle f, \xi \rangle^n. \qquad (7.6)$$

From the generating function in equation (5.2), we have

$$\sum_{n=0}^{\infty} \frac{\langle f, \xi \rangle^n}{n!|f|_0^{2n}} : y^n :_{|f|_0^2} = e^{\frac{1}{|f|_0^2}(y\langle f, \xi \rangle - \frac{1}{2}\langle f, \xi \rangle^2)}. \qquad (7.7)$$

Equations (7.6) and (7.7) yield the S-transform of $F(\langle \cdot, f \rangle)$.   □

Example 7.4. (*Donsker's delta function*) Let $F = \delta_a$ and $f = 1_{[0,t]}$. Note that $\langle \cdot, 1_{[0,t]} \rangle = B(t)$ is a Brownian motion. Thus $F(\langle \cdot, f \rangle) = \delta_a(B(t))$ is Donsker's delta function. By Theorem 7.3, $\delta_a(B(t)) \in (\mathcal{S})^*$. It has the following Wiener-Itô decomposition

$$\delta_a(B(t)) = \frac{1}{\sqrt{2\pi t}} e^{-\frac{a^2}{2t}} \sum_{n=0}^{\infty} \frac{1}{n!t^n} : a^n :_t \langle : \cdot^{\otimes n} :, 1_{[0,t]}^{\otimes n} \rangle.$$

Its S-transform is given by

$$(S\delta_a(B(t)))(\xi) = \frac{1}{\sqrt{2\pi t}} \exp\left[-\frac{1}{2t}\left(a - \int_0^t \xi(u)\,du \right)^2 \right], \quad \xi \in \mathcal{S}_c(\mathbb{R}).$$

Note that $\langle : \cdot^{\otimes n} :, 1_{[0,t]}^{\otimes n} \rangle = : B(t)^n :_t$. Hence $\delta_a(B(t))$ can be rewritten as

$$\delta_a(B(t)) = \frac{1}{\sqrt{2\pi t}} e^{-\frac{a^2}{2t}} \sum_{n=0}^{\infty} \frac{1}{n!t^n} : a^n :_t : B(t)^n :_t .$$

Example 7.5. We can use equation (7.1) to derive the Wiener-Itô decomposition of $|B(t)|$ as follows. Note that Hermite polynomials with parameter σ^2 satisfy the following identities:

$$: y^{n+1} :_{\sigma^2} - y : y^n :_{\sigma^2} + \sigma^2 n : y^{n-1} :_{\sigma^2} = 0, \quad \frac{d}{dy} : y^n :_{\sigma^2} = n : y^{n-1} :_{\sigma^2} .$$

By using these formulas, we can easily derive that

$$
\int_0^\infty :y^n:_{\sigma^2} e^{-\frac{y^2}{2\sigma^2}} \, dy = \begin{cases} 2^{-1}\sqrt{2\pi}\,\sigma, & \text{if } n = 0; \\ 0, & \text{if } n = 2k \geq 2; \\ (-1)^k (2k-1)!!\sigma^{2(k+1)}, & \text{if } n = 2k+1, \end{cases}
$$

where we use the same convention $(-1)!! = 1$ as in §5.1. These formulas can be used to derive

$$
\int_{-\infty}^\infty |y| :y^n:_{\sigma^2} e^{-\frac{y^2}{2\sigma^2}} \, dy = \begin{cases} 2\sigma^2, & \text{if } n = 0; \\ 2(-1)^{k+1}\sigma^{2(k+1)}(2k-3)!!, & \text{if } n = 2k \geq 2; \\ 0, & \text{if } n = 2k+1. \end{cases}
$$

From this formula we get the following expansion for $|x|$

$$
|x| = \sqrt{\frac{2}{\pi}}\,\sigma \sum_{n=0}^\infty \frac{(-1)^{n+1}}{(2n-1)2^n n!\sigma^{2n}} :x^{2n}:_{\sigma^2}.
$$

Now, put $x = \langle \cdot, 1_{[0,t]} \rangle = B(t)$ and $\sigma^2 = t$ to obtain the Wiener-Itô decomposition of $|B(t)| \in (L^2)$

$$
|B(t)| = \sqrt{\frac{2t}{\pi}} \sum_{n=0}^\infty \frac{(-1)^{n+1}}{(2n-1)2^n n! t^n} :B(t)^{2n}:_t.
$$

The S-transform of $|B(t)|$ is given by

$$
(S|B(t)|)(\xi) = \frac{ic}{\sqrt{\pi}} \int_0^{ic/\sqrt{2t}} \left(1 - u^{-2}(e^{u^2} - 1)\right) du,
$$

where $c = \int_0^t \xi(v) \, dv$.

We mention that equation (7.1) can also be used to derive the expansions of the distribution derivatives of $F \in \mathcal{S}'_c(\mathbb{R})$. By using the formula for Hermite polynomials $\frac{d}{dx} :x^n:_{\sigma^2} = n :x^{n-1}:_{\sigma^2}$, we can easily derive the k-th order distribution derivative $F^{(k)}$ of F, $k \geq 1$,

$$
F^{(k)}(x) = \frac{1}{\sqrt{2\pi}\,\sigma} \sum_{n=0}^\infty \frac{1}{n!\sigma^{2(n+k)}} :x^n:_{\sigma^2} \int_{\mathbb{R}} F(y) :y^{n+k}:_{\sigma^2} e^{-\frac{y^2}{2\sigma^2}} \, dy.
$$

Let $F = \delta_a$ and $x = \langle \cdot, 1_{[0,t]} \rangle = B(t)$. Then we get the Wiener-Itô decomposition of $\delta_a^{(k)}(B(t))$

$$
\delta_a^{(k)}(B(t)) = \frac{1}{\sqrt{2\pi t}} e^{-\frac{a^2}{2t}} \sum_{n=0}^\infty \frac{1}{n! t^{n+k}} :a^{n+k}:_t \langle :\cdot^{\otimes n}:, 1_{[0,t]}^{\otimes n} \rangle
$$

$$
= \frac{1}{\sqrt{2\pi t}} e^{-\frac{a^2}{2t}} \sum_{n=0}^\infty \frac{1}{n! t^{n+k}} :a^{n+k}:_t :B(t)^n:_t.
$$

Being motivated by equation (7.2), we can also consider the following series for any integer $k \geq 2$

$$\Phi_{F,f,k} \equiv \frac{1}{\sqrt{2\pi}\,|f|_0} \sum_{n=0}^{\infty} \frac{1}{n!\,|f|_0^{2n}}$$

$$\times \left(\int_{\mathbb{R}} F(y) : y^n :_{|f|_0^2} e^{-\frac{y^2}{2|f|_0^2}} \, dy \right) \langle : \cdot^{\otimes kn} :, f^{\otimes kn} \rangle. \qquad (7.8)$$

Observe that $: (x^k)^n :_{\sigma^2} \neq : x^{kn} :_{\sigma^2}$. Hence $\Phi_{F,f,k} \neq F(\langle \cdot, f \rangle^k)$. In fact, $F(\langle \cdot, f \rangle^k)$ has no meaning for a general $F \in S_c'(\mathbb{R})$.

Theorem 7.6. *Let $F \in S_c'(\mathbb{R})$ and $f \in L^2(\mathbb{R})$, $f \neq 0$. Then for any integer $k \geq 2$, the function $\Phi_{F,f,k}$ in equation (7.8) is a generalized function in $(S)_{\frac{k-1}{k}}^*$ and for all $\xi \in S_c(\mathbb{R})$*

$$(S\Phi_{F,f,k})(\xi) = \frac{1}{\sqrt{2\pi}\,|f|_0} \int_{\mathbb{R}} F(y) \exp\left[-\frac{1}{2|f|_0^2} (y - \langle f, \xi \rangle^k)^2 \right] dy,$$

where the integral is understood to be the bilinear pairing of $S_c'(\mathbb{R})$ and $S_c(\mathbb{R})$.

Proof. By Lemma 7.2

$$\|\Phi_{F,f,k}\|_{-p,-\beta}^2 \leq \frac{1}{2\pi|f|_0^2} \sum_{n=0}^{\infty} ((kn)!)^{1-\beta}$$

$$\times \frac{1}{(n!)^2|f|_0^{4n}} K_p^2 \, n! \, |f|_0^{2n+2} (n+1)^{2p} |f^{\otimes kn}|_{-p}^2.$$

Note that $|f^{\otimes kn}|_{-p} \leq 2^{-knp}|f|_0^{kn}$. Hence we get the following estimate

$$\|\Phi_{F,f,k}\|_{-p,-\beta}^2 \leq \frac{1}{2\pi} K_p^2 \sum_{n=0}^{\infty} \frac{((kn)!)^{1-\beta}}{n!} (n+1)^{2p} 2^{-2knp} |f|_0^{2(k-1)n}. \qquad (7.9)$$

Now, by using the Stirling formula, we can derive that

$$\frac{((kn)!)^{1-\beta}}{n!} \approx (2\pi)^{-\frac{\beta}{2}} k^{\frac{1-\beta}{2}} \frac{n^{(k(1-\beta)-1)n} k^{k(1-\beta)n}}{n^{\frac{\beta}{2}} e^{(k(1-\beta)-1)n}}.$$

In view of this asymptotic value, the smallest number β such that the series in (7.9) may converge is $\beta = \frac{k-1}{k}$. With this $\beta = \frac{k-1}{k}$, we choose p so large that $k|f|_0^{2(k-1)} < 2^{2kp}$. Then the series in equation (7.9) does converge. Hence $\Phi_{F,f,k} \in (S)_{\frac{k-1}{k}}^*$. Finally the S-transform of $\Phi_{F,f,k}$ can be derived easily in the same way as in the proof of Theorem 7.3. $\qquad \square$

Example 7.7. Let $F = \delta_a$ and $f = 1_{[0,t]}$ as in Example 7.4. Then we get the following generalized function in $(\mathcal{S})^*_{\frac{k-1}{k}}$

$$\Phi = \frac{1}{\sqrt{2\pi t}} e^{-\frac{a^2}{2t}} \sum_{n=0}^{\infty} \frac{1}{n! t^n} :a^n:_t \langle :\cdot^{\otimes kn}:, 1_{[0,t]}^{\otimes kn} \rangle.$$

The S-transform of Φ is given by

$$(S\Phi)(\xi) = \frac{1}{\sqrt{2\pi t}} \exp\left\{ -\frac{1}{2t}\left[a - \left(\int_0^t \xi(u)\, du \right)^k \right]^2 \right\}, \quad \xi \in \mathcal{S}_c(\mathbb{R}).$$

7.2 Kubo-Yokoi delta function

The analogue of the finite dimensional Dirac's delta function in white noise distribution theory is the Kubo-Yokoi delta function $\widetilde{\delta}_x$ at $x \in \mathcal{E}'$ defined by

$$\langle\langle \widetilde{\delta}_x, \varphi \rangle\rangle = \int_{\mathcal{E}'} \varphi(y)\, d\delta_x(y) = \varphi(x), \quad \varphi \in (\mathcal{E}),$$

where δ_x is the delta measure on $\mathcal{E}'$ at x. Recall that continuous versions of test functions are assumed (see the end of §6.1) so that $\varphi(x)$ is meaningful. By equation (6.16) the linear functional $\widetilde{\delta}_x$ is continuous on $(\mathcal{E})$ and so $\widetilde{\delta}_x \in (\mathcal{E})^*$ for each $x \in \mathcal{E}'$.

Theorem 7.8. *The Kubo-Yokoi delta function $\widetilde{\delta}_x$ at $x \in \mathcal{E}'$ has the following Wiener-Itô decomposition*

$$\widetilde{\delta}_x = \sum_{n=0}^{\infty} \frac{1}{n!} \langle :\cdot^{\otimes n}:, :x^{\otimes n}: \rangle.$$

Its S-transform is given by

$$S(\widetilde{\delta}_x)(\xi) = e^{\langle x, \xi \rangle - \frac{1}{2}\langle \xi, \xi \rangle}, \quad \xi \in \mathcal{E}_c.$$

Proof. Since $\widetilde{\delta}_x \in (\mathcal{E})^*$, it has the Wiener-Itô decomposition by §5.1:

$$\widetilde{\delta}_x = \sum_{n=0}^{\infty} \langle :\cdot^{\otimes n}:, F_n \rangle,$$

where $F_n \in (\mathcal{E}_c')^{\widehat{\otimes} n}$ and there exists some $p \geq 0$ such that

$$\sum_{n=0}^{\infty} n! |F_n|^2_{-p} < \infty.$$

Now for each fixed n, consider $\varphi = \langle : \cdot^{\otimes n} :, f \rangle$, $f \in \mathcal{E}_c^{\widehat{\otimes} n}$. On the one hand, by the definition of $\widetilde{\delta}_x$, we have

$$\langle\!\langle \widetilde{\delta}_x, \varphi \rangle\!\rangle = \varphi(x) = \langle : x^{\otimes n} :, f \rangle. \tag{7.10}$$

On the other hand, by equation (5.8),

$$\langle\!\langle \widetilde{\delta}_x, \varphi \rangle\!\rangle = n! \langle F_n, f \rangle. \tag{7.11}$$

It follows from equations (7.10) and (7.11) that $n! \langle F_n, f \rangle = \langle : x^{\otimes n} :, f \rangle$ for all $f \in \mathcal{E}_c^{\widehat{\otimes} n}$. This implies that $F_n = \frac{1}{n!} : x^{\otimes n} :$ and we get the Wiener-Itô decomposition of $\widetilde{\delta}_x$. To obtain the S-transform, we use Definition 5.8

$$
\begin{aligned}
S\big(\widetilde{\delta}_x\big)(\xi) &= e^{-\frac{1}{2}\langle \xi, \xi \rangle} \langle\!\langle \widetilde{\delta}_x, e^{\langle \cdot, \xi \rangle} \rangle\!\rangle \\
&= e^{\langle x, \xi \rangle - \frac{1}{2}\langle \xi, \xi \rangle}.
\end{aligned}
$$

$\square$

We can use the inequality in (6.16) to estimate the norm $\|\widetilde{\delta}_x\|_{-p, -\beta}$. However, we can get better estimates by using the representation of $\widetilde{\delta}_x$ given in Theorem 7.8.

Theorem 7.9. *For any $x \in \mathcal{E}'$ and $p > 0$, the following equality holds:*

$$\|\widetilde{\delta}_x\|_{-p} = \left(\|\Gamma(A)^{-2p}\|_{HS} \right)^{1/2} \exp\left[\frac{1}{2} |(I + A^{-2p})^{-1/2} x|^2_{-p} \right],$$

where $\Gamma(A)$ is the second quantization operator of A given in §4.2. In particular,

$$\left(\|\Gamma(A)^{-2p}\|_{HS} \right)^{1/2} e^{\frac{1}{2}(1+\lambda_1^{-2p})^{-1/2} |x|^2_{-p}}$$

$$\leq \|\widetilde{\delta}_x\|_{-p} \leq \left(\|\Gamma(A)^{-2p}\|_{HS} \right)^{1/2} e^{\frac{1}{2}|x|^2_{-p}},$$

where λ_1 is the smallest eigenvalue of A.

Remark. Let α be the positive number given in the assumptions for the operator A in §4.2. Recall from §4.2 that $\Gamma(A)^{-\alpha/2}$ is a Hilbert-Schmidt operator. Hence for $p \geq \alpha/4$ we see that $\widetilde{\delta}_x \in (\mathcal{E}_p)^*$ (i.e., $\|\widetilde{\delta}_x\|_{-p} < \infty$) if and only if $x \in \mathcal{E}_p'$ (i.e., $|x|_{-p} < \infty$).

Proof. By the representation of $\widetilde{\delta}_x$ in Theorem 7.8,

$$
\begin{aligned}
\|\widetilde{\delta}_x\|^2_{-p} &= \sum_{n=0}^{\infty} n! \frac{1}{(n!)^2} |: x^{\otimes n} :|^2_{-p} \\
&= \sum_{n=0}^{\infty} \frac{1}{n!} |: x^{\otimes n} :|^2_{-p}.
\end{aligned}
\tag{7.12}
$$

Let $\{\zeta_j; j = 1, 2, \ldots\}$ be the orthonormal basis for E given in §4.2. Then the following set is an orthonormal basis for $E^{\widetilde{\otimes} n}$

$$\left\{ \zeta_{n_1, n_2, \ldots} \equiv \sqrt{\frac{n!}{n_1! n_2! \cdots}} \; \zeta_1^{\otimes n_1} \widehat{\otimes} \zeta_2^{\otimes n_2} \widehat{\otimes} \cdots ; \; n_1 + n_2 + \cdots = n \right\}.$$

Therefore

$$A^{-p} : x^{\otimes n} : = \sum_{n_1 + n_2 + \cdots = n} \langle : x^{\otimes n} :, A^{-p} \zeta_{n_1, n_2, \ldots} \rangle \zeta_{n_1, n_2, \ldots}$$

$$= \sum_{n_1 + n_2 + \cdots = n} \prod_{j=1}^{\infty} \lambda_j^{-n_j p} \langle : x^{\otimes n} :, \zeta_{n_1, n_2, \ldots} \rangle \zeta_{n_1, n_2, \ldots}$$

$$= \sqrt{n!} \sum_{n_1 + n_2 + \cdots = n} \left(\prod_{j=1}^{\infty} \lambda_j^{-n_j p} \frac{1}{\sqrt{n_j!}} : \langle x, \zeta_j \rangle^{n_j} :_1 \right) \zeta_{n_1, n_2, \ldots}.$$

Here we have used Corollary 5.3 to get the last inequality. Thus

$$|: x^{\otimes n} :|_{-p}^2 = |A^{-p} : x^{\otimes n} :|_0^2$$

$$= n! \sum_{n_1 + n_2 + \cdots = n} \prod_{j=1}^{\infty} \lambda_j^{-2 n_j p} \frac{1}{n_j!} : \langle x, \zeta_j \rangle^{n_j} :_1^2. \qquad (7.13)$$

From equations (7.12) and (7.13), we obtain

$$\|\widetilde{\delta}_x\|_{-p}^2 = \sum_{n=0}^{\infty} \sum_{n_1 + n_2 + \cdots = n} \prod_{j=1}^{\infty} \lambda_j^{-2 n_j p} \frac{1}{n_j!} : \langle x, \zeta_j \rangle^{n_j} :_1^2$$

$$= \prod_{j=1}^{\infty} \sum_{n=0}^{\infty} \lambda_j^{-2np} \frac{1}{n!} : \langle x, \zeta_j \rangle^n :_1^2.$$

But Hermite polynomials $: x^n :_1$, $n \geq 0$, with parameter 1 satisfy the following identity:

$$\sum_{n=0}^{\infty} \frac{t^n}{n!} : x^n :_1^2 = (1 - t^2)^{-1/2} \exp\left[\frac{t}{1+t} x^2 \right], \quad |t| < 1.$$

Hence by using this identity, we get

$$\|\widetilde{\delta}_x\|_{-p}^2 = \prod_{j=1}^{\infty} \left(1 - \lambda_j^{-4p}\right)^{-1/2} \exp\left[\frac{\lambda_j^{-2p}}{1 + \lambda_j^{-2p}} \langle x, \zeta_j \rangle^2 \right]$$

$$= \left(\prod_{j=1}^{\infty} (1 - \lambda_j^{-4p}) \right)^{-1/2} \exp\left[\sum_{j=1}^{\infty} \frac{\lambda_j^{-2p}}{1 + \lambda_j^{-2p}} \langle x, \zeta_j \rangle^2 \right]. \qquad (7.14)$$

Recall that from §4.2 the Hilbert-Schmidt operator norm of $\Gamma(A)^{-2p}$ is given by

$$\|\Gamma(A)^{-2p}\|_{HS} = \left(\prod_{j=1}^{\infty} (1 - \lambda_j^{-4p}) \right)^{-1/2}. \tag{7.15}$$

On the other hand, we have

$$\left| (I + A^{-2p})^{-1/2} x \right|_{-p}^2 = \left| (I + A^{-2p})^{-1/2} A^{-p} x \right|_0^2$$

$$= \sum_{j=1}^{\infty} \frac{\lambda_j^{-2p}}{1 + \lambda_j^{-2p}} \langle x, \zeta_j \rangle^2. \tag{7.16}$$

Thus by equations (7.14)–(7.16) we obtain the equality in the theorem. The inequalities in the theorem are obvious in view of equation (7.16) and the following inequalities:

$$\frac{1}{1 + \lambda_1^{-2p}} \leq \frac{1}{1 + \lambda_j^{-2p}} \leq 1. \qquad \qquad \square$$

Next we estimate the norm $\|\tilde{\delta}_x\|_{-p,-\beta}$. Obviously we have $\|\tilde{\delta}_x\|_{-p,-\beta} \leq \|\tilde{\delta}_x\|_{-p}$ for any $0 \leq \beta < 1$. However, we can get a better estimate when $0 < \beta < 1$. In order to do so, we need to estimate the norm $|:x^{\otimes n}:|_{-p}$.

Lemma 7.10. *For any $n \geq 1$ and $x \in \mathcal{E}'_p$,*

$$|:x^{\otimes n}:|_{-p} \leq \sqrt{n!} \left(|x|_{-p} + |\tau|_{-p}^{1/2} \right)^n,$$

where τ is the trace operator given in §5.1.

Proof. Recall from Definition 5.1

$$:x^{\otimes n}: = \sum_{k=0}^{[n/2]} \binom{n}{2k} (2k-1)!! \, (-1)^k x^{\otimes(n-2k)} \widehat{\otimes} \tau^{\otimes k}.$$

Hence we get

$$|:x^{\otimes n}:|_{-p} \leq \sum_{k=0}^{[n/2]} \binom{n}{2k} (2k-1)!! \, |x|_{-p}^{n-2k} \, |\tau|_{-p}^k.$$

Note that $(2k-1)!! \leq (n-1)!! \leq \sqrt{n!}$ for all $0 \leq k \leq [n/2]$. Therefore

$$|:x^{\otimes n}:|_{-p} \leq \sqrt{n!} \sum_{k=0}^{[n/2]} \binom{n}{2k} |x|_{-p}^{n-2k} \left(|\tau|_{-p}^{1/2} \right)^{2k}$$

$$\leq \sqrt{n!} \sum_{k=0}^{n} \binom{n}{k} |x|_{-p}^{n-k} \left(|\tau|_{-p}^{1/2} \right)^{k}$$

$$= \sqrt{n!} \left(|x|_{-p} + |\tau|_{-p}^{1/2} \right)^n. \qquad \square$$

Observe that if we use the estimate in Lemma 7.10 for the norm $\|\tilde{\delta}_x\|_{-p}$ in equation (7.12), then we get

$$\|\tilde{\delta}_x\|_{-p}^2 = \sum_{n=0}^{\infty} n! \frac{1}{(n!)^2} |:x^{\otimes n}:|_{-p}^2$$

$$\leq \sum_{n=0}^{\infty} (|x|_{-p} + |\tau|_{-p}^{1/2})^{2n}.$$

Recall from the inequality in (5.1) that $|\tau|_{-p} \to 0$ as $p \to \infty$. On the other hand, for each $x \in \mathcal{E}'$, we also have $|x|_{-p} \to 0$ as $p \to \infty$. Thus for large p we have $|x|_{-p} + |\tau|_{-p}^{1/2} < 1$ and so we get

$$\|\tilde{\delta}_x\|_{-p} \leq (1 - |x|_{-p} - |\tau|_{-p}^{1/2})^{-1/2}.$$

Of course, this estimate is not so precise as the one given in Theorem 7.9. However, the idea can be used to get a better estimate for the norm $\|\tilde{\delta}_x\|_{-p,-\beta}$.

Theorem 7.11. If $x \in \mathcal{E}'_p$, then $\tilde{\delta}_x \in (\mathcal{E}_p)^*_\beta$ for all $0 < \beta < 1$ and

$$\|\tilde{\delta}_x\|_{-p,-\beta} \leq 2^{\frac{1-\beta}{2}} \exp\left[\beta 2^{\frac{1-2\beta}{\beta}} (|x|_{-p} + |\tau|_{-p}^{1/2})^{2/\beta}\right].$$

Proof. By Theorem 7.8 and Lemma 7.10 we have

$$\|\tilde{\delta}_x\|_{-p,-\beta}^2 = \sum_{n=0}^{\infty} (n!)^{1-\beta} \frac{1}{(n!)^2} |:x^{\otimes n}:|_{-p}^2$$

$$\leq \sum_{n=0}^{\infty} \frac{1}{(n!)^\beta} (|x|_{-p} + |\tau|_{-p}^{1/2})^{2n}.$$

In the proof of Theorem 5.7 we have an estimate for the following series

$$\sum_{n=0}^{\infty} \frac{1}{(n!)^{1-\beta}} a^{2n}.$$

Obviously we can simply replace β by $1 - \beta$ in that estimate to get

$$\|\tilde{\delta}_x\|_{-p,-\beta}^2 \leq 2^{1-\beta} \exp\left[\beta 2^{\frac{1-\beta}{\beta}} (|x|_{-p} + |\tau|_{-p}^{1/2})^{2/\beta}\right].$$

This implies the estimate in the theorem immediately. $\qquad\square$

7.3 Continuity of the delta functions

In this section we will prove the continuity of Donsker's delta function $\delta_a(B(t))$ on $a \in \mathbb{R}$ and the Kubo-Yokoi delta function $\widetilde{\delta}_x$ on $x \in \mathcal{E}'$.

First we prove some properties about Hermite polynomials $: x^n :_{\sigma^2}$ defined in §5.1. They can be used to obtain the corresponding properties for Wick tensors.

Lemma 7.12. $: (x + y)^n :_{\sigma^2} = \displaystyle\sum_{k=0}^{n} \binom{n}{k} : x^{n-k} :_{\sigma^2} y^k.$

Proof. By the generating function in equation (5.2) we have

$$e^{t(x+y) - \frac{1}{2}\sigma^2 t^2} = \sum_{n=0}^{\infty} \frac{t^n}{n!} : (x + y)^n :_{\sigma^2} . \tag{7.17}$$

On the other hand,

$$e^{t(x+y) - \frac{1}{2}\sigma^2 t^2} = e^{tx - \frac{1}{2}\sigma^2 t^2} e^{ty}$$

$$= \Big(\sum_{m=0}^{\infty} \frac{t^m}{m!} : x^m :_{\sigma^2} \Big) \Big(\sum_{k=0}^{\infty} \frac{t^k}{k!} y^k \Big)$$

$$= \sum_{n=0}^{\infty} \frac{t^n}{n!} \sum_{k=0}^{n} \binom{n}{k} : x^{n-k} :_{\sigma^2} y^k. \tag{7.18}$$

The identity in the lemma follows from equations (7.17) and (7.18) by comparing the coefficients of t^n. $\square$

Lemma 7.13. $\big| : x^n :_{\sigma^2} \big| \leq \sqrt{n!} \, (|x| + \sigma)^n.$

Proof. By equation (5.3)

$$\big| : x^n :_{\sigma^2} \big| \leq \sum_{k=0}^{[n/2]} \binom{n}{2k} (2k - 1)!! \, |x|^{n-2k} \sigma^{2k}.$$

Then apply the same argument as in the proof of Lemma 7.10 to get the inequality in this lemma. $\square$

Lemma 7.14. *The following inequality holds:*

$$\big| : y^n :_{\sigma^2} - : x^n :_{\sigma^2} \big| \leq n \sqrt{(n-1)!} \, |y - x| \, (|x| + |y - x| + \sigma)^{n-1}.$$

Proof. Let $\widetilde{y} = y - x$. Then by Lemmas 7.12 and 7.13

$$\left|{:}y^n{:}_{\sigma^2} - {:}x^n{:}_{\sigma^2}\right|$$

$$\leq \sum_{k=1}^{n} \binom{n}{k} \sqrt{(n-k)!} \, (|x| + \sigma)^{n-k} |\widetilde{y}|^k$$

$$= n|\widetilde{y}| \sum_{m=0}^{n-1} \frac{\sqrt{(n-1-m)!}}{m+1} \binom{n-1}{m} (|x| + \sigma)^{n-1-m} |\widetilde{y}|^m.$$

This proves the lemma since the inequality $\frac{\sqrt{(n-1-m)!}}{m+1} \leq \sqrt{(n-1)!}$ holds for all $0 \leq m \leq n-1$. $\qquad\square$

Theorem 7.15. *Donsker's delta function $\delta_a(B(t))$ depends continuously on a, i.e., the function $a \mapsto \delta_a(B(t))$ is continuous from $\mathbb{R}$ into $(\mathcal{S})^*$ with the inductive limit topology for $(\mathcal{S})^*$.*

Proof. In view of the representation of $\delta_a(B(t))$ in Example 7.4, we need only to prove the continuity of the following function

$$g(a) \equiv \sum_{n=0}^{\infty} \frac{1}{n! t^n} {:}a^n{:}_t \, \langle {:}{\cdot}^{\otimes n}{:}, 1_{[0,t]}^{\otimes n} \rangle.$$

For any two real numbers a and b, we have

$$\|g(b) - g(a)\|_{-p}^2 = \sum_{n=1}^{\infty} \frac{1}{n! t^{2n}} |{:}b^n{:}_t - {:}b^n{:}_t|^2 \left|1_{[0,t]}^{\otimes n}\right|_{-p}^2.$$

Note that

$$\left|1_{[0,t]}^{\otimes n}\right|_{-p} = \left|(A^{-p})^{\otimes n} 1_{[0,t]}^{\otimes n}\right|_0 \leq \lambda_1^{-np} t^{n/2}.$$

This estimate and Lemma 7.14 yield that

$$\|g(b) - g(a)\|_{-p}^2$$

$$\leq |b - a|^2 (|a| + |b - a| + \sqrt{t})^{-2} \sum_{n=1}^{\infty} n \left(\frac{|a| + |b - a| + \sqrt{t}}{\sqrt{t}\,\lambda_1^p}\right)^{2n}.$$

Note that $\sum_{n=1}^{\infty} n x^n = x(1-x)^{-2}$ for $|x| < 1$. Thus by choosing p large enough such that $\sqrt{t}\,\lambda_1^p > |a| + |b - a| + \sqrt{t}$, we get

$$\|g(b) - g(a)\|_{-p} \leq |b - a| \sqrt{t}\,\lambda_1^p \left(t\lambda_1^{2p} - (|a| + |b - a| + \sqrt{t})^2\right)^{-1}.$$

From this inequality we can check easily the continuity of the function $g \colon \mathbb{R} \to (\mathcal{S})^*$ with the inductive limit topology for $(\mathcal{S})^*$. $\qquad\square$

Next, we prove the continuity of the Kubo-Yokoi delta function $\widetilde{\delta}_x$ on $x \in \mathcal{E}'$. We will use the same arguments as in the proof of Theorem 6.4 due to the similarity in the following representations:

$$:e^{\langle \cdot, x \rangle}: = \sum_{n=0}^{\infty} \frac{1}{n!} \langle : \cdot^{\otimes n} :, x^{\otimes n} \rangle, \qquad \widetilde{\delta}_x = \sum_{n=0}^{\infty} \frac{1}{n!} \langle : \cdot^{\otimes n} :, : x^{\otimes n} : \rangle.$$

Lemma 7.16. *For any $x, y \in \mathcal{E}'$ and any $n \geq 1$*

$$:(x + y)^{\otimes n}: = \sum_{k=0}^{n} \binom{n}{k} :x^{\otimes (n-k)}: \widehat{\otimes} y^{\otimes k}.$$

Proof. Let $\xi \in \mathcal{E}$. By Lemma 5.2

$$\langle :(x+y)^{\otimes n}:, \xi^{\otimes n} \rangle = :\langle x+y, \xi \rangle^n :_{|\xi|_0^2}$$
$$= :\left(\langle x, \xi \rangle + \langle y, \xi \rangle \right)^n :_{|\xi|_0^2}.$$

Then apply Lemma 7.12 to get

$$\langle :(x+y)^{\otimes n}:, \xi^{\otimes n} \rangle = \sum_{k=0}^{n} \binom{n}{k} :\langle x, \xi \rangle^{n-k}:_{|\xi|_0^2} \langle y, \xi \rangle^k$$

$$= \sum_{k=0}^{n} \binom{n}{k} \langle :x^{\otimes (n-k)}:, \xi^{\otimes (n-k)} \rangle \langle y^{\otimes k}, \xi^{\otimes k} \rangle$$

$$= \sum_{k=0}^{n} \binom{n}{k} \langle :x^{\otimes (n-k)}: \widehat{\otimes} y^{\otimes k}, \xi^{\otimes n} \rangle$$

$$= \left\langle \sum_{k=0}^{n} \binom{n}{k} :x^{\otimes (n-k)}: \widehat{\otimes} y^{\otimes k}, \xi^{\otimes n} \right\rangle.$$

This proves the lemma in view of the polarization identity for $F \in (\mathcal{E}')^{\otimes n}$ given in the proof of Proposition 5.10. $\qquad \square$

Lemma 7.17. *Let $x \in \mathcal{E}'_q$ be fixed and let W be a neighborhood of zero given by the balanced convex hull of the sets $\{ w \in \mathcal{E}'_p; \; |w|_{-p} < \delta_p \}$ with $p \geq q$, $0 < \delta_p \leq 1$. Then for any $y \in x + W$, there exist positive numbers $\alpha_p, q \leq p \leq N$, with $\sum_{p=q}^{N} \alpha_p = 1$ such that for any n,*

$$:y^{\otimes n}: = :x^{\otimes n}: + \sum_{p=q}^{N} \alpha_p v_{n,p},$$

where $v_{n,p} \in (\mathcal{E}'_p)^{\widehat{\otimes}n}$ and $|v_{n,p}|_{-p} \le n\sqrt{(n-1)!}\left(1 + |x|_{-p} + |\tau|_{-p}^{1/2}\right)^{n-1}\delta_p$.

Proof. The proof is identical to the proof of Lemma 6.3 except for some estimates. We will point out the differences only. By using Lemma 7.16 we see that $:y^{\otimes n}: = :x^{\otimes n}: + \sum_{p=q}^{N}\alpha_p v_{n,p}$, where $v_{n,p}$ is given by

$$v_{n,p} \equiv \alpha_p^{-1}\sum_{k=1}^{n}\binom{n}{k}\sum_{p_1\vee\cdots\vee p_k=p}\alpha_{p_1}\cdots\alpha_{p_k} :x^{\otimes(n-k)}: \widehat{\otimes} w_{p_1}\widehat{\otimes}\cdots\widehat{\otimes}w_{p_k}.$$

Instead of inequality (6.6), we have the following inequality by using Lemma 7.10:

$$|v_{n,p}|_{-p} \le \alpha_p^{-1}\sum_{k=1}^{n}\binom{n}{k}\sum_{p_1\vee\cdots\vee p_k=p}$$
$$\alpha_{p_1}\cdots\alpha_{p_k}\sqrt{(n-k)!}\left(|x|_{-p} + |\tau|_{-p}^{1/2}\right)^{n-k}\delta_{p_1}\cdots\delta_{p_k}.$$

Note that $(n-k)! \le (n-1)!$ for $1 \le k \le n$. We can use the same arguments for inequalities (6.7)–(6.9) to get the estimate for this lemma. $\qquad\square$

Theorem 7.18. *The Kubo-Yokoi delta function $x \mapsto \tilde{\delta}_x$ is continuous from $\mathcal{E}'$ into $(\mathcal{E})^*$ with the inductive limit topology (equivalently, the strong topology) for both $\mathcal{E}'$ and $(\mathcal{E})^*$.*

Proof. We need only to modify the arguments in the proof of Theorem 6.4. Suppose $x \in \mathcal{E}'_q$. Choose large r such that $\lambda_1^{-r}(1 + |x|_{-q} + |\tau|_{-q}^{1/2}) < 1$. Let V be a balanced convex hull of 0 in $(\mathcal{E})^*$ of the form

$$V = \text{conv}\left(\bigcup_{p\ge q+r}\{\Phi \in (\mathcal{E}_p)^*; \|\Phi\|_{-p} < \epsilon_p\}\right).$$

Define a neighborhood W of 0 in $\mathcal{E}'$ by

$$W = \text{balanced convex hull of the sets }\{w \in \mathcal{E}'_p; |w|_{-p} < \delta_p\}, \ p \ge q,$$

where $\delta_p = \min\left\{1, \epsilon_{p+r}\lambda_1^r\left(1 - \lambda_1^{-2r}(1 + |x|_{-p} + |\tau|_{-p}^{1/2})\right)\right\}$.

For any $y \in x + W$, let $0 < \alpha_p \le 1$ and $v_{n,p} \in (\mathcal{E}'_p)^{\widehat{\otimes}n}$ be associated with y as given in Lemma 7.17. Then for any $n \ge 1$

$$:y^{\otimes n}: = :x^{\otimes n}: + \sum_{p=q}^{N}\alpha_p v_{n,p}.$$

Thus similar to equation (6.10), we have $\tilde{\delta}_y - \tilde{\delta}_x = \sum_{p=q}^{N}\alpha_p\Phi_p$ with Φ_p given by

$$\Phi_p = \sum_{n=1}^{\infty}\left\langle :\cdot^{\otimes n}:, \frac{1}{n!}v_{n,p}\right\rangle.$$

Note that by Lemma 7.17 we have

$$|v_{n,p}|_{-(p+r)} \leq \lambda_1^{-rn} |v_{n,p}|_{-p}$$

$$\leq \lambda_1^{-rn} n \sqrt{(n-1)!} \, \big(1 + |x|_{-p} + |\tau|_{-p}^{1/2}\big)^{n-1} \delta_p.$$

Therefore

$$\|\Phi_p\|_{-(p+r)}^2 = \sum_{n=1}^{\infty} n! \frac{1}{(n!)^2} |v_{n,p}|_{-(p+r)}^2$$

$$\leq \delta_p^2 \big(1 + |x|_{-p} + |\tau|_{-p}^{1/2}\big)^{-2} \sum_{n=1}^{\infty} n \Big(\lambda_1^{-2r} \big(1 + |x|_{-p} + |\tau|_{-p}^{1/2}\big)^2\Big)^n.$$

Since $\lambda_1^{-r}(1 + |x|_{-q} + |\tau|_{-q}^{1/2}) < 1$, the last series is convergent. Moreover, it is easy to check that

$$\|\Phi_p\|_{-(p+r)}^2 \leq \delta_p^2 \lambda_1^{-2r} \Big(1 - \lambda_1^{-2r}\big(1 + |x|_{-p} + |\tau|_{-p}^{1/2}\big)^2\Big)^{-2}.$$

Hence by the choice of δ_p, we have $\|\Phi_p\|_{-(p+r)} < \epsilon_{p+r}$ for any $q \leq p \leq N$. This shows that $\widetilde{\delta}_y \in \widetilde{\delta}_x + V$ for any $y \in x + W$, i.e., $\widetilde{\delta}_{(\cdot)}$ is continuous at x. This completes the proof. $\qquad\square$

Finally we estimate the norm $\|\widetilde{\delta}_y - \widetilde{\delta}_x\|_{-p}$. In order to do so, we first need to estimate the norm $|:y^{\otimes n}: - :x^{\otimes n}:|_{-p}$.

Lemma 7.19. *For any $x, y \in \mathcal{E}'$ and $n \geq 1$, the following inequality holds:*

$$|:y^{\otimes n}: - :x^{\otimes n}:|_{-p}$$

$$\leq n\sqrt{(n-1)!} \, |y - x|_{-p} \big(|x|_{-p} + |y - x|_{-p} + |\tau|_{-p}^{1/2}\big)^{n-1}.$$

Proof. The proof is similar to Lemma 7.14 except that Lemmas 7.10 and 7.16 (instead of Lemmas 7.12 and 7.13) should be used. $\qquad\square$

Theorem 7.20. *Let $x, y \in \mathcal{E}'_p$. Then the inequality*

$$\|\widetilde{\delta}_y - \widetilde{\delta}_x\|_{-(p+r)}$$

$$\leq |y - x|_{-p} \lambda_1^{-r} \Big(1 - \lambda_1^{-2r}\big(|x|_{-p} + |y - x|_{-p} + |\tau|_{-q}^{1/2}\big)^2\Big)^{-1/2}$$

holds for any large r such that $\lambda_1^{-r}(|x|_{-p} + |y - x|_{-p} + |\tau|_{-q}^{1/2}) < 1$.

Proof. Use Lemma 7.19 and the same estimation in Theorem 7.18. $\qquad\square$

8

Characterization Theorems

A fundamental tool in white noise distribution theory is the S-transform defined in §5.2. Often we specify a generalized function (or Hida distribution) by its S-transform. Thus it is very important to characterize those functions which are S-transforms of generalized functions. Similarly, it is desirable to characterize those functions which are S-transforms of test functions.

8.1 Characterization of generalized functions

Let $\Phi \in (\mathcal{E})^*_\beta$. Its S-transform $F(\xi) \equiv S\Phi$ is a function on $\mathcal{E}_c$ given by

$$F(\xi) = \langle\!\langle \Phi, :e^{\langle \cdot, \xi\rangle} :\rangle\!\rangle = e^{-\frac{1}{2}\langle \xi, \xi\rangle}\langle\!\langle \Phi, e^{\langle \cdot, \xi\rangle}\rangle\!\rangle, \quad \xi \in \mathcal{E}_c.$$

In order to characterize those functions on $\mathcal{E}_c$ which are S-transforms of generalized functions in $(\mathcal{E})^*_\beta$, we first need to find the necessary conditions satisfied by F.

Lemma 8.1. *Let* $\Phi \in (\mathcal{E})^*_\beta$ *and* $F = S\Phi$. *Then for any* ξ *and* η *in* $\mathcal{E}_c$, *the function* $F(z\xi + \eta)$ *is an entire function of* $z \in \mathbb{C}$.

Proof. Let Φ be represented by

$$\Phi = \sum_{n=0}^{\infty} \langle :\cdot^{\otimes n}:, f_n\rangle, \quad f_n \in (\mathcal{E}'_{p,c})^{\widehat{\otimes}n}.$$

Since $(\mathcal{E})^*_\beta = \cup_{p\geq 0}(\mathcal{E}_p)^*_\beta$, $\Phi \in (\mathcal{E}_p)^*_\beta$ for some $p \geq 0$, i.e.,

$$\|\Phi\|^2_{-p,-\beta} = \sum_{n=0}^{\infty}(n!)^{1-\beta}|f_n|^2_{-p} < \infty. \tag{8.1}$$

77

By Proposition 5.9 the S-transform $F = S\Phi$ is given by

$$F(\xi) = \sum_{n=0}^{\infty} \langle f_n, \xi^{\otimes n} \rangle, \quad \xi \in \mathcal{E}_c.$$

Hence for any $\xi, \eta \in \mathcal{E}_c$ and any $z \in \mathbb{C}$,

$$F(z\xi + \eta) = \sum_{n=0}^{\infty} \sum_{k=0}^{n} \binom{n}{k} z^{n-k} \langle f_n, \xi^{\otimes(n-k)} \widehat{\otimes} \eta^{\otimes k} \rangle.$$

By changing the order of summations (which can be justified by the calculation below) we get easily that

$$F(z\xi + \eta) = \sum_{n=0}^{\infty} A_n z^n, \tag{8.2}$$

where A_n is given by

$$A_n = \sum_{k=0}^{\infty} \binom{n+k}{k} \langle f_{n+k}, \xi^{\otimes n} \widehat{\otimes} \eta^{\otimes k} \rangle.$$

Obviously we have

$$|A_n| \leq \frac{1}{n!} |\xi|_p^n \sum_{k=0}^{\infty} \frac{(n+k)!}{k!} |f_{n+k}|_{-p} |\eta|_p^k.$$

Now, write the general term in the last series as

$$\frac{(n+k)!}{k!} |f_{n+k}|_{-p} |\eta|_p^k = \left(\frac{1}{k!} [(n+k)!]^{\frac{1+\beta}{2}} |\eta|_p^k \right) \left([(n+k)!]^{\frac{1-\beta}{2}} |f_{n+k}|_{-p} \right)$$

and then apply the Schwarz inequality to get

$$|A_n| \leq \frac{1}{n!} |\xi|_p^n \left(\sum_{k=0}^{\infty} \frac{[(n+k)!]^{1+\beta}}{(k!)^2} |\eta|_p^{2k} \right)^{1/2} \left(\sum_{k=0}^{\infty} [(n+k)!]^{1-\beta} |f_{n+k}|_{-p}^2 \right)^{1/2}.$$

Hence by equation (8.1) we obtain

$$|A_n| \leq \|\Phi\|_{-p,-\beta} \frac{1}{n!} |\xi|_p^n \left(\sum_{k=0}^{\infty} \frac{[(n+k)!]^{1+\beta}}{(k!)^2} |\eta|_p^{2k} \right)^{1/2}. \tag{8.3}$$

We can use the inequality $(n+k)! \leq n!k!2^{n+k}$ to estimate the last summation

$$\sum_{k=0}^{\infty} \frac{[(n+k)!]^{1+\beta}}{(k!)^2} |\eta|_p^{2k} \leq (n!)^{1+\beta} 2^{n(1+\beta)} \sum_{k=0}^{\infty} \frac{2^{k(1+\beta)}}{(k!)^{1-\beta}} |\eta|_p^{2k}. \tag{8.4}$$

Then by the same trick as in the proof of Theorem 5.7, i.e., write the general term of the last series as

$$\left(\frac{1}{2^{k\beta}}\right)\left(\frac{2^{k\beta}2^{k(1+\beta)}}{(k!)^{1-\beta}}|\eta|_p^{2k}\right),$$

we can apply the Hölder inequality with the pair $\{\frac{1}{\beta}, \frac{1}{1-\beta}\}$ to get

$$\sum_{k=0}^{\infty}\frac{2^{k(1+\beta)}}{(k!)^{1-\beta}}|\eta|_p^{2k} \leq 2^{\beta}\exp\left[(1-\beta)2^{\frac{1+2\beta}{1-\beta}}|\eta|_p^{\frac{2}{1-\beta}}\right]. \tag{8.5}$$

From the inequalities (8.3)–(8.5), we can derive that

$$|A_n| \leq \|\Phi\|_{-p,-\beta}\frac{1}{(n!)^{\frac{1-\beta}{2}}}\left(2^{\frac{1+\beta}{2}}|\xi|_p\right)^n 2^{\frac{\beta}{2}}\exp\left[(1-\beta)2^{\frac{3\beta}{1-\beta}}|\eta|_p^{\frac{2}{1-\beta}}\right]. \tag{8.6}$$

The inequality (8.6) implies that the series in (8.2) converges for all $z \in \mathbb{C}$. Thus the function $F(z\xi + \eta)$ is an entire function. $\qquad\square$

Another necessary condition is the growth condition. Note that

$$F(\xi) = \langle\!\langle \Phi, :e^{\langle\cdot,\xi\rangle}:\rangle\!\rangle, \quad \xi \in \mathcal{E}_c.$$

Therefore, in view of Theorem 5.7, $F = S\Phi$ with $\Phi \in (\mathcal{E}_p)_{\beta}^*$ should satisfy the following growth condition

$$|F(\xi)| \leq c\|\Phi\|_{-p,-\beta}\exp\left[a|\xi|_p^{\frac{2}{1-\beta}}\right], \quad \xi \in \mathcal{E}_c,$$

where $c = 2^{\beta/2}$ and $a = (1-\beta)2^{\frac{2\beta-1}{1-\beta}}$.

It turns out this growth condition and the entire function property in Lemma 8.1 are also sufficient in order for F to be the S-transform of some generalized functions. That is, we have the next characterization theorem.

Theorem 8.2. *Let $\Phi \in (\mathcal{E})_{\beta}^*$. Then its S-transform $F = S\Phi$ satisfies the conditions:*

(a) *For any ξ and η in $\mathcal{E}_c$, the function $F(z\xi + \eta)$ is an entire function of $z \in \mathbb{C}$.*

(b) *There exist nonnegative constants K, a, and p such that*

$$|F(\xi)| \leq K\exp\left[a|\xi|_p^{\frac{2}{1-\beta}}\right], \quad \forall \xi \in \mathcal{E}_c.$$

Conversely, suppose a function F defined on $\mathcal{E}_c$ satisfies the above two conditions. Then there exists a unique $\Phi \in (\mathcal{E})_{\beta}^*$ such that $F = S\Phi$ and for any q satisfying the condition that $e^2\left(\frac{2a}{1-\beta}\right)^{1-\beta}\|A^{-(q-p)}\|_{HS}^2 < 1$, the following inequality holds:

$$\|\Phi\|_{-q,-\beta} \le K\left(1 - e^2\left(\frac{2a}{1-\beta}\right)^{1-\beta}\|A^{-(q-p)}\|_{HS}^2\right)^{-1/2}.$$

Remarks. (1) This characterization theorem was first proved by Potthoff and Streit [150] for the space $(\mathcal{S})^*$ (see also the book by Hida et al. [58]). It has been extended in Kondratiev and Streit [83] [84] to the spaces $(\mathcal{S})_\beta^*$. The proof below for the case $\beta = 0$ is given in Kubo and Kuo [91]. (2) The condition (a) is not equivalent to the condition that for any $\xi \in \mathcal{E}_c$, the function $F(z\xi)$ is an entire function of $z \in \mathbb{C}$. For example, take any $x \ne y$ in $\mathcal{E}'$ and define

$$F(\xi) = \begin{cases} \dfrac{\langle x,\xi\rangle^2\langle y,\xi\rangle^2}{\langle x,\xi\rangle^2 + \langle y,\xi\rangle^2}, & \text{if } \langle x,\xi\rangle^2 + \langle y,\xi\rangle^2 \ne 0; \\ 0, & \text{if } \langle x,\xi\rangle = \langle y,\xi\rangle = 0. \end{cases}$$

Obviously, for any $\xi \in \mathcal{E}_c$, the function $F(z\xi)$ is an entire function. However, it does not satisfy condition (a).

Proof. The first assertion has already been shown above. So, suppose F is a function defined on $\mathcal{E}_c$ satisfying conditions (a) and (b). The uniqueness of Φ with $F = S\Phi$ is a consequence of Proposition 5.10. For the existence, we need to find $f_n \in (\mathcal{E}_c')^{\widehat{\otimes}n}$ such that

$$\Phi = \sum_{n=0}^{\infty} \langle :\cdot^{\otimes n}:, f_n\rangle \tag{8.7}$$

defines a generalized function in $(\mathcal{E})_\beta^*$ and $F = S\Phi$, i.e.,

$$F(\xi) = \sum_{n=0}^{\infty} \langle f_n, \xi^{\otimes n}\rangle, \quad \xi \in \mathcal{E}_c.$$

This implies that $f_0 = F(0)$ and for $n \ge 1$, f_n is given by

$$\langle f_n, \xi_1\widehat{\otimes}\cdots\widehat{\otimes}\xi_n\rangle = \frac{1}{n!}\frac{\partial}{\partial z_1}\cdots\frac{\partial}{\partial z_n}F(z_1\xi_1 + \cdots + z_n\xi_n)\Big|_{z_1=\cdots=z_n=0}.$$

Note that condition (a) holds if and only if for any $n \ge 1$ and any $\xi_1,\ldots,\xi_n \in \mathcal{E}_c$, the function $F(z_1\xi_1 + \cdots + z_n\xi_n)$ is entire in each variable separately. In fact, by Hartogs' theorem (see Bochner and Martin [13]), it is holomorphic on $\mathbb{C}^n$. For $n \ge 1$, define a symmetric n-linear functional J_n on $\mathcal{E}_c \times \cdots \times \mathcal{E}_c$ by

$$J_n(\xi_1,\ldots,\xi_n) = \frac{1}{n!}\frac{\partial}{\partial z_1}\cdots\frac{\partial}{\partial z_n}F(z_1\xi_1 + \cdots + z_n\xi_n)\Big|_{z_1=\cdots=z_n=0}. \tag{8.8}$$

By the Cauchy formula, we have

$$J_n(\xi_1, \ldots, \xi_n)$$

$$= \frac{1}{n!} \frac{1}{(2\pi i)^n} \int_{|z_1|=r_1} \cdots \int_{|z_n|=r_n} \frac{F(z_1\xi_1 + \cdots + z_n\xi_n)}{z_1^2 \cdots z_n^2} dz_1 \cdots dz_n.$$

First suppose $|\xi_1|_p = \cdots = |\xi_n|_p = 1$. Then by condition (b)

$$|J_n(\xi_1, \ldots, \xi_n)|$$

$$\leq \frac{1}{n!} \frac{1}{(2\pi)^n} \int_{|z_1|=r_1} \cdots \int_{|z_n|=r_n} \frac{|F(z_1\xi_1 + \cdots + z_n\xi_n)|}{|z_1|^2 \cdots |z_n|^2} |dz_1 \cdots dz_n|$$

$$\leq \frac{1}{n!} \frac{1}{(2\pi)^n} \int_0^{2\pi} \cdots \int_0^{2\pi} \frac{K e^{a(r_1 + \cdots + r_n)^{\frac{2}{1-\beta}}}}{r_1^2 \cdots r_n^2} r_1 \cdots r_n \, d\theta_1 \cdots d\theta_n$$

$$= K \frac{1}{n!} \frac{e^{a(r_1 + \cdots + r_n)^{\frac{2}{1-\beta}}}}{r_1 \cdots r_n}.$$

We can check easily that the minimum of the right hand side occurs at

$$r_1 = \cdots = r_n = n^{-\frac{1+\beta}{2}} \left(\frac{1-\beta}{2a} \right)^{\frac{1-\beta}{2}}.$$

With this choice of r_j's, we get

$$|J_n(\xi_1, \ldots, \xi_n)| \leq K \frac{1}{n!} n^{\frac{n(1+\beta)}{2}} \left(\frac{2ae}{1-\beta} \right)^{\frac{n(1-\beta)}{2}}.$$

Thus for any $\xi_1, \ldots, \xi_n \in \mathcal{E}_c$, we have

$$|J_n(\xi_1, \ldots, \xi_n)| \leq K \frac{1}{n!} n^{\frac{n(1+\beta)}{2}} \left(\frac{2ae}{1-\beta} \right)^{\frac{n(1-\beta)}{2}} |\xi_1|_p \cdots |\xi_n|_p. \qquad (8.9)$$

Take the eigenvectors $\{\zeta_n\}$ of A (see §4.2) and define

$$f_n = \sum_{j_1, \ldots, j_n} J_n(A^{-q}\zeta_{j_1}, \ldots, A^{-q}\zeta_{j_n})(A^q\zeta_{j_1}) \otimes \cdots \otimes (A^q\zeta_{j_n}).$$

By equation (8.9), we have for any $n \geq 1$

$$|f_n|^2_{-q} = \sum_{j_1, \ldots, j_n} |J_n(A^{-q}\zeta_{j_1}, \ldots, A^{-q}\zeta_{j_n})|^2$$

$$\leq K^2 \frac{1}{(n!)^2} n^{n(1+\beta)} \left(\frac{2ae}{1-\beta} \right)^{n(1-\beta)} \sum_{j_1, \ldots, j_n} |A^{-q}\zeta_{j_1}|_p^2 \cdots |A^{-q}\zeta_{j_n}|_p^2$$

$$= K^2 \frac{1}{(n!)^2} n^{n(1+\beta)} \left(\frac{2ae}{1-\beta} \right)^{n(1-\beta)} \left(\sum_{j=1}^{\infty} |A^{-q}\zeta_j|_p^2 \right)^n.$$

But $|A^{-q}\zeta_j|_p = |A^{-(q-p)}\zeta_j|_0$. Hence

$$|f_n|^2_{-q} \leq K^2 \frac{1}{(n!)^2} n^{n(1+\beta)} \left(\frac{2ae}{1-\beta}\right)^{n(1-\beta)} \left(\sum_{j=1}^{\infty} |A^{-(q-p)}\zeta_j|^2_0\right)^n$$

$$= K^2 \frac{1}{(n!)^2} n^{n(1+\beta)} \left(\frac{2ae}{1-\beta}\right)^{n(1-\beta)} \|A^{-(q-p)}\|^{2n}_{HS}.$$

Now, let $f_0 = F(0)$ and define Φ by equation (8.7). Note that $|f_0| \leq K$. Hence we get

$$\|\Phi\|^2_{-q,-\beta} = \sum_{n=0}^{\infty} (n!)^{1-\beta} |f_n|^2_{-q}$$

$$\leq K^2 \sum_{n=0}^{\infty} \frac{1}{(n!)^{1+\beta}} n^{n(1+\beta)} \left(\frac{2ae}{1-\beta}\right)^{n(1-\beta)} \|A^{-(q-p)}\|^{2n}_{HS},$$

where we have used the convention $0^0 = 1$. To estimate the sum of the last series, note that for $n \geq 1$,

$$n! \geq \sqrt{2\pi n} \left(\frac{n}{e}\right)^n \geq \left(\frac{n}{e}\right)^n.$$

Therefore

$$\|\Phi\|^2_{-q,-\beta} \leq K^2 \sum_{n=0}^{\infty} e^{2n} \left(\frac{2a}{1-\beta}\right)^{n(1-\beta)} \|A^{-(q-p)}\|^{2n}_{HS}.$$

Hence for q satisfying the condition that $e^2 \left(\frac{2a}{1-\beta}\right)^{1-\beta} \|A^{-(q-p)}\|^2_{HS} < 1$, we have

$$\|\Phi\|^2_{-q,-\beta} \leq K^2 \left(1 - e^2 \left(\frac{2a}{1-\beta}\right)^{1-\beta} \|A^{-(q-p)}\|^2_{HS}\right)^{-1}.$$

Thus $\Phi \in (\mathcal{E})^*_\beta$ and we have a norm estimate for $\|\Phi\|_{-q,-\beta}$. To find the S-transform of Φ, note that from equation (8.8) we have

$$J_n(\xi, \ldots, \xi) = \frac{1}{n!} \frac{d^n}{dz^n} F(z\xi) \Big|_{z=0}.$$

Thus for any $\xi \in \mathcal{E}_c$,

$$S\Phi(\xi) = \sum_{n=0}^{\infty} \langle f_n, \xi^{\otimes n} \rangle$$

$$= \sum_{n=0}^{\infty} J_n(\xi, \ldots, \xi)$$

$$= \sum_{n=0}^{\infty} \frac{1}{n!} \frac{d^n}{dz^n} F(z\xi) \Big|_{z=0}.$$

But the Taylor series of $F(z\xi)$ at $z = 1$ is given by

$$F(\xi) = \sum_{n=0}^{\infty} \frac{1}{n!} \frac{d^n}{dz^n} F(z\xi)\Big|_{z=0}.$$

Hence we have $S\Phi(\xi) = F(\xi)$ for any $\xi \in \mathcal{E}_c$, i.e., $S\Phi = F$. $\square$

Example 8.3. *Gaussian white noise functions*

For any $x \in \mathcal{E}_c'$ and $c \in \mathbb{C}, c \neq -1$, consider the function

$$F(\xi) = \exp\left[\frac{1}{1+c}\langle x, \xi\rangle - \frac{1}{2(1+c)}\langle \xi, \xi\rangle\right], \quad \xi \in \mathcal{E}_c.$$

It is easy to check that F satisfies conditions (a) and (b) in Theorem 8.2 with $\beta = 0$. Hence there exists a unique $g_{x,c} \in (\mathcal{E})^*$ such that

$$Sg_{x,c}(\xi) = \exp\left[\frac{1}{1+c}\langle x, \xi\rangle - \frac{1}{2(1+c)}\langle \xi, \xi\rangle\right], \quad \xi \in \mathcal{E}_c. \tag{8.10}$$

This generalized function $g_{x,c}$ is called the *Gaussian white noise function* with mean x and (generalized) variance c. When $x = 0$, $g_{x,c}$ will be denoted simply by g_c.

Example 8.4. *Generalized functions induced by Gaussian measures*

Let $\mu_{x,t}$ be the Gaussian measure on $\mathcal{E}'$ with mean $x \in \mathcal{E}'$ and variance $t > 0$, i.e., its characteristic function is given by

$$\int_{\mathcal{E}'} e^{i\langle y, \xi\rangle} \, d\mu_{x,t}(y) = \exp\left[i\langle x, \xi\rangle - \frac{1}{2} t |\xi|_0^2\right], \quad \xi \in \mathcal{E}.$$

Define a function F on $\mathcal{E}_c$ by

$$F(\xi) \equiv e^{-\frac{1}{2}\langle \xi, \xi\rangle} \int_{\mathcal{E}'} e^{\langle y, \xi\rangle} \, d\mu_{x,t}(y)$$

$$= \exp\left[\langle x, \xi\rangle - \frac{1}{2}(1-t)\langle \xi, \xi\rangle\right].$$

Obviously this function F satisfies conditions (a) and (b) in Theorem 8.2 with $\beta = 0$. Hence there exists a unique generalized function $\widetilde{\mu}_{x,t} \in (\mathcal{E})^*$ with S-transform

$$S\widetilde{\mu}_{x,t}(\xi) = \exp\left[\langle x, \xi\rangle - \frac{1}{2}(1-t)\langle \xi, \xi\rangle\right], \quad \xi \in \mathcal{E}_c.$$

When $x = 0$, $\mu_{x,t}$ and $\widetilde{\mu}_{x,t}$ will be simply denoted by μ_t and $\widetilde{\mu}_t$, respectively. Observe that $\widetilde{\mu}_{x,t}$ is related to the Gaussian white noise functions by

$$g_{x,c} = \widetilde{\mu}_{\frac{1}{1+c}x, \frac{c}{1+c}}, \quad x \in \mathcal{E}', \ c < -1 \text{ or } 0 \leq c \leq \infty,$$

$$\widetilde{\mu}_{x,t} = g_{\frac{1}{1-t}x, \frac{t}{1-t}}, \quad x \in \mathcal{E}', \ t > 0, \ t \neq 1.$$

Example 8.5. *Grey noise measure* (see Kondratiev and Streit [84] and Schneider [161])

Let $0 < \lambda \leq 1$. Consider the Mittag-Leffler function defined by

$$L_\lambda(t) = \sum_{n=0}^{\infty} \frac{(-t)^n}{\Gamma(1 + \lambda n)}, \quad t \geq 0, \tag{8.11}$$

where Γ is the Gamma function. It is easy to check that this series converges for all $t \geq 0$ by using the Stirling formula:

$$\Gamma(w) \approx \sqrt{2\pi}\, e^{-w} w^{w-\frac{1}{2}}, \quad |\arg w| < \pi. \tag{8.12}$$

By Feller [34, p.453], the function L_λ is the Laplace transform of a distribution function G_λ, namely,

$$L_\lambda(t) = \int_0^\infty e^{-tu}\, dG_\lambda(u), \quad t \geq 0. \tag{8.13}$$

In fact, the distribution function G_λ is given as follows. When $\lambda = 1$, $G_\lambda = 1_{[1,\infty)}$. When $0 < \lambda < 1$, $G_\lambda(u) = 1 - F_\lambda(u^{-1/\lambda}), u \geq 0$, where F_λ is the λ-stable distribution with Laplace transform given by

$$\int_0^\infty e^{-tu}\, dF_\lambda(u) = \exp\left[-t^\lambda\right], \quad t \geq 0.$$

Define a function J_λ on $\mathcal{E}$ by

$$J_\lambda(\xi) = L_\lambda(|\xi|_0^2), \quad \xi \in \mathcal{E}.$$

Then $J_\lambda(0) = 1$ and by equation (8.13) we have

$$J_\lambda(\xi) = \int_0^\infty e^{-u|\xi|_0^2}\, dG_\lambda(u), \quad \xi \in \mathcal{E}. \tag{8.14}$$

Observe that the function $e^{-u|\xi|_0^2}, \xi \in \mathcal{E}$, is the characteristic function of the probability measure $\mu(\cdot/\sqrt{2u})$ (μ is the Gaussian measure on $\mathcal{E}'$). Thus the function J_λ is positive definite and continuous. Hence by the Minlos theorem there exists a unique probability measure ν_λ such that

$$\int_{\mathcal{E}'} e^{i\langle x, \xi \rangle}\, d\nu_\lambda(x) = J_\lambda(\xi) = L_\lambda(|\xi|_0^2), \quad \xi \in \mathcal{E}. \tag{8.15}$$

This probability measure is called a *grey noise measure* in Schneider [161].

Now, observe that the function $L_\lambda(t), t \geq 0$, in equation (8.11) has an entire extension given by

$$L_\lambda(z) = \sum_{n=0}^{\infty} \frac{(-z)^n}{\Gamma(1 + \lambda n)}, \quad z \in \mathbb{C}.$$

The convergence of this series can also be checked by using the Stirling formula in equation (8.12). In view of this entire extension, it is reasonable to expect that the grey noise measure ν_λ induces a generalized function $\Phi_\lambda \in (\mathcal{E})^*_\beta$, i.e.,

$$\langle\!\langle \Phi_\lambda, \varphi \rangle\!\rangle = \int_{\mathcal{E}'} \varphi(x) \, d\nu_\lambda(x), \quad \varphi \in (\mathcal{E})_\beta, \tag{8.16}$$

where β is to be determined later. By equation (8.15) we have

$$\int_{\mathcal{E}'} e^{i\langle x, \xi \rangle} \, d\nu_\lambda(x) = L_\lambda(\langle \xi, \xi \rangle), \quad \xi \in \mathcal{E}_c. \tag{8.17}$$

Thus by equations (8.16) and (8.17) (with ξ being replaced by $-i\xi$), the S-transform of Φ_λ is given by

$$S\Phi_\lambda(\xi) = e^{-\frac{1}{2}\langle \xi, \xi \rangle} \langle\!\langle \Phi_\lambda, e^{\langle \cdot, \xi \rangle} \rangle\!\rangle$$

$$= e^{-\frac{1}{2}\langle \xi, \xi \rangle} \int_{\mathcal{E}'} e^{\langle x, \xi \rangle} \, d\nu_\lambda(x)$$

$$= e^{-\frac{1}{2}\langle \xi, \xi \rangle} L_\lambda(-\langle \xi, \xi \rangle), \quad \xi \in \mathcal{E}_c. \tag{8.18}$$

Finally, we use the characterization theorem to show that Φ_λ is a generalized function in $(\mathcal{E})^*_{1-\lambda}$. First of all, it is obvious that Φ_λ satisfies condition (a) of Theorem 8.2. To check condition (b), note that equation (8.12) implies that

$$\Gamma(1 + \lambda n) \approx \lambda^{1/2} (2\pi)^{(1-\lambda)/2} n^{(1-\lambda)/2} (n!)^\lambda.$$

Hence there exists some constant C_λ depending only on λ such that

$$|L_\lambda(z)| \leq C_\lambda \sum_{n=0}^{\infty} \frac{|z|^n}{(n!)^\lambda}, \quad z \in \mathbb{C}.$$

Then by Lemma 6.6 we have

$$|L_\lambda(z)| \leq C_\lambda \exp\left[\lambda |z|^{1/\lambda}\right], \quad z \in \mathbb{C}.$$

This inequality and equation (8.18) yield the growth condition for $S\Phi_\lambda(\xi)$:

$$|S\Phi_\lambda(\xi)| \le C_\lambda \exp\left[\frac{1}{2}|\xi|_0^2\right] \exp\left[\lambda|\xi|_0^{2/\lambda}\right].$$

Since $2/\lambda \ge 2$, we can find constants a_λ and K_λ depending only on λ such that

$$|S\Phi_\lambda(\xi)| \le K_\lambda \exp\left[a_\lambda|\xi|_0^{2/\lambda}\right].$$

Hence Φ_λ satisfies condition (b) of Theorem 8.2 with $\beta = 1 - \lambda$. Thus by Theorem 8.2, Φ_λ is a generalized function in $(\mathcal{E})_{1-\lambda}^*$. In fact, we have a nice representation of Φ_λ in terms of G_λ and the generalized functions $\tilde{\mu}_t$ in Example 8.4:

$$\Phi_\lambda = \int_0^\infty \tilde{\mu}_{2u}\, dG_\lambda(u).$$

This equality can be proved easily by using equation (8.14) to check that both sides have the same S-transform.

8.2 Convergence of generalized functions

From the previous section we see that generalized functions can be characterized by the S-transform. Thus it is natural to ask whether the convergence of generalized functions can be determined in terms of the S-transform. The following theorem has been proved in Potthoff and Streit [150] (see also the book by Hida et al. [58]) for the special case $(\mathcal{S})^*$.

Theorem 8.6. *Let* $\Phi_n \in (\mathcal{E})_\beta^*$ *and* $F_n = S\Phi_n$. *Then* Φ_n *converges strongly in* $(\mathcal{E})_\beta^*$ *if and only if the following conditions are satisfied:*
(a) $\lim_{n\to\infty} F_n(\xi)$ *exists for each* $\xi \in \mathcal{E}_c$.
(b) *There exist nonnegative constants* $K, a,$ *and* p, *independent of* n, *such that*

$$|F_n(\xi)| \le K \exp\left[a|\xi|_p^{\frac{2}{1-\beta}}\right], \quad \forall n \in \mathbf{N}, \xi \in \mathcal{E}_c.$$

Proof. Suppose Φ_n converges strongly to, say, Φ in $(\mathcal{E})_\beta^*$. Let $F = S\Phi$. Then for any $\xi \in \mathcal{E}_c$,

$$\lim_{n\to\infty} F_n(\xi) = \lim_{n\to\infty} \langle\!\langle \Phi_n, :e^{\langle\cdot,\xi\rangle}: \rangle\!\rangle$$
$$= \langle\!\langle \Phi, :e^{\langle\cdot,\xi\rangle}: \rangle\!\rangle$$
$$= F(\xi).$$

Moreover, since Φ_n converges strongly to Φ in $(\mathcal{E})^*_\beta$, there exist nonnegative numbers p and L such that $\sup_n \|\Phi_n\|_{-p,-\beta} \leq L$. Hence by the argument preceding Theorem 8.2

$$|F_n(\xi)| \leq K \exp\left[a|\xi|_p^{\frac{2}{1-\beta}}\right], \quad \forall n \in \mathbb{N}, \xi \in \mathcal{E}_c,$$

where $K = L2^{\beta/2}$ and $a = (1-\beta)2^{\frac{2\beta-1}{1-\beta}}$. Thus condition (b) is satisfied.

Conversely, let $\Phi_n \in (\mathcal{E})^*_\beta$ and $F_n = S\Phi_n$. Assume that the sequence $\{F_n; n \geq 1\}$ satisfies conditions (a) and (b). By condition (a) the following limit exists for each $\xi \in \mathcal{E}_c$

$$F(\xi) \equiv \lim_{n\to\infty} F_n(\xi). \tag{8.19}$$

With conditions (a) and (b), we can apply Morera's theorem to show that for any $\xi, \eta \in \mathcal{E}_c$, the function $F(z\xi + \eta)$ is an entire function of $z \in \mathbb{C}$. Moreover,

$$|F(\xi)| \leq K \exp\left[a|\xi|_p^{\frac{2}{1-\beta}}\right], \quad \forall \xi \in \mathcal{E}_c.$$

Thus by Theorem 8.2, there exists a unique $\Phi \in (\mathcal{E})^*_\beta$ such that $F = S\Phi$. By equation (8.19) we have

$$\langle\!\langle \Phi, :e^{\langle \cdot, \xi \rangle}: \rangle\!\rangle = \lim_{n\to\infty} \langle\!\langle \Phi_n, :e^{\langle \cdot, \xi \rangle}: \rangle\!\rangle, \quad \forall \xi \in \mathcal{E}_c.$$

This implies that

$$\langle\!\langle \Phi, \varphi \rangle\!\rangle = \lim_{n\to\infty} \langle\!\langle \Phi_n, \varphi \rangle\!\rangle, \quad \forall \varphi \in V,$$

where V is the linear span of $\{: e^{\langle \cdot, \xi \rangle}:; \xi \in \mathcal{E}_c\}$. On the other hand, by the uniform bound for F_n and F, we conclude from Theorem 8.2 that $\sup_n \|\Phi_n\|_{-q,-\beta} < \infty$, $\|\Phi\|_{-q,-\beta} < \infty$ for some $q \geq 0$.

Now, recall that V is dense in $(\mathcal{E})_\beta$. For any φ in $(\mathcal{E})_\beta$, choose a sequence $\{\varphi_k\} \subset V$ such that $\varphi_k \to \varphi$ in $(\mathcal{E})_\beta$. By writing $\langle\!\langle \Phi, \varphi \rangle\!\rangle - \langle\!\langle \Phi_n, \varphi \rangle\!\rangle$ as

$$\langle\!\langle \Phi, \varphi \rangle\!\rangle - \langle\!\langle \Phi_n, \varphi \rangle\!\rangle = (\langle\!\langle \Phi, \varphi \rangle\!\rangle - \langle\!\langle \Phi, \varphi_k \rangle\!\rangle) +$$

$$(\langle\!\langle \Phi, \varphi_k \rangle\!\rangle - \langle\!\langle \Phi_n, \varphi_k \rangle\!\rangle) + (\langle\!\langle \Phi_n, \varphi_k \rangle\!\rangle - \langle\!\langle \Phi_n, \varphi \rangle\!\rangle),$$

we see clearly that

$$\langle\!\langle \Phi, \varphi \rangle\!\rangle = \lim_{n\to\infty} \langle\!\langle \Phi_n, \varphi \rangle\!\rangle, \quad \forall \varphi \in (\mathcal{E})_\beta.$$

Hence Φ_n converges weakly to Φ. But the weak convergence of a sequence in $(\mathcal{E})^*_\beta$ is equivalent to strong convergence (see §2.2). Hence Φ_n converges strongly in $(\mathcal{E})^*_\beta$ to Φ. $\qquad\square$

Example 8.7. *Gaussian white noise functions by renormalization*

The white noise analogue of the finite dimensional Gaussian function with mean $x \in \mathcal{E}'$ and variance $c > 0$ is the "informal expression"

$$\exp\left[-\frac{1}{2c}|y - x|_0^2\right], \quad y \in \mathcal{E}'.$$

This expression is meaningless as a function on $\mathcal{E}'$. It requires a renormalization in order to get a generalized function. Take a sequence $\{P_n\}$ of finite dimensional orthogonal projections of E with range in $\mathcal{E}$ such that $P_n \to I$ strongly on each space $\mathcal{E}_p$, $p \geq 0$. For instance, take P_n to be the orthogonal projection onto the space spanned by the first n eigenvectors $\zeta_1, \ldots, \zeta_n$ of the operator A (see §4.2). Define

$$\varphi_n(y) \equiv \exp\left[-\frac{1}{2c}|P_n(y - x)|_0^2\right], \quad y \in \mathcal{E}'.$$

Then $\varphi_n \in (L^2)$ with expectation

$$E\varphi_n = \left(\frac{1 + c}{c}\right)^{-r_n/2} \exp\left[-\frac{1}{2(1 + c)}|P_n x|_0^2\right],$$

where r_n is the dimension of $P_n E$. Define $\Phi_n \equiv \varphi_n / E\varphi_n$. Then $\Phi_n \in (L^2)$ with S-transform given by

$$S\Phi_n(\xi) = \exp\left[\frac{1}{1 + c}\langle x, P_n\xi\rangle - \frac{1}{2(1 + c)}\langle P_n\xi, \xi\rangle\right], \quad \xi \in \mathcal{E}_c.$$

Note that by the uniform boundedness principle $\sup_n \|P_n\|_{\mathcal{E}_p} < \infty$ for each $p \geq 0$. This implies that condition (b) in Theorem 8.6 with $\beta = 0$ is satisfied. On the other hand, we have

$$\lim_{n \to \infty} S\Phi_n(\xi) = \exp\left[\frac{1}{1 + c}\langle x, \xi\rangle - \frac{1}{2(1 + c)}\langle \xi, \xi\rangle\right], \quad \xi \in \mathcal{E}_c.$$

Hence by Theorem 8.6 Φ_n converges strongly in $(\mathcal{E})^*$. The limit, in view of the above procedure, is a multiplicative renormalization and is denoted by $\mathcal{N}\exp\left[-\frac{1}{2c}|\cdot - x|_0^2\right]$. Its S-transform is given by

$$S\mathcal{N}\exp\left[-\frac{1}{2c}|\cdot - x|_0^2\right](\xi) = \exp\left[\frac{1}{1 + c}\langle x, \xi\rangle - \frac{1}{2(1 + c)}\langle \xi, \xi\rangle\right], \quad \xi \in \mathcal{E}_c.$$

Note that the renormalization procedure works only when $c > 0$. However, the S-transform of this renormalization makes sense for any complex number $c \neq -1$. This extension is, in fact, the Gaussian white noise function in Example 8.3, i.e.,

$$g_{x,c} = \mathcal{N}\exp\left[-\frac{1}{2c}|\cdot - x|_0^2\right].$$

Example 8.8. *Convergence of $g_{x,c}$ to $\widetilde{\delta}_x$ as $c \to 0$*

Let $x \in \mathcal{E}'$ be fixed. Then for each $\xi \in \mathcal{E}_c$,

$$Sg_{x,c}(\xi) = \exp\left[\frac{1}{1+c}\langle x, \xi\rangle - \frac{1}{2(1+c)}\langle\xi,\xi\rangle\right]$$

$$\longrightarrow \exp\left[\langle x, \xi\rangle - \frac{1}{2}\langle\xi,\xi\rangle\right] \quad \text{as } c \to 0.$$

Thus by Theorem 7.8, $\lim_{c \to 0} Sg_{x,c}(\xi) = S\widetilde{\delta}_x(\xi)$ for each $\xi \in \mathcal{E}_c$. Moreover, it is easy to check that $Sg_{x,c}$ satisfies the uniform bound in Theorem 8.6 with $\beta = 0$ for $|c| < \frac{1}{2}$. Hence $g_{x,c} \to \widetilde{\delta}_x$ strongly in $(\mathcal{E})^*$ as $c \to 0$. This is the white noise analogue of a well-known property of Gaussian measures in finite dimensional spaces.

8.3 Characterization of test functions

The test functions in $(\mathcal{E})_\beta$ can also be characterized by the S-transform. First we prove a norm estimate similar to the one in Theorem 8.2, but for positive indices.

Theorem 8.9. *Let F be a function on $\mathcal{E}_c$ satisfying the conditions:*

(a) *For any ξ and η in $\mathcal{E}_c$, the function $F(z\xi + \eta)$ is an entire function of $z \in \mathbb{C}$.*

(b) *There exist positive constants K, a, and p such that*

$$|F(\xi)| \leq K \exp\left[a|\xi|_{-p}^{\frac{2}{1+\beta}}\right], \quad \forall\xi \in \mathcal{E}_c.$$

Then there exists a unique $\varphi \in (\mathcal{E})_\beta^*$ such that $F = S\varphi$. In fact, $\varphi \in (\mathcal{E}_q)_\beta$ for any $q \in [0,p)$ satisfying the condition $e^2\left(\frac{2a}{1+\beta}\right)^{1+\beta}\|A^{-(p-q)}\|_{HS}^2 < 1$ and

$$\|\varphi\|_{q,\beta} \leq K\left(1 - e^2\left(\frac{2a}{1+\beta}\right)^{1+\beta}\|A^{-(p-q)}\|_{HS}^2\right)^{-1/2}.$$

Proof. We need only to modify the proof of Theorem 8.2. Define J_n's in the same way. Then for $|\xi_1|_{-p} = \cdots = |\xi_n|_{-p} = 1$, we have

$$|J_n(\xi_1, \ldots, \xi_n)| \leq K\frac{1}{n!}\frac{e^{a(r_1 + \cdots + r_n)^{\frac{2}{1+\beta}}}}{r_1 \cdots r_n}.$$

Choose $r_1 = \cdots = r_n = n^{-\frac{1-\beta}{2}}\left(\frac{1+\beta}{2a}\right)^{\frac{1+\beta}{2}}$. Then

$$|J_n(\xi_1, \ldots, \xi_n)| \leq K\frac{1}{n!}n^{\frac{n(1-\beta)}{2}}\left(\frac{2ae}{1+\beta}\right)^{\frac{n(1+\beta)}{2}}.$$

Hence for any $\xi_1, \ldots, \xi_n \in \mathcal{E}_c$, we have

$$|J_n(\xi_1, \ldots, \xi_n)| \leq K \frac{1}{n!} \, n^{\frac{n(1-\beta)}{2}} \left(\frac{2ae}{1+\beta}\right)^{\frac{n(1+\beta)}{2}} |\xi_1|_{-p} \cdots |\xi_n|_{-p}.$$

Now we define f_n by

$$f_n = \sum_{j_1, \ldots, j_n} J_n(A^q \zeta_{j_1}, \ldots, A^q \zeta_{j_n})(A^{-q}\zeta_{j_1}) \otimes \cdots \otimes (A^{-q}\zeta_{j_n}).$$

(This f_n is actually the same as f_n in Theorem 8.2 although it appears slightly different.) Then by the same argument as in the proof of Theorem 8.2, we can easily derive that

$$|f_n|_q^2 \leq K^2 \frac{1}{(n!)^2} n^{n(1-\beta)} \left(\frac{2ae}{1+\beta}\right)^{n(1+\beta)} \|A^{-(p-q)}\|_{HS}^{2n}.$$

With these functions f_n, $n \geq 1$, and $f_0 = F(0)$, we define

$$\varphi = \sum_{n=0}^{\infty} \langle : \cdot^{\otimes n} :, f_n \rangle.$$

Then $S\varphi = F$ and

$$\|\varphi\|_{q,\beta}^2 \leq K^2 \sum_{n=0}^{\infty} e^{2n} \left(\frac{2a}{1+\beta}\right)^{n(1+\beta)} \|A^{-(p-q)}\|_{HS}^{2n}.$$

Thus for $0 \leq q < p$ such that $e^2 \left(\frac{2a}{1+\beta}\right)^{1+\beta} \|A^{-(p-q)}\|_{HS}^2 < 1$, we have

$$\|\varphi\|_{q,\beta}^2 \leq K^2 \left(1 - e^2 \left(\frac{2a}{1+\beta}\right)^{1+\beta} \|A^{-(p-q)}\|_{HS}^2\right)^{-1}. \qquad \square$$

Now, we are ready to state the characterization theorem for test functions in $(\mathcal{E})_\beta$. Let $\varphi \in (\mathcal{E})_\beta$ and $F = S\varphi$. Then by regarding φ as a generalized function, the function $F(z\xi + \eta)$ is an entire function of $z \in \mathbb{C}$ for any $\xi, \eta \in \mathcal{E}_c$. However, $F(\xi)$ has a different growth condition than the one in Theorem 8.2. By Lemma 6.7 we have

$$|F(\xi)| \leq \|\varphi\|_{q,\beta} \exp\left[\frac{1}{2}(1+\beta)|\xi|_{-q}^{\frac{2}{1+\beta}}\right], \quad \xi \in \mathcal{E}_c.$$

Recall that $\|A^{-1}\| = \lambda_1^{-1}$ (see §4.2) and so $|\xi|_{-q} \leq \lambda_1^{-(q-p)}|\xi|_{-p}$ for $q \geq p$. Hence

$$|F(\xi)| \leq \|\varphi\|_{q,\beta} \exp\left[\frac{1}{2}(1+\beta)\lambda_1^{-(q-p)\frac{2}{1+\beta}} |\xi|_{-p}^{\frac{2}{1+\beta}}\right], \quad \xi \in \mathcal{E}_c.$$

For any $a > 0$ and $p \geq 0$, choose large q such that $\frac{1}{2}(1+\beta)\lambda_1^{-(q-p)\frac{2}{1+\beta}} < a$. Then we get the growth condition

$$|F(\xi)| \leq \|\varphi\|_{q,\beta} \exp\left[a|\xi|_{-p}^{\frac{2}{1+\beta}}\right], \quad \xi \in \mathcal{E}_c.$$

It turns out that the above conditions are also sufficient for a function on $\mathcal{E}_c$ to be the S-transform of a test function in $(\mathcal{E})_\beta$.

Theorem 8.10. *Let $\varphi \in (\mathcal{E})_\beta$. Then its S-transform $F = S\varphi$ satisfies the conditions:*

(a) *For any ξ and η in $\mathcal{E}_c$, the function $F(z\xi + \eta)$ is an entire function of $z \in \mathbb{C}$.*

(b) *For any constants $a > 0$ and $p \geq 0$, there exists a constant $K > 0$ such that*

$$|F(\xi)| \leq K \exp\left[a|\xi|_{-p}^{\frac{2}{1+\beta}}\right], \quad \forall \xi \in \mathcal{E}_c.$$

Conversely, suppose a function F defined on $\mathcal{E}_c$ satisfies the above two conditions. Then there exists a unique $\varphi \in (\mathcal{E})_\beta$ such that $F = S\varphi$.

Remark. This characterization theorem was first proved in Kuo et al. [115] for the space $(\mathcal{S})$ (see also the book by Hida et al. [58]). It has been extended to the spaces $(\mathcal{S})_\beta$ in Kondratiev and Streit [84]. A similar result appeared in Kondratiev [80].

Proof. The first assertion has already been shown above. To prove the second assertion, let F be a function on $\mathcal{E}_c$ satisfying conditions (a) and (b). Given any a and p, we get K from condition (b) such that the growth condition holds. Hence by Theorem 8.9, there exists a unique $\varphi \in (\mathcal{E})_\beta^*$ such that $F = S\varphi$. We need only to show that φ is in fact in $(\mathcal{E})_\beta$. To do so, suppose q is any positive number. Choose $p > q$ such that $A^{-(p-q)}$ is a Hilbert-Schmidt operator of E (e.g., $p = q + \frac{\alpha}{2}$, where α is given in §4.2). Then we choose a positive number a such that

$$e^2 \left(\frac{2a}{1+\beta}\right)^{1+\beta} \|A^{-(p-q)}\|_{HS}^2 < 1.$$

Now, with the numbers a and p, we apply condition (b) to get a constant K such that

$$|F(\xi)| \leq K \exp\left[a|\xi|_{-p}^{\frac{2}{1+\beta}}\right], \quad \forall \xi \in \mathcal{E}_c.$$

Thus by Theorem 8.9,

$$\|\varphi\|_{q,\beta} \leq K \left(1 - e^2 \left(\frac{2a}{1+\beta}\right)^{1+\beta} \|A^{-(p-q)}\|_{HS}^2\right)^{-1/2}.$$

Hence $\|\varphi\|_{q,\beta} < \infty$ for any positive number q and so $\varphi \in (\mathcal{E})_\beta$. $\qquad\square$

8.4 Wick product and convolution

Let Φ and Ψ be two generalized functions in $(\mathcal{E})^*_\beta$. Let $F = S\Phi$ and $G = S\Psi$. Obviously the product FG satisfies conditions (a) and (b) in Theorem 8.2. Hence there exists a unique $Y \in (\mathcal{E})^*_\beta$ such that $SY = FG$.

Definition 8.11. The *Wick product* of two generalized functions Φ and Ψ in $(\mathcal{E})^*_\beta$, denoted by $\Phi \diamond \Psi$, is the unique generalized function in $(\mathcal{E})^*_\beta$ such that $S(\Phi \diamond \Psi) = (S\Phi)(S\Psi)$.

Recall that $S\langle :\cdot^{\otimes m}:, f_m\rangle(\xi) = \langle f_m, \xi^{\otimes n}\rangle$. This implies that

$$\langle :\cdot^{\otimes m}:, f_m\rangle \diamond \langle :\cdot^{\otimes n}:, g_n\rangle = \langle :\cdot^{\otimes(m+n)}:, f_m\widehat{\otimes}g_n\rangle.$$

We can use this relationship to find the Wick product in terms of the Wiener-Itô decomposition.

Theorem 8.12. *If $\varphi, \psi \in (\mathcal{E})_\beta$, then $\varphi \diamond \psi \in (\mathcal{E})_\beta$. In fact, for any $p \geq 0$, $c \geq 1$, and $q \geq p + (2\ln\lambda_1)^{-1}\left[\ln\frac{c+1}{c} + (1+\beta)\ln 2\right]$, the following inequality holds:*

$$\|\varphi \diamond \psi\|_{p,\beta} \leq c\|\varphi\|_{q,\beta}\|\psi\|_{q,\beta}, \quad \forall \varphi, \psi \in (\mathcal{E})_\beta. \tag{8.20}$$

Remarks. (1) The first assertion follows from Theorem 8.10. But it also follows from inequality (8.20). Moreover, this inequality implies that the mapping $(\varphi, \psi) \mapsto \varphi \diamond \psi$ is continuous from $(\mathcal{E})_\beta \times (\mathcal{E})_\beta$ into $(\mathcal{E})_\beta$.
(2) The proof below can be easily modified to show that for any $p \geq 0, c \geq 1$, and $q \geq p + (2\ln\lambda_1)^{-1}\left(\ln\frac{c+1}{c} + (1-\beta)\ln 2\right)$, the following inequality holds for all Φ and Ψ in $(\mathcal{E}_p)^*_\beta$,

$$\|\Phi \diamond \Psi\|_{-q,-\beta} \leq c\|\Phi\|_{-p,-\beta}\|\Psi\|_{-p,-\beta}.$$

This implies that the mapping $(\Phi, \Psi) \mapsto \Phi \diamond \Psi$ is continuous from $(\mathcal{E})^*_\beta \times (\mathcal{E})^*_\beta$ into $(\mathcal{E})^*_\beta$.

Proof. Let $\varphi, \psi \in (\mathcal{E})_\beta$ be represented by

$$\varphi(x) = \sum_{n=0}^\infty \langle :x^{\otimes n}:, f_n\rangle, \quad \psi(x) = \sum_{n=0}^\infty \langle :x^{\otimes n}:, g_n\rangle.$$

Define ϑ by

$$\vartheta(x) = \sum_{n=0}^\infty \langle :x^{\otimes n}:, \textstyle\sum_{k=0}^n f_{n-k}\widehat{\otimes}g_k\rangle.$$

First we show that $\vartheta \in (\mathcal{E})_\beta$. To do so, note that

$$\Big| \sum_{k=0}^{n} f_{n-k} \widehat{\otimes} g_k \Big|_p^2 \leq \Big(\sum_{k=0}^{n} |f_{n-k}|_p \, |g_k|_p \Big)^2$$

$$\leq (n+1) \sum_{k=0}^{n} |f_{n-k}|_p^2 \, |g_k|_p^2.$$

Then we use the inequality $n+1 \leq c\big(\frac{c+1}{c}\big)^n$ for any $n \geq 0$ and $c \geq 1$ to get

$$\Big| \sum_{k=0}^{n} f_{n-k} \widehat{\otimes} g_k \Big|_p^2 \leq c\Big(\frac{c+1}{c}\Big)^n \sum_{k=0}^{n} |f_{n-k}|_p^2 \, |g_k|_p^2.$$

Therefore

$$\|\vartheta\|_{p,\beta}^2 = \sum_{n=0}^{\infty} (n!)^{1+\beta} \Big| \sum_{k=0}^{n} f_{n-k} \widehat{\otimes} g_k \Big|_p^2$$

$$\leq \sum_{n=0}^{\infty} c\Big(\frac{c+1}{c}\Big)^n \sum_{k=0}^{n} \binom{n}{k}^{1+\beta} ((n-k)!)^{1+\beta} |f_{n-k}|_p^2 (k!)^{1+\beta} |g_k|_p^2$$

$$\leq c \sum_{n=0}^{\infty} \Big(\frac{c+1}{c}\Big)^n 2^{n(1+\beta)} \sum_{k=0}^{n} ((n-k)!)^{1+\beta} |f_{n-k}|_p^2 (k!)^{1+\beta} |g_k|_p^2.$$

But for $q \geq p$, we have $|f_n|_p \leq \lambda_1^{-n(q-p)} |f_n|_q$ and similarly for $|g_n|_p$. Hence we get

$$\|\vartheta\|_{p,\beta}^2 \leq c \sum_{n=0}^{\infty} \Big(\frac{c+1}{c}\Big)^n 2^{n(1+\beta)} \lambda_1^{-2n(q-p)}$$

$$\times \sum_{k=0}^{n} ((n-k)!)^{1+\beta} |f_{n-k}|_q^2 (k!)^{1+\beta} |g_k|_q^2.$$

If q satisfies $\frac{c+1}{c} 2^{(1+\beta)} \lambda_1^{-2(q-p)} \leq 1$, i.e., $q \geq p + (2\ln\lambda_1)^{-1} \big[\ln \frac{c+1}{c} + (1+\beta)\ln 2\big]$, then

$$\|\vartheta\|_{p,\beta}^2 \leq c \sum_{n=0}^{\infty} \sum_{k=0}^{n} ((n-k)!)^{1+\beta} |f_{n-k}|_q^2 (k!)^{1+\beta} |g_k|_q^2$$

$$= c \sum_{k=0}^{\infty} \Big(\sum_{n=k}^{\infty} ((n-k)!)^{1+\beta} |f_{n-k}|_q^2 \Big) (k!)^{1+\beta} |g_k|_q^2$$

$$= c \|\varphi\|_{q,\beta}^2 \, \|\psi\|_{q,\beta}^2.$$

This implies that $\vartheta \in (\mathcal{E})_\beta$. The S-transform of ϑ is given by

$$S\vartheta(\xi) = \sum_{n=0}^{\infty} \langle \sum_{k=0}^{n} f_{n-k} \widehat{\otimes} g_k, \xi^{\otimes n} \rangle$$

$$= \sum_{n=0}^{\infty} \sum_{k=0}^{n} \langle f_{n-k}, \xi^{\otimes(n-k)} \rangle \langle g_k, \xi^{\otimes k} \rangle$$

$$= \sum_{k=0}^{\infty} \Big(\sum_{n=k}^{\infty} \langle f_{n-k}, \xi^{\otimes(n-k)} \rangle \Big) \langle g_k, \xi^{\otimes k} \rangle$$

$$= S\varphi(\xi) S\psi(\xi).$$

Hence $\vartheta = \varphi \diamond \psi$ and we have completed the proof.   □

When Hida introduced the theory of white noise in [45], he used the $\mathcal{T}$-transform. The S-transform is convenient to use because $: e^{\langle \cdot, \xi \rangle} :$ has a nice Wiener-Itô decomposition. However, it is sometimes better to use the $\mathcal{T}$-transform.

Definition 8.13. The *$\mathcal{T}$-transform* $\mathcal{T}\Phi$ of a generalized function $\Phi \in (\mathcal{E})_\beta^*$ is defined to be the function

$$\mathcal{T}\Phi(\xi) = \langle\!\langle \Phi, e^{i\langle \cdot, \xi \rangle} \rangle\!\rangle, \quad \xi \in \mathcal{E}_c.$$

Similar to the S-transform, the function $\mathcal{T}\Phi(z\xi + \eta)$ is an entire function of $z \in \mathbb{C}$ for any $\xi, \eta \in \mathcal{E}_c$. Moreover, it is easy to check that the $\mathcal{T}$-transform and the S-transform are related by the formulas:

$$S\Phi(\xi) = e^{-\frac{1}{2}\langle \xi, \xi \rangle} \mathcal{T}\Phi(-i\xi), \quad \mathcal{T}\Phi(\xi) = e^{-\frac{1}{2}\langle \xi, \xi \rangle} S\Phi(i\xi), \quad \xi \in \mathcal{E}_c. \quad (8.21)$$

Thus Theorems 8.2, 8.6, 8.9, and 8.10 remain valid if the S-transform is replaced with the $\mathcal{T}$-transform (except for the norm estimates which can be modified).

Definition 8.14. A complex measure ν on $\mathcal{E}'$ is called a *Hida complex measure* if $\varphi \in L^1(|\nu|)$ for all $\varphi \in (\mathcal{E})_\beta$ and the linear functional

$$\varphi \mapsto \int_{\mathcal{E}'} \varphi(x) \, d\nu(x)$$

is continuous on $(\mathcal{E})_\beta$. Here $|\nu|$ denotes the total variation of ν.

If ν is a Hida complex measure, then it induces a generalized function $\widetilde{\nu}$ in $(\mathcal{E})_\beta^*$:

$$\langle\!\langle \widetilde{\nu}, \varphi \rangle\!\rangle = \int_{\mathcal{E}'} \varphi(x) \, d\nu(x), \quad \varphi \in (\mathcal{E})_\beta.$$

A Hida complex measure is called a *Hida measure* if it is a measure. From Example 8.4, the Gaussian measure $\mu_{x,t}$ is a Hida measure for any $x \in \mathcal{E}'$ and $t > 0$.

Let ν_1 and ν_2 be two Hida complex measures. Consider the convolution $\nu_1 * \nu_2$ defined by

$$\nu_1 * \nu_2(\cdot) = \int_{\mathcal{E}'} \nu_1(\cdot - x) \, d\nu_2(x).$$

It is easy to check that

$$\int_{\mathcal{E}'} e^{i\langle x,\xi \rangle} \, d(\nu_1 * \nu_2)(x) = (\mathcal{T}\tilde{\nu}_1)(\xi)(\mathcal{T}\tilde{\nu}_2)(\xi).$$

Therefore, by the $\mathcal{T}$-transform version of Theorem 8.2, the convolution $\nu_1 * \nu_2$ is also a Hida complex measure and we have $\mathcal{T}(\nu_1 * \nu_2)^{\sim} = (\mathcal{T}\tilde{\nu}_1)(\mathcal{T}\tilde{\nu}_2)$. This is the motivation for the following definition of the convolution of two generalized functions.

Definition 8.15. The *convolution* $\Phi * \Psi$ of two generalized functions Φ and Ψ in $(\mathcal{E})_\beta^*$ is the unique generalized function in $(\mathcal{E})_\beta^*$ such that $\mathcal{T}(\Phi * \Psi) = (\mathcal{T}\Phi)(\mathcal{T}\Psi)$.

By using the formulas in (8.21) we can derive the following relationships between convolution and the Wick product (Kuo [111]):

$$\Phi * \Psi = \Phi \diamond \Psi \diamond g_{-2}, \quad \Phi \diamond \Psi = \Phi * \Psi * g_{-1/2}.$$

Next, we consider the pointwise product of two test functions in $(\mathcal{E})_\beta$. Note that in general the product $\Phi\Psi$ has no meaning for $\Phi, \Psi \in (\mathcal{E})_\beta^*$. For example $\langle \cdot, \delta_t \rangle^2$ is not a generalized function. However, we will show below (Theorem 8.17) that the pointwise product $\varphi\psi$ of two test functions φ and ψ in $(\mathcal{E})_\beta$ is also a test function in $(\mathcal{E})_\beta$. This is an important property of the space $(\mathcal{E})_\beta$ of test functions.

Recall that the transformation $\Theta : (\mathcal{E})_\beta \to (\mathcal{E})_\beta$ is defined in §6.1 by

$$\varphi = \sum_{n=0}^{\infty} \langle :\cdot^{\otimes n}:, f_n \rangle \longmapsto \Theta\varphi = \sum_{n=0}^{\infty} \langle :\cdot^{\otimes n}:, g_n \rangle,$$

where g_n is the following function

$$g_n = \sum_{k=0}^{\infty} \binom{n+2k}{2k} (2k-1)!!(-1)^k \langle \tau^{\otimes k}, f_{n+2k} \rangle. \tag{8.22}$$

On the other hand, let $\varphi \in (\mathcal{E})_\beta$ and consider the restriction of $S\varphi$ to $\mathcal{E}$, i.e.,

$$S\varphi(\xi) = \langle\!\langle \varphi, :e^{\langle \cdot, \xi \rangle}: \rangle\!\rangle, \quad \xi \in \mathcal{E}.$$

Observe that $:e^{\langle \cdot, x \rangle}: \in (\mathcal{E})^*$ for any $x \in \mathcal{E}'$. Thus the function $S\varphi(\xi)$, $\xi \in \mathcal{E}$, extends to $S\varphi(x)$, $x \in \mathcal{E}'$. We will use $\check{S}\varphi$ to denote this extension, i.e., for $\varphi \in (\mathcal{E})_\beta$, define

$$\check{S}\varphi(x) = \langle\!\langle :e^{\langle \cdot, x \rangle}:, \varphi \rangle\!\rangle, \quad x \in \mathcal{E}'.$$

By Proposition 5.9, we see that if $\varphi(x) = \sum_{n=0}^{\infty} \langle :x^{\otimes n}:, f_n \rangle$, then

$$\check{S}\varphi(x) = \sum_{n=0}^{\infty} \langle x^{\otimes n}, f_n \rangle.$$

But from §5.1, we have

$$x^{\otimes n} = \sum_{k=0}^{[n/2]} \binom{n}{2k} (2k-1)!! :x^{\otimes(n-2k)}: \widehat{\otimes} \tau^{\otimes k}.$$

Thus by a result similar to Proposition 6.1, we get

$$\check{S}\varphi(x) = \sum_{n=0}^{\infty} \langle :x^{\otimes n}:, h_n \rangle,$$

where h_n is given by

$$h_n = \sum_{k=0}^{\infty} \binom{n+2k}{2k} (2k-1)!! \langle \tau^{\otimes k}, f_{n+2k} \rangle. \tag{8.23}$$

By comparing equations (8.22) and (8.23), we see clearly the similarity between Θ and $\check{S}$. In particular, Theorem 6.2 is valid when Θ is replaced with $\check{S}$. Thus $\check{S}$ is a continuous linear operator from $(\mathcal{E})_\beta$ into itself.

Theorem 8.16. $\check{S}\Theta = \Theta\check{S} = I$ on $(\mathcal{E})_\beta$ for any $0 \leq \beta < 1$.

Proof. For simplicity, let $\varphi_\eta \equiv :e^{\langle \cdot, \eta \rangle}:$. Recall that $\varphi_\eta \in (\mathcal{E})_\beta$ for any $\eta \in \mathcal{E}$ and

$$\varphi_\eta = \sum_{n=0}^{\infty} \left\langle :\cdot^{\otimes n}:, \frac{1}{n!}\eta^{\otimes n} \right\rangle.$$

Hence by the definition of Θ,

$$\Theta\varphi_\eta = \sum_{n=0}^{\infty} \langle :\cdot^{\otimes n}:, g_n \rangle, \tag{8.24}$$

where g_n is given by equation (8.22) with $f_{n+2k} = \frac{1}{(n+2k)!}\eta^{\otimes(n+2k)}$, i.e.,

$$g_n = \sum_{k=0}^{\infty} \binom{n+2k}{2k} (2k-1)!!(-1)^k \left\langle \tau^{\otimes k}, \frac{1}{(n+2k)!}\eta^{\otimes(n+2k)} \right\rangle$$

$$= \frac{1}{n!}\sum_{k=0}^{\infty} \frac{1}{2^k k!}(-1)^k |\eta|_0^{2k} \eta^{\otimes n}$$

$$= \frac{1}{n!}e^{-\frac{1}{2}|\eta|_0^2}\eta^{\otimes n}.$$

Put this g_n into equation (8.24) to get

$$\Theta\varphi_\eta = \sum_{n=0}^{\infty} \left\langle :\cdot^{\otimes n}:, \frac{1}{n!} e^{-\frac{1}{2}|\eta|_0^2} \eta^{\otimes n} \right\rangle$$

$$= e^{-\frac{1}{2}|\eta|_0^2} \sum_{n=0}^{\infty} \left\langle :\cdot^{\otimes n}:, \frac{1}{n!} \eta^{\otimes n} \right\rangle$$

$$= e^{-\frac{1}{2}|\eta|_0^2} \varphi_\eta.$$

Thus we have shown that for any $\eta \in \mathcal{E}$,

$$\Theta\varphi_\eta = e^{-\frac{1}{2}|\eta|_0^2} \varphi_\eta.$$

Similarly, by using equation (8.23), we get

$$\check{S}\varphi_\eta = e^{\frac{1}{2}|\eta|_0^2} \varphi_\eta.$$

Obviously this implies that $\check{S}\Theta\varphi_\eta = \Theta\check{S}\varphi_\eta = \varphi_\eta$ for all $\eta \in \mathcal{E}$. But the linear span of $\{\varphi_\eta; \eta \in \mathcal{E}\}$ is dense in $(\mathcal{E})_\beta$, and the operators Θ and $\check{S}$ are continuous on $(\mathcal{E})_\beta$. Hence $\check{S}\Theta = \Theta\check{S} = I$ on $(\mathcal{E})_\beta$. □

Let $\varphi, \psi \in (\mathcal{E})_\beta$. By Theorems 8.12, $\varphi \diamond \psi \in (\mathcal{E})_\beta$ and we have

$$S(\varphi \diamond \psi)(\xi) = S\varphi(\xi)S\psi(\xi), \quad \forall \xi \in \mathcal{E}.$$

But from the definition of $\check{S}$, we see that $S\varphi(\xi) = \check{S}\varphi(\xi)$ for any $\varphi \in (\mathcal{E})_\beta$ and $\xi \in \mathcal{E}$. Thus we have

$$\check{S}(\varphi \diamond \psi)(\xi) = \check{S}\varphi(\xi)\check{S}\psi(\xi), \quad \forall \xi \in \mathcal{E}.$$

Note that $\check{S}$ is a linear operator from $(\mathcal{E})_\beta$ into itself and $\mathcal{E}$ is dense in $\mathcal{E}'$. Hence we can conclude that $\check{S}(\varphi \diamond \psi)(x) = \check{S}\varphi(x)\check{S}\psi(x)$ for all $x \in \mathcal{E}'$, i.e.,

$$\check{S}(\varphi \diamond \psi) = (\check{S}\varphi)(\check{S}\psi), \tag{8.25}$$

or equivalently,

$$\varphi \diamond \psi = \Theta\big((\check{S}\varphi)(\check{S}\psi)\big).$$

Theorem 8.17. *If $\varphi, \psi \in (\mathcal{E})_\beta$, then $\varphi\psi \in (\mathcal{E})_\beta$. Moreover, $\varphi\psi = \check{S}\big((\Theta\varphi) \diamond (\Theta\psi)\big)$, or equivalently, $\Theta(\varphi\psi) = (\Theta\varphi) \diamond (\Theta\psi)$.*

Proof. Let $\varphi, \psi \in (\mathcal{E})_\beta$. Since Θ is a linear operator from $(\mathcal{E})_\beta$ into itself, we have $\Theta\varphi, \Theta\psi \in (\mathcal{E})_\beta$. Then, by Theorem 8.12, $(\Theta\varphi) \diamond (\Theta\psi) \in (\mathcal{E})_\beta$. But

$\check{S}$ is a linear operator from $(\mathcal{E})_\beta$ into itself. Hence $\check{S}((\Theta\varphi) \diamond (\Theta\psi)) \in (\mathcal{E})_\beta$. Now, by equation (8.25) and Theorem 8.16,

$$\check{S}((\Theta\varphi) \diamond (\Theta\psi)) = (\check{S}\Theta\varphi)(\check{S}\Theta\psi) = \varphi\psi.$$

Thus we conclude that $\varphi\psi \in (\mathcal{E})_\beta$ and $\varphi\psi = \check{S}((\Theta\varphi) \diamond (\Theta\psi))$. By applying Θ to both sides, we get the other equivalent identity. $\qquad\square$

Before we state the next theorem, recall that the constants $\lambda_j, j \geq 1$, and α are associated with the operator A in §4.2.

Theorem 8.18. *There exist positive constants C and r depending only on $\lambda_j, j \geq 1, \alpha$, and β such that for any $p \geq \alpha/4$ the inequality holds*

$$\|\varphi\psi\|_{p,\beta} \leq C\|\varphi\|_{p+r,\beta}\|\psi\|_{p+r,\beta}, \quad \forall \varphi, \psi \in (\mathcal{E})_\beta.$$

Remark. It follows from this theorem that the pointwise multiplication is continuous from $(\mathcal{E})_\beta \times (\mathcal{E})_\beta$ into $(\mathcal{E})_\beta$.

Proof. By Theorem 8.17 $\varphi\psi = \check{S}((\Theta\varphi) \diamond (\Theta\psi))$. We can apply Theorem 6.2 for $\check{S}$ (see the remark preceding Theorem 8.16) to get

$$\|\varphi\psi\|_{p,\beta} \leq C_1 \|(\Theta\varphi) \diamond (\Theta\psi)\|_{u,\beta}, \qquad (8.26)$$

where $u > p + r_1$ and the constants C_1 and r_1 depend only on λ_j's, α, and β (as given in Theorem 6.2 for the operator $\check{S}$). Next by Theorem 8.12 with $c = 1$, we have

$$\|(\Theta\varphi) \diamond (\Theta\psi)\|_{u,\beta} \leq \|\Theta\varphi\|_{v,\beta}\|\Theta\psi\|_{v,\beta}, \qquad (8.27)$$

where $v > u + r_2$ with r_2 depending only on λ_1 and β. But by Theorem 6.2 we also have

$$\|\Theta\varphi\|_{v,\beta}\|\Theta\psi\|_{v,\beta} \leq C_2^2\|\varphi\|_{q,\beta}\|\psi\|_{q,\beta}, \qquad (8.28)$$

where $q > v + r_3$ and the constants C_2 and r_3 depend only on λ_j's, α, and β. Therefore, by the inequalities in (8.26)-(8.28),

$$\|\varphi\psi\|_{p,\beta} \leq C_1 C_2^2 \|\varphi\|_{q,\beta}\|\psi\|_{q,\beta}.$$

To finish the proof, simply let $r = r_1 + r_2 + r_3 + 1$. $\qquad\square$

Corollary 8.19. *There exist positive constants C and r depending only on $\lambda_j, j \geq 1, \alpha$, and β such that for any $n \geq 1$ the inequality holds*

$$\|\varphi^n\|_{0,\beta} \leq C^{n-1}\|\varphi\|_{(n-1)r,\beta}^n, \quad \forall \varphi \in (\mathcal{E})_\beta.$$

Proof. By applying Theorem 8.18 repeatedly, we get

$$\|\varphi^n\|_{0,\beta} \le C\|\varphi\|_{r,\beta}\|\varphi^{n-1}\|_{r,\beta}$$
$$\le C^2\|\varphi\|_{r,\beta}\|\varphi\|_{2r,\beta}\|\varphi^{n-2}\|_{2r,\beta}$$
$$\le C^{n-1}\|\varphi\|_{r,\beta}\|\varphi\|_{2r,\beta}\cdots\|\varphi\|_{(n-2)r,\beta}\|\varphi\|^2_{(n-1)r,\beta}.$$

Note that $\|\varphi\|_{p,\beta} \le \|\varphi\|_{q,\beta}$ for any $p \le q$. Hence we obtain the inequality in the lemma immediately. □

At the end of this section, we introduce the concept of the Wick composition of an entire function and a generalized function.

Definition 8.20. Let h be an entire function and let $\Phi \in (\mathcal{S})^*_\beta$ for some β. Suppose the function $h(S\Phi(\cdot))$ is the S-transform of some generalized function. Then this generalized function is defined to be the *Wick composition*, denoted by $h \circ \Phi$, of h and Φ, i.e.,

$$S(h \circ \Phi)(\xi) = h(S\Phi(\xi)), \quad \xi \in \mathcal{S}_c.$$

For example, $\exp \circ \langle \cdot, y \rangle = :e^{\langle \cdot, y \rangle}:$, $y \in \mathcal{E}'_c$ (see Theorem 5.13). The Wick composition can be used to generate new generalized functions. If h is a polynomial, then $h \circ \Phi \in (\mathcal{E})^*_\beta$ for all $\Phi \in (\mathcal{E})^*_\beta$. On the other hand, suppose g is an entire function and there exist nonnegative constants C and a such that

$$|g(z)| \le Ce^{a|z|^2}, \quad \forall z \in \mathbb{C}.$$

Let $\Phi_k = \langle :\cdot^{\otimes k}:, F \rangle$, $F \in (\mathcal{E}'_c)^{\otimes k}$. Then the Wick composition $g \circ \Phi_k$ belongs to $(\mathcal{E})^*_{\frac{k-1}{k}}$. Note that $\Phi_k \in (\mathcal{E})^*$ for all k, while the $g \circ \Phi_k$'s belong to spaces with different indices. Some interesting examples of generalized functions of this type are $\sin \circ \dot{B}(t)$ and $\cos \circ \dot{B}(t)$ which have S-transforms $\sin(\xi(t))$ and $\cos(\xi(t))$, respectively.

8.5 Integrable functions

In finite dimensional distribution theory all functions in $L^p(\mathbb{R}^r), 1 \le p \le \infty$, induce generalized functions. Thus it is natural to ask whether this property is still true in white noise distribution theory. More precisely, suppose $f \in L^p(\mu)$. We want to know whether the following linear functional is defined and continuous on $(\mathcal{E})$:

$$\Phi_f : \varphi \longrightarrow \int_{\mathcal{E}'} \varphi(x)f(x)\, d\mu(x).$$

We will show that this is indeed true for $1 < p \leq \infty$. The derivation of this property is essentially an application of the inequality in Corollary 8.19. Our presentation in this section follows the idea in Obata [142].

First we point out that when $f \in L^1(\mu)$, Φ_f may not be defined on $(\mathcal{E})$. For example, take $\zeta \in \mathcal{E}$ with $|\zeta|_0 = 1$ and let

$$f(x) = \frac{1}{\langle x, \zeta \rangle^2 + 1} e^{\frac{1}{2}\langle x, \zeta \rangle^2}.$$

It is easy to check that $f \in L^1(\mu)$, but Φ_f is not defined at every $\varphi \in (\mathcal{E})$.

Theorem 8.21. *Suppose $f \in L^p(\mu)$ for some $1 < p \leq \infty$. Then $\Phi_f(\varphi)$ is defined for every $\varphi \in (\mathcal{E})$ and Φ_f is a continuous linear functional on $(\mathcal{E})$. Moreover,*

$$\|\Phi_f\|_{-(n-1)r} \leq C^{1-\frac{1}{n}} \|f\|_{L^p(\mu)},$$

where n is any integer such that $n \geq \frac{p}{p-1}$, and the positive constants C and r depend only on $\lambda_j, j \geq 1$, and α.

Proof. Let $q = \frac{p}{p-1}$ ($q = 1$ if $p = \infty$). Then by the Hölder inequality,

$$|\Phi_f(\varphi)| = \left| \int_{\mathcal{E}'} \varphi(x) f(x) \, d\mu(x) \right|$$

$$\leq \|\varphi\|_{L^q(\mu)} \|f\|_{L^p(\mu)}.$$

Take an integer $n \geq q$. Apply the Hölder inequality again and use Corollary 8.19 (with $\beta = 0$) to get

$$\|\varphi\|_{L^q(\mu)} = \left(\int_{\mathcal{E}'} |\varphi(x)|^q \, d\mu(x) \right)^{1/q}$$

$$\leq \left(\int_{\mathcal{E}'} |\varphi(x)|^{2n} \, d\mu(x) \right)^{1/(2n)}$$

$$= \|\varphi^n\|_0^{1/n}$$

$$\leq C^{1-\frac{1}{n}} \|\varphi\|_{(n-1)r}.$$

Hence for any $\varphi \in (\mathcal{E})$,

$$|\Phi_f(\varphi)| \leq C^{1-\frac{1}{n}} \|\varphi\|_{(n-1)r} \|f\|_{L^p(\mu)}.$$

Thus $\Phi_f(\varphi)$ is defined for every $\varphi \in (\mathcal{E})$. Furthermore, Φ_f is a continuous linear functional on $(\mathcal{E})$, i.e., $\Phi_f \in (\mathcal{E})^*$ and

$$\|\Phi_f\|_{-(n-1)r} \leq C^{1-\frac{1}{n}} \|f\|_{L^p(\mu)}. \qquad \square$$

Proposition 8.22. *Suppose $f \in L^1(\mu)$ satisfies the condition that*

$$\int_{\mathcal{E}'} f(x)e^{i\langle x,\xi\rangle} \, d\mu(x) = 0, \quad \forall \xi \in \mathcal{E}.$$

Then $f = 0$, μ-a.e.

Proof. Recall that ζ_j's are the eigenvectors of the operator A in §4.2. Let $\mathcal{B}_n$ be the σ-field generated by $\{\langle \cdot, \zeta_j \rangle; 1 \leq j \leq n\}$. Let V_n be the linear space spanned by $\zeta_j, 1 \leq j \leq n$ and P_n the orthogonal projection of E onto V_n. By the martingale convergence theorem the conditional expectation $E(f|\mathcal{B}_n)$ converges to f, μ-a.e. On the other hand, it follows from the assumption that for any $n \geq 1$

$$\int_{\mathcal{E}'} f(x)e^{i\langle x,\xi\rangle} \, d\mu(x) = 0, \quad \forall \xi \in V_n.$$

By taking the conditional expectation with respect to $\mathcal{B}_n$, we get

$$\int_{\mathcal{E}'} E(f|\mathcal{B}_n)(x)e^{i\langle x,\xi\rangle} \, d\mu(x) = 0, \quad \forall \xi \in V_n.$$

Let g_n be the restriction of $E(f|\mathcal{B}_n)$ on V_n. Then

$$\int_{V_n} g_n(x)e^{i\langle x,\xi\rangle} \, d\mu_n(x) = \int_{\mathcal{E}'} E(f|\mathcal{B}_n)(x)e^{i\langle x,\xi\rangle} \, d\mu(x) = 0, \quad \forall \xi \in V_n,$$

where μ_n is the standard Gaussian measure on V_n. Hence $g_n = 0$, μ_n-a.e. But $E(f|\mathcal{B}_n) = g_n \circ P_n$. Thus $E(f|\mathcal{B}_n) = 0$, μ-a.e. for all $n \geq 1$. Since $E(f|\mathcal{B}_n)$ converges to f, μ-a.e., we conclude that $f = 0$, μ-a.e. $\quad\square$

Now, by Theorem 8.21 and Proposition 8.22 the linear mapping $f \mapsto \Phi_f$ is injective and continuous. Thus $L^p(\mu)$ can be imbedded in $(\mathcal{E})^*$ for any $1 < p \leq \infty$. Moreover, if we endow $\cup_{p>1}L^p(\mu)$ with the inductive limit topology, then the inclusion $\cup_{p>1}L^p(\mu) \subset (\mathcal{E})^*$ is continuous. This fact is stated as the next theorem.

Theorem 8.23. *The inclusion $\cup_{p>1}L^p(\mu) \subset (\mathcal{E})^*$ is continuous.*

9

Differential Operators

In finite dimensional distribution theory, there are several important continuous linear operators acting on the space of test functions; e.g., differentiation, translation, dilation, multiplication, convolution, and the Fourier transform. We can use duality to define adjoint operators. But often they turn out to be the extensions of continuous operators on the space of test functions to the space of generalized functions because of the nature of the Lebesgue measure. For instance, the adjoint D_x^* of the differential operator D_x is related to the extension $\widetilde{D}_x$ of D_x to the space of generalized functions by $D_x^* = -\widetilde{D}_x$. However the situation is quite different in white noise distribution theory since the Gaussian measure is used. We will see that adjoint operators play an important role.

9.1 Differential operators

Consider a simple test function $\varphi(x) = \langle :x^{\otimes n}:, f\rangle \in (\mathcal{E})_\beta$. By using the formula in Lemma 7.16, we can easily check that

$$\lim_{\epsilon \to 0} \frac{\varphi(x + \epsilon y) - \varphi(x)}{\epsilon} = n\langle :x^{\otimes(n-1)}: \widehat{\otimes} y, f\rangle$$

$$= n\langle :x^{\otimes(n-1)}:, \langle y, f\rangle\rangle.$$

This shows that the function φ has Gâteaux derivative $D_y\varphi$ in any direction $y \in \mathcal{E}'$. Being motivated by this example, we define an operator D_y on $(\mathcal{E})_\beta$ as follows. For $\varphi(x) = \sum_{n=0}^{\infty} \langle :x^{\otimes n}:, f_n\rangle \in (\mathcal{E})_\beta$, define

$$D_y\varphi(x) \equiv \sum_{n=1}^{\infty} n\langle :x^{\otimes(n-1)}:, \langle y, f_n\rangle\rangle.$$

First we will show that the operator D_y is continuous from $(\mathcal{E})_\beta$ into itself. Then we will justify that $D_y\varphi$ is the derivative of φ in the y-direction.

Theorem 9.1. *For any $y \in \mathcal{E}'$, the operator D_y is continuous from $(\mathcal{E})_\beta$ into itself. In fact, if $y \in \mathcal{E}_q'$ and $p > q$ is such that $2^{1-\beta}\lambda_1^{-2(p-q)} \leq 1$, then for any $\varphi \in (\mathcal{E})_\beta$,*

$$\|D_y\varphi\|_{q,\beta} \leq \lambda_1^{-(p-q)}|y|_{-q}\|\varphi\|_{p,\beta}.$$

Remark. The inequality shows that the operator D_y can be extended by continuity to a continuous linear operator from $(\mathcal{E}_p)_\beta$ into $(\mathcal{E}_q)_\beta$.

Proof. Let $\varphi \in (\mathcal{E})_\beta$ be represented by $\varphi(x) = \sum_{n=0}^\infty \langle :x^{\otimes n}:, f_n\rangle$. Then

$$D_y\varphi(x) = \sum_{n=1}^\infty n\langle :x^{\otimes(n-1)}:, \langle y, f_n\rangle\rangle$$

$$= \sum_{n=0}^\infty (n+1)\langle :x^{\otimes n}:, \langle y, f_{n+1}\rangle\rangle.$$

Therefore we get

$$\|D_y\varphi\|_{q,\beta}^2 = \sum_{n=0}^\infty (n!)^{1+\beta}(n+1)^2|\langle y, f_{n+1}\rangle|_q^2$$

$$\leq \sum_{n=0}^\infty (n!)^{1+\beta}(n+1)^2|y|_{-q}^2|f_{n+1}|_q^2$$

$$= |y|_{-q}^2 \sum_{n=0}^\infty (n+1)^{1-\beta}((n+1)!)^{1+\beta}|f_{n+1}|_q^2.$$

Note that $|g_n|_q \leq \lambda_1^{-(p-q)n}|g_n|_p$ for any $p > q \geq 0$ and $g_n \in \mathcal{E}_c^{\widehat{\otimes}n}$. Also note that $n+1 \leq 2^n$ for any $n \geq 0$. Hence

$$\|D_y\varphi\|_{q,\beta}^2 \leq |y|_{-q}^2 \sum_{n=0}^\infty 2^{n(1-\beta)}\lambda_1^{-2(p-q)(n+1)}((n+1)!)^{1+\beta}|f_{n+1}|_p^2.$$

Thus if $p > q$ is chosen such that $2^{1-\beta}\lambda_1^{-2(p-q)} \leq 1$, then we have

$$\|D_y\varphi\|_{q,\beta}^2 \leq |y|_{-q}^2\lambda_1^{-2(p-q)} \sum_{n=0}^\infty ((n+1)!)^{1+\beta}|f_{n+1}|_p^2$$

$$= |y|_{-q}^2\lambda_1^{-2(p-q)}\|\varphi\|_{p,\beta}^2.$$

Obviously this inequality implies that $D_y \varphi \in (\mathcal{E})_\beta$ for any $\varphi \in (\mathcal{E})_\beta$ and that D_y is a continuous linear operator from $(\mathcal{E})_\beta$ into itself. $\qquad \square$

Theorem 9.2. *Suppose $\varphi \in (\mathcal{E})_\beta$ and $y \in \mathcal{E}'_p$. Let*

$$\psi(\cdot, y) \equiv \varphi(\cdot + y) - \varphi(\cdot) - D_y \varphi(\cdot).$$

Then for any $p > q \geq 0$ such that $2^{1-\beta} \lambda_1^{-2(p-q)} \leq 1$, the inequality holds:

$$\|\psi(\cdot, y)\|_{q,\beta} \leq$$

$$2^{1-\beta} \left(1 - 2^{1-\beta} \lambda_1^{-2(p-q)}\right)^{-1/2} \|\varphi\|_{p,\beta} |y|_{-p}^2 \exp\left[(1+\beta) 2^{-\frac{2\beta}{1+\beta}} |y|_{-p}^{\frac{2}{1+\beta}}\right].$$

Remark. From Theorem 6.13 we know that φ has an analytic extension $\varphi(x), x \in \mathcal{E}'_c$. Therefore it is possible to get a pointwise estimate for $|\psi(x, y)|$, i.e., for any fixed $x \in \mathcal{E}'$, we have $\psi(x, y) = o(|y|_{-p})$ for some p as $|y|_{-p} \to 0$. But this pointwise estimate is much more complicated than the one in the lemma.

Proof. Let φ be represented by $\varphi = \sum_{n=0}^\infty \langle : \cdot^{\otimes n} :, f_n \rangle$. Then by Lemma 7.16

$$\psi(x, y) = \sum_{n=2}^\infty \Big\langle \sum_{k=2}^\infty \binom{n}{k} : x^{\otimes(n-k)} :, \langle y^{\otimes k}, f_n \rangle \Big\rangle.$$

By changing the order of summation (which is justified by the calculation below), we get

$$\psi(x, y) = \sum_{n=0}^\infty \Big\langle : x^{\otimes n} :, \sum_{k=0}^\infty \binom{n+k+2}{k+2} \langle y^{\otimes(k+2)}, f_{n+k+2} \rangle \Big\rangle.$$

To estimate the norm $\|\psi(\cdot, y)\|_{q,\beta}$, note that $|g_n|_q \leq \lambda_1^{-(p-q)n} |g_n|_p$ for any $p > q \geq 0$ and $g_n \in \mathcal{E}_c^{\widehat{\otimes} n}$. Therefore

$$\|\psi(\cdot, y)\|_{q,\beta}^2$$

$$= \sum_{n=0}^\infty (n!)^{1+\beta} \Big| \sum_{k=0}^\infty \binom{n+k+2}{k+2} \langle y^{\otimes(k+2)}, f_{n+k+2} \rangle \Big|_q^2$$

$$\leq \sum_{n=0}^\infty (n!)^{1+\beta} \lambda_1^{-2(p-q)n} \Big| \sum_{k=0}^\infty \binom{n+k+2}{k+2} \langle y^{\otimes(k+2)}, f_{n+k+2} \rangle \Big|_p^2$$

$$\leq \sum_{n=0}^\infty (n!)^{1+\beta} \lambda_1^{-2(p-q)n} \Big(\sum_{k=0}^\infty \binom{n+k+2}{k+2} |y|_{-p}^{k+2} |f_{n+k+2}|_p \Big)^2$$

$$= \sum_{n=0}^\infty \frac{1}{(n!)^{1-\beta}} \lambda_1^{-2(p-q)n} \Big(\sum_{k=0}^\infty \frac{(n+k+2)!}{(k+2)!} |y|_{-p}^{k+2} |f_{n+k+2}|_p \Big)^2. \qquad (9.1)$$

For simplicity, let

$$L \equiv \sum_{k=0}^{\infty} \frac{(n+k+2)!}{(k+2)!} |y|_{-p}^{k+2} |f_{n+k+2}|_p. \qquad (9.2)$$

By writing the general term in the last series as

$$\frac{(n+k+2)!}{(k+2)!} |y|_{-p}^{k+2} |f_{n+k+2}|_p$$

$$= \left(\frac{((n+k+2)!)^{\frac{1-\beta}{2}}}{(k+2)!} |y|_{-p}^{k+2} \right) \left(((n+k+2)!)^{\frac{1+\beta}{2}} |f_{n+k+2}|_p \right)$$

and then applying the Schwarz inequality to the series in equation (9.2), we get

$$L^2 \leq \|\varphi\|_{p,\beta}^2 \sum_{k=0}^{\infty} \frac{((n+k+2)!)^{1-\beta}}{((k+2)!)^2} |y|_{-p}^{2(k+2)}. \qquad (9.3)$$

It follows from (9.1)–(9.3) that

$$\|\psi(\cdot, y)\|_{q,\beta}^2 \leq \|\varphi\|_{p,\beta}^2 \sum_{n,k=0}^{\infty} \lambda_1^{-2(p-q)n}$$

$$\times \left(\frac{(n+k+2)!}{n!(k+2)!} \right)^{1-\beta} \frac{1}{((k+2)!)^{1+\beta}} |y|_{-p}^{2(k+2)}.$$

Now we use the fact that $\binom{n+k+2}{k+2} \leq 2^{n+k+2}$. When $2^{1-\beta}\lambda_1^{-2(p-q)} < 1$, we can sum up the series in n to get

$$\|\psi(\cdot, y)\|_{q,\beta}^2$$

$$\leq \|\varphi\|_{p,\beta}^2 (1 - 2^{1-\beta}\lambda_1^{-2(p-q)})^{-1} \sum_{k=0}^{\infty} \frac{1}{((k+2)!)^{1+\beta}} (2^{1-\beta}|y|_{-p}^2)^{k+2}.$$

Finally we apply Lemma 6.6 to get the inequality in the theorem.   □

We have defined the Gâteaux derivative $D_y\varphi$ for $\varphi \in (\mathcal{E})_\beta$ in the direction $y \in \mathcal{E}'$. The last theorem justifies the term "derivative" for $D_y\varphi$. However, we need to show something more, i.e., the continuity of $D_y\varphi$ in the y variable.

Theorem 9.3. *For any fixed $\varphi \in (\mathcal{E})_\beta$, the linear operator $y \mapsto D_y\varphi$ is continuous from $\mathcal{E}'$ (with the strong topology) into $(\mathcal{E})_\beta$.*

Proof. Suppose $\{y_\lambda\}$ is a net converging strongly to y in $\mathcal{E}'$. Then there exists $p_0 \geq 0$ such that $y_\lambda \to y$ in $\mathcal{E}'_{p_0}$. This implies that $y_\lambda \to y$ in $\mathcal{E}'_q$

for all $q \geq p_0$. Now to show that $D_{y_\lambda}\varphi$ converges to $D_y\varphi$ in $(\mathcal{E})_\beta$, we need only to show that $\|D_{y_\lambda}\varphi - D_y\varphi\|_{q,\beta} \to 0$ for any $q \geq p_0$ (since the norm $\|\cdot\|_p$ increases as p increases). Note that $D_{y_\lambda}\varphi - D_y\varphi = D_{y_\lambda - y}\varphi$. Hence by Theorem 9.1 we can choose large p such that

$$\|D_{y_\lambda - y}\varphi\|_{q,\beta} \leq \lambda_1^{-(p-q)}|y_\lambda - y|_{-q}\|\varphi\|_{p,\beta}.$$

Then $\|D_{y_\lambda - y}\varphi\|_{q,\beta} \to 0$ for any $q \geq p_0$ and so $D_{y_\lambda}\varphi \to D_y\varphi$ in $(\mathcal{E})_\beta$. $\qquad \square$

As a matter of fact, we have the following pointwise differentiation of test functions.

Corollary 9.4. *For any fixed $\varphi \in (\mathcal{E})_\beta$ and $x \in \mathcal{E}'$, the linear functional $y \mapsto D_y\varphi(x)$ is continuous on $\mathcal{E}'$.*

Proof. Let $\widetilde{\delta}_x$ be the Kubo-Yokoi delta function. Then

$$D_y\varphi(x) = \langle\!\langle \widetilde{\delta}_x, D_y\varphi \rangle\!\rangle.$$

Hence by Theorem 9.3 the function $y \mapsto D_y\varphi(x)$ is continuous on $\mathcal{E}'$. $\qquad \square$

From the above discussion we can call D_y a differential operator acting on the space $(\mathcal{E})_\beta$. An important case is for the Gel'fand triples $\mathcal{S} \subset L^2(T, \nu) \subset \mathcal{S}'$ (see the introductory part of Chapter 5).

Definition 9.5. Let T be a topological space and ν a σ-finite Borel measure on T. The *white noise differential operator* ∂_t is defined to be the operator D_{δ_t} acting on the space $(\mathcal{S})_\beta$ of test functions, where $\delta_t \in \mathcal{S}'$ is the delta function at $t \in T$.

From Theorems 9.1 and 9.3 we get the following corollary about the white noise differential operator ∂_t.

Corollary 9.6. *For each $t \in T$, the operator ∂_t is a continuous linear operator from $(\mathcal{S})_\beta$ into itself. For each $\varphi \in (\mathcal{S})_\beta$, the function $t \mapsto \partial_t\varphi$ is continuous from T into $(\mathcal{S})_\beta$.*

Next we show that the S-transform of $D_y\varphi$ can be expressed in terms of the derivative of the $\check{S}$-transform of φ. Recall from Theorem 6.13 that any test function $\varphi \in (\mathcal{E})_\beta$ has a unique extension $\varphi(x)$ to $x \in \mathcal{E}'_c$ such that for any $x, y \in \mathcal{E}'_c$, the function $\varphi(zx + y)$ is an entire function of $z \in \mathbb{C}$.

Theorem 9.7. *For any $\varphi \in (\mathcal{E})_\beta$ and any $y \in \mathcal{E}'$, the following equality holds:*

$$(SD_y\varphi)(\xi) = \frac{d}{dz}(\check{S}\varphi)(\xi + zy)\big|_{z=0}, \quad \xi \in \mathcal{E}_c.$$

Remark. In fact the equality has an obvious extension, i.e., for any $y \in \mathcal{E}'_c$,

$$(\check{S}D_y\varphi)(x) = \frac{d}{dz}(\check{S}\varphi)(x + zy)\big|_{z=0}, \quad x \in \mathcal{E}'_c.$$

Proof. Let φ be represented by $\varphi = \sum_{n=0}^{\infty} \langle :\cdot^{\otimes n}:, f_n \rangle$. Then by Proposition 5.9 and the definition of D_y, we have

$$(SD_y\varphi)(\xi) = \sum_{n=1}^{\infty} n \langle y \widehat{\otimes} \xi^{\otimes(n-1)}, f_n \rangle, \quad \xi \in \mathcal{E}_c.$$

On the other hand,

$$\check{S}\varphi(\xi + zy) = \sum_{n=0}^{\infty} \langle (\xi + zy)^{\otimes n}, f_n \rangle.$$

It is straightforward to check that

$$\frac{d}{dz}(\check{S}\varphi)(\xi + zy)\big|_{z=0} = \sum_{n=1}^{\infty} n \langle y \widehat{\otimes} \xi^{\otimes(n-1)}, f_n \rangle, \quad \xi \in \mathcal{E}_c.$$

Hence we get the equality in the theorem. $\qquad\qquad\qquad\qquad\qquad\square$

Observe that $D_y\varphi$ can be regarded as a bilinear mapping of y and φ. We have seen that if $y \in \mathcal{E}'$, then D_y is a continuous linear operator from $(\mathcal{E})_\beta$ into itself. On the other hand, if $\eta \in \mathcal{E}$, then it is reasonable to expect that the operator D_η has an extension to the space $(\mathcal{E})_\beta^*$ of generalized functions. This is indeed the case. But first we prove that the operator D_y has the derivation property with respect to the Wick product and pointwise multiplication.

Theorem 9.8. *Let $y \in \mathcal{E}'$. The following identities hold for all φ and ψ in $(\mathcal{E})_\beta$:*

$$D_y(\varphi \diamond \psi) = (D_y\varphi) \diamond \psi + \varphi \diamond (D_y\psi),$$
$$D_y(\varphi\psi) = (D_y\varphi)\psi + \varphi(D_y\psi).$$

Proof. Let $\varphi, \psi \in (\mathcal{E})_\beta$. By Theorem 9.7

$$SD_y(\varphi \diamond \psi)(\xi) = \frac{d}{dz}\left(\check{S}(\varphi \diamond \psi)(\xi + zy)\right)\big|_{z=0}$$

$$= \frac{d}{dz}\left((\check{S}\varphi)(\xi + zy)\,(\check{S}\psi)(\xi + zy)\right)\big|_{z=0}.$$

Obviously this yields that $SD_y(\varphi \diamond \psi)(\xi) = S\left((D_y\varphi) \diamond \psi + \varphi \diamond (D_y\psi)\right)(\xi)$ for all $\xi \in \mathcal{E}_c$. Therefore we get the first identity. To prove the second identity, note that by Theorem 8.17, $\varphi\psi = \check{S}\left((\Theta\varphi) \diamond (\Theta\psi)\right)$. Hence we have

$$D_y(\varphi\psi) = \check{S}D_y\left((\Theta\varphi) \diamond (\Theta\psi)\right)$$

$$= \check{S}\left((D_y\Theta\varphi) \diamond (\Theta\psi) + (\Theta\varphi) \diamond (D_y\Theta\psi)\right)$$

$$= (D_y\varphi)\psi + \varphi(D_y\psi). \qquad\qquad\qquad\square$$

Lemma 9.9. *Let $\eta \in \mathcal{E}$. For any $q > p \geq 0$ such that $2^{1+\beta}\lambda_1^{-2(q-p)} \leq 1$, the inequality holds for all $\varphi \in (\mathcal{E})_\beta$,*

$$\|D_\eta\varphi\|_{-q,-\beta} \leq \lambda_1^{-(q-p)}|\eta|_q\|\varphi\|_{-p,-\beta}.$$

Proof. We need only to modify the argument in the proof of Theorem 9.1. Suppose $\varphi = \sum_{n=0}^{\infty}\langle :\cdot^{\otimes n}:, f_n\rangle \in (\mathcal{E})_\beta$. Then

$$\|D_\eta\varphi\|_{-q,-\beta}^2 \leq \sum_{n=0}^{\infty}(n!)^{1-\beta}(n+1)^2|\eta|_q^2|f_{n+1}|_{-q}^2$$

$$= |\eta|_q^2 \sum_{n=0}^{\infty}(n+1)^{1+\beta}\big((n+1)!\big)^{1-\beta}|f_{n+1}|_{-q}^2.$$

But $|g_n|_{-q} \leq \lambda_1^{-(q-p)n}|g_n|_{-p}$ for any $q > p \geq 0$ and $g_n \in (\mathcal{E}_c')^{\widehat{\otimes}n}$, and $n+1 \leq 2^n$ for any $n \geq 0$. Hence

$$\|D_\eta\varphi\|_{-q,-\beta}^2 \leq |\eta|_q^2 \sum_{n=0}^{\infty}2^{n(1+\beta)}\lambda_1^{-2(q-p)(n+1)}\big((n+1)!\big)^{1-\beta}|f_{n+1}|_{-p}^2.$$

By assumption $2^{1+\beta}\lambda_1^{-2(q-p)} \leq 1$. Therefore

$$\|D_\eta\varphi\|_{-q,-\beta}^2 \leq |\eta|_q^2\lambda_1^{-2(q-p)} \sum_{n=0}^{\infty}\big((n+1)!\big)^{1-\beta}|f_{n+1}|_{-p}^2$$

$$= |\eta|_q^2\lambda_1^{-2(q-p)}\|\varphi\|_{-p,-\beta}^2. \qquad \square$$

Theorem 9.10. *Let $\eta \in \mathcal{E}$. Then the differential operator D_η from $(\mathcal{E})_\beta$ into itself has a unique extension by continuity to a continuous linear operator $\widetilde{D}_\eta$ from $(\mathcal{E})_\beta^*$ into itself. Moreover, for any $\Phi, \Psi \in (\mathcal{E})_\beta^*$,*

$$\widetilde{D}_\eta(\Phi \diamond \Psi) = (\widetilde{D}_\eta\Phi) \diamond \Psi + \Phi \diamond (\widetilde{D}_\eta\Psi).$$

Proof. Note that the space $(\mathcal{E})_\beta$ is dense in $(\mathcal{E})_\beta^*$. Let $\Phi \in (\mathcal{E})_\beta^*$. Then there exists a sequence $\{\varphi_n\} \subset (\mathcal{E})_\beta$ such that $\varphi_n \to \Phi$ strongly in $(\mathcal{E})_\beta^*$. Hence there is some $p \geq 0$ such that $\varphi_n, \Phi \in (\mathcal{E}_p)_\beta^*$ and $\|\varphi_n - \Phi\|_{-p,-\beta} \to 0$. With this p, we choose $q > p$ such that $2^{1+\beta}\lambda_1^{-2(q-p)} \leq 1$. Then by Lemma 9.9

$$\|D_\eta\varphi_n - D_\eta\varphi_m\|_{-q,-\beta} \leq \lambda_1^{-(q-p)}|\eta|_q\|\varphi_n - \varphi_m\|_{-p,-\beta}.$$

Therefore $\{D_\eta\varphi_n\}$ is a Cauchy sequence in $(\mathcal{E}_q)_\beta^*$. Let

$$\widetilde{D}_\eta\Phi \equiv \lim_{n\to\infty} D_\eta\varphi_n \text{ in } (\mathcal{E}_q)_\beta^*.$$

Obviously $\tilde{D}_\eta \Phi$ is well-defined and we have

$$\|\tilde{D}_\eta \Phi\|_{-q,-\beta} \le \lambda_1^{-(q-p)} |\eta|_q \|\Phi\|_{-p,-\beta}.$$

This shows that $\tilde{D}_\eta$ is a continuous linear operator from $(\mathcal{E})_\beta^*$ into itself. The identity $\tilde{D}_\eta(\Phi \diamond \Psi) = (\tilde{D}_\eta \Phi) \diamond \Psi + \Phi \diamond (\tilde{D}_\eta \Psi)$ follows from the extension and Theorem 9.8.   □

9.2 Adjoint operators

Suppose J is a continuous linear operator from $(\mathcal{E})_\beta$ into itself. Its adjoint operator J^* is defined by the duality of $(\mathcal{E})_\beta^*$ and $(\mathcal{E})_\beta$, i.e.,

$$\langle\langle J^* \Phi, \varphi \rangle\rangle = \langle\langle \Phi, J\varphi \rangle\rangle, \quad \Phi \in (\mathcal{E})_\beta^*, \ \varphi \in (\mathcal{E})_\beta.$$

Theorem 9.11. *Suppose J is a continuous linear operator from $(\mathcal{E})_\beta$ into itself. Then its adjoint J^* is a continuous linear operator from $(\mathcal{E})_\beta^*$ into itself.*

Proof. Obviously J^* is a linear operator from $(\mathcal{E})_\beta^*$ into itself. We need to show that for any $p \ge 0$, there exist two constants $C, q \ge 0$ such that

$$\|J^* \Phi\|_{-q,-\beta} \le C\|\Phi\|_{-p,-\beta}, \quad \forall \Phi \in (\mathcal{E}_p)_\beta^*.$$

So let $p \ge 0$ be given. Since J is a continuous linear operator from $(\mathcal{E})_\beta$ into itself, there exist two constants $C, q \ge 0$ such that

$$\|J\varphi\|_{p,\beta} \le C\|\varphi\|_{q,\beta}, \quad \forall \varphi \in (\mathcal{E})_\beta.$$

Now for any $\Phi \in (\mathcal{E}_p)_\beta^*$, we have

$$\begin{aligned}
|\langle\langle J^* \Phi, \varphi \rangle\rangle| &= |\langle\langle \Phi, J\varphi \rangle\rangle| \\
&\le \|\Phi\|_{-p,-\beta} \|J\varphi\|_{p,\beta} \\
&\le C\|\Phi\|_{-p,-\beta} \|\varphi\|_{q,\beta}, \quad \forall \varphi \in (\mathcal{E})_\beta.
\end{aligned}$$

This implies that $\|J^* \Phi\|_{-q,-\beta} \le C\|\Phi\|_{-p,-\beta}$ for any $\Phi \in (\mathcal{E}_p)_\beta^*$ and so the theorem is proved.   □

By using the above theorem and the properties of differential operators from the previous section we get the next theorem. The proof is routine and thus omitted.

Theorem 9.12. *The following properties hold:*

(a) *For any $y \in \mathcal{E}'$, the operator D_y^* is a continuous linear operator from $(\mathcal{E})_\beta^*$ into itself.*

(b) *For any $\Phi \in (\mathcal{E})_\beta^*$, the linear operator $y \mapsto D_y^*\Phi$ is continuous from $\mathcal{E}'$ (with the strong topology) into $(\mathcal{E})_\beta^*$.*

(c) *For any $t \in T$, the operator ∂_t^* is a continuous linear operator from $(\mathcal{S})_\beta^*$ into itself.*

(d) *For any $\Phi \in (\mathcal{S})_\beta^*$, the function $t \mapsto \partial_t^*\Phi$ is continuous from T into $(\mathcal{S})_\beta^*$.*

There is a nice relationship between the S-transform and the adjoint operators. This will be useful in later chapters.

Theorem 9.13. *Let $y \in \mathcal{E}'$ and $\Phi \in (\mathcal{E})_\beta^*$. Then the S-transform of $D_y^*\Phi$ is given by*

$$(SD_y^*\Phi)(\xi) = \langle y, \xi \rangle S\Phi(\xi), \quad \xi \in \mathcal{E}_c.$$

In particular, for $t \in T$ and $\Phi \in (\mathcal{S})_\beta^*$,

$$(S\partial_t^*\Phi)(\xi) = \xi(t)S\Phi(\xi), \quad \xi \in \mathcal{S}_c.$$

Proof. Recall that

$$:e^{\langle \cdot, \xi \rangle} := \sum_{n=0}^{\infty} \frac{1}{n!}\langle :\cdot^{\otimes n}:, \xi^{\otimes n} \rangle.$$

Hence we have

$$D_y :e^{\langle \cdot, \xi \rangle} := \sum_{n=1}^{\infty} n\frac{1}{n!}\langle :\cdot^{\otimes(n-1)}:, \langle y, \xi^{\otimes n} \rangle\rangle$$

$$= \langle y, \xi \rangle \sum_{n=1}^{\infty} \frac{1}{(n-1)!}\langle :\cdot^{\otimes(n-1)}:, \xi^{\otimes(n-1)} \rangle$$

$$= \langle y, \xi \rangle :e^{\langle \cdot, \xi \rangle}:.$$

Therefore for any $\xi \in \mathcal{E}_c$,

$$(SD_y^*\Phi)(\xi) = \langle\!\langle D_y^*\Phi, :e^{\langle \cdot, \xi \rangle}: \rangle\!\rangle$$

$$= \langle\!\langle \Phi, D_y :e^{\langle \cdot, \xi \rangle}: \rangle\!\rangle$$

$$= \langle y, \xi \rangle\langle\!\langle \Phi, :e^{\langle \cdot, \xi \rangle}: \rangle\!\rangle$$

$$= \langle y, \xi \rangle S\Phi(\xi).$$

In particular, when $y = \delta_t$ we get the second assertion. $\qquad\square$

Corollary 9.14. *If $y \in \mathcal{E}'$, then $D_y^* \Phi = \langle \cdot, y \rangle \diamond \Phi$ for any $\Phi \in (\mathcal{E})_\beta^*$.
Moreover, if $\eta \in \mathcal{E}$, then the restriction of the operator D_η^* to $(\mathcal{E})_\beta$ is a
continuous linear operator from $(\mathcal{E})_\beta$ into itself.*

Proof. The first assertion follows from Theorem 9.13 since $S \langle \cdot, y \rangle (\xi) = \langle y, \xi \rangle$. Note that for fixed $\eta \in \mathcal{E}$, the function $\langle \cdot, \eta \rangle$ is a test function in $(\mathcal{E})_\beta$. Therefore the second assertion follows from the first assertion and Theorem 8.12. $\square$

We can express $D_y^* \Phi$ in terms of the Wiener-Itô decomposition of Φ.
Notice that $\langle \cdot, y \rangle \diamond \langle : \cdot^{\otimes n} :, F_n \rangle = \langle : \cdot^{\otimes (n+1)} :, y \widehat{\otimes} F_n \rangle$. Thus if $\Phi \in (\mathcal{E})_\beta^*$ is
represented by

$$\Phi = \sum_{n=0}^{\infty} \langle : \cdot^{\otimes n} :, F_n \rangle,$$

then $D_y^* \Phi$ is represented by

$$D_y^* \Phi = \sum_{n=0}^{\infty} \langle : \cdot^{\otimes (n+1)} :, y \widehat{\otimes} F_n \rangle.$$

In particular, for the adjoint operator ∂_t^*, we have

$$\partial_t^* \Phi = \sum_{n=0}^{\infty} \langle : \cdot^{\otimes (n+1)} :, \delta_t \widehat{\otimes} F_n \rangle.$$

The next theorem gives the commutation relationships for the differential
operators and their adjoints. The commutator $[A, B]$ of two operators A
and B is defined to be $[A, B] = AB - BA$.

Theorem 9.15. *The following commutation relationships hold:*
(a) $[D_x, D_y] = 0$ *on $(\mathcal{E})_\beta$ for all $x, y \in \mathcal{E}'$.*
(b) $[D_x^*, D_y^*] = 0$ *on $(\mathcal{E})_\beta^*$ for all $x, y \in \mathcal{E}'$.*
(c) $[\widetilde{D}_\xi, \widetilde{D}_\eta] = 0$ *on $(\mathcal{E})_\beta^*$ for all $\xi, \eta \in \mathcal{E}$.*
(d) $[\widetilde{D}_\eta, D_y^*] = \langle y, \eta \rangle I$ *on $(\mathcal{E})_\beta^*$ for all $\eta \in \mathcal{E}$ and $y \in \mathcal{E}'$.*
(e) $[D_y, D_\eta^*] = \langle y, \eta \rangle I$ *on $(\mathcal{E})_\beta$ for all $y \in \mathcal{E}'$ and $\eta \in \mathcal{E}$.*

Remark. For the white noise differential operator and its adjoint, we have

$$[\partial_s, \partial_t] = 0, \quad [\partial_s^*, \partial_t^*] = 0, \quad [\partial_s, \partial_t^*] = \delta_s(t) I. \tag{9.4}$$

Proof. The identity (a) is obvious from the definition of D_y. By taking the
adjoint of (a), we get (b). The identity (c) follows from (a) and Theorem

9.10. The identities (d) and (e) can be proved by similar arguments. To show (d) we use Theorem 9.10 and Corollary 9.14, i.e., for any $\Phi \in (\mathcal{E})^*_\beta$,

$$(\tilde{D}_\eta D^*_y - D^*_y \tilde{D}_\eta)\Phi = \tilde{D}_\eta(\langle \cdot, y \rangle \diamond \Phi) - \langle \cdot, y \rangle \diamond (\tilde{D}_\eta \Phi)$$

$$= (\langle y, \eta \rangle \diamond \Phi + \langle \cdot, y \rangle \diamond (\tilde{D}_\eta \Phi)) - \langle \cdot, y \rangle \diamond (\tilde{D}_\eta \Phi)$$

$$= \langle y, \eta \rangle \Phi. \qquad \square$$

9.3 Multiplication operators

Let φ be a test function in $(\mathcal{E})_\beta$. From Theorem 8.17 we know that multiplication by φ is a continuous linear operator from $(\mathcal{E})_\beta$ into itself. In fact, we can extend this multiplication to the space $(\mathcal{E})^*_\beta$ of generalized functions as follows. For $\Phi \in (\mathcal{E})^*_\beta$, we define $\varphi\Phi$ by

$$\langle\langle \varphi\Phi, \psi \rangle\rangle \equiv \langle\langle \Phi, \varphi\psi \rangle\rangle, \quad \psi \in (\mathcal{E})_\beta. \tag{9.5}$$

It can be shown that the mapping $\Phi \mapsto \varphi\Phi$ is a continuous linear operator from $(\mathcal{E})^*_\beta$ into itself. A special case of this multiplication is given by the test function $\varphi = \langle \cdot, \xi \rangle$ for $\xi \in \mathcal{E}$. We will use Q_ξ to denote this multiplication on $(\mathcal{E})_\beta$ and $\tilde{Q}_\xi$ its extension to $(\mathcal{E})^*_\beta$, i.e.,

$$Q_\xi\varphi = \langle \cdot, \xi \rangle \varphi, \quad \varphi \in (\mathcal{E})_\beta; \qquad \tilde{Q}_\xi\Phi = \langle \cdot, \xi \rangle \Phi, \quad \Phi \in (\mathcal{E})^*_\beta. \tag{9.6}$$

The multiplication operator can be expressed in terms of the differential operator and its adjoint. In order to prove this fact, we need two lemmas.

Lemma 9.16. *For any $\eta, \zeta \in \mathcal{E}$, the equality holds:*

$$\int_{\mathcal{E}'} \langle x, \eta \rangle e^{\langle x, \zeta \rangle} \, d\mu(x) = \langle \eta, \zeta \rangle e^{\frac{1}{2}|\zeta|_0^2}.$$

Proof. Note that we have the identity for $t \in \mathbb{R}$,

$$\int_{\mathcal{E}'} e^{\langle x, \zeta + t\eta \rangle} \, d\mu(x) = e^{\frac{1}{2}|\zeta + t\eta|_0^2}.$$

By differentiating both sides of this identity and then letting $t = 0$, we get immediately the equality in the lemma. $\square$

Lemma 9.17. *Let $\eta \in \mathcal{E}$. Then the equality holds for all φ and ψ in $(\mathcal{E})_\beta$:*

$$\langle\langle D_\eta\varphi, \psi \rangle\rangle + \langle\langle \varphi, D_\eta\psi \rangle\rangle = \langle\langle \langle \cdot, \eta \rangle \varphi, \psi \rangle\rangle.$$

Proof. By the continuity of the differential and multiplication operators and the fact that the linear span of the set $\{:e^{\langle \cdot, \xi \rangle}:; \xi \in \mathcal{E}\}$ is dense in $(\mathcal{E})_\beta$, it suffices to prove the equality for $\varphi =: e^{\langle \cdot, \xi \rangle}:$ and $\psi =: e^{\langle \cdot, \zeta \rangle}:$, $\xi, \zeta \in \mathcal{E}$. From the proof of Theorem 9.13, we get

$$D_\eta \varphi = \langle \eta, \xi \rangle :e^{\langle \cdot, \xi \rangle}:, \quad D_\eta \psi = \langle \eta, \zeta \rangle :e^{\langle \cdot, \zeta \rangle}:.$$

But by Theorem 5.13, we have $\langle\!\langle :e^{\langle \cdot, \xi \rangle}:, :e^{\langle \cdot, \eta \rangle}: \rangle\!\rangle = e^{\langle \zeta, \xi \rangle}$. Therefore

$$\langle\!\langle D_\eta \varphi, \psi \rangle\!\rangle + \langle\!\langle \varphi, D_\eta \psi \rangle\!\rangle = \langle \eta, \xi + \zeta \rangle e^{\langle \zeta, \xi \rangle}. \tag{9.7}$$

On the other hand,

$$\langle\!\langle \langle \cdot, \eta \rangle \varphi, \psi \rangle\!\rangle = e^{-\frac{1}{2}(|\xi|_0^2 + |\zeta|_0^2)} \int_{\mathcal{E}'} \langle x, \eta \rangle e^{\langle x, \zeta + \xi \rangle} \, d\mu(x).$$

Then we use Lemma 9.16 to get

$$\langle\!\langle \langle \cdot, \eta \rangle \varphi, \psi \rangle\!\rangle = \langle \eta, \zeta + \xi \rangle e^{\langle \zeta, \xi \rangle}. \tag{9.8}$$

The conclusion follows from equations (9.7) and (9.8). $\qquad\square$

Theorem 9.18. *For any $\eta \in \mathcal{E}$, the equalities hold:*
(a) $Q_\eta = D_\eta + D_\eta^*$ *as continuous linear operators from $(\mathcal{E})_\beta$ into itself.*
(b) $\tilde{Q}_\eta = \tilde{D}_\eta + D_\eta^*$ *as continuous linear operators from $(\mathcal{E})_\beta^*$ into itself.*

Remark. By equality (a) the following integral holds for all $\varphi, \psi \in (\mathcal{E})_\beta$:

$$\int_{\mathcal{E}'} (D_\eta \varphi)(x) \psi(x) \, d\mu(x) = - \int_{\mathcal{E}'} \varphi(x) \left[(D_\eta \psi)(x) - \langle x, \eta \rangle \psi(x) \right] d\mu(x).$$

This is the integration by parts formula for the Gaussian measure μ on $\mathcal{E}'$.

Proof. Assertion (a) follows right away from Lemma 9.17. Assertion (b) is the extension of assertion (a) by continuity to $(\mathcal{E})_\beta^*$. $\qquad\square$

Now we consider the multiplication operator Q_y for $y \in \mathcal{E}'$. Suppose $\varphi \in (\mathcal{E})_\beta$. Then $Q_y \varphi = \langle \cdot, y \rangle \varphi$ is a generalized function in $(\mathcal{E})_\beta^*$ defined by equation (9.5). In view of the last theorem we expect that $Q_y \varphi = D_y \varphi + D_y^* \varphi$. This is indeed the case. We need a lemma to prove this fact.

Lemma 9.19. *For any $x \in \mathcal{E}'$, the equality holds*

$$:x^{\otimes(n+1)}: = :x^{\otimes n}: \hat{\otimes} x - n :x^{\otimes(n-1)}: \hat{\otimes} \tau.$$

Proof. Note that Hermite polynomials satisfy the recursion formula

$$:x^{n+1}:_{\sigma^2} - x :x^n:_{\sigma^2} + \sigma^2 n :x^{n-1}:_{\sigma^2} = 0.$$

Hence for any $\xi \in \mathcal{E}$, it follows from Lemma 5.2 that

$$\langle :x^{\otimes(n+1)}:, \xi^{\otimes(n+1)} \rangle = :\langle x, \xi \rangle^{n+1}: |\xi|_0^2$$
$$= \langle x, \xi \rangle :\langle x, \xi \rangle^n: |\xi|_0^2 - n|\xi|_0^2 :\langle x, \xi \rangle^{n-1}: |\xi|_0^2$$
$$= \langle x, \xi \rangle \langle :x^{\otimes n}:, \xi^{\otimes n} \rangle - n|\xi|_0^2 \langle :x^{\otimes(n-1)}:, \xi^{\otimes(n-1)} \rangle$$
$$= \langle :x^{\otimes n}: \widehat{\otimes} x, \xi^{\otimes(n+1)} \rangle - n \langle :x^{\otimes(n-1)}: \widehat{\otimes} \tau, \xi^{\otimes(n+1)} \rangle.$$

Thus by the polarization identity we get the equality in the lemma. $\square$

Theorem 9.20. For any $y \in \mathcal{E}'$, the equality $Q_y = D_y + D_y^*$ holds as continuous linear operators from $(\mathcal{E})_\beta$ into $(\mathcal{E})_\beta^*$.

Remark. An important special case of this theorem is the white noise multiplication for the Gel'fand triple $(\mathcal{S})_\beta \subset (L^2) \subset (\mathcal{S})_\beta^*$,

$$\dot{B}(t)\varphi = \partial_t\varphi + \partial_t^*\varphi, \quad \varphi \in (\mathcal{S})_\beta, \tag{9.9}$$

where both sides are regarded as generalized functions in the space $(\mathcal{S})_\beta^*$.

Proof. Let $y \in \mathcal{E}'$ be fixed. We need only to prove that

$$\langle \cdot, y \rangle \varphi = D_y\varphi + D_y^*\varphi, \quad \forall \varphi \in (\mathcal{E})_\beta. \tag{9.10}$$

For any $n \geq 0$ and any $f_n \in \mathcal{E}^{\widehat{\otimes} n}$, we can apply Lemma 9.19 to get

$$\langle\langle \langle \cdot, y \rangle \varphi, \langle :\cdot^{\otimes n}:, f_n \rangle \rangle\rangle$$
$$= \langle\langle \langle \cdot, y \rangle \langle :\cdot^{\otimes n}:, f_n \rangle, \varphi \rangle\rangle$$
$$= \langle\langle n \langle :\cdot^{\otimes(n-1)}:, \langle y, f_n \rangle \rangle + \langle :\cdot^{\otimes(n+1)}:, y \widehat{\otimes} f_n \rangle, \varphi \rangle\rangle. \tag{9.11}$$

On the other hand, we have

$$D_y \langle :\cdot^{\otimes n}:, f_n \rangle = n \langle :\cdot^{\otimes(n-1)}:, \langle y, f_n \rangle \rangle \tag{9.12}$$
$$D_y^* \langle :\cdot^{\otimes n}:, f_n \rangle = \langle :\cdot^{\otimes(n+1)}:, y \widehat{\otimes} f_n \rangle. \tag{9.13}$$

It follows from equations (9.11)–(9.13) that

$$\langle\langle \langle \cdot, y \rangle \varphi, \langle :\cdot^{\otimes n}:, f_n \rangle \rangle\rangle = \langle\langle (D_y + D_y^*) \langle :\cdot^{\otimes n}:, f_n \rangle, \varphi \rangle\rangle$$
$$= \langle\langle (D_y + D_y^*)\varphi, \langle :\cdot^{\otimes n}:, f_n \rangle \rangle\rangle.$$

Now observe that the linear span of the set $\{\langle :\cdot^{\otimes n}:, f_n \rangle; n \geq 0, f_n \in \mathcal{E}^{\widehat{\otimes} n}\}$ is dense in $(\mathcal{E})_\beta$. Hence $\langle \cdot, y \rangle \varphi = D_y\varphi + D_y^*\varphi$ and this proves equation (9.10). $\square$

The product rule for the adjoint operator D_y^* is given in the next theorem. Note that $\varphi\Psi$ as defined in equation (9.5) is a generalized function in $(\mathcal{E})_\beta^*$ for any $\varphi \in (\mathcal{E})_\beta$ and $\Psi \in (\mathcal{E})_\beta^*$.

Theorem 9.21. *Let* $y \in \mathcal{E}'$. *Then for any* $\varphi \in (\mathcal{E})_\beta$ *and* $\Psi \in (\mathcal{E})_\beta^*$,

$$D_y^*(\varphi\Psi) = \varphi(D_y^*\Psi) - (D_y\varphi)\Psi. \tag{9.14}$$

Remark. In particular, for $y = \delta_t$ we have

$$\partial_t^*(\varphi\Psi) = \varphi(\partial_t^*\Psi) - (\partial_t\varphi)\Psi. \tag{9.15}$$

This equality will be used in Chapter 13 when we study white noise integration.

Proof. First let $\varphi, \psi \in (\mathcal{E})_\beta$. Then by Theorem 9.20,

$$\begin{aligned}
D_y^*(\varphi\psi) &= (Q_y - D_y)(\varphi\psi) \\
&= Q_y(\varphi\psi) - (D_y\varphi)\psi - \varphi(D_y\psi) \\
&= \varphi(Q_y\psi - D_y\psi) - (D_y\varphi)\psi \\
&= \varphi(D_y^*\psi) - (D_y\varphi)\psi.
\end{aligned}$$

Hence equation (9.14) holds for any $\varphi, \psi \in (\mathcal{E})_\beta$. Note that the operator D_y^* and the multiplication $\Psi \mapsto \varphi\Psi$ are continuous from $(\mathcal{E})_\beta^*$ into itself. Therefore we can use the continuity to show that equation (9.14) holds for any $\varphi \in (\mathcal{E})_\beta$ and $\Psi \in (\mathcal{E})_\beta^*$. $\qquad\square$

9.4 Gross differentiation and gradient

We have studied the differential operators D_y and D_ξ for $y \in \mathcal{E}'$ and $\xi \in \mathcal{E}$. Recall that the operator D_y is continuous from $(\mathcal{E})_\beta$ into itself, while the operator D_ξ has an extension $\widetilde{D}_\xi$ which is continuous from $(\mathcal{E})_\beta^*$ into itself. It is natural to ask about the operator D_h for $h \in E$; e.g., whether it is continuous from (L^2) into itself. The derivative $D_h\varphi$ for $h \in E$ is often called the *Gross derivative* of φ. Let $h \in E$ be fixed. It is easy to see that D_h is an unbounded operator on (L^2) for nonzero h. The domain of the operator D_h is closely related to the number operator N.

Definition 9.22. For $\varphi = \sum_{n=0}^\infty \langle :\cdot^{\otimes n}:, f_n \rangle$, define

$$N\varphi = \sum_{n=1}^\infty n \langle :\cdot^{\otimes n}:, f_n \rangle.$$

This operator N is called the *number operator*.

Theorem 9.23. *The number operator N is a continuous linear operator from $(\mathcal{E})_\beta$ into itself and from $(\mathcal{E})_\beta^*$ into itself.*

Proof. First we prove that N is a continuous linear operator from $(\mathcal{E})_\beta$ into itself. Let $p \geq 0$ be given. Then for any $\varphi = \sum_{n=0}^{\infty} \langle : \cdot^{\otimes n} :, f_n \rangle \in (\mathcal{E})_\beta$ and $q \geq p$, we have

$$\|N\varphi\|_{p,\beta}^2 = \sum_{n=1}^{\infty} (n!)^{1+\beta} n^2 |f_n|_p^2$$

$$\leq \sum_{n=1}^{\infty} (n!)^{1+\beta} n^2 \lambda_1^{-2(q-p)n} |f_n|_q^2.$$

Note that $n^2 \leq 2^{2n}$ for $n \geq 1$. Choose $q \geq p$ such that $\lambda_1^{q-p} \geq 2$. Then

$$\|N\varphi\|_{p,\beta}^2 \leq \sum_{n=1}^{\infty} (n!)^{1+\beta} 2^{2n} \lambda_1^{-2(q-p)n} |f_n|_q^2$$

$$\leq \sum_{n=1}^{\infty} (n!)^{1+\beta} |f_n|_q^2$$

$$= \|\varphi\|_{q,\beta}^2.$$

This shows that $N\varphi \in (\mathcal{E})_\beta$ for any $\varphi \in (\mathcal{E})_\beta$ and $\|N\varphi\|_{p,\beta} \leq \|\varphi\|_{q,\beta}$. The linearity of N is obvious. Hence N is a continuous linear operator from $(\mathcal{E})_\beta$ into itself.

To prove that N is a continuous linear operator from $(\mathcal{E})_\beta^*$ into itself we need to show that for any given $p \geq 0$, there exist $q > p$ and $C \geq 0$ such that

$$\|N\Phi\|_{-q,-\beta} \leq C\|\Phi\|_{-p,-\beta}, \quad \forall \Phi \in (\mathcal{E})_\beta^*.$$

With the given p, we choose $q > p$ such that $\lambda_1^{q-p} \geq 2$. Then for any generalized function $\Phi = \sum_{n=0}^{\infty} \langle : \cdot^{\otimes n} :, F_n \rangle \in (\mathcal{E})_\beta^*$, we have

$$\|N\varphi\|_{-q,-\beta} = \sum_{n=1}^{\infty} (n!)^{1-\beta} n^2 |F_n|_{-q}^2$$

$$\leq \sum_{n=1}^{\infty} (n!)^{1-\beta} n^2 \lambda_1^{-2(q-p)n} |F_n|_{-p}^2$$

$$\leq \sum_{n=1}^{\infty} (n!)^{1-\beta} 2^{2n} \lambda_1^{-2(q-p)n} |F_n|_{-p}^2$$

$$\leq \sum_{n=1}^{\infty} (n!)^{1-\beta} |F_n|_{-p}^2$$

$$= \|\Phi\|_{-p,-\beta}^2.$$

Hence N is a continuous linear operator from $(\mathcal{E})^*_\beta$ into itself.   □

The power $N^r, r \in \mathbb{R}$, of the number operator can be defined in an obvious way, i.e., for $\varphi = \sum_{n=0}^\infty \langle :\cdot^{\otimes n}:, f_n \rangle$, define

$$N^r\varphi = \sum_{n=1}^\infty n^r \langle :\cdot^{\otimes n}:, f_n \rangle.$$

Theorem 9.24. *For any* $r \in \mathbb{R}$, *the operator* N^r *is a continuous linear operator from* $(\mathcal{E})_\beta$ *into itself and from* $(\mathcal{E})^*_\beta$ *into itself.*

Proof. The conclusion is obvious if $r \le 0$. When $r > 0$, just use the same argument as in the proof of Theorem 9.23.   □

Theorem 9.25. *For any* $r \in \mathbb{R}$, *the equality* $\langle\!\langle N^r\Phi, \varphi \rangle\!\rangle = \langle\!\langle \Phi, N^r\varphi \rangle\!\rangle$ *holds for all* Φ *in* $(\mathcal{E})^*_\beta$ *and* $\varphi \in (\mathcal{E})_\beta$.

Proof. Let $\Phi \in (\mathcal{E})^*_\beta$ and $\varphi \in (\mathcal{E})_\beta$ be represented by

$$\Phi = \sum_{n=0}^\infty \langle :\cdot^{\otimes n}:, F_n \rangle, \qquad \varphi = \sum_{n=0}^\infty \langle :\cdot^{\otimes n}:, f_n \rangle.$$

Then we have

$$N^r\Phi = \sum_{n=0}^\infty n^r \langle :\cdot^{\otimes n}:, F_n \rangle, \qquad N^r\varphi = \sum_{n=0}^\infty n^r \langle :\cdot^{\otimes n}:, f_n \rangle.$$

Hence by equation (5.8), the equality $\langle\!\langle N^r\Phi, \varphi \rangle\!\rangle = \langle\!\langle \Phi, N^r\varphi \rangle\!\rangle$ holds.   □

Theorem 9.26. *Let* φ *be such that* $N^{1/2}\varphi \in (L^2)$ *and* $h \in E$. *Then* $D_h\varphi \in (L^2)$ *and* $\|D_h\varphi\|_0 \le |h|_0\|N^{1/2}\varphi\|_0$.

Proof. Let $\varphi = \sum_{n=0}^\infty \langle :\cdot^{\otimes n}, f_n \rangle$. Then for $h \in E$,

$$D_h\varphi = \sum_{n=1}^\infty n\langle :\cdot^{\otimes(n-1)}, \langle h, f_n \rangle \rangle.$$

Therefore

$$\|D_h\varphi\|_0^2 \le \sum_{n=1}^\infty (n-1)!n^2|h|_0^2|f_n|_0^2 = |h|_0^2\sum_{n=1}^\infty n!n|f_n|_0^2.$$

On the other hand, it is easy to see that

$$\|N^{1/2}\varphi\|_0^2 = \sum_{n=1}^\infty n!n|f_n|_0^2. \tag{9.16}$$

Hence we get immediately that $\|D_h\varphi\|_0 \le |h|_0 \|N^{1/2}\varphi\|_0$. $\qquad\square$

For the rest of this section we consider the special Gel'fand triple $\mathcal{S} \subset L^2(T,\nu) \subset \mathcal{S}'$ (see the introductory part of Chapter 5).

Theorem 9.27. Let φ be such that $N^{1/2}\varphi \in (L^2)$ and $h \in L^2(T,\nu)$. Then

$$\int_T \|\partial_t\varphi\|_0^2 \, d\nu(t) = \|N^{1/2}\varphi\|_0^2, \quad D_h\varphi = \int_T h(t)\partial_t\varphi \, d\nu(t).$$

Proof. Let $\varphi = \sum_{n=0}^\infty \langle :\cdot^{\otimes n}, f_n \rangle$. Then for almost all $t \in T$ with respect to ν, we have

$$\partial_t\varphi = \sum_{n=1}^\infty n\langle :\cdot^{\otimes(n-1)}, f_n(t,\cdot) \rangle.$$

Therefore

$$\|\partial_t\varphi\|_0^2 = \sum_{n=1}^\infty (n-1)! n^2 |f_n(t,\cdot)|_0^2.$$

This yields that

$$\int_T \|\partial_t\varphi\|_0^2 \, d\nu(t) = \sum_{n=1}^\infty n! n |f_n|_0^2.$$

Hence, in view of equation (9.16), we get the first equality in the theorem. To prove the second equality, note that $\|\partial_t\varphi\|_0 \in L^2(T,\nu)$, ν-a.e., by the Fubini theorem and the first equality. Thus we get

$$\int_T h(t)\partial_t\varphi \, d\nu(t) = \sum_{n=1}^\infty n\langle :\cdot^{\otimes(n-1)}, \langle h, f_n \rangle \rangle = D_h\varphi. \qquad\square$$

Finally we make some remarks about Theorem 9.27. We can view $\partial_t\varphi(x)$ in another way, i.e., let

$$(\nabla\varphi)(t,x) \equiv \partial_t\varphi(x).$$

We will call $\nabla\varphi$ the *gradient* of φ. Then for φ such that $N^{1/2}\varphi \in (L^2)$, we have

$$\int_T \int_{\mathcal{S}'} |(\nabla\varphi)(t,x)|^2 \, d\mu(x) d\nu(t) = \|N^{1/2}\varphi\|_0^2.$$

Thus the gradient operator ∇ is a linear operator from the domain $\mathcal{D}(N^{1/2})$ of $N^{1/2}$ (regarded as densely defined in (L^2)) into $L^2(T \times \mathcal{S}')$. However, $\|N^{1/2}\varphi\|_0$ does not define a norm for $\varphi \in \mathcal{D}(N^{1/2})$ since $N^{1/2}$ is not injective. To overcome this difficulty, we use the operator $(N+1)^{1/2}$ instead of $N^{1/2}$. Define

$$|\varphi|_{1/2} \equiv \|(N+1)^{1/2}\varphi\|_0, \quad \mathcal{W}^{1/2} \equiv \{\varphi \in (L^2); |\varphi|_{1/2} < \infty\}.$$

Then $\mathcal{W}^{1/2}$ is a Hilbert space with the norm $\| \cdot \|_{1/2}$. It is a Sobolev space for the Gel'fand triple $(\mathcal{S}) \subset (L^2) \subset (\mathcal{S})^*$. By Theorem 9.27 the gradient operator

$$\nabla : \mathcal{W}^{1/2} \longrightarrow L^2(T \times \mathcal{S}')$$

is a continuous linear operator. If $\varphi(x) = \sum_{n=0}^{\infty} \langle :x^{\otimes n}:, f_n \rangle$, then

$$(\nabla\varphi)(t, x) = \sum_{n=1}^{\infty} n \langle :x^{\otimes(n-1)}:, f_n(t, \cdot) \rangle. \tag{9.17}$$

The adjoint operator ∇^* is a continuous linear operator from $L^2(T \times \mathcal{S}')$ into $\mathcal{W}^{1/2}$. The next theorem shows that it can be expressed in terms of the adjoint operator ∂_t^*.

Theorem 9.28. *The equality* $\nabla^* X = \int_T \partial_t^* X(t) \, d\nu(t)$ *holds for all* $X \in L^2(T \times \mathcal{S}')$.

Proof. Let $\varphi_\xi =: e^{\langle \cdot, \xi \rangle}:, \xi \in \mathcal{E}$. Then we have $\varphi_\xi = \sum_{n=0}^{\infty} \frac{1}{n!} \langle :\cdot^{\otimes n}:, \xi^{\otimes n} \rangle$. Hence by equation (9.17)

$$(\nabla\varphi_\xi)(t, x) = \sum_{n=1}^{\infty} \frac{1}{(n-1)!} \langle :x^{\otimes(n-1)}:, \xi(t)\xi^{\otimes(n-1)} \rangle = \xi(t)\varphi_\xi(x). \tag{9.18}$$

Now for $X \in L^2(T \times \mathcal{S}')$, we can use equation (9.18) to find the S-transform of $\nabla^* X$

$$
\begin{aligned}
S(\nabla^* X)(\xi) &= \langle\!\langle \nabla^* X, \varphi_\xi \rangle\!\rangle \\
&= \langle\!\langle X, \nabla\varphi_\xi \rangle\!\rangle \\
&= \int_T \int_{\mathcal{S}'} X(t, x)\xi(t)\varphi_\xi(x) \, d\mu(x) d\nu(t) \\
&= \int_T \xi(t) S(X(t))(\xi) \, d\nu(t).
\end{aligned} \tag{9.19}
$$

On the other hand, we have

$$S\left(\int_T \partial_t^* X(t) \, d\nu(t) \right)(\xi) = \int_T \xi(t) S(X(t))(\xi) \, d\nu(t). \tag{9.20}$$

The conclusion of the theorem follows from equations (9.19) and (9.20). $\square$

10

Integral Kernel Operators

In an interesting development, Hida, Obata, and Saitô have introduced in [60] integral kernel operators of the form

$$\int_{T^{j+k}} \theta(s_1, \ldots, s_j, t_1, \ldots, t_k) \partial_{s_1}^* \cdots \partial_{s_j}^* \partial_{t_1} \cdots \partial_{t_k} \, ds_1 \cdots ds_j dt_1 \cdots dt_k,$$

where $\theta \in (\mathcal{S}')^{\otimes(j+k)}$. These integral kernel operators are continuous from the space $(\mathcal{S})$ into its dual space $(\mathcal{S})^*$ (see the book by Hida et al. [58], Hida et al. [60], and Obata [142]). They have important applications to quantum probability theory (see Obata [140] [143]).

In this chapter we will study this type of integral kernel operator for the Gel'fand triple $(\mathcal{S})_\beta \subset (L^2) \subset (\mathcal{S})^*_\beta$ which is associated with the Gel'fand triple $\mathcal{S} \subset L^2(T, \nu) \subset \mathcal{S}'$ (see the introductory part of Chapter 5).

Notations. For simplicity we will use the following notations. Let $\mathbf{t} = (t_1, \ldots, t_n) \in T^n$. Define

(1) $\partial_{\mathbf{t}} = \partial_{t_1} \cdots \partial_{t_n}$.
(2) $\partial_{\mathbf{t}}^* = \partial_{t_1}^* \cdots \partial_{t_n}^*$.
(3) $d\mathbf{t} = d\nu(t_1) \cdots d\nu(t_n)$ (ν is suppressed).

10.1 Heuristic discussion

In the previous chapter we have defined the differential operator D_y and its adjoint operator D_y^* for $y \in \mathcal{S}'$. Recall that D_y is a continuous linear operator from $(\mathcal{S})_\beta$ into itself and D_y^* is a continuous linear operator from $(\mathcal{S})^*_\beta$ into itself. In view of Theorem 9.27 these operators can be written informally as

$$D_y = \int_T y(t) \partial_t \, dt, \quad D_y^* = \int_T y(s) \partial_s^* \, ds.$$

If we consider the products of two operators of this kind, we get the following cases:

Case 1. $D_x D_y$ *for* $x, y \in \mathcal{S}'$

The symbolic expression is given by

$$D_x D_y = \int_T \int_T x(t_1) y(t_2) \partial_{t_1} \partial_{t_2} \, dt_1 dt_2,$$

where the kernel function $\theta(t_1, t_2) = x(t_1) y(t_2)$ is in $(\mathcal{S}')^{\otimes 2}$. This operator is continuous from $(\mathcal{S})_\beta$ into itself. But for $\xi, \eta \in \mathcal{S}$, the operator $D_\xi D_\eta$ extends by continuity to a continuous operator from $(\mathcal{S})^*_\beta$ into itself.

Case 2. $D_x^* D_y^*$ *for* $x, y \in \mathcal{S}'$

The symbolic expression is given by

$$D_x^* D_y^* = \int_T \int_T x(s_1) y(s_2) \partial_{s_1}^* \partial_{s_2}^* \, ds_1 ds_2,$$

where the kernel function $\theta(s_1, s_2) = x(s_1) y(s_2)$ is in $(\mathcal{S}')^{\otimes 2}$. This operator is continuous from $(\mathcal{S})^*_\beta$ into itself. But for $\xi, \eta \in \mathcal{S}$, the operator $D_\xi^* D_\eta^*$, by restriction, is a continuous operator from $(\mathcal{S})_\beta$ into itself.

Case 3. $D_x^* D_y$ *for* $x, y \in \mathcal{S}'$

The symbolic expression is given by

$$D_x^* D_y = \int_T \int_T x(s) y(t) \partial_s^* \partial_t \, dsdt,$$

where the kernel function $\theta(s, t) = x(s) y(t)$ belongs to $(\mathcal{S}')^{\otimes 2}$. This operator is continuous from $(\mathcal{S})_\beta$ into $(\mathcal{S})^*_\beta$. But for $x \in \mathcal{S}'$ and $\eta \in \mathcal{S}$, the operator $D_x^* D_\eta$ extends by continuity to a continuous operator from $(\mathcal{S})^*_\beta$ into itself.

Case 4. $D_x D_y^*$ *for* $x, y \in \mathcal{S}'$

In general this product has no meaning since D_x may not extend to $(\mathcal{S})^*_\beta$. We need to put a restriction on x, i.e., we consider the operator $\tilde{D}_\eta D_y^*$ for $\eta \in \mathcal{S}, y \in \mathcal{S}'$. The symbolic expression is given by

$$\tilde{D}_\eta D_y^* = \int_T \int_T \eta(t) y(s) \partial_t \partial_s^* \, dtds,$$

where the kernel function $\theta(t, s) = \eta(t) y(s)$ belongs to $\mathcal{S} \otimes \mathcal{S}'$. This operator is continuous from $(\mathcal{S})^*_\beta$ into itself. But by Theorem 9.15, we have $\tilde{D}_\eta D_y^* = \langle y, \eta \rangle I + D_y^* \tilde{D}_\eta$. Thus the expression for this operator can be rewritten as

$$\tilde{D}_\eta D_y^* = \langle y, \eta \rangle I + \int_T \int_T y(s) \eta(t) \partial_s^* \partial_t \, dsdt.$$

In view of the above different cases, it is natural to consider operators of the forms

$$\Xi_{0,2}(\theta) = \int_T \int_T \theta(t_1, t_2) \partial_{t_1} \partial_{t_2} \, dt_1 dt_2,$$

$$\Xi_{2,0}(\theta) = \int_T \int_T \theta(s_1, s_2) \partial_{s_1}^* \partial_{s_2}^* \, ds_1 ds_2,$$

$$\Xi_{1,1}(\theta) = \int_T \int_T \theta(s, t) \partial_s^* \partial_t \, dsdt,$$

where the kernel function $\theta \in (\mathcal{S}')^{\otimes 2}$. We would expect these linear operators to be continuous from $(\mathcal{S})_\beta$ into $(\mathcal{S})_\beta^*$. They may extend by continuity to continuous linear operators from $(\mathcal{S})_\beta^*$ into itself (e.g., $\Xi_{0,2}(\theta)$ with $\theta = \xi \otimes \eta$, $\xi, \eta \in \mathcal{S}$). On the other hand, they may be restricted to $(\mathcal{S})_\beta$ to be continuous linear operators from $(\mathcal{S})_\beta$ into itself (e.g., $\Xi_{2,0}(\theta)$ with $\theta = \xi \otimes \eta$, $\xi, \eta \in \mathcal{S}$).

More generally, we can consider integral kernel operators of the form

$$\Xi_{j,k}(\theta) = \int_{\mathbf{s} \in T^j} \int_{\mathbf{t} \in T^k} \theta(\mathbf{s}, \mathbf{t}) \partial_{\mathbf{s}}^* \partial_{\mathbf{t}} \, d\mathbf{s} d\mathbf{t}, \tag{10.1}$$

where $\theta \in (\mathcal{S}')^{\otimes (j+k)}$. Of course, we may consider multiple products of the operators $\partial_{\mathbf{s}}^*$ and $\partial_{\mathbf{t}}$ in any mixed order. But by the discussion in Case 4 above, we can always group the differential operators and the adjoint operators separately. Thus without loss of generality we need to study only integral kernel operators of the form in equation (10.1). We will show that this informal expression $\Xi_{j,k}(\theta)$ defines a continuous linear operator from $(\mathcal{S})_\beta$ into $(\mathcal{S})_\beta^*$. Moreover, we will find conditions on θ so that $\Xi_{j,k}(\theta)$ is a continuous linear operator from $(\mathcal{S})_\beta$ into itself. On the other hand, we will also find conditions on θ so that $\Xi_{j,k}(\theta)$ extends by continuity to a continuous linear operator from $(\mathcal{S})_\beta^*$ into itself.

10.2 Integral kernel operators

Suppose $\varphi = \sum_{n=0}^{\infty} \langle : \cdot^{\otimes n} :, f_n \rangle \in (\mathcal{S})_\beta$. By Corollary 9.5, the white noise derivative $\partial_t \varphi$ is a continuous function of $t \in T$. However, recall that $\partial_t \varphi$ is given by

$$\partial_t \varphi = \sum_{n=1}^{\infty} n \langle : \cdot^{\otimes (n-1)} :, f_n(t, \cdot) \rangle.$$

Since $f_n \in \mathcal{S}_c^{\otimes n}$, it is reasonable to expect that $\langle\!\langle \Phi, \partial_t \varphi \rangle\!\rangle$ as a function of t is in $\mathcal{S}$ for any $\Phi \in (\mathcal{S})_\beta^*$. Thus we expect the integral in equation (10.1) to be meaningful as the pairing of a generalized function θ and a test function. We will show that this is indeed true. But first we need a lemma.

Lemma 10.1. Let $\varphi, \psi \in (\mathcal{S})_\beta$ and let

$$h(\mathbf{s}, \mathbf{t}) \equiv \langle\!\langle \partial_{\mathbf{t}} \varphi, \partial_{\mathbf{s}} \psi \rangle\!\rangle, \quad \mathbf{s} \in T^j, \ \mathbf{t} \in T^k.$$

Then $h \in \mathcal{S}_c^{\otimes(j+k)}$. In fact, for any $p \geq 0$ such that $\lambda_1^{2p} \geq 2^{1-\beta}$, the following inequality holds:

$$|h|_p \leq (j!k!2^{j+k})^{\frac{1-\beta}{2}} \|\varphi\|_{p,\beta} \|\psi\|_{p,\beta}.$$

Proof. Let φ and ψ be represented by

$$\varphi = \sum_{n=0}^{\infty} \langle :\cdot^{\otimes n}:, f_n \rangle, \quad \psi = \sum_{n=0}^{\infty} \langle :\cdot^{\otimes n}:, g_n \rangle.$$

It is easy to check that

$$\partial_{\mathbf{t}} \varphi = \sum_{n=0}^{\infty} \frac{(n+k)!}{n!} \langle :\cdot^{\otimes n}:, f_{n+k}(\mathbf{t}, \cdot) \rangle,$$

$$\partial_{\mathbf{s}} \psi = \sum_{n=0}^{\infty} \frac{(n+j)!}{n!} \langle :\cdot^{\otimes n}:, g_{n+j}(\mathbf{s}, \cdot) \rangle.$$

Thus we get right away

$$h(\mathbf{s}, \mathbf{t}) = \sum_{n=0}^{\infty} \frac{(n+k)!(n+j)!}{n!} \langle f_{n+k}(\mathbf{t}, \cdot), g_{n+j}(\mathbf{s}, \cdot) \rangle. \tag{10.2}$$

Let $\rho_n(\mathbf{s}, \mathbf{t}) \equiv \langle f_{n+k}(\mathbf{t}, \cdot), g_{n+j}(\mathbf{s}, \cdot) \rangle$. Then

$$|\rho_n|_p = |(A^p)^{\otimes(j+k)} \langle f_{n+k}(\mathbf{t}, \cdot), g_{n+j}(\mathbf{s}, \cdot) \rangle|_0$$

$$\leq |((A^p)^{\otimes k} \otimes I^{\otimes n}) f_{n+k}|_0 |((A^p)^{\otimes j} \otimes I^{\otimes n}) g_{n+j}|_0.$$

Now we use the fact that $\|A^{-1}\| = \lambda_1$ to obtain

$$|((A^p)^{\otimes k} \otimes I^{\otimes n}) f_{n+k}|_0 = |(I^{\otimes k} \otimes (A^{-p})^{\otimes n})(A^p)^{\otimes(n+k)} f_{n+k}|_0$$

$$\leq \lambda_1^{-np} |f_{n+k}|_p.$$

Similarly, we get

$$|((A^p)^{\otimes j} \otimes I^{\otimes n}) g_{n+j}|_0 \leq \lambda_1^{-np} |g_{n+j}|_p.$$

Therefore

$$|\rho_n|_p \leq \lambda_1^{-2np} |f_{n+k}|_p |g_{n+j}|_p. \tag{10.3}$$

It follows from equations (10.2) and (10.3) that

$$|h|_p \leq \sum_{n=0}^{\infty} \frac{(n+k)!(n+j)!}{n!} \lambda_1^{-2np} |f_{n+k}|_p |g_{n+j}|_p$$

$$= \sum_{n=0}^{\infty} C_n ((n+k)!)^{\frac{1+\beta}{2}} |f_{n+k}|_p ((n+j)!)^{\frac{1+\beta}{2}} |g_{n+j}|_p, \tag{10.4}$$

where $C_n = \frac{1}{n!} ((n+k)!(n+j)!)^{\frac{1-\beta}{2}} \lambda_1^{-2np}$. By using the fact that $(n+k)! \leq 2^{n+k} n! k!$ (similarly for j), we can easily check that

$$C_n \leq (j!k!2^{j+k})^{\frac{1-\beta}{2}} \frac{1}{(n!)^{\beta}} (2^{1-\beta} \lambda_1^{-2p})^n.$$

But $2^{1-\beta} \lambda_1^{-2p} \leq 1$ by assumption and $(n!)^{\beta} \geq 1$. Thus we have

$$C_n \leq (j!k!2^{j+k})^{\frac{1-\beta}{2}}. \tag{10.5}$$

From the inequalities in (10.4) and (10.5), we get immediately that

$$|h|_p \leq (j!k!2^{j+k})^{\frac{1-\beta}{2}} \sum_{n=0}^{\infty} ((n+k)!)^{\frac{1+\beta}{2}} |f_{n+k}|_p ((n+j)!)^{\frac{1+\beta}{2}} |g_{n+j}|_p.$$

Thus by the Schwarz inequality, we have

$$|h|_p^2 \leq (j!k!2^{j+k})^{1-\beta} \left(\sum_{n=0}^{\infty} ((n+k)!)^{1+\beta} |f_{n+k}|_p^2 \right) \left(\sum_{n=0}^{\infty} ((n+j)!)^{1+\beta} |g_{n+j}|_p^2 \right)$$

$$\leq (j!k!2^{j+k})^{1-\beta} \|\varphi\|_{p,\beta}^2 \|\psi\|_{p,\beta}^2. \qquad \square$$

Theorem 10.2. Let $\theta \in (S_c')^{\otimes(j+k)}$. Then there exists a unique continuous linear operator $\Xi_{j,k}(\theta)$ from $(S)_{\beta}$ into $(S)_{\beta}^*$ such that

$$\langle\!\langle \Xi_{j,k}(\theta)\varphi, \psi \rangle\!\rangle = \int_{\mathbf{s}\in T^j} \int_{\mathbf{t}\in T^k} \theta(\mathbf{s}, \mathbf{t}) \langle\!\langle \partial_{\mathbf{s}}^* \partial_{\mathbf{t}} \varphi, \psi \rangle\!\rangle \, d\mathbf{s} d\mathbf{t}, \quad \forall \varphi, \psi \in (S)_{\beta},$$

where the integral is interpreted as the bilinear pairing of $(S_c')^{\otimes(j+k)}$ and $S_c^{\otimes(j+k)}$. For any $p \geq 0$ such that $\lambda_1^{2p} \geq 2^{1-\beta}$, the inequality holds for all $\varphi \in (S)_{\beta}$:

$$\|\Xi_{j,k}(\theta)\varphi\|_{-p,-\beta} \leq (j!k!2^{j+k})^{\frac{1-\beta}{2}} |\theta|_{-p} \|\varphi\|_{p,\beta}.$$

Remark. Let $S_{j,k}\theta$ denote the symmetrization of θ with respect to the first j variables and the other k variables separately, i.e.,

$$S_{j,k}\theta = \frac{1}{j!k!}\sum_{\sigma,\rho}\theta_{\sigma,\rho},$$

where the summation is over all permutations σ of $\{1,\ldots,j\}$ and ρ of $\{1,\ldots,k\}$ and $\theta_{\sigma,\rho}$ is given by

$$\theta_{\sigma,\rho}(s_1,\ldots,s_j,t_1,\ldots,t_k) = \theta(s_{\sigma(1)},\ldots,s_{\sigma(j)},t_{\rho(1)},\ldots,t_{\rho(k)}).$$

In view of the fact that $[\partial_s,\partial_t] = [\partial_s^*,\partial_t^*] = 0$ (see the remark of Theorem 9.15), we see immediately that

$$\Xi_{j,k}(\theta) = \Xi_{j,k}(S_{j,k}\theta).$$

Proof. Let $\varphi,\psi \in (\mathcal{S})_\beta$. Then by Lemma 10.1 the function $h(s,t) \equiv \langle\!\langle\partial_t\varphi,\partial_s\psi\rangle\!\rangle$ belongs to $\mathcal{S}_c^{\otimes(j+k)}$. Thus we can define a bilinear functional B on $(\mathcal{S})_\beta \times (\mathcal{S})_\beta$ by

$$B(\varphi,\psi) = \int_{s\in T^j}\int_{t\in T^k}\theta(s,t)\langle\!\langle\partial_s^*\partial_t\varphi,\psi\rangle\!\rangle\,dsdt, \quad \varphi,\psi \in (\mathcal{E})_\beta,$$

where the integral is interpreted as the bilinear pairing of $(\mathcal{S}_c')^{\otimes(j+k)}$ and $\mathcal{S}_c^{\otimes(j+k)}$. By Lemma 10.1 we have

$$|B(\varphi,\psi)| \le (j!k!2^{j+k})^{\frac{1-\beta}{2}}|\theta|_{-p}\|\varphi\|_{p,\beta}\|\psi\|_{p,\beta}. \tag{10.6}$$

This shows that the bilinear functional B is continuous. Thus by the abstract kernel theorem (Fact 2.5) there exists a continuous linear operator, denoted by $\Xi_{j,k}(\theta)$, such that

$$B(\varphi,\psi) = \langle\!\langle\Xi_{j,k}(\theta)\varphi,\psi\rangle\!\rangle, \quad \varphi,\psi \in (\mathcal{S})_\beta.$$

This operator $\Xi_{j,k}(\theta)$ is obviously unique. Moreover, by the inequality in (10.6)

$$\|\Xi_{j,k}(\theta)\varphi\|_{-p,-\beta} \le (j!k!2^{j+k})^{\frac{1-\beta}{2}}|\theta|_{-p}\|\varphi\|_{p,\beta}. \qquad \square$$

Definition 10.3. The operator $\Xi_{j,k}(\theta)$ is called the *integral kernel operator* associated with $\theta \in (\mathcal{S}_c')^{\otimes(j+k)}$. It is symbolically written as

$$\Xi_{j,k}(\theta) = \int_{s\in T^j}\int_{t\in T^k}\theta(s,t)\partial_s^*\partial_t\,dsdt.$$

For any $\theta \in (S'_c)^{\otimes(j+k)}$, the associated integral kernel operator $\Xi_{j,k}(\theta)$ is continuous from $(S)_\beta$ into $(S)^*_\beta$. In fact, there exist $p, q \geq 0$ such that $\Xi_{j,k}(\theta)$ is a Hilbert-Schmidt operator from $(S_p)_\beta$ into $(S_q)^*_\beta$ (see the abstract kernel theorem in Fact 2.5).

Now, we raise the following two questions:

Question 1. When is $\Xi_{j,k}(\theta)$ a continuous linear operator from $(S)_\beta$ into itself?

Question 2. When does $\Xi_{j,k}(\theta)$ extend by continuity to a continuous linear operator from $(S)^*_\beta$ into itself?

The answers to these questions depend on the kernel function θ. As a clue to find the answers, recall that from Cases 1 and 2 in §10.1:

(1) For $\theta(t_1, t_2) = x(t_1)y(t_2) \in S'(T^{0+2})$ and $\theta(s_1, s_2) = \xi(s_1)\eta(s_2) \in S(T^{2+0})$, the associated operators are continuous from $(S)_\beta$ into itself.
(2) For $\theta(t_1, t_2) = \xi(t_1)\eta(t_2) \in S(T^{0+2})$ and $\theta(s_1, s_2) = x(s_1)y(s_2) \in S'(T^{2+0})$, the associated operators extend by continuity to continuous linear operators from $(S)^*_\beta$ into itself.

Thus we can expect that when the **s**-variable part of θ is a "test" function, then the associated integral kernel operator is continuous from $(S)_\beta$ into itself. On the other hand, when the **t**-variable part of θ is a "test" function, then the associated integral kernel operator extends by continuity to a continuous linear operator from $(S)^*_\beta$ into itself.

Notation. For convenience we will use $|u|_{\{p,q\}}$ to denote the following norm of a function $u(\mathbf{s}, \mathbf{t}), \mathbf{s} \in T^j, \mathbf{t} \in T^k$ with indices $p, q \in \mathbb{R}$:

$$|u|_{\{p,q\}} \equiv |(A^p)^{\otimes j} \otimes (A^q)^{\otimes k} u|_0.$$

Lemma 10.4. Let $h(\mathbf{s}, \mathbf{t}) \equiv \langle\!\langle \partial_t \varphi, \partial_s \psi \rangle\!\rangle$, $\mathbf{s} \in T^j$, $\mathbf{t} \in T^k$. Then for any $p \geq 0$ and $q > p$ such that $\lambda_1^{q-p} \geq 2$, the inequality holds for all $\varphi, \psi \in (S)_\beta$:

$$|h|_{\{-p,q\}} \leq (j! 2^j)^{\frac{1+\beta}{2}} (k! 2^k)^{\frac{1-\beta}{2}} \|\varphi\|_{q,\beta} \|\psi\|_{-p,-\beta}.$$

Proof. We need only to modify the proof of Lemma 10.1. Suppose

$$\varphi = \sum_{n=0}^{\infty} \langle :\cdot^{\otimes n}:, f_n \rangle, \quad \psi = \sum_{n=0}^{\infty} \langle :\cdot^{\otimes n}:, g_n \rangle.$$

Let $\rho_n(\mathbf{s}, \mathbf{t}) \equiv \langle f_{n+k}(\mathbf{t}, \cdot), g_{n+j}(\mathbf{s}, \cdot) \rangle$. Then

$$|\rho_n|_{\{-p,q\}} = |(A_t^q)^{\otimes k} \otimes (A_s^{-p})^{\otimes j} \langle f_{n+k}(\mathbf{t}, \cdot), g_{n+j}(\mathbf{s}, \cdot) \rangle|_0$$

$$= |(A_t^q)^{\otimes k} \otimes (A_s^{-p})^{\otimes j} \langle (A^p)^{\otimes n} f_{n+k}(\mathbf{t}, \cdot), (A^{-p})^{\otimes n} g_{n+j}(\mathbf{s}, \cdot) \rangle|_0$$

$$= |\langle (A^q)^{\otimes k} \otimes (A^p)^{\otimes n} f_{n+k}, (A^{-p})^{\otimes(n+j)} g_{n+j} \rangle|_0,$$

where we have used subscripts s and t to indicate the variables that the operator A is being applied to. Therefore, by the Schwarz inequality and the fact that $\|A^{-1}\| = \lambda_1$, we get

$$
\begin{aligned}
|\rho_n|_{\{-p,q\}} &\leq |(A^q)^{\otimes k} \otimes (A^p)^{\otimes n} f_{n+k}|_0 \, |(A^{-p})^{\otimes(n+j)} g_{n+j}|_0 \\
&= |(I^{\otimes k} \otimes (A^{-(q-p)})^{\otimes n})(A^q)^{\otimes(n+k)} f_{n+k}|_0 \, |g_{n+j}|_{-p} \\
&\leq \lambda_1^{-(q-p)n} |f_{n+k}|_q |g_{n+j}|_{-p}.
\end{aligned}
\tag{10.7}
$$

It follows from equation (10.2) and inequality (10.7) that

$$
\begin{aligned}
|h|_{\{-p,q\}} &\leq \sum_{n=0}^{\infty} \frac{(n+k)!(n+j)!}{n!} \lambda_1^{-(q-p)n} |f_{n+k}|_q \, |g_{n+j}|_{-p} \\
&= \sum_{n=0}^{\infty} C_n ((n+k)!)^{\frac{1+\beta}{2}} |f_{n+k}|_q ((n+j)!)^{\frac{1-\beta}{2}} |g_{n+j}|_{-p},
\end{aligned}
$$

where the constant C_n is given by

$$
C_n = \frac{1}{n!} ((n+k)!)^{\frac{1-\beta}{2}} ((n+j)!)^{\frac{1+\beta}{2}} \lambda_1^{-(q-p)n}.
$$

But $(n+k)! \leq 2^{n+k} n! k!$ (similarly for j) and q satisfies the condition $\lambda_1^{q-p} \geq 2$. Hence

$$
\begin{aligned}
C_n &\leq (j! 2^j)^{\frac{1+\beta}{2}} (k! 2^k)^{\frac{1-\beta}{2}} (2\lambda_1^{-(q-p)})^n \\
&\leq (j! 2^j)^{\frac{1+\beta}{2}} (k! 2^k)^{\frac{1-\beta}{2}}.
\end{aligned}
$$

Thus as in the proof of Lemma 10.1 we get right away that

$$
|h|_{\{-p,q\}} \leq (j! 2^j)^{\frac{1+\beta}{2}} (k! 2^k)^{\frac{1-\beta}{2}} \|\varphi\|_{q,\beta} \|\psi\|_{-p,-\beta}. \qquad \square
$$

Theorem 10.5. *If* $\theta \in \mathcal{S}_c^{\otimes j} \otimes (\mathcal{S}_c')^{\otimes k}$, *then the operator* $\Xi_{j,k}(\theta)$ *is a continuous linear operator from* $(\mathcal{S})_\beta$ *into itself. In fact, for any* $p \geq 0$ *and any* $q > p$ *such that* $\lambda_1^{q-p} \geq 2$, *the inequality holds for all* $\varphi \in (\mathcal{S})_\beta$

$$
\|\Xi_{j,k}(\theta)\varphi\|_{p,\beta} \leq (j! 2^j)^{\frac{1+\beta}{2}} (k! 2^k)^{\frac{1-\beta}{2}} |\theta|_{\{p,-q\}} \|\varphi\|_{q,\beta}.
$$

Conversely, if $\theta \in (\mathcal{S}_c')^{\otimes(j+k)}$ *and the operator* $\Xi_{j,k}(\theta)$ *is continuous from* $(\mathcal{S})_\beta$ *into itself, then* $\theta \in \mathcal{S}_c^{\otimes j} \otimes (\mathcal{S}_c')^{\otimes k}$.

Proof. Suppose $\theta \in \mathcal{S}_c^{\otimes j} \otimes (\mathcal{S}_c')^{\otimes k}$. Then by Lemma 10.4 and the argument in the proof of Theorem 10.2,

$$
\begin{aligned}
&|\langle\!\langle \Xi_{j,k}(\theta)\varphi, \psi \rangle\!\rangle| \\
&\leq (j! 2^j)^{\frac{1+\beta}{2}} (k! 2^k)^{\frac{1-\beta}{2}} |\theta|_{\{p,-q\}} \|\varphi\|_{q,\beta} \|\psi\|_{-p,-\beta}, \quad \forall \varphi, \psi \in (\mathcal{S})_\beta.
\end{aligned}
$$

Thus for any $\varphi \in (\mathcal{S})_\beta$,

$$\|\Xi_{j,k}(\theta)\varphi\|_{p,\beta} \leq (j!2^j)^{\frac{1+\beta}{2}}(k!2^k)^{\frac{1-\beta}{2}}|\theta|_{\{p,-q\}}\|\varphi\|_{q,\beta}.$$

This shows that $\Xi_{j,k}(\theta)\varphi \in (\mathcal{S})_\beta$ for all $\varphi \in (\mathcal{S})_\beta$ and that the operator $\Xi_{j,k}(\theta)$ is continuous from $(\mathcal{S})_\beta$ into itself.

Conversely, suppose $\Xi_{j,k}(\theta)$ is a continuous linear operator from $(\mathcal{S})_\beta$ into itself. Then for any $p \geq 0$, there exist nonnegative constants C and q such that

$$\|\Xi_{j,k}(\theta)\varphi\|_{p,\beta} \leq C\|\varphi\|_{q,\beta}, \quad \forall \varphi \in (\mathcal{S})_\beta.$$

Therefore

$$|\langle\!\langle \Xi_{j,k}(\theta)\varphi, \psi \rangle\!\rangle| \leq C\|\varphi\|_{q,\beta}\|\psi\|_{-p,-\beta}, \quad \forall \varphi, \psi \in (\mathcal{S})_\beta. \tag{10.8}$$

On the other hand, for $\varphi = \langle :\!\cdot^{\otimes k}\!:, \xi\rangle$ and $\psi = \langle :\!\cdot^{\otimes j}\!:, \eta\rangle$, we have

$$\langle\!\langle \partial_s^* \partial_t \varphi, \psi \rangle\!\rangle = j!k!(\eta \otimes \xi)(\mathbf{s}, \mathbf{t}).$$

Thus for this particular φ and ψ,

$$\langle\!\langle \Xi_{j,k}(\theta)\varphi, \psi \rangle\!\rangle = j!k!\langle \theta, \eta \otimes \xi \rangle. \tag{10.9}$$

Note that $\|\varphi\|_{q,\beta} = (k!)^{\frac{1+\beta}{2}}|\xi|_q$ and $\|\psi\|_{-p,-\beta} = (j!)^{\frac{1-\beta}{2}}|\eta|_{-p}$. Hence from (10.8) and (10.9), we get

$$|\langle \theta, \eta \otimes \xi \rangle| \leq C(j!)^{-\frac{1+\beta}{2}}(k!)^{-\frac{1-\beta}{2}}|\xi|_q\,|\eta|_{-p}.$$

This implies that $\theta \in \mathcal{S}_c^{\otimes j} \otimes (\mathcal{S}_c')^{\otimes k}$ and so the theorem is proved. $\qquad\square$

The last theorem provides an answer to Question 1 (preceding Lemma 10.4). As for Question 2, the answer is given in the next theorem.

Theorem 10.6. *If $\theta \in (\mathcal{S}_c')^{\otimes j} \otimes \mathcal{S}_c^{\otimes k}$, then the operator $\Xi_{j,k}(\theta)$ extends by continuity to a continuous linear operator from $(\mathcal{S})_\beta^*$ into itself. Conversely, if $\theta \in (\mathcal{S}_c')^{\otimes(j+k)}$ and the operator $\Xi_{j,k}(\theta)$ extends by continuity to a continuous operator from $(\mathcal{S})_\beta^*$ into itself, then $\theta \in (\mathcal{S}_c')^{\otimes j} \otimes \mathcal{S}_c^{\otimes k}$.*

Proof. Suppose $\theta \in (\mathcal{S}_c')^{\otimes j} \otimes \mathcal{S}_c^{\otimes k}$. Define

$$\gamma(\mathbf{t}, \mathbf{s}) = \theta(\mathbf{s}, \mathbf{t}), \quad \mathbf{t} \in T^k, \mathbf{s} \in T^j.$$

It is easy to check that $\Xi_{j,k}(\theta) = \Xi_{k,j}(\gamma)^*$. The assumption $\theta \in (\mathcal{S}_c')^{\otimes j} \otimes \mathcal{S}_c^{\otimes k}$ implies that $\gamma \in \mathcal{S}_c^{\otimes k} \otimes (\mathcal{S}_c')^{\otimes j}$. Thus by Theorem 10.5, the operator $\Xi_{k,j}(\gamma)$ is continuous from $(\mathcal{S})_\beta$ into itself. Hence the operator $\Xi_{j,k}(\theta)$, being equal to the adjoint $\Xi_{k,j}(\gamma)^*$, is continuous from $(\mathcal{S})_\beta^*$ into itself.

Conversely, suppose $\theta \in (S_c')^{\otimes(j+k)}$ and the operator $\Xi_{j,k}(\theta)$ extends by continuity to a continuous operator from $(S)_\beta^*$ into itself. Since $\Xi_{j,k}(\theta) = \Xi_{k,j}(\gamma)^*$, we can check that $\Xi_{k,j}(\gamma)$ is continuous from $(S)_\beta$ into itself. Hence by Theorem 10.5, $\gamma \in S_c^{\otimes k} \otimes (S_c')^{\otimes j}$, or equivalently, $\theta \in (S_c')^{\otimes j} \otimes S_c^{\otimes k}$. This completes the proof. $\square$

Obviously we get the following corollary from Theorems 10.5 and 10.6.

Corollary 10.7. *If* $\theta \in S_c^{\otimes(j+k)}$, *then the operator* $\Xi_{j,k}(\theta)$ *is continuous from* $(S)_\beta$ *into itself and extends by continuity to a continuous linear operator from* $(S)_\beta^*$ *into itself.*

Remark. The converse is not true. For instance, take θ such that $\theta(\mathbf{s}, \mathbf{t}) = \theta(\mathbf{t}, \mathbf{s})$, $\mathbf{s}, \mathbf{t} \in T^j$ and $\theta \in S_c^{\otimes j} \otimes (S_c')^{\otimes j}$. Then the operator $\Xi_{j,j}(\theta)$ is continuous from $(S)_\beta$ into itself and extends by continuity to a continuous linear operator from $(S)_\beta^*$ into itself.

10.3 Gross Laplacian and number operator

For $\theta \in S_c'$, we have integral kernel operators $\Xi_{0,1}(\theta)$ and $\Xi_{1,0}(\theta)$. In view of the informal expressions for D_y and D_y^* with $y \in S'$ in §10.1, we have $\Xi_{0,1}(\theta) = D_\theta$ and $\Xi_{1,0}(\theta) = D_\theta^*$. (Note that the differential operator D_w for $w \in \mathcal{E}_c'$ can be defined on $(\mathcal{E})_\beta$ as in §9.1 by taking analytic extensions of test functions.)

Suppose $\theta \in (S_c')^{\otimes 2}$. Then there are three associated integral kernel operators, i.e., $\Xi_{0,2}(\theta)$, $\Xi_{2,0}(\theta)$, and $\Xi_{1,1}(\theta)$. An important case of such a generalized function θ is the trace operator τ. The corresponding integral kernel operators can be identified with the Gross Laplacian, its adjoint, and the number operator. We will study these operators in this section.

First we define the Gross Laplacian Δ_G for the Gel'fand triple $(\mathcal{E})_\beta \subset (L^2) \subset (\mathcal{E})_\beta^*$ which is constructed from the Gel'fand triple $\mathcal{E} \subset E \subset \mathcal{E}'$.

Definition 10.8. A bilinear functional B on $E \times E$ is said to have *finite trace* if for any orthonormal basis $\{e_k; k \geq 1\}$ for E, the series $\sum_{k=1}^{\infty} B(e_k, e_k)$ converges and the sum is independent of the orthonormal basis $\{e_k; k \geq 1\}$. The sum is denoted by $\text{trace}_E B$.

Definition 10.9. A function φ on $\mathcal{E}'$ is said to have *Gross Laplacian* at $x \in \mathcal{E}'$ if the function $\varphi(x + h)$ is twice differentiable on $h \in E$ and its second derivative at 0, denoted by $D^2\varphi(x)$, has finite trace. In this case, the Gross Laplacian $\Delta_G\varphi(x)$ at x is defined by

$$\Delta_G\varphi(x) = \text{trace}_E D^2\varphi(x).$$

As can be expected, a test function in $(\mathcal{E})_\beta$ has Gross Laplacian at every $x \in \mathcal{E}'$. To prove this fact, we need a lemma.

Lemma 10.10. *Let* $\varphi(x) = \sum_{n=0}^{\infty} \langle :x^{\otimes n}:, f_n \rangle \in (\mathcal{E})_\beta$. *Then for any* $x \in \mathcal{E}'$ *and any orthonormal basis* $\{e_k; k \geq 1\}$ *for* E,

$$\sum_{k=1}^{\infty} \sum_{n=0}^{\infty} (n+2)(n+1) |\langle :x^{\otimes n}:, \langle e_k \otimes e_k, f_{n+2} \rangle \rangle| < \infty.$$

Proof. First note that for any $p \geq 0$,

$$\begin{aligned}
|\langle e_k \otimes e_k, f_{n+2} \rangle|_p &= |(A^p)^{\otimes n} \langle e_k \otimes e_k, f_{n+2} \rangle|_0 \\
&= |(A^p)^{\otimes n} \langle (A^{-p})^{\otimes 2} e_k \otimes e_k, (A^p)^{\otimes 2} f_{n+2} \rangle|_0 \\
&\leq |A^{-p} e_k|_0^2 |f_{n+2}|_p.
\end{aligned} \tag{10.10}$$

For any fixed $x \in \mathcal{E}'$, there exists $p > \frac{\alpha}{2}$ such that $|x|_{-p} < \infty$. Then by Lemma 7.10 and inequality (10.10)

$$|\langle :x^{\otimes n}:, \langle e_k \otimes e_k, f_{n+2} \rangle \rangle| \leq \sqrt{n!} \left(|x|_{-p} + |\tau|_{-p}^{1/2} \right)^n |A^{-p} e_k|_0^2 |f_{n+2}|_p.$$

Let L denote the sum of the series in the lemma. Then for any $q > p$,

$$\begin{aligned}
L &\leq \sum_{k=1}^{\infty} \sum_{n=0}^{\infty} (n+2)(n+1) \sqrt{n!} \left(|x|_{-p} + |\tau|_{-p}^{1/2} \right)^n |A^{-p} e_k|_0^2 |f_{n+2}|_p \\
&= \|A^{-p}\|_{HS}^2 \lambda_1^{-2(q-p)} \sum_{n=0}^{\infty} (n+2)(n+1) \sqrt{n!} \, C^n |f_{n+2}|_q,
\end{aligned}$$

where $C = \left(|x|_{-p} + |\tau|_p^{1/2} \right) \lambda_1^{-(q-p)}$. By using the inequality $(n+2)(n+1) \leq 2^{n+2}$, we get

$$(n+2)(n+1) \sqrt{n!} \, C^n \leq 2^{1-\beta} \left(2^{\frac{1-\beta}{2}} C \right)^n \left((n+2)! \right)^{\frac{1+\beta}{2}}.$$

Thus for q large enough such that $2^{\frac{1-\beta}{2}} C < 1$,

$$\begin{aligned}
L &\leq \|A^{-p}\|_{HS}^2 \lambda_1^{-2(q-p)} 2^{1-\beta} \sum_{n=0}^{\infty} \left(2^{\frac{1-\beta}{2}} C \right)^n \left((n+2)! \right)^{\frac{1+\beta}{2}} |f_{n+2}|_q \\
&\leq \|A^{-p}\|_{HS}^2 \lambda_1^{-2(q-p)} 2^{1-\beta} \left[1 - \left(2^{\frac{1-\beta}{2}} C \right)^2 \right]^{-1/2} \|\varphi\|_{q,\beta} \\
&< \infty.
\end{aligned}$$

$\square$

Theorem 10.11. Let $\varphi(x) = \sum_{n=0}^{\infty} \langle : x^{\otimes n} :, f_n \rangle \in (\mathcal{E})_\beta$. Then the Gross Laplacian $\Delta_G \varphi(x)$ exists at every $x \in \mathcal{E}'$ and

$$\Delta_G \varphi(x) = \sum_{n=0}^{\infty} (n+2)(n+1) \langle : x^{\otimes n} :, \langle \tau, f_{n+2} \rangle \rangle.$$

Proof. It is easy to check that for any $u, v \in E$,

$$D^2 \varphi(x)(u, v) = \sum_{n=0}^{\infty} (n+2)(n+1) \langle : x^{\otimes n} :, \langle u \otimes v, f_{n+2} \rangle \rangle.$$

Let $\{e_k; k \geq 1\}$ be any orthonormal basis for E. Then by Lemma 10.10, we can interchange the order of summation to get

$$\text{trace}_E \, D^2 \varphi(x) = \sum_{k=1}^{\infty} \sum_{n=0}^{\infty} (n+2)(n+1) \langle : x^{\otimes n} :, \langle e_k \otimes e_k, f_{n+2} \rangle \rangle$$

$$= \sum_{n=0}^{\infty} (n+2)(n+1) \langle : x^{\otimes n} :, \langle \tau, f_{n+2} \rangle \rangle.$$

Thus $\text{trace}_E \, D^2 \varphi(x)$ exists and is independent of the orthonormal basis $\{e_k; k \geq 1\}$. Hence $\Delta_G \varphi(x)$ exists at any $x \in \mathcal{E}'$ and is given by

$$\Delta_G \varphi(x) = \text{trace}_E \, D^2 \varphi(x)$$

$$= \sum_{n=0}^{\infty} (n+2)(n+1) \langle : x^{\otimes n} :, \langle \tau, f_{n+2} \rangle \rangle. \qquad \square$$

Theorem 10.12. *The Gross Laplacian Δ_G is a continuous linear operator from $(\mathcal{E})_\beta$ into itself. In fact, for any $p \geq 0$ and $q > p$ such that $\lambda_1^{2(q-p)} \geq 2^{1-\beta}$, the inequality holds for all $\varphi \in (\mathcal{E})_\beta$:*

$$\|\Delta_G \varphi\|_{p,\beta} \leq |\tau|_{-p} 2^{1-\beta} \lambda_1^{-2(q-p)} \|\varphi\|_{q,\beta}.$$

Proof. By Theorem 10.11, we have

$$\|\Delta_G \varphi\|_{p,\beta}^2 = \sum_{n=0}^{\infty} (n!)^{1+\beta} (n+2)^2 (n+1)^2 |\langle \tau, f_{n+2} \rangle|_p^2.$$

Note that for any $q > p$

$$|\langle \tau, f_{n+2} \rangle|_p = |(A^p)^{\otimes n} \langle \tau, f_{n+2} \rangle|_0$$

$$= |(A^p)^{\otimes n} \langle (A^{-p})^{\otimes 2} \tau, (A^p)^{\otimes 2} f_{n+2} \rangle|_0$$

$$\leq |\tau|_{-p} |f_{n+2}|_p$$

$$\leq |\tau|_{-p} \lambda_1^{-(q-p)(n+2)} |f_{n+2}|_q.$$

Hence, by using the inequality $(n + 2)(n + 1) \leq 2^{n+2}$, we get immediately that

$$\|\Delta_G \varphi\|_{p,\beta}^2$$

$$\leq |\tau|_{-p}^2 \sum_{n=0}^{\infty} (n!)^{1+\beta}(n+2)^2(n+1)^2 \lambda_1^{-2(q-p)(n+2)} |f_{n+2}|_q^2$$

$$\leq |\tau|_{-p}^2 2^{2(1-\beta)} \lambda_1^{-4(q-p)} \sum_{n=0}^{\infty} \left(2^{1-\beta} \lambda_1^{-2(q-p)}\right)^n \left((n+2)!\right)^{1+\beta} |f_{n+2}|_q^2.$$

Therefore, if $q > p$ such that $\lambda_1^{2(q-p)} \geq 2^{1-\beta}$, then

$$\|\Delta_G \varphi\|_{p,\beta}^2 \leq |\tau|_{-p}^2 2^{2(1-\beta)} \lambda_1^{-4(q-p)} \sum_{n=0}^{\infty} \left((n+2)!\right)^{1+\beta} |f_{n+2}|_q^2$$

$$\leq |\tau|_{-p}^2 2^{2(1-\beta)} \lambda_1^{-4(q-p)} \|\varphi\|_{q,\beta}^2. \qquad \square$$

Now, take the Gel'fand triple $(\mathcal{S})_\beta \subset (L^2) \subset (\mathcal{S})_\beta^*$. In this case, the Gross Laplacian can be identified as an integral kernel operator. First note that $\tau \in (\mathcal{S}_c')^{\otimes 2}$. Hence by Theorem 10.5 the integral kernel operator $\Xi_{0,2}(\tau)$ is a continuous linear operator from $(\mathcal{S})_\beta$ into itself. This operator will be denoted symbolically as

$$\Xi_{0,2}(\tau) = \int_T \int_T \tau(t_1, t_2) \partial_{t_1} \partial_{t_2} \, dt_1 dt_2 = \int_T \partial_t^2 \, dt.$$

Theorem 10.13. *The Gross Laplacian* Δ_G *coincides with the integral kernel operator* $\Xi_{0,2}(\tau)$ *as an operator from* $(\mathcal{S})_\beta$ *into itself, i.e.,*

$$\Delta_G = \int_T \partial_t^2 \, dt.$$

Proof. Let $\varphi_\xi =: e^{\langle \cdot, \xi \rangle} :, \xi \in \mathcal{S}_c$. Then $\varphi_\xi = \sum_{n=0}^{\infty} \frac{1}{n!} \langle :\cdot^{\otimes n}:, \xi^{\otimes n} \rangle$ and by Theorem 10.11

$$\Delta_G \varphi_\xi = \sum_{n=0}^{\infty} (n+2)(n+1) \frac{1}{(n+2)!} \langle :\cdot^{\otimes n}:, \langle \tau, \xi^{\otimes(n+2)} \rangle \rangle$$

$$= \langle \xi, \xi \rangle \varphi_\xi.$$

Thus for any $\xi, \eta \in \mathcal{S}_c$,

$$\langle\!\langle \Delta_G \varphi_\xi, \varphi_\eta \rangle\!\rangle = \langle \xi, \xi \rangle \langle\!\langle \varphi_\xi, \varphi_\eta \rangle\!\rangle = \langle \xi, \xi \rangle e^{\langle \eta, \xi \rangle}. \qquad (10.11)$$

On the other hand, it is easy to check that $\partial_{t_1}\partial_{t_2}\varphi_\xi = \xi(t_1)\xi(t_2)\varphi_\xi$. This yields that

$$
\begin{aligned}
\langle\!\langle \Xi_{0,2}(\tau)\varphi_\xi, \varphi_\eta \rangle\!\rangle &= \int_T \int_T \tau(t_1, t_2)\langle\!\langle \partial_{t_1}\partial_{t_2}\varphi_\xi, \varphi_\eta \rangle\!\rangle \, dt_1 dt_2 \\
&= \int_T \int_T \tau(t_1, t_2)\xi(t_1)\xi(t_2)e^{\langle \eta, \xi \rangle} \, dt_1 dt_2 \\
&= \langle \xi, \xi \rangle e^{\langle \eta, \xi \rangle}.
\end{aligned}
\tag{10.12}
$$

By equations (10.11) and (10.12), $\langle\!\langle \Delta_G\varphi_\xi, \varphi_\eta \rangle\!\rangle = \langle\!\langle \Xi_{0,2}(\tau)\varphi_\xi, \varphi_\eta \rangle\!\rangle$ for all $\xi, \eta \in \mathcal{S}_c$. But the linear span of $\{\varphi_\eta; \eta \in \mathcal{S}_c\}$ is dense in $(\mathcal{S})^*_\beta$. Hence $\Delta_G\varphi_\xi = \Xi_{0,2}(\tau)\varphi_\xi$ for all $\xi \in \mathcal{S}_c$. Thus $\Delta_G = \Xi_{0,2}(\tau)$ since both operators are continuous from $(\mathcal{S})_\beta$ into itself. $\qquad\square$

Now, consider the operator $\Xi_{2,0}(\tau)$. This operator can be written symbolically as

$$
\Xi_{2,0}(\tau) = \int_T \int_T \tau(s_1, s_2)\partial^*_{t_1}\partial^*_{t_2} \, dt_1 dt_2 = \int_T (\partial^*_s)^2 \, ds.
$$

By Theorem 10.6 the operator $\Xi_{2,0}(\tau)$ extends by continuity to a continuous linear operator from $(\mathcal{S})^*_\beta$ into itself. This extension is simply the adjoint Δ^*_G of the Gross Laplacian as given in the next corollary, an immediate consequence of Theorem 10.13.

Corollary 10.14. *The adjoint Δ^*_G of Δ_G coincides with the extension of the integral kernel operator $\Xi_{2,0}(\tau)$ as an operator from $(\mathcal{S})^*_\beta$ into itself, i.e.,*

$$
\Delta^*_G = \int_T (\partial^*_t)^2 \, dt.
$$

Finally we consider the integral kernel operator $\Xi_{1,1}(\tau)$. This operator is written symbolically as

$$
\Xi_{1,1}(\tau) = \int_T \int_T \tau(s, t)\partial^*_s\partial_t \, dsdt = \int_T \partial^*_t\partial_t \, dt.
$$

Lemma 10.15. *The trace operator τ belongs to $\mathcal{S}_c \otimes \mathcal{S}'_c$ and $\mathcal{S}'_c \otimes \mathcal{S}_c$.*

Proof. To prove $\tau \in \mathcal{S}_c \otimes \mathcal{S}'_c$, we need to show that for any $p \geq 0$ there exists some $q \geq 0$ such that $|\tau|_{p, -q} < \infty$. Take the orthonormal basis $\{\zeta_n\}$ for $L^2(T, \nu)$ (see §4.2 for the case $E = L^2(T, \nu)$). Then $\tau = \sum_{n=1}^\infty \zeta_n \otimes \zeta_n$

and so we have

$$|\tau|^2_{p,-q} = \left|(A^p \otimes A^{-q})\tau\right|^2_0$$

$$= \sum_{n=1}^{\infty} \left|(A^p\zeta_n) \otimes (A^{-q}\zeta_n)\right|^2_0$$

$$= \sum_{n=1}^{\infty} \left|\lambda_n^{p-q}(\zeta_n \otimes \zeta_n)\right|^2_0$$

$$= \sum_{n=1}^{\infty} \lambda_n^{2(p-q)}.$$

Thus for any $p \geq 0$, there exists some $q > p$ (e.g., $q = p + \frac{\alpha}{2}$) such that $|\tau|_{p,-q} < \infty$. Hence $\tau \in \mathcal{S}_c \otimes \mathcal{S}'_c$. Note that τ is symmetric. Hence we also have $\tau \in \mathcal{S}'_c \otimes \mathcal{S}_c$. □

Now, it follows from Lemma 10.15 and Theorem 10.5 that the integral kernel operator $\Xi_{1,1}(\tau)$ is continuous from $(\mathcal{S})_\beta$ into itself. On the other hand, by Lemma 10.15 and Theorem 10.6, the operator $\Xi_{1,1}(\tau)$ extends by continuity to a continuous linear operator from $(\mathcal{S})^*_\beta$ into itself.

Theorem 10.16. *The number operator N coincides with the integral kernel operator $\Xi_{1,1}(\tau)$ as an operator from $(\mathcal{S})_\beta$ into itself and as an operator from $(\mathcal{S})^*_\beta$ itself, i.e.,*

$$N = \int_T \partial_t^* \partial_t \, dt.$$

Proof. Let $\varphi_\xi =: e^{\langle \cdot, \xi \rangle} :, \xi \in \mathcal{E}$. We have $\partial_t \varphi_\xi = \xi(t)\varphi_\xi$. Hence for any $\xi, \eta \in \mathcal{S}$.

$$\langle\!\langle \Xi_{1,1}(\tau)\varphi_\xi, \varphi_\eta \rangle\!\rangle = \int_T \int_T \tau(s,t)\langle\!\langle \partial_t\varphi_\xi, \partial_s\varphi_\eta \rangle\!\rangle \, dsdt$$

$$= \langle \eta, \xi \rangle e^{\langle \eta, \xi \rangle}.$$

On the other hand, we have

$$\varphi_\xi = \sum_{n=0}^{\infty} \frac{1}{n!}\langle :\cdot^{\otimes n}:, \xi^{\otimes n}\rangle, \quad N\varphi_\xi = \sum_{n=1}^{\infty} \frac{1}{(n-1)!}\langle :\cdot^{\otimes n}:, \xi^{\otimes n}\rangle.$$

Therefore

$$\langle\!\langle N\varphi_\xi, \varphi_\eta \rangle\!\rangle = \sum_{n=1}^{\infty} n! \frac{1}{n!(n-1)!}\langle \xi^{\otimes n}, \eta^{\otimes n}\rangle$$

$$= \sum_{n=1}^{\infty} \frac{1}{(n-1)!}\langle \eta, \xi \rangle^n$$

$$= \langle \eta, \xi \rangle e^{\langle \eta, \xi \rangle}.$$

Thus we have shown that $\langle\langle N\varphi_\xi, \varphi_\eta\rangle\rangle = \langle\langle \Xi_{1,1}(\tau)\varphi_\xi, \varphi_\eta\rangle\rangle$ for all $\xi, \eta \in \mathcal{S}$. Hence by the same arguments as in the proofs of Theorem 10.13 and Corollary 10.14, we get the assertions of the theorem. $\qquad\qquad\square$

10.4 Lambda operator

There is a relationship between the Gross Laplacian and the number operator. Recall from §9.1 that for any $\varphi = \sum_{n=0}^\infty \langle :\cdot^{\otimes n}:, f_n\rangle \in (\mathcal{E})_\beta$ and $y \in \mathcal{E}'$,

$$D_y\varphi(x) = \sum_{n=1}^\infty n\langle :x^{\otimes(n-1)}:, \langle y, f_n\rangle\rangle$$

$$= \sum_{n=1}^\infty n\langle :x^{\otimes(n-1)}: \widehat\otimes y, f_n\rangle.$$

Thus $D_x\varphi(x)$ is meaningful for any $\varphi \in (\mathcal{E})_\beta$. Note that the finite dimensional version of this expression is $x_1\partial/\partial x_1 + \cdots + x_r\partial/\partial x_r$.

Definition 10.17. For $\varphi(x) = \sum_{n=0}^\infty \langle :x^{\otimes n}:, f_n\rangle \in (\mathcal{E})_\beta$, define

$$\Lambda\varphi(x) = \sum_{n=1}^\infty n\langle :x^{\otimes(n-1)}: \widehat\otimes x, f_n\rangle.$$

The operator Λ acting on $(\mathcal{E})_\beta$ is called the *lambda operator*.

Theorem 10.18. *The lambda operator Λ is a continuous linear operator from $(\mathcal{E})_\beta$ into itself. In fact, $\Lambda = N + \Delta_G$.*

Proof. We can check easily from Definition 5.1 that

$$:x^{\otimes n}: = :x^{\otimes(n-1)}: \widehat\otimes x - (n-1) :x^{\otimes(n-2)}: \widehat\otimes \tau.$$

Let $\varphi = \sum_{n=0}^\infty \langle :\cdot^{\otimes n}:, f_n\rangle \in (\mathcal{E})_\beta$. Then from the above identity we get

$$(\Lambda\varphi)(x) = \sum_{n=1}^\infty n\langle :x^{\otimes n}:, f_n\rangle + \sum_{n=2}^\infty n(n-1)\langle :x^{\otimes(n-2)}:, \langle \tau, f_n\rangle\rangle.$$

Thus by the definition of number operator and Theorem 10.11, we get

$$\Lambda\varphi(x) = N\varphi(x) + \Delta_G\varphi(x).$$

Note that by Theorems 9.23 and 10.12 the number operator N and the Gross Laplacian Δ_G are continuous from $(\mathcal{E})_\beta$ into itself. Hence the lambda operator Λ is also continuous from $(\mathcal{E})_\beta$ into itself. $\qquad\square$

It follows from the above theorem that the adjoint Λ^* of Λ is a continuous linear operator from $(\mathcal{E})_\beta^*$ into itself and we have $\Lambda^* = N + \Delta_G^*$.

Recall from equation (9.5) that $\varphi\Phi$ is defined for any $\varphi \in (\mathcal{E})_\beta$ and $\Phi \in (\mathcal{E})_\beta^*$. For any fixed Φ, the linear operator $\varphi \mapsto \varphi\Phi$ is continuous from $(\mathcal{E})_\beta$ into $(\mathcal{E})_\beta^*$. An important special case is given by $\Phi = \langle :\cdot^{\otimes 2}:, \tau \rangle \in (\mathcal{E})_\beta^*$. We will use Q_τ to denote this operator, i.e.,

$$Q_\tau\varphi = \langle :\cdot^{\otimes 2}:, \tau \rangle \varphi, \quad \varphi \in (\mathcal{E})_\beta.$$

This operator Q_τ is related to the lambda operator Λ and its adjoint Λ^* as continuous linear operators from $(\mathcal{E})_\beta$ into $(\mathcal{E})_\beta^*$ in the next theorem.

Theorem 10.19. *For any $\varphi \in (\mathcal{E})_\beta$, the equality holds:*

$$Q_\tau\varphi = \Lambda\varphi + \Lambda^*\varphi.$$

Proof. It suffices to show that for any $\varphi, \psi \in (\mathcal{E})_\beta$,

$$\langle\!\langle \Lambda\varphi, \psi \rangle\!\rangle + \langle\!\langle \varphi, \Lambda\psi \rangle\!\rangle = \langle\!\langle Q_\tau\varphi, \psi \rangle\!\rangle. \tag{10.13}$$

But by the continuity of the operators Λ and Q_τ, we need only to prove equation (10.13) for $\varphi = \varphi_\xi$ and $\psi = \varphi_\eta$ for any $\xi, \eta \in \mathcal{E}$. Here $\varphi_\xi =\, :e^{\langle \cdot, \xi \rangle}: = \sum_{n=0}^\infty \frac{1}{n!} \langle :x^{\otimes n}:, \xi^{\otimes n} \rangle$. From the definition of Λ, we get

$$\Lambda\varphi_\xi = \sum_{n=1}^\infty n \langle :x^{\otimes(n-1)}: \,\widehat{\otimes}\, x, \frac{1}{n!}\xi^{\otimes n} \rangle$$

$$= \langle x, \xi \rangle \sum_{n=1}^\infty \frac{1}{(n-1)!} \langle :x^{\otimes(n-1)}:, \xi^{\otimes(n-1)} \rangle$$

$$= \langle x, \xi \rangle \varphi_\xi.$$

Thus for any $\xi, \eta \in \mathcal{E}$,

$$\langle\!\langle \Lambda\varphi, \psi \rangle\!\rangle + \langle\!\langle \varphi, \Lambda\psi \rangle\!\rangle = \langle\!\langle \langle \cdot, \xi \rangle \varphi_\xi, \varphi_\eta \rangle\!\rangle + \langle\!\langle \varphi_\xi, \langle \cdot, \xi\eta \rangle \varphi_\eta \rangle\!\rangle$$

$$= \langle\!\langle \langle \cdot, \xi + \eta \rangle \varphi_\xi, \varphi_\eta \rangle\!\rangle$$

$$= e^{-\frac{1}{2}(|\xi|_0^2 + |\eta|_0^2)} \int_{\mathcal{E}'} \langle x, \xi + \eta \rangle e^{\langle x, \xi+\eta \rangle} \, d\mu(x).$$

Then, by Lemma 9.16, we can derive easily that

$$\langle\!\langle \Lambda\varphi_\xi, \varphi_\eta \rangle\!\rangle + \langle\!\langle \varphi_\xi, \Lambda\varphi_\eta \rangle\!\rangle = |\xi + \eta|_0^2 \, e^{\langle \xi, \eta \rangle}. \tag{10.14}$$

On the other hand,

$$\langle\!\langle Q_\tau\varphi_x, \varphi_\eta \rangle\!\rangle = \langle\!\langle \langle :\cdot^{\otimes 2}:, \tau \rangle, \varphi_\xi\varphi_\eta \rangle\!\rangle$$

$$= e^{\langle \xi, \eta \rangle} \langle\!\langle \langle :\cdot^{\otimes 2}:, \tau \rangle, \varphi_{\xi+\eta} \rangle\!\rangle.$$

Hence by equation (5.8) we get

$$\langle\!\langle Q_\tau \varphi_x, \varphi_\eta \rangle\!\rangle = |\xi + \eta|_0^2 \, e^{\langle \xi, \eta \rangle}. \tag{10.15}$$

It follows from equations (10.14) and (10.15) that equation (10.13) holds for all $\varphi = \varphi_\xi$ and $\psi = \varphi_\eta$, $\xi, \eta \in \mathcal{E}$, and the proof is completed. $\square$

The following corollary (for $\beta = 0$) has been proved in Obata [139] without using the lambda operator Λ and its adjoint Λ^*.

Corollary 10.20. *The following equality holds as operators from $(\mathcal{E})_\beta$ into $(\mathcal{E})_\beta^*$*

$$Q_\tau = 2N + \Delta_G + \Delta_G^*.$$

Proof. By Theorem 10.18 we have $\Lambda = N + \Delta_G$. This implies that $\Lambda^* = N + \Delta_G^*$. Hence by Theorem 10.19 $Q_\tau = 2N + \Delta_G + \Delta_G^*$. $\square$

It is worthwhile to point out that by this corollary the operator $Q_\tau - \Delta_G^*$ is a continuous linear operator from $(\mathcal{E})_\beta$ into itself.

10.5 Translation operators

We have studied several operators acting on the space $(\mathcal{E})_\beta$ of test functions. For instance, the differential operator D_y for $y \in \mathcal{E}'$ and the operator D_ξ^* for $\xi \in \mathcal{E}'$. We now introduce another operator, i.e., the translation operator T_y with $y \in \mathcal{E}'$. For $\varphi \in (\mathcal{E})_\beta$, the *translation* $T_y \varphi$ of φ by y is defined by

$$T_y \varphi(x) = \varphi(x + y).$$

Theorem 10.21. *For any $y \in \mathcal{E}'$, the translation operator T_y is continuous from $(\mathcal{E})_\beta$ into itself. In fact, if $y \in \mathcal{E}_p'$ and $q > p$ satisfies $\lambda_1^{2(q-p)} \geq 2^{1-\beta}$, then for all $\varphi \in (\mathcal{E})_\beta$,*

$$\|T_y \varphi\|_{p,\beta}$$
$$\leq \|\varphi\|_{q,\beta} \left(1 - 2^{1-\beta} \lambda_1^{-2(q-p)}\right)^{-1/2} \exp\left[(1+\beta)2^{-\frac{2\beta}{1+\beta}} \lambda_1^{-\frac{2(q-p)}{1+\beta}} |y|_{-p}^{\frac{2}{1+\beta}}\right].$$

Proof. Let $\varphi \in (\mathcal{E})_\beta$ be represented by

$$\varphi(x) = \sum_{n=0}^{\infty} \langle :x^{\otimes n}:, f_n \rangle.$$

Then by using Lemma 7.16 and changing the order of summation (which can be justified by the calculation below), we get

$$\varphi(x+y) = \sum_{n=0}^{\infty} \Big\langle :x^{\otimes n}:, \sum_{k=0}^{\infty} \binom{n+k}{k} \langle y^{\otimes k}, f_{n+k} \rangle \Big\rangle.$$

Thus for any $p \geq 0$ and $q > p$, we have

$$\Big| \sum_{k=0}^{\infty} \frac{(n+k)!}{k!n!} \langle y^{\otimes k}, f_{n+k} \rangle \Big|_p$$

$$\leq \sum_{k=0}^{\infty} \frac{(n+k)!}{k!n!} |y|_{-p}^k |f_{n+k}|_p$$

$$\leq \sum_{k=0}^{\infty} \frac{(n+k)!}{k!n!} |y|_{-p}^k \lambda_1^{-(q-p)(n+k)} |f_{n+k}|_q$$

$$= \sum_{k=0}^{\infty} \Big(((n+k)!)^{\frac{1-\beta}{2}} \frac{1}{k!n!} |y|_{-p}^k \lambda_1^{-(q-p)(n+k)} \Big) \Big(((n+k)!)^{\frac{1+\beta}{2}} |f_{n+k}|_q \Big).$$

Hence by the Schwarz inequality

$$\Big| \sum_{k=0}^{\infty} \frac{(n+k)!}{k!n!} \langle y^{\otimes k}, f_{n+k} \rangle \Big|_p^2$$

$$\leq \|\varphi\|_{q,\beta}^2 \sum_{k=0}^{\infty} ((n+k)!)^{1-\beta} \frac{1}{(k!n!)^2} |y|_{-p}^{2k} \lambda_1^{-2(q-p)(n+k)}.$$

Now we use the inequality $(n+k)! \leq 2^{n+k} k!n!$ to derive that

$$\|T_y\varphi\|_{p,\beta}^2 = \sum_{n=0}^{\infty} (n!)^{1+\beta} \Big| \sum_{k=0}^{\infty} \frac{(n+k)!}{k!n!} \langle y^{\otimes k}, f_{n+k} \rangle \Big|_p^2$$

$$\leq \|\varphi\|_{q,\beta}^2 \sum_{n=0}^{\infty} \sum_{k=0}^{\infty} (n!)^{1+\beta} ((n+k)!)^{1-\beta} \frac{1}{(k!n!)^2} |y|_{-p}^{2k} \lambda_1^{-2(q-p)(n+k)}$$

$$\leq \|\varphi\|_{q,\beta}^2 \sum_{n=0}^{\infty} \sum_{k=0}^{\infty} \big(2^{1-\beta} \lambda_1^{-2(q-p)} \big)^n \frac{1}{(k!)^{1+\beta}} \big(2^{1-\beta} \lambda_1^{-2(q-p)} |y|_{-p}^2 \big)^k.$$

Since q satisfies the condition that $\lambda_1^{2(q-p)} \geq 2^{1-\beta}$, we can sum up the above series in n. Moreover, we can use Lemma 6.6 to estimate the series in k. This yields the inequality in the theorem immediately. $\qquad\square$

It follows from Theorem 10.21 that the adjoint operator T_y^* is continuous from $(\mathcal{E})_\beta^*$ into itself. The next theorem states that the operator T_ξ has a continuous extension to $(\mathcal{E})_\beta^*$ if $\xi \in \mathcal{E}$.

Theorem 10.22. *If $\xi \in \mathcal{E}$, then the translation operator T_ξ extends by continuity to a continuous linear operator $\widetilde{T}_\xi$ from $(\mathcal{E})_\beta^*$ into itself. In fact,*

$$\widetilde{T}_\xi \Phi = T_{-\xi}^* \big(:e^{\langle \cdot, \xi \rangle} : \Phi \big), \quad \Phi \in (\mathcal{E})_\beta^*.$$

Proof. Let $\varphi, \psi \in (\mathcal{E})_\beta$. By the translation formula for μ (see §2.1), we have

$$\langle\!\langle T_\xi \varphi, \psi \rangle\!\rangle = \int_{\mathcal{E}'} \varphi(x + \xi) \psi(x) \, d\mu(x)$$

$$= \int_{\mathcal{E}'} \varphi(y) \psi(y - \xi) \, d\mu(y - \xi)$$

$$= \int_{\mathcal{E}'} \varphi(y) (T_{-\xi} \psi)(y) :e^{\langle y, \xi \rangle} : \, d\mu(y)$$

$$= \langle\!\langle \varphi, (T_{-\xi} \psi) :e^{\langle \cdot, \xi \rangle} : \rangle\!\rangle$$

$$= \langle\!\langle T_{-\xi}^* \big(:e^{\langle \cdot, \xi \rangle} : \varphi \big), \psi \rangle\!\rangle.$$

Thus we have shown that $\langle\!\langle T_\xi \varphi, \psi \rangle\!\rangle = \langle\!\langle T_{-\xi}^* (:e^{\langle \cdot, \xi \rangle} : \varphi), \psi \rangle\!\rangle$ for all $\varphi, \psi \in (\mathcal{E})_\beta$. This implies that $T_\xi \varphi = T_{-\xi}^* (:e^{\langle \cdot, \xi \rangle} : \varphi)$ for all $\varphi \in (\mathcal{E})_\beta^*$. But the operator $T_{-\xi}^*$ and the multiplication operator by $:e^{\langle \cdot, \xi \rangle} :$ are continuous operators from $(\mathcal{E})_\beta^*$ into itself. Thus T_ξ extends by continuity to a continuous linear operator $\widetilde{T}_\xi$ from $(\mathcal{E})_\beta^*$ into itself and $\widetilde{T}_\xi \Phi = T_{-\xi}^* (:e^{\langle \cdot, \xi \rangle} : \Phi)$ for all $\Phi \in (\mathcal{E})_\beta^*$. □

Now we consider the case when $E = L^2(T, \nu)$, i.e., the Gel'fand triple $(\mathcal{S})_\beta \subset (L^2) \subset (\mathcal{S})_\beta^*$. The next theorem shows that the translation operator T_y can be expressed as a series of integral kernel operators. But in order to show the convergence of the series, we need to introduce the following operator norm.

Notation. For a linear operator L from $(\mathcal{S})_\beta$ into itself, define $\|L\|_{\{q,p\}}$ by

$$\|L\|_{\{q,p\}} = \sup\{\|L\varphi\|_{p,\beta}; \|\varphi\|_{q,\beta} \leq 1\}, \quad p, q \in \mathbb{R}.$$

Note that L is a continuous linear operator from $(\mathcal{S})_\beta$ into itself if and only if for any $p \geq 0$, there exists $q \geq 0$ such that $\|L\|_{\{q,p\}} < \infty$. Moreover, if $\|L\|_{\{q,p\}} < \infty$, then the operator L extends by continuity to a continuous linear operator from $(\mathcal{S}_q)_\beta$ into $(\mathcal{S}_p)_\beta$.

Theorem 10.23. *For any $y \in \mathcal{E}'$, the equality holds*

$$T_y = \sum_{n=0}^{\infty} \frac{1}{n!} \Xi_{0,n}(y^{\otimes n}), \qquad (10.16)$$

where the series converges in the sense that for any $p \geq 0$, there exists $q \geq 0$ such that the series converges in the operator norm $\|\cdot\|_{\{q,p\}}$.

Proof. By Theorem 10.5, for any $p \geq 0$, there exists $q \geq 0$ such that

$$\|\Xi_{0,n}(y^{\otimes n})\|_{\{q,p\}} \leq (n!2^n)^{\frac{1-\beta}{2}} |y|_{-q}^n.$$

This implies that the series in equation (10.16) converges in the operator norm $\|\cdot\|_{\{q,p\}}$. To prove the equality, let $\varphi_\xi = :e^{\langle \cdot, \xi \rangle}:$, $\xi \in \mathcal{S}$. Then for any $\xi, \eta \in \mathcal{S}$,

$$\langle\langle T_y \varphi_\xi, \varphi_\eta \rangle\rangle = \langle\langle e^{\langle \cdot + y, \xi \rangle - \frac{1}{2}|\xi|_0^2}, \varphi_\eta \rangle\rangle$$

$$= e^{\langle y, \xi \rangle} \langle\langle \varphi_\xi, \varphi_\eta \rangle\rangle$$

$$= e^{\langle y, \xi \rangle + \langle \eta, \xi \rangle}. \qquad (10.17)$$

On the other hand, recall that $\partial_t \varphi_\xi = \xi(t) \varphi_\xi$. Hence

$$\Xi_{0,n}(y^{\otimes n}) \varphi_\xi = \int_{T^n} y(t_1) \cdots y(t_n) \xi(t_1) \cdots \xi(t_n) \varphi_\xi \, dt_1 \cdots dt_n$$

$$= \langle y, \xi \rangle^n \varphi_\xi.$$

Therefore

$$\left\langle\!\left\langle \sum_{n=0}^{\infty} \frac{1}{n!} \Xi_{0,n}(y^{\otimes n}) \varphi_\xi, \varphi_\eta \right\rangle\!\right\rangle = \langle\langle \varphi_\xi, \varphi_\eta \rangle\rangle \sum_{n=0}^{\infty} \frac{1}{n!} \langle y, \xi \rangle^n$$

$$= e^{\langle \eta, \xi \rangle + \langle y, \xi \rangle}. \qquad (10.18)$$

Obviously equations (10.17) and (10.18) imply equation (10.16). $\qquad \square$

From Theorem 10.23 we get immediately the next theorem about the adjoint T_y^* of the translation operator T_y.

Theorem 10.24. *For any $y \in \mathcal{E}'$, the equality holds*

$$T_y^* = \sum_{n=0}^{\infty} \frac{1}{n!} \Xi_{n,0}(y^{\otimes n}),$$

where the series converges in the sense that for any $p \geq 0$, there exists $q \geq 0$ such that the series converges in the operator norm $\|\cdot\|_{\{-p,-q\}}$.

Next, we consider multiplications. Let Φ_0 be a fixed generalized function in $(\mathcal{E})^*_\beta$. We know from §8.4 that the Wick multiplication $\Phi \mapsto \Phi \diamond \Phi_0$ is continuous from $(\mathcal{E})^*_\beta$ into itself. Moreover, from §9.3 the multiplication $\varphi \mapsto \varphi \Phi_0$ is continuous from $(\mathcal{S})_\beta$ into $(\mathcal{S})^*_\beta$. An interesting special case is when $\Phi_0 = {:}e^{\langle \cdot, y \rangle}{:}$ for $y \in \mathcal{S}'$.

Theorem 10.25. *Let $y \in \mathcal{E}'$. Then the equality holds:*

$$ {:}e^{\langle \cdot, y \rangle}{:} \diamond \Phi = T^*_y \Phi, \quad \forall \Phi \in (\mathcal{E})^*_\beta. $$

Proof. Let $\varphi_\xi = {:}e^{\langle \cdot, \xi \rangle}{:}$. Then for any $\xi, \eta \in \mathcal{E}$,

$$ \langle\!\langle {:}e^{\langle \cdot, y \rangle}{:} \diamond \varphi_\xi, \varphi_\eta \rangle\!\rangle = \langle\!\langle {:}e^{\langle \cdot, y+\xi \rangle}{:}, \varphi_\eta \rangle\!\rangle $$
$$ = \langle\!\langle \varphi_{y+\xi}, \varphi_\eta \rangle\!\rangle $$
$$ = e^{\langle \eta, y+\xi \rangle}. $$

On the other hand, we have

$$ \langle\!\langle T^*_y \varphi_\xi, \varphi_\eta \rangle\!\rangle = \langle\!\langle \varphi_\xi, T_y \varphi_\eta \rangle\!\rangle $$
$$ = \langle\!\langle \varphi_\xi, {:}e^{\langle \cdot + y, \eta \rangle}{:} \rangle\!\rangle $$
$$ = e^{\langle y, \eta \rangle} \langle\!\langle \varphi_\xi, \varphi_\eta \rangle\!\rangle $$
$$ = e^{\langle y, \eta \rangle + \langle \eta, \xi \rangle}. $$

Thus $\langle\!\langle {:}e^{\langle \cdot, y \rangle}{:} \diamond \varphi_\xi, \varphi_\psi \rangle\!\rangle = \langle\!\langle T^*_y \varphi_\xi, \varphi_\eta \rangle\!\rangle$ for all $\xi, \eta \in \mathcal{E}$. This implies that ${:}e^{\langle \cdot, y \rangle}{:} \diamond \Phi = T^*_y \Phi$ for all $\Phi \in (\mathcal{E})^*_\beta$. $\square$

Theorem 10.26. *Let $y \in \mathcal{E}'$. Then the equality holds:*

$$ {:}e^{\langle \cdot, y \rangle}{:} \varphi = T^*_y T_y \varphi, \quad \forall \varphi \in (\mathcal{E})_\beta. $$

Proof. As before, let $\varphi_\xi = {:}e^{\langle \cdot, \xi \rangle}{:}$. Then $T_y \varphi_\xi = e^{\langle y, \xi \rangle} \varphi_\xi$. Hence for any $\xi, \eta \in \mathcal{E}$,

$$ \langle\!\langle T^*_y T_y \varphi_\xi, \varphi_\eta \rangle\!\rangle = \langle\!\langle T_y \varphi_\xi, T_y \varphi_\eta \rangle\!\rangle $$
$$ = \langle\!\langle e^{\langle y, \xi \rangle} \varphi_\xi, e^{\langle y, \eta \rangle} \varphi_\eta \rangle\!\rangle $$
$$ = e^{\langle y, \xi+\eta \rangle} e^{\langle \xi, \eta \rangle}. $$

On the other hand, it is straightforward to check that

$$ \varphi_\xi \varphi_\eta = e^{\langle \xi, \eta \rangle} \varphi_{\xi+\eta}. $$

Therefore

$$\langle\!\langle :e^{\langle\cdot,y\rangle}: \varphi_\xi, \varphi_\eta\rangle\!\rangle = \langle\!\langle :e^{\langle\cdot,y\rangle}:, \varphi_\xi\varphi_\eta\rangle\!\rangle$$

$$= \langle\!\langle :e^{\langle\cdot,y\rangle}:, e^{\langle\xi,\eta\rangle} :e^{\langle\cdot,\eta\xi+\xi\rangle}:\rangle\!\rangle$$

$$= e^{\langle\xi,\eta\rangle} e^{\langle y,\xi+\eta\rangle}.$$

Thus we have shown that $\langle\!\langle :e^{\langle\cdot,y\rangle}: \varphi_\xi, \varphi_\eta\rangle\!\rangle = \langle\!\langle T_y^* T_y \varphi_\xi, \varphi_\eta\rangle\!\rangle$ for all $\xi, \eta \in \mathcal{E}$. This implies that $:e^{\langle\cdot,y\rangle}: \varphi = T_y^* T_y \varphi$ for all $\varphi \in (\mathcal{E})_\beta$. $\square$

10.6 Representation theorem

In the last section we have seen several examples of continuous linear operators from $(\mathcal{S})_\beta$ into $(\mathcal{S})_\beta^*$ that can be represented as series of integral kernel operators. Thus it is natural to ask whether this kind of representation is still true for any continuous linear operator from $(\mathcal{S})_\beta$ into $(\mathcal{S})_\beta^*$. A representation theorem for such an operator has been obtained in Huang [66] and Obata [139] and [140] for the case $\beta = 0$. In this section we will follow the same idea as in Obata [142] to describe a representation theorem for the general case.

Let Ξ be a continuous linear operator from $(\mathcal{S})_\beta$ into $(\mathcal{S})_\beta^*$. In the proofs of Theorems 10.13, 10.16, 10.23, 10.25, and 10.26, we have often used the function $\langle\!\langle \Xi :e^{\langle\cdot,\xi\rangle}:, :e^{\langle\cdot,\eta\rangle}:\rangle\!\rangle$, $\xi, \eta \in \mathcal{E}_c$. In view of its importance we make the following definition.

Definition 10.27. The *symbol* of a continuous linear operator Ξ from $(\mathcal{S})_\beta$ into $(\mathcal{S})_\beta^*$ is defined to be function

$$F(\xi,\eta) = \langle\!\langle \Xi :e^{\langle\cdot,\xi\rangle}:, :e^{\langle\cdot,\eta\rangle}:\rangle\!\rangle, \quad \xi, \eta \in \mathcal{S}_c.$$

Obviously a continuous linear operator from $(\mathcal{S})_\beta$ into $(\mathcal{S})_\beta^*$ is uniquely determined by its symbol. Observe that the symbol of Ξ is analogous to the S-transform of a generalized function $\Phi \in (\mathcal{S})_\beta^*$.

The idea to obtain a representation theorem for Ξ is essentially the same as the one we used in the proof of Theorem 8.2 to characterize the generalized functions. First we show that the symbol F of Ξ satisfies conditions similar to those for the S-transform:

(a) For any $\xi_1, \xi_2, \eta_1, \eta_2 \in \mathcal{S}_c$, the function $F(z\xi_1 + \xi_2, w\eta_1 + \eta_2)$ is a holomorphic function of $(z,w) \in \mathbb{C}^2$.

(b) There exist nonnegative constants K, a, and p such that

$$|F(\xi,\eta)| \le K \exp\left[a\left(|\xi|_p^{\frac{2}{1-\beta}} + |\eta|_p^{\frac{2}{1-\beta}}\right)\right], \quad \forall \xi, \eta \in \mathcal{S}_c.$$

The proof of (a) is similar to Lemma 8.1. It is straightforward to check that

$$F(z\xi_1 + \xi_2, w\eta_1 + \eta_2) = \sum_{m,n=0}^{\infty} A_{mn} z^m w^n, \tag{10.19}$$

where A_{mn} is given by

$$A_{mn} = \frac{1}{m!n!} \sum_{j,k=0}^{\infty} \frac{1}{j!k!} \langle\!\langle \Xi\langle : \cdot^{\otimes(m+j)} :, \xi_1^{\otimes m} \widehat{\otimes} \xi_2^{\otimes j}\rangle, \langle : \cdot^{\otimes(n+k)} :, \eta_1^{\otimes n} \widehat{\otimes} \eta_2^{\otimes k}\rangle\rangle\!\rangle.$$

We can derive estimates similar to those in the proof of Lemma 8.1 and show that the series in equation (10.19) converges for all $(z, w) \in \mathbb{C}^2$. Thus condition (a) holds.

To derive condition (b), note that the bilinear functional $B(\varphi, \psi) \equiv \langle\!\langle \Xi\varphi, \psi\rangle\!\rangle$, $\varphi, \psi \in (\mathcal{S})_\beta$, is continuous since Ξ is a continuous linear operator from $(\mathcal{S})_\beta$ into $(\mathcal{S})_\beta^*$. Hence there exist nonnegative constants C and p such that

$$|\langle\!\langle \Xi\varphi, \psi\rangle\!\rangle| \le C\|\varphi\|_{p,\beta}\|\psi\|_{p,\beta}, \quad \forall \varphi, \psi \in (\mathcal{S})_\beta.$$

But recall from Theorem 5.7 that for any $\xi \in \mathcal{E}_c$,

$$\|:e^{\langle \cdot, \xi\rangle}:\|_{p,\beta} \le 2^{\beta/2} \exp\left[(1-\beta)2^{\frac{2\beta-1}{1-\beta}} |\xi|_p^{\frac{2}{1-\beta}}\right].$$

Thus for any $\xi, \eta \in \mathcal{E}_c$,

$$|F(\xi, \eta)| = |\langle\!\langle \Xi :e^{\langle \cdot, \xi\rangle}:, :e^{\langle \cdot, \eta\rangle}:\rangle\!\rangle|$$

$$\le K \exp\left[a\left(|\xi|_p^{\frac{2}{1-\beta}} + |\eta|_p^{\frac{2}{1-\beta}}\right)\right],$$

where $K = 2^\beta C$ and $a = (1-\beta)2^{\frac{2\beta-1}{1-\beta}}$.

Conversely, suppose F is a function on $\mathcal{S}_c \times \mathcal{S}_c$ satisfying conditions (a) and (b). Then there exists a unique continuous linear operator Ξ from $(\mathcal{S})_\beta$ into $(\mathcal{S})_\beta^*$ such that F is the symbol of Ξ. For the proof, see Obata [142].

Now, we explain the idea for representing a continuous linear operator Ξ from $(\mathcal{S})_\beta$ into $(\mathcal{S})_\beta^*$ as a series of integral kernel operators,

$$\Xi\varphi = \sum_{j,k=0}^{\infty} \Xi_{j,k}(\theta_{j,k})\varphi, \quad \forall \varphi \in (\mathcal{S})_\beta, \tag{10.20}$$

where $\theta_{j,k} \in (\mathcal{S}_c')^{\otimes(j+k)}$ and the series on the right hand side converges in $(\mathcal{S})_\beta^*$. Observe that the representation is unique if we require that $\theta_{j,k} \in (\mathcal{S}_c')^{\otimes(j+k)}_{\text{symm}(j,k)}$ (symmetric for the first j variables in s and for the other k

variables in $\mathbf{t}$). First we need to find the $\theta_{j,k}$'s from the symbol F of Ξ. In order to do so, assume that equation (10.20) is valid. Let φ_ξ denote the function $\varphi_\xi \equiv :e^{\langle \cdot, \xi \rangle}:$ as before. Then the symbol of Ξ is given by

$$F(\xi, \eta) = \langle\!\langle \Xi\varphi_\xi, \varphi_\eta \rangle\!\rangle$$

$$= \sum_{j,k=0}^{\infty} \int_{\mathbf{s} \in T^j} \int_{\mathbf{t} \in T^k} \theta_{j,k}(\mathbf{s}, \mathbf{t}) \langle\!\langle \partial_{\mathbf{t}}\varphi_\xi, \partial_{\mathbf{s}}\varphi_\eta \rangle\!\rangle \, d\mathbf{s} d\mathbf{t}$$

$$= \sum_{j,k=0}^{\infty} \int_{\mathbf{s} \in T^j} \int_{\mathbf{t} \in T^k} \theta_{j,k}(\mathbf{s}, \mathbf{t})(\eta^{\otimes j} \otimes \xi^{\otimes k})(\mathbf{s}, \mathbf{t}) \langle\!\langle \varphi_\xi, \varphi_\eta \rangle\!\rangle \, d\mathbf{s} d\mathbf{t}$$

$$= e^{\langle \xi, \eta \rangle} \sum_{j,k=0}^{\infty} \langle \theta_{j,k}, \eta^{\otimes j} \otimes \xi^{\otimes k} \rangle.$$

Thus we get the following identity

$$e^{-\langle \xi, \eta \rangle} F(\xi, \eta) = \sum_{j,k=0}^{\infty} \langle \theta_{j,k}, \eta^{\otimes j} \otimes \xi^{\otimes k} \rangle. \tag{10.21}$$

For simplicity, let $f(\xi, \eta) \equiv e^{-\langle \xi, \eta \rangle} F(\xi, \eta)$. Then by condition (a) on F, we see that the function f also satisfies condition (a). In fact, for any $\xi_1, \ldots, \xi_k, \eta_1, \ldots, \eta_j \in S_c$, the function

$$f(z_1\xi_1 + \cdots + z_k\xi_k, w_1\eta_1 + \cdots + w_j\eta_j)$$

is holomorphic on $\mathbb{C}^{j+k}$. Then we apply the same arguments as in the proof of Theorem 8.2 to show that there exists $\theta_{j,k} \in (S_c')^{\otimes(j+k)}_{\mathrm{symm}(j,k)}$ such that

$$\langle \theta_{j,k}, \eta_1 \otimes \cdots \otimes \eta_j \otimes \xi_1 \otimes \cdots \otimes \xi_k \rangle$$
$$= \frac{1}{j!k!} \frac{\partial}{\partial \mathbf{z}} \frac{\partial}{\partial \mathbf{w}} f(z_1\xi_1 + \cdots + z_k\xi_k, w_1\eta_1 + \cdots + w_j\eta_j)\Big|_{\mathbf{z}=\mathbf{w}=0}, \tag{10.22}$$

where $\mathbf{u} = (u_1, \ldots, u_k)$ and $\frac{\partial}{\partial \mathbf{u}} = \frac{\partial}{\partial u_1} \cdots \frac{\partial}{\partial u_k}$, $\mathbf{u} = \mathbf{z}, \mathbf{w}$. Then we use similar arguments (but much more complicated estimates) to show that the following series

$$\sum_{j,k=0}^{\infty} \Xi_{j,k}(\theta_{j,k})\varphi$$

converges in $(S)^*_\beta$ for each $\varphi \in (S)_\beta$. Finally we can use equation (10.21) to show that the symbol of the operator given by the above series coincides with the symbol F of the operator Ξ. Thus we get the representation of Ξ in equation (10.20).

Suppose the operator Ξ is a continuous linear operator from $(\mathcal{S})_\beta$ into itself. Then for any j and k, $\theta_{j,k}$ as given by equation (10.22) belongs to $\mathcal{S}_c^{\widehat{\otimes}j} \otimes (\mathcal{S}_c')^{\widehat{\otimes}k}$.

We summarize the above discussion as the next representation theorem for continuous linear operators from $(\mathcal{S})_\beta$ into $(\mathcal{S})_\beta^*$.

Theorem 10.28. *Let Ξ be a continuous linear operator from $(\mathcal{S})_\beta$ into $(\mathcal{S})_\beta^*$. Then Ξ can be represented uniquely as a series of integral kernel operators*

$$\Xi\varphi = \sum_{j,k=0}^{\infty} \Xi_{j,k}(\theta_{j,k})\varphi, \quad \forall \varphi \in (\mathcal{S})_\beta,$$

where $\theta_{j,k} \in (\mathcal{S}_c')^{\otimes(j+k)}$ and the series on the right hand side converges in $(\mathcal{S})_\beta^*$. If Ξ is a continuous linear operator from $(\mathcal{S})_\beta$ into itself, then $\theta_{j,k} \in \mathcal{S}_c^{\widehat{\otimes}j} \otimes (\mathcal{S}_c')^{\widehat{\otimes}k}$ for all $j, k \geq 0$.

11

Fourier Transforms

In the previous two chapters we have studied several continuous linear operators on the space of test functions and on the space of generalized functions. In this chapter we will study the Fourier transform on the space of generalized functions. This Fourier transform is the white noise analogue of the finite dimensional Fourier transform. It was first introduced in Kuo [103]. However, the definition in Kuo [103] is quite informal and can serve only as a heuristic motivation. A rigorous definition has been given in the book by Hida et al. [58], Kuo [108] [109] [110], and Obata [142]. Other approaches to define the Fourier transform have appeared in Y. Ito et al. [73], Y.-J. Lee [122], and Yan [178]. In this chapter we will study the Fourier transform as a continuous linear operator from the space $(\mathcal{E})_\beta^*$ into itself.

11.1 Definition of the Fourier transform

First we recall how the finite dimensional Fourier transform is defined on the space $\mathcal{S}(\mathbb{R}^r)$ of test functions and on the space $\mathcal{S}'(\mathbb{R}^r)$ of generalized functions. The Fourier transform $\widehat{f}$ of a function $f \in \mathcal{S}(\mathbb{R}^r)$ is defined by

$$\widehat{f}(y) = (2\pi)^{-r/2} \int_{\mathbb{R}^r} e^{-i\langle x, y \rangle} f(x) \, dx. \qquad (11.1)$$

The Fourier transform is a continuous linear operator from $\mathcal{S}(\mathbb{R}^r)$ into itself. On the other hand, the Fourier transform $\widehat{F}$ of a generalized function $F \in \mathcal{S}'(\mathbb{R}^r)$ is defined to be the generalized function given by

$$\langle \widehat{F}, f \rangle = \langle F, \widehat{f} \rangle, \quad f \in \mathcal{S}(\mathbb{R}^r). \qquad (11.2)$$

Note that the extension of the Fourier transform to $\mathcal{S}'(\mathbb{R}^r)$ is actually the adjoint of the Fourier transform on $\mathcal{S}(\mathbb{R}^r)$ and hence it is continuous from $\mathcal{S}'(\mathbb{R}^r)$ into itself.

Now, observe that we can not use equation (11.1) to define the Fourier transform of a test function in $(\mathcal{E})_\beta$ since the Lebesgue measure does not exist on $\mathcal{E}'$. On the other hand, assuming that we can define the Fourier transform in some way, this Fourier transform can not be from $(\mathcal{E})_\beta$ into itself since the function $\varphi \equiv 1$ is in $(\mathcal{E})_\beta$ while its Fourier transform is the Kubo-Yokoi delta function $\tilde{\delta}_0$, which is outside $(\mathcal{E})_\beta$. Thus the white noise Fourier transform has to be defined on the space of generalized functions, i.e., we will define the Fourier transform directly on $(\mathcal{E})_\beta^*$.

The main idea to define the Fourier transform on $(\mathcal{E})_\beta^*$ is to rewrite equation (11.1) in a dimension free form. Being motivated by the renormalized exponential function in §5.2, we define

$$:e^{-i\langle x,y\rangle}:_y = e^{-i\langle x,y\rangle+\frac{1}{2}|x|^2},$$

where the $:\cdot:_y$ denotes the renormalization with respect to the y variable. Then equation (11.1) can be written as

$$\widehat{f}(y) = (2\pi)^{-r/2} \int_{\mathbb{R}^r} :e^{-i\langle x,y\rangle}:_y f(x)e^{-\frac{1}{2}|x|^2}\,dx$$

$$= \int_{\mathbb{R}^r} :e^{-i\langle x,y\rangle}:_y f(x)\,d\mu_r(x),$$

where μ_r is the standard Gaussian measure on $\mathbb{R}^r$. Thus it is reasonable to define the Fourier transform $\widehat{\Phi}$ of a generalized function $\Phi \in (\mathcal{E})_\beta^*$ by

$$\widehat{\Phi}(y) = \int_{\mathcal{E}'} :e^{-i\langle x,y\rangle}:_y \Phi(x)\,d\mu(x).$$

But this integral is purely symbolic since the function $:e^{-i\langle x,y\rangle}:_y$ is not a test function in the x variable. Furthermore, we have not yet defined the integral of an $(\mathcal{E})_\beta^*$-valued function on $\mathcal{E}'$. However, if we take the S-transform of $\widehat{\Phi}$ informally, then we get

$$(S\widehat{\Phi})(\xi) = \int_{\mathcal{E}'} e^{-i\langle x,\xi\rangle}\Phi(x)\,d\mu(x), \quad \xi \in \mathcal{E}_c.$$

Note that for any $\xi \in \mathcal{E}_c$, the function $e^{-i\langle\cdot,\xi\rangle}$ belongs to $(\mathcal{E})_\beta$. Therefore

$$(S\widehat{\Phi})(\xi) = \langle\!\langle \Phi, e^{-i\langle\cdot,\xi\rangle}\rangle\!\rangle, \quad \xi \in \mathcal{E}_c.$$

On the other hand, it is easy to check that

$$\langle\!\langle \Phi, e^{-i\langle\cdot,\xi\rangle}\rangle\!\rangle = (S\Phi)(-i\xi)e^{-\frac{1}{2}\langle\xi,\xi\rangle}, \quad \xi \in \mathcal{E}_c. \tag{11.3}$$

If $\Phi \in (\mathcal{E})_\beta^*$, then the function in equation (11.3) satisfies the conditions (a) and (b) in Theorem 8.2. Hence it is the S-transform of a unique element in $\Psi \in (\mathcal{E})_\beta^*$. This is the motivation for the definition of the Fourier transform.

Definition 11.1. The *Fourier transform* $\widehat{\Phi}$ of a generalized function $\Phi \in (\mathcal{E})_\beta^*$ is defined to be the unique element in $(\mathcal{E})_\beta^*$ with S-transform given by

$$(S\widehat{\Phi})(\xi) = \langle\!\langle \Phi, e^{-i\langle\cdot,\xi\rangle} \rangle\!\rangle = (S\Phi)(-i\xi)e^{-\frac{1}{2}\langle\xi,\xi\rangle}, \quad \xi \in \mathcal{E}_c.$$

Example 11.2. Let $\widetilde{\delta}_x$ be the Kubo-Yokoi delta function at $x \in \mathcal{E}_c'$ (see §7.2). Then

$$(\widetilde{\delta}_x)^\wedge = :e^{-i\langle\cdot,x\rangle}:, \quad (:e^{\langle\cdot,x\rangle}:)^\wedge = \widetilde{\delta}_{-ix}.$$

In particular, when $x = 0$, we have $(\widetilde{\delta}_0)^\wedge = 1$ and $\widehat{1} = \widetilde{\delta}_0$. To check this, simply use Theorems 5.13 and 7.8, i.e.,

$$S(:e^{-i\langle\cdot,x\rangle}:)(\xi) = e^{-i\langle x,\xi\rangle}, \quad S(\widetilde{\delta}_x)(\xi) = e^{\langle x,\xi\rangle - \frac{1}{2}\langle\xi,\xi\rangle}, \quad \xi \in \mathcal{E}_c.$$

Example 11.3. Let $g_{x,c}$ be the Gaussian white noise function with S-transform given by equation (8.10). By checking the S-transform it is easy to see that $\widehat{g}_{x,c} = g_{-icx,c^{-1}}$. In particular, we have $\widehat{g}_c = g_{c^{-1}}$.

We will also use $\mathcal{F}\Phi$ to denote the Fourier transform of Φ. Thus $\mathcal{F}$ is a linear operator from $(\mathcal{E})_\beta^*$ into itself. As expected, the Fourier transform $\mathcal{F}$ is continuous. There are several ways to prove this property. We will give an elementary proof by using the Wiener-Itô decomposition.

We first give an informal derivation of the Wiener-Itô decomposition of $\widehat{\Phi}$. Let $\Phi = \sum_{n=0}^\infty \langle :\cdot^{\otimes n}:, F_n \rangle$. Then we have

$$S\Phi(\xi) = \sum_{n=0}^\infty \langle F_n, \xi^{\otimes n} \rangle.$$

Therefore

$$(S\widehat{\Phi})(\xi) = (S\Phi)(-i\xi)e^{-\frac{1}{2}\langle\xi,\xi\rangle}$$

$$= \left(\sum_{n=0}^\infty \langle F_n, (-i\xi)^{\otimes n} \rangle \right) \left(\sum_{n=0}^\infty \frac{(-1)^n}{2^n n!} \langle\xi,\xi\rangle^n \right)$$

$$= \left(\sum_{n=0}^\infty (-i)^n \langle F_n, \xi^{\otimes n} \rangle \right) \left(\sum_{n=0}^\infty \frac{(-1)^n}{2^n n!} \langle \tau^{\otimes n}, \xi^{\otimes 2n} \rangle \right).$$

From the last equality, it is easy to check that

$$(S\widehat{\Phi})(\xi) = \sum_{n=0}^\infty (-i)^n \Big\langle \sum_{m=0}^{[n/2]} \frac{1}{2^m m!} F_{n-2m} \widehat{\otimes} \tau^{\otimes m}, \xi^{\otimes n} \Big\rangle.$$

Thus informally we must have

$$\hat{\Phi} = \sum_{n=0}^{\infty} \left\langle : \cdot^{\otimes n} :, (-i)^n \sum_{m=0}^{[n/2]} \frac{1}{2^m m!} F_{n-2m} \widehat{\otimes} \tau^{\otimes m} \right\rangle. \qquad (11.4)$$

We will show that the right hand side of equation (11.4) defines a generalized function in $(\mathcal{E})^*_\beta$ and that it is indeed the Fourier transform of Φ. In doing so, we will show at the same time that the Fourier transform $\mathcal{F}$ is continuous from $(\mathcal{E})^*_\beta$ into itself.

Lemma 11.4. *Suppose* $\Phi = \sum_{n=0}^{\infty} \langle : \cdot^{\otimes n} :, F_n \rangle \in (\mathcal{E})^*_\beta$. *Then the function*

$$\Psi = \sum_{n=0}^{\infty} \left\langle : \cdot^{\otimes n} :, (-i)^n \sum_{m=0}^{[n/2]} \frac{1}{2^m m!} F_{n-2m} \widehat{\otimes} \tau^{\otimes m} \right\rangle$$

defines a generalized function in $(\mathcal{E})^*_\beta$. *In fact, if* $\Phi \in (\mathcal{E}_p)^*_\beta$, *then* $\Psi \in (\mathcal{E}_q)^*_\beta$ *for any* $q > p$ *such that* $2^{2-\beta} \lambda_1^{-2(q-p)} \leq 1$ *and* $2^{2-\beta} |\tau|_{-q} < 1$. *Moreover, the inequality holds*

$$\|\Psi\|_{-q,-\beta} \leq \left(1 - 2^{2(2-\beta)} |\tau|_{-q}^2\right)^{-1/2} \|\Phi\|_{-p,-\beta}.$$

Proof. For simplicity, let

$$G_n \equiv (-i)^n \sum_{m=0}^{[n/2]} \frac{1}{2^m m!} F_{n-2m} \widehat{\otimes} \tau^{\otimes m}.$$

Then we have $\Psi = \sum_{n=0}^{\infty} \langle : \cdot^{\otimes n} :, G_n \rangle$ and

$$|G_n|_{-q}^2 \leq ([n/2]+1) \sum_{m=0}^{[n/2]} \frac{1}{2^{2m}(m!)^2} |F_{n-2m}|_{-q}^2 \, |\tau|_{-q}^{2m}.$$

But $[n/2]+1 \leq 2^n$ for $n \geq 0$ and $|F_k|_{-q} \leq \lambda_1^{-(q-p)k} |F_k|_{-p}$ for any $q \geq p$. Hence

$$|G_n|_{-q}^2 \leq 2^n \sum_{m=0}^{[n/2]} \frac{1}{2^{2m}(m!)^2} \lambda_1^{-2(q-p)(n-2m)} |F_{n-2m}|_{-p}^2 \, |\tau|_{-q}^{2m}.$$

Therefore

$$\|\Psi\|_{-q,-\beta}^2 = \sum_{n=0}^{\infty} (n!)^{1-\beta} |G_n|_{-q}^2$$

$$\leq \sum_{n=0}^{\infty} \sum_{m=0}^{[n/2]} (n!)^{1-\beta} 2^n \frac{1}{2^{2m}(m!)^2} \lambda_1^{-2(q-p)(n-2m)} |F_{n-2m}|_{-p}^2 \, |\tau|_{-q}^{2m}.$$

By changing the order of summation, we get

$$
\|\Psi\|^2_{-q,-\beta}
$$

$$
\leq \sum_{m=0}^{\infty}\sum_{n=2m}^{\infty}(n!)^{1-\beta}2^n\frac{1}{2^{2m}(m!)^2}\lambda_1^{-2(q-p)(n-2m)}|F_{n-2m}|^2_{-p}\,|\tau|^{2m}_{-q}
$$

$$
= \sum_{m=0}^{\infty}\sum_{k=0}^{\infty}((2m+k)!)^{1-\beta}2^{2m+k}\frac{1}{2^{2m}(m!)^2}\lambda_1^{-2(q-p)k}|F_k|^2_{-p}\,|\tau|^{2m}_{-q}.
$$

But $(2m+k)! \leq 2^{2m+k}(2m)!k! \leq 2^{2m+k}2^{2m}(m!)^2 k!$. Hence we can derive that

$$
\|\Psi\|^2_{-q,-\beta} \leq \sum_{m=0}^{\infty}\sum_{k=0}^{\infty}2^{(2m+k)(2-\beta)}(k!)^{1-\beta}\frac{1}{\left(2^{2m}(m!)\right)^\beta}\lambda_1^{-2(q-p)k}|F_k|^2_{-p}|\tau|^{2m}_{-q}
$$

$$
\leq \left(\sum_{m=0}^{\infty}\left(2^{2(2-\beta)}|\tau|^2_{-q}\right)^m\right)\left(\sum_{k=0}^{\infty}\left(2^{2-\beta}\lambda_1^{-2(q-p)}\right)^k(k!)^{1-\beta}|F_k|^2_{-p}\right).
$$

Suppose $\Phi \in (\mathcal{E}_p)^*_\beta$ and $q > p$ satisfies $2^{2-\beta}\lambda_1^{-2(q-p)} \leq 1$ and $2^{2-\beta}|\tau|_{-q} < 1$. Then

$$
\|\Psi\|^2_{-q,-\beta} \leq \left(\sum_{m=0}^{\infty}\left(2^{2(2-\beta)}|\tau|^2_{-q}\right)^m\right)\left(\sum_{k=0}^{\infty}(k!)^{1-\beta}|F_k|^2_{-p}\right)
$$

$$
= \left(1 - 2^{2(2-\beta)}|\tau|^2_{-q}\right)^{-1}\|\Phi\|^2_{-p,-\beta}. \qquad \square
$$

Lemma 11.5. *If* $\Phi = \sum_{n=0}^{\infty}\langle :\cdot^{\otimes n}:, F_n\rangle \in (\mathcal{E})^*_\beta$, *then its Fourier transform is given by*

$$
\hat{\Phi} = \sum_{n=0}^{\infty}\left\langle :\cdot^{\otimes n}:, (-i)^n\sum_{m=0}^{[n/2]}\frac{1}{2^m m!}F_{n-2m}\widehat{\otimes}\tau^{\otimes m}\right\rangle.
$$

Proof. Define $\Psi \in (\mathcal{E})^*_\beta$ by

$$
\Psi \equiv \sum_{n=0}^{\infty}\left\langle :\cdot^{\otimes n}:, (-i)^n\sum_{m=0}^{[n/2]}\frac{1}{2^m m!}F_{n-2m}\widehat{\otimes}\tau^{\otimes m}\right\rangle.
$$

By Lemma 11.4, Ψ is a generalized function in $(\mathcal{E})^*_\beta$. Its S-transform is given by

$$
S\Psi(\xi) = \sum_{n=0}^{\infty}\sum_{m=0}^{[n/2]}(-i)^n\frac{1}{2^m m!}\langle F_{n-2m}\widehat{\otimes}\tau^{\otimes m}, \xi^{\otimes n}\rangle, \quad \xi \in \mathcal{E}_c.
$$

By the proof of Lemma 11.4 we can change the order of summation to get

$$
S\Psi(\xi) = \sum_{m=0}^{\infty} \sum_{n=2m}^{\infty} (-i)^n \frac{1}{2^m m!} \langle F_{n-2m} \widehat{\otimes} \tau^{\otimes m}, \xi^{\otimes n} \rangle
$$

$$
= \sum_{m=0}^{\infty} \sum_{n=2m}^{\infty} (-i)^n \frac{1}{2^m m!} \langle F_{n-2m}, \xi^{\otimes(n-2m)} \rangle \langle \xi, \xi \rangle^m
$$

$$
= \sum_{m=0}^{\infty} \frac{1}{2^m m!} \langle \xi, \xi \rangle^m \sum_{k=0}^{\infty} (-i)^{2m+k} \langle F_k, \xi^{\otimes k} \rangle
$$

$$
= \sum_{m=0}^{\infty} \frac{(-1)^m}{2^m m!} \langle \xi, \xi \rangle^m \sum_{k=0}^{\infty} \langle F_k, (-i\xi)^{\otimes k} \rangle
$$

$$
= e^{-\frac{1}{2}\langle \xi, \xi \rangle} (S\Phi)(-i\xi).
$$

Hence by the definition of the Fourier transform, we have $S\Psi = S\widehat{\Phi}$ and so $\Psi = \widehat{\Phi}$. □

Theorem 11.6. *The Fourier transform $\mathcal{F}$ is a continuous linear operator from $(\mathcal{E})_\beta^*$ into itself. In fact, for any $p \geq 0$, if $q > p$ satisfies $2^{2-\beta}\lambda_1^{-2(q-p)} \leq 1$ and $2^{2-\beta}|\tau|_{-q} < 1$, then*

$$
\|\widehat{\Phi}\|_{-q,-\beta} \leq \left(1 - 2^{2(2-\beta)}|\tau|_{-q}^2\right)^{-1/2} \|\Phi\|_{-p,-\beta}, \quad \forall \Phi \in (\mathcal{E}_p)_\beta^*.
$$

Proof. The inequality follows from Lemmas 11.4 and 11.5. On the other hand, the inequality implies the continuity of the Fourier transform $\mathcal{F}$. □

Theorem 11.7. *The Fourier transform $\mathcal{F}$ satisfies $\mathcal{F}^4 = I$ on $(\mathcal{E})_\beta^*$.*

Proof. By the definition of the Fourier transform,

$$
S(\mathcal{F}^2 \Phi)(\xi) = S(\mathcal{F}\Phi)(-i\xi)e^{-\frac{1}{2}\langle \xi, \xi \rangle}
$$

$$
= S\Phi(-\xi)e^{\frac{1}{2}\langle \xi, \xi \rangle}e^{-\frac{1}{2}\langle \xi, \xi \rangle}
$$

$$
= S\Phi(-\xi).
$$

Thus for any $\xi \in \mathcal{E}_c$,

$$
S(\mathcal{F}^4 \Phi)(\xi) = S(\mathcal{F}^2 \Phi)(-\xi) = S\Phi(\xi).
$$

This implies that $\mathcal{F}^4 \Phi = \Phi$ for all $\Phi \in (\mathcal{E})_\beta^*$. □

Corollary 11.8. *The Fourier transform $\mathcal{F} : (\mathcal{E})_\beta^* \to (\mathcal{E})_\beta^*$ is injective and onto.*

Proof. Suppose $\mathcal{F}\Phi = \mathcal{F}\Psi$. Then by applying $\mathcal{F}^3$ to both sides of the equality, we get $\Phi = \Psi$. Hence $\mathcal{F}$ is injective. On the other hand, for any $\Phi \in (\mathcal{E})^*_\beta$, let $\Psi = \mathcal{F}^3\Phi$. Then $\Psi \in (\mathcal{E})^*_\beta$ and $\mathcal{F}\Psi = \Phi$. Hence $\mathcal{F}$ is onto. $\square$

Obviously the Fourier transform $\mathcal{F} : (\mathcal{E})^*_\beta \to (\mathcal{E})^*_\beta$ is invertible and $\mathcal{F}^{-1} = \mathcal{F}^3$. As a matter of fact, it is easy to check that the S-transform of $\mathcal{F}^{-1}\Phi$ is given by

$$S(\mathcal{F}^{-1}\Phi)(\xi) = \langle\!\langle \Phi, e^{i\langle\cdot,\xi\rangle} \rangle\!\rangle = (S\Phi)(i\xi)e^{-\frac{1}{2}\langle\xi,\xi\rangle}, \quad \xi \in \mathcal{E}_c. \tag{11.5}$$

Moreover, if $\Phi = \sum_{n=0}^\infty \langle :\cdot^{\otimes n}:, F_n \rangle$, then

$$\mathcal{F}^{-1}\Phi = \sum_{n=0}^\infty \left\langle :\cdot^{\otimes n}:, i^n \sum_{m=0}^{[n/2]} \frac{1}{2^m m!} F_{n-2m} \widehat{\otimes} \tau^{\otimes m} \right\rangle. \tag{11.6}$$

11.2 Representations of the Fourier transform

In this section we will show that the Fourier transform can be represented in terms of other transformations. Recall that the S- and $\mathcal{T}$-transforms are defined by

$$S\Phi(\xi) = \langle\!\langle \Phi, :e^{\langle\cdot,\xi\rangle}: \rangle\!\rangle, \quad \mathcal{T}\Phi(\xi) = \langle\!\langle \Phi, e^{i\langle\cdot,\xi\rangle} \rangle\!\rangle, \quad \xi \in \mathcal{E}_c.$$

Theorem 11.9. *The Fourier transform $\mathcal{F}$ can be represented by $\mathcal{F} = \mathcal{T}^{-1}S$.*

Proof. Recall from the second equation in (8.21) that for any $\Phi \in (\mathcal{E})^*_\beta$,

$$\mathcal{T}\Phi(\xi) = e^{-\frac{1}{2}\langle\xi,\xi\rangle} S\Phi(i\xi).$$

Thus by the definition of the Fourier transform, we have

$$\begin{aligned}
\mathcal{T}\widehat{\Phi}(\xi) &= e^{-\frac{1}{2}\langle\xi,\xi\rangle} S\widehat{\Phi}(i\xi) \\
&= e^{-\frac{1}{2}\langle\xi,\xi\rangle} S\Phi(-i^2\xi)e^{-\frac{1}{2}\langle i\xi,i\xi\rangle} \\
&= S\Phi(\xi).
\end{aligned}$$

This shows that $\mathcal{T}\widehat{\Phi} = S\Phi$, i.e., $\widehat{\Phi} = \mathcal{T}^{-1}S\Phi$ for all $\Phi \in (\mathcal{E})^*_\beta$. Hence $\mathcal{F} = \mathcal{T}^{-1}S$. $\square$

To obtain another representation of the Fourier transform, recall from Theorem 7.8 that the Kubo-Yokoi delta function $\widetilde{\delta}_0$ at 0 has the following S-transform:

$$S(\widetilde{\delta}_0)(\xi) = e^{-\frac{1}{2}\langle\xi,\xi\rangle}, \quad \xi \in \mathcal{E}_c. \tag{11.7}$$

This S-transform is a factor in the S-transform $S\widehat{\Phi}$. For the other factor in $S\widehat{\Phi}$, we consider the second quantization $\Gamma(-iI)$ of the operator $-iI$. Let $\Phi = \sum_{n=0}^{\infty} \langle :\cdot^{\otimes n}:, F_n \rangle$. Then $\Gamma(-iI)\Phi$ is given by

$$\Gamma(-iI)\Phi = \sum_{n=0}^{\infty} \langle :\cdot^{\otimes n}:, (-iI)^{\otimes n} F_n \rangle$$

$$= \sum_{n=0}^{\infty} (-i)^n \langle :\cdot^{\otimes n}:, F_n \rangle.$$

Obviously for any $p \geq 0$

$$\|\Gamma(-iI)\Phi\|_{-p,-\beta} = \|\Phi\|_{-p,-\beta}. \tag{11.8}$$

Thus $\Gamma(-iI)$ is a continuous linear operator from $(\mathcal{E})^*_\beta$ into itself. Moreover, we have

$$S(\Gamma(-iI)\Phi)(\xi) = \sum_{n=0}^{\infty} (-i)^n \langle F_n, \xi^{\otimes n} \rangle$$

$$= \sum_{n=0}^{\infty} \langle F_n, (-i\xi)^{\otimes n} \rangle$$

$$= S\Phi(-i\xi). \tag{11.9}$$

Theorem 11.10. *The following equalities hold for any* $\Phi \in (\mathcal{E})^*_\beta$

$$\widehat{\Phi} = (\Gamma(-iI)\Phi) \diamond \widetilde{\delta}_0, \quad \widehat{\Phi} \diamond g_{-2} = \Gamma(-iI)\Phi.$$

Remark. It follows from (a) Theorem 7.9, (b) the second remark for Theorem 8.12, and (c) equation (11.8) that for any $p \geq 0$ and $c \geq 1$, there exists some $q \geq p$ such that

$$\|\widehat{\Phi}\|_{-q,-\beta} \leq c\|\Phi\|_{-p,-\beta}\|\widetilde{\delta}_0\|_{-p}, \quad \forall \Phi \in (\mathcal{E}_p)^*_\beta.$$

Thus we have another proof for the continuity of the Fourier transform.

Proof. By equations (11.7) and (11.9) we have

$$S\big((\Gamma(-iI)\Phi) \diamond \widetilde{\delta}_0\big)(\xi) = S\Phi(-i\xi)e^{-\frac{1}{2}\langle \xi, \xi \rangle}.$$

But the right hand side is the S-transform of $\widehat{\Phi}$. Hence $\widehat{\Phi} = (\Gamma(-iI)\Phi) \diamond \widetilde{\delta}_0$. The other equality is obvious since $\widetilde{\delta}_0 \diamond g_{-2} = 1$. $\qquad\square$

11.3 Basic properties

The Fourier transform $\mathcal{F} : (\mathcal{E})^*_\beta \to (\mathcal{E})^*_\beta$ has many properties similar to those of the finite dimensional Fourier transform. We first consider differentiation and multiplication. Recall these facts from Chapter 9:

(1) For any $\eta \in \mathcal{E}$, the differential operator D_η has a unique extension to a continuous operator $\widetilde{D}_\eta : (\mathcal{E})^*_\beta \to (\mathcal{E})^*_\beta$ (Theorem 9.10).
(2) For any $y \in \mathcal{E}'$, the adjoint operator $D^*_y : (\mathcal{E})^*_\beta \to (\mathcal{E})^*_\beta$ is continuous (Theorem 9.12).
(3) For any $\eta \in \mathcal{E}$, the multiplication operator Q_η has a unique extension to a continuous operator $\widetilde{Q}_\eta : (\mathcal{E})^*_\beta \to (\mathcal{E})^*_\beta$ (Theorem 9.18).

Theorem 11.11. *For any $\eta \in \mathcal{E}$ and $x \in \mathcal{E}'$, the equalities hold:*

$$\mathcal{F}\widetilde{D}_\eta = i\widetilde{Q}_\eta\mathcal{F},$$
$$\mathcal{F}D^*_x = -iD^*_x\mathcal{F},$$
$$\mathcal{F}\widetilde{Q}_\eta = i\widetilde{D}_\eta\mathcal{F}.$$

Proof. For simplicity, let $\varphi_\xi =: e^{\langle \cdot, \xi \rangle}:$ for $\xi \in \mathcal{E}$. It is easy to check that

$$\widetilde{D}_\eta \varphi_\xi = \langle \xi, \eta \rangle \varphi_\xi.$$

Hence by Example 11.2

$$\mathcal{F}\widetilde{D}_\eta \varphi_\xi = \langle \xi, \eta \rangle \widetilde{\delta}_{-i\xi}. \tag{11.10}$$

On the other hand,

$$i\widetilde{Q}_\eta \mathcal{F}\varphi_\xi = i\widetilde{Q}_\eta \widetilde{\delta}_{-i\xi} = i\langle \cdot, \eta \rangle \widetilde{\delta}_{-i\xi}. \tag{11.11}$$

But for any test function $\psi \in (\mathcal{E})_\beta$, we have

$$\langle\!\langle i\langle \cdot, \eta \rangle \widetilde{\delta}_{-i\xi}, \psi \rangle\!\rangle = i\langle\!\langle \widetilde{\delta}_{-i\xi}, \langle \cdot, \eta \rangle \psi \rangle\!\rangle$$
$$= i\langle -i\xi, \eta \rangle \psi(-i\xi)$$
$$= \langle\!\langle \langle \xi, \eta \rangle \widetilde{\delta}_{-i\xi}, \psi \rangle\!\rangle.$$

This shows that $i\langle \cdot, \eta \rangle \widetilde{\delta}_{-i\xi} = \langle \xi, \eta \rangle \widetilde{\delta}_{-i\xi}$. Thus in view of equation (11.11),

$$i\widetilde{Q}_\eta \mathcal{F}\varphi_\xi = \langle \xi, \eta \rangle \widetilde{\delta}_{-i\xi}. \tag{11.12}$$

From equations (11.10) and (11.12) we get

$$\mathcal{F}\widetilde{D}_\eta \varphi_\xi = i\widetilde{Q}_\eta \mathcal{F}\varphi_\xi, \quad \forall \xi \in \mathcal{E}. \tag{11.13}$$

Now, note that the linear span of the set $\{\varphi_\xi; \xi \in \mathcal{E}\}$ is dense in $(\mathcal{E})_\beta^*$ and that the operators $\mathcal{F}, \widetilde{D}_\eta, \widetilde{Q}_\eta$ are continuous from $(\mathcal{E})_\beta^*$ into itself. Hence equation (11.13) implies that $\mathcal{F}\widetilde{D}_\eta = i\widetilde{Q}_\eta \mathcal{F}$.

To prove the second equality, note that by Theorem 9.13 and the definition of the Fourier transform

$$S(D_x^* \mathcal{F}\Phi)(\xi) = \langle x, \xi \rangle S(\mathcal{F}\Phi)(\xi)$$
$$= \langle x, \xi \rangle \langle\!\langle \Phi, e^{-i\langle \cdot, \xi \rangle} \rangle\!\rangle. \tag{11.14}$$

On the other hand, we have $D_x e^{-i\langle \cdot, \xi \rangle} = -i\langle x, \xi \rangle e^{-i\langle \cdot, \xi \rangle}$ and so

$$S(\mathcal{F}D_x^* \Phi)(\xi) = \langle\!\langle D_x^* \Phi, e^{-i\langle \cdot, \xi \rangle} \rangle\!\rangle$$
$$= \langle\!\langle \Phi, D_x e^{-i\langle \cdot, \xi \rangle} \rangle\!\rangle$$
$$= \langle\!\langle \Phi, -i\langle x, \xi \rangle e^{-i\langle \cdot, \xi \rangle} \rangle\!\rangle$$
$$= -i\langle x, \xi \rangle \langle\!\langle \Phi, e^{-i\langle \cdot, \xi \rangle} \rangle\!\rangle. \tag{11.15}$$

Equations (11.14) and (11.15) imply that $S(\mathcal{F}D_x^*\Phi)(\xi) = -iS(D_x^*\mathcal{F}\Phi)(\xi)$ for all $\xi \in \mathcal{E}_c$ and $\Phi \in (\mathcal{E})_\beta^*$. Hence $\mathcal{F}D_x^* = -iD_x^*\mathcal{F}$.

Finally, note that $\widetilde{Q}_\eta = \widetilde{D}_\eta + D_\eta^*$ by Theorem 9.18. Hence

$$\mathcal{F}\widetilde{Q}_\eta = \mathcal{F}(\widetilde{D}_\eta + D_\eta^*)$$
$$= i\widetilde{Q}_\eta \mathcal{F} - iD_\eta^* \mathcal{F}$$
$$= i\widetilde{D}_\eta \mathcal{F}. \qquad \square$$

The last theorem can be extended to some integral kernel operators. Recall the following facts from Chapter 10:

(1) Let $\zeta \in \mathcal{S}_c^{\otimes n}$. Then the integral kernel operator $\Xi_{0,n}(\zeta)$ is continuous from $(\mathcal{S})_\beta$ into itself and extends by continuity to a continuous linear operator $\widetilde{\Xi}_{0,n}(\zeta) : (\mathcal{S})_\beta^* \to (\mathcal{S})_\beta^*$ (Corollary 10.7).

(2) Let $y \in (\mathcal{S}_c')^{\otimes n}$. Then the integral kernel operator $\Xi_{n,0}(y)$ extends by continuity to a continuous linear operator $\widetilde{\Xi}_{n,0}(y) : (\mathcal{S})_\beta^* \to (\mathcal{S})_\beta^*$ (Theorem 10.6).

Furthermore, suppose $\zeta \in \mathcal{S}_c^{\otimes n}$. Let Q_ζ denote the multiplication by $\langle \cdot^{\otimes n}, \zeta \rangle$. It is easy to check that the operator $Q_\zeta : (\mathcal{S})_\beta \to (\mathcal{S})_\beta$ is continuous and extends by continuity to a continuous linear operator $\widetilde{Q}_\zeta : (\mathcal{S})_\beta^* \to (\mathcal{S})_\beta^*$.

The same argument as in the proof of Theorem 11.11 can be used to prove the next theorem.

Theorem 11.12. *For any $\zeta \in \mathcal{S}_c^{\otimes n}$ and $y \in (\mathcal{S}_c')^{\otimes n}$, the equalities hold:*

$$\mathcal{F}\widetilde{\Xi}_{0,n}(\zeta) = i^n \widetilde{Q}_\zeta \mathcal{F},$$

$$\mathcal{F}\widetilde{\Xi}_{n,0}(y) = (-i)^n \widetilde{\Xi}_{n,0}(y)\mathcal{F},$$

$$\mathcal{F}\widetilde{Q}_\zeta = i^n \widetilde{\Xi}_{0,n}(\zeta)\mathcal{F}.$$

Next we will derive the relationship between the Fourier transform and translation operators. Recall that from §10.5 we have the facts:

(1) For any $x \in \mathcal{E}'$, the translation operator T_x is continuous from $(\mathcal{E})_\beta$ into itself and its adjoint $T_x^* : (\mathcal{E})_\beta^* \to (\mathcal{E})_\beta^*$ is continuous (Theorem 10.21).

(2) For any $\xi \in \mathcal{E}$, the translation operator T_ξ has a unique extension to a continuous operator $\widetilde{T}_\xi : (\mathcal{E})_\beta^* \to (\mathcal{E})_\beta^*$.

Theorem 11.13. *For any $\eta \in \mathcal{E}$, $x \in \mathcal{E}'$, and $\Phi \in (\mathcal{E})_\beta^*$, the equalities hold:*

$$\mathcal{F}\widetilde{T}_\eta\Phi = e^{i\langle\cdot,\eta\rangle}\mathcal{F}\Phi, \quad \mathcal{F}\big(e^{-i\langle\cdot,\eta\rangle}\Phi\big) = \widetilde{T}_\eta\mathcal{F}\Phi,$$

$$\mathcal{F}T_x^*\Phi =: e^{-i\langle\cdot,x\rangle} : \diamond (\mathcal{F}\Phi), \quad \mathcal{F}\big(:e^{i\langle\cdot,x\rangle}: \diamond\Phi\big) = T_x^*\mathcal{F}\Phi,$$

where $\diamond$ denotes the Wick product.

Proof. Let $\varphi_\xi \equiv :e^{\langle\cdot,\xi\rangle}:$, $\xi \in \mathcal{E}$. It is easy to check that

$$\widetilde{T}_\eta\varphi_\xi = e^{\langle\xi,\eta\rangle}\varphi_\xi.$$

Hence by Example 11.2,

$$\mathcal{F}\widetilde{T}_\eta\varphi_\xi = e^{\langle\xi,\eta\rangle}\widetilde{\delta}_{-i\xi}. \tag{11.16}$$

On the other hand, by the same argument as in the proof of Theorem 11.11, we have

$$e^{i\langle\cdot,\eta\rangle}\widetilde{\delta}_{-i\xi} = e^{\langle\xi,\eta\rangle}\widetilde{\delta}_{-i\xi}.$$

By using this fact, we get

$$e^{i\langle\cdot,\eta\rangle}\mathcal{F}\varphi_\xi = e^{i\langle\cdot,\eta\rangle}\widetilde{\delta}_{-i\xi} = e^{\langle\xi,\eta\rangle}\widetilde{\delta}_{-i\xi}. \tag{11.17}$$

It follows from equations (11.16) and (11.17) that $\mathcal{F}\widetilde{T}_\eta\varphi_\xi = e^{i\langle\cdot,\eta\rangle}\mathcal{F}\varphi_\xi$ for all $\xi \in \mathcal{E}$. Thus by the same argument as in the proof of Theorem 11.11 we can conclude that $\mathcal{F}\widetilde{T}_\eta = e^{i\langle\cdot,\eta\rangle}\mathcal{F}$. Similarly, we can prove the second equality. The other two equalities can be derived directly by checking the S-transforms of both sides.   □

The Fourier transform $\mathcal{F}$ is related to convolution and the Wick product by formulas similar to those for the finite dimensional Fourier transform. Recall from §8.4 that the convolution and the Wick product of Φ and Ψ in $(\mathcal{E})_\beta^*$ are defined by

$$\Phi * \Psi = \mathcal{T}^{-1}\big((\mathcal{T}\Phi)(\mathcal{T}\Psi)\big), \quad \Phi \diamond \Psi = S^{-1}\big((S\Phi)(S\Psi)\big).$$

Moreover, we have the identities

$$\Phi * \Psi = \Phi \diamond \Psi \diamond g_{-2}, \quad \Phi \diamond \Psi = \Phi * \Psi * g_{-1/2},$$

where g_c is the Gaussian white noise function with parameter c (Example 8.3).

Theorem 11.14. *For any Φ and Ψ in $(\mathcal{E})_\beta^*$, the equalities hold:*

$$(\Phi \diamond \Psi)\widehat{} = \widehat{\Phi} * \widehat{\Psi}, \quad (\Phi * \Psi)\widehat{} = \widehat{\Phi} \diamond \widehat{\Psi}.$$

Proof. By using the fact that $Sg_{-2}(\xi) = \exp[\frac{1}{2}\langle \xi, \xi \rangle]$, we can check easily that

$$S(\Phi \diamond \Psi)\widehat{}\,(\xi) = S\widehat{\Phi}(\xi)\, S\widehat{\Psi}(\xi)\, Sg_{-2}(\xi), \quad \xi \in \mathcal{E}_c.$$

Hence $(\Phi \diamond \Psi)\widehat{} = \widehat{\Phi} \diamond \widehat{\Psi} \diamond g_{-2} = \widehat{\Phi} * \widehat{\Psi}$. To prove the other equality, note that $\widehat{g}_{-2} = g_{-1/2}$. Thus by the first equality of this theorem

$$\begin{aligned}
(\Phi * \Psi)\widehat{} &= (\Phi \diamond \Psi \diamond g_{-2})\widehat{} \\
&= \widehat{\Phi} * \widehat{\Psi} * g_{-1/2} \\
&= \widehat{\Phi} \diamond \widehat{\Psi}.
\end{aligned}$$

□

Recall from §6.3 that every test function $\varphi \in (\mathcal{E})_\beta$ has a unique analytic extension $\varphi(x), x \in \mathcal{E}_c'$. We will use the same notation φ to denote its analytic extension.

Definition 11.15. Let each $\varphi \in (\mathcal{E})_\beta$. The *scaling* of φ by $z \in \mathbb{C}$ is defined to be the function $S_z\varphi(x) = \varphi(zx), x \in \mathcal{E}'$.

We will show that the scaling operator S_z is a continuous linear operator from $(\mathcal{E})_\beta$ into itself. For the proof, we need to derive identities for the scaling of the Hermite polynomial $: x^n :_{\sigma^2}$ and the Wick tensor $: x^{\otimes n} :$ defined in §5.1.

Lemma 11.16. *For any $z \in \mathbb{C}$, the identity holds for all $x \in \mathbb{R}$*

$$:(zx)^n:_{\sigma^2} = \sum_{k=0}^{[n/2]} \binom{n}{2k}(2k-1)!!(-\sigma^2)^k z^{n-2k}(1-z^2)^k :x^{n-2k}:_{\sigma^2}.$$

Proof. From the generating function in equation (5.2),

$$\sum_{n=0}^{\infty} \frac{t^n}{n!} :(zx)^n:_{\sigma^2} = e^{zxt - \frac{1}{2}\sigma^2 t^2}$$

$$= e^{-\frac{1}{2}\sigma^2(1-z^2)t^2} e^{x(zt) - \frac{1}{2}\sigma^2(zt)^2}$$

$$= e^{-\frac{1}{2}\sigma^2(1-z^2)t^2} \sum_{n=0}^{\infty} \frac{(zt)^n}{n!} :x^n:_{\sigma^2}.$$

By comparing the coefficients of t^n on both sides, we get the identity in the lemma. $\square$

Lemma 11.17. *For any $z \in \mathbb{C}$, the identity holds for all $x \in \mathcal{E}'$*

$$:(zx)^{\otimes n}: = \sum_{k=0}^{[n/2]} \binom{n}{2k}(2k-1)!! z^{n-2k}(z^2-1)^k :x^{\otimes(n-2k)}: \widehat{\otimes} \tau^{\otimes k}.$$

Proof. For any $\xi \in \mathcal{E}$, we use Lemmas 5.2 and 11.16 to get

$$\langle :(zx)^{\otimes n}:, \xi^{\otimes n} \rangle$$

$$= :(z\langle x, \xi \rangle)^n:_{|\xi|_0^2}$$

$$= \sum_{k=0}^{[n/2]} \binom{n}{2k}(2k-1)!!(-1)^k |\xi|_0^{2k} z^{n-2k}(1-z^2)^k :\langle x, \xi \rangle^{n-2k}:_{|\xi|_0^2}$$

$$= \sum_{k=0}^{[n/2]} \binom{n}{2k}(2k-1)!! z^{n-2k}(z^2-1)^k \langle :x^{\otimes(n-2k)}\widehat{\otimes}\tau^{\otimes k}, \xi^{\otimes n} \rangle$$

$$= \Big\langle \sum_{k=0}^{[n/2]} \binom{n}{2k}(2k-1)!! z^{n-2k}(z^2-1)^k :x^{\otimes(n-2k)}\widehat{\otimes}\tau^{\otimes k}, \xi^{\otimes n} \Big\rangle.$$

Hence by the polarization identity, we obtain the equality in the lemma. $\square$

Now we are ready to prove the continuity of the scaling operator. The proof is very similar to the one for the continuity of the operator Θ in Theorem 6.2.

Theorem 11.18. *For any* $z \in \mathbb{C}$, *the scaling operator* S_z *is continuous from* $(\mathcal{E})_\beta$ *into itself. In fact, for any* $p \geq 0$, *if* $q > p$ *satisfies*

$$\lambda_1^{2(q-p)} > \max\{2^{1-\beta}|z^2 - 1||\tau|_{-p}, \, 2^{1-\beta}|z|^2\},$$

then the following inequality holds for all $\varphi \in (\mathcal{E})_\beta$

$$\|S_z\varphi\|_{p,\beta} \leq K_{p,q,\beta}\|\varphi\|_{q,\beta},$$

where $K_{p,q,\beta}$ *is the constant given by*

$$K_{p,q,\beta} = \left[1 - (2^{1-\beta}\lambda_1^{-2(q-p)}|z^2 - 1||\tau|_{-p})^2\right]^{-1/2}$$
$$\times \left(1 - 2^{1-\beta}|z|^2\lambda_1^{-2(q-p)}\right)^{-1/2}.$$

Proof. The proof is similar to Theorem 6.2. Let $\varphi \in (\mathcal{E})_\beta$ be represented by

$$\varphi = \sum_{n=0}^{\infty} \langle :\cdot^{\otimes n}:, f_n \rangle.$$

We can use Lemma 11.17 together with changing the order of summation (which can be justified by the same argument as in the proof of Proposition 6.1) to derive that

$$S_z\varphi = \sum_{n=0}^{\infty} \langle :\cdot^{\otimes n}:, u_n \rangle,$$

where u_n is given by

$$u_n = z^n \sum_{k=0}^{\infty} \binom{n+2k}{2k} (2k-1)!!(z^2-1)^k \langle \tau^{\otimes k}, f_{n+2k} \rangle.$$

By the same argument as in estimating $|g_n|_p$ in the proof of Theorem 6.2, we get the following estimate

$$|u_n|_p^2 \leq \|\varphi\|_{q,\beta}^2 |z|^{2n} \frac{1}{(n!)^{1+\beta}} \lambda_1^{-2(q-p)n}$$
$$\times 2^{(1-\beta)n} \left[1 - (2^{1-\beta}\lambda_1^{-2(q-p)}|z^2-1||\tau|_{-p})^2\right]^{-1}.$$

With this estimate for $|u_n|_p$, we can derive the inequality for $\|S_z\varphi\|_{p,\beta}$ in this lemma by the same argument for $\|\Theta\varphi\|_{p,\beta}$ as in the proof of Theorem 6.2. $\square$

It follows from Theorem 11.18 that the adjoint operator S_z^* is continuous from $(\mathcal{E})_\beta^*$ into itself. However, the operator S_z has no continuous extension to $(\mathcal{E})_\beta^*$ unless $z = \pm 1$ ($S_1 = I$ and $S_{-1} = \Gamma(-I)$).

Theorem 11.19. *For $z \in \mathbb{C}$, the equalities hold*

$$S_z^* \mathcal{F} = \left(\mathcal{F} S_z^*\right) \diamond g_{\frac{z^2}{1-z^2}}, \quad \text{if } z \neq \pm 1,$$

$$S_z^* \mathcal{F} = \mathcal{F} S_z^*, \quad \text{if } z = \pm 1,$$

where g_c is the Gaussian white noise function with parameter c.

Proof. It is straightforward to check that

$$S(\mathcal{F} S_z^* \Phi)(\xi) = (S\Phi)(-iz\xi)e^{-\frac{1}{2}z^2\langle\xi,\xi\rangle}, \quad \xi \in \mathcal{E}_c,$$

$$S(S_z^* \mathcal{F}\Phi)(\xi) = (S\Phi)(-iz\xi)e^{-\frac{1}{2}\langle\xi,\xi\rangle}, \quad \xi \in \mathcal{E}_c.$$

Thus if $z = \pm 1$, we have $\mathcal{F} S_z^* \Phi = S_z^* \mathcal{F}\Phi$ for all $\Phi \in (\mathcal{E})_\beta^*$, i.e., $\mathcal{F} S_z^* = S_z^* \mathcal{F}$. On the other hand, if $z \neq \pm 1$, we have

$$Sg_{\frac{z^2}{1-z^2}}(\xi) = e^{-\frac{1}{2}(1-z^2)\langle\xi,\xi\rangle}, \quad \xi \in \mathcal{E}_c.$$

It is easy to see that

$$S(S_z^* \mathcal{F}\Phi)(\xi) = S\left((\mathcal{F} S_z^* \Phi) \diamond g_{\frac{z^2}{1-z^2}}\right)(\xi), \quad \xi \in \mathcal{E}_c.$$

Hence $S_z^* \mathcal{F}\Phi = (\mathcal{F} S_z^* \Phi) \diamond g_{\frac{z^2}{1-z^2}}$ for all $\Phi \in (\mathcal{E})_\beta^*$, i.e.,

$$S_z^* \mathcal{F} = \left(\mathcal{F} S_z^*\right) \diamond g_{\frac{z^2}{1-z^2}}. \qquad \square$$

In finite dimensional distribution theory, the Fourier transform is a mapping from the space $\mathcal{S}(\mathbb{R}^r)$ of test functions onto itself. However, this is not the case for the Fourier transform on $(\mathcal{E})_\beta$. For instance, $1 \in (\mathcal{E})_\beta$, but $\widehat{1} = \widetilde{\delta}_0 \notin (\mathcal{E})_\beta$.

Theorem 11.20. *If $\varphi \in (\mathcal{E})_\beta$, then $\widehat{\varphi} \diamond g_{-2} \in (\mathcal{E})_\beta$ and for any $p \geq 0$,*

$$\|\widehat{\varphi} \diamond g_{-2}\|_{p,\beta} = \|\varphi\|_{p,\beta}.$$

Conversely, if $\widehat{\varphi} \diamond g_{-2} \in (\mathcal{E})_\beta$, then $\varphi \in (\mathcal{E})_\beta$ and the above equality holds.

Remark. In fact, the proof below also shows that $\Phi \in (\mathcal{E}_p)_\beta^*$ if and only if $\widehat{\Phi} \diamond g_{-2} \in (\mathcal{E}_p)_\beta^*$ for any $p \geq 0$ and

$$\|\widehat{\Phi} \diamond g_{-2}\|_{-p,-\beta} = \|\Phi\|_{-p,-\beta}.$$

Proof. By Theorem 11.10 we have $\widehat{\varphi} \diamond g_{-2} = \Gamma(-iI)\varphi$. Note that if $\varphi = \sum_{n=0}^{\infty} \langle :\cdot^{\otimes n}:, f_n \rangle$, then

$$\Gamma(-iI)\varphi = \sum_{n=0}^{\infty} \langle :\cdot^{\otimes n}:, (-i)^n f_n \rangle.$$

Obviously $\|\Gamma(-iI)\varphi\|_{p,\beta} = \|\varphi\|_{p,\beta}$. Hence $\|\widehat{\varphi} \diamond g_{-2}\|_{p,\beta} = \|\varphi\|_{p,\beta}$ for any $p \geq 0$. Moreover, this implies that $\varphi \in (\mathcal{E})_\beta$ if and only if $\widehat{\varphi} \diamond g_{-2} \in (\mathcal{E})_\beta$. □

At the end of this section we consider Fourier transforms of generalized functions induced by functions in $L^p(\mu)$. Recall from §8.5 that if $f \in L^p(\mu)$, $1 < p \leq \infty$, then it induces a generalized function $\Phi_f \in (\mathcal{E})^*$ defined by

$$\langle\langle \Phi_f, \varphi \rangle\rangle = \int_{\mathcal{E}'} \varphi(x) f(x)\, d\mu(x), \quad \varphi \in (\mathcal{E}).$$

Theorem 11.21. *There exist positive constants C and r depending only on $\lambda_j, j \geq 1, \alpha$ such that for any $f \in L^p(\mu), 1 < p \leq \infty$,*

$$\|\widehat{\Phi}_f \diamond g_{-2}\|_{-(n-1)r} \leq C^{1-\frac{1}{n}} \|f\|_{L^p(\mu)},$$

where n is any integer such that $n \geq \frac{p}{p-1}$. (For the constants λ_j's and α, see §4.2.)

Proof. Let C and r be the constants from Theorem 8.18. Then by Theorem 8.21 we have

$$\|\Phi_f\|_{-(n-1)r} \leq C^{1-\frac{1}{n}} \|f\|_{L^p(\mu)},$$

where n is any integer such that $n \geq \frac{p}{p-1}$. On the other hand, by equation (11.8)

$$\|\Gamma(-iI)\Phi_f\|_{-(n-1)r} = \|\Phi_f\|_{-(n-1)r}.$$

Therefore

$$\|\Gamma(-iI)\Phi_f\|_{-(n-1)r} \leq C^{1-\frac{1}{n}} \|f\|_{L^p(\mu)}.$$

But by Theorem 11.10, $\widehat{\Phi}_f \diamond g_{-2} = \Gamma(-iI)\Phi_f$. Hence we get the inequality in the lemma immediately. □

11.4 Decomposition of the Fourier transform

In this section we consider the Gel'fand triple $(\mathcal{S})_\beta \subset (L^2) \subset (\mathcal{S})_\beta^*$. The Fourier transform $\mathcal{F}$ is a continuous linear operator from $(\mathcal{S})_\beta^*$ into itself. Hence it is also continuous from $(\mathcal{S})_\beta$ into $(\mathcal{S})_\beta^*$. Thus by Theorem 10.28 it can be decomposed into a series of integral kernel operators. We will find the associated kernel functions of these integral kernel operators.

Lemma 11.22. *The symbol of the Fourier transform $\mathcal{F}$ is given by*

$$F(\xi, \eta) = e^{-i\langle \xi, \eta \rangle - \frac{1}{2}\langle \eta, \eta \rangle}, \quad \xi, \eta \in \mathcal{S}_c.$$

Proof. From Example 11.2 we have $\mathcal{F} : e^{\langle \cdot, \xi \rangle} := \widetilde{\delta}_{-i\xi}$. Hence by Definition 10.27 the symbol F of $\mathcal{F}$ is given by

$$F(\xi, \eta) = \langle\!\langle \mathcal{F} : e^{\langle \cdot, \xi \rangle} :, : e^{\langle \cdot, \eta \rangle} : \rangle\!\rangle$$

$$= \langle\!\langle \widetilde{\delta}_{-i\xi}, : e^{\langle \cdot, \eta \rangle} : \rangle\!\rangle$$

$$= e^{-i\langle \xi, \eta \rangle - \frac{1}{2}\langle \eta, \eta \rangle}.$$
$\square$

We will find the kernel function $\theta_{j,k}$ from equation (10.21) (instead of equation (10.22)) by expanding the following function into a series

$$f(\xi, \eta) \equiv e^{-\langle \xi, \eta \rangle} F(\xi, \eta) = e^{-(1+i)\langle \xi, \eta \rangle - \frac{1}{2}\langle \eta, \eta \rangle}.$$

Note that

$$e^{-(1+i)\langle \xi, \eta \rangle} = \sum_{n=0}^{\infty} \frac{(-1)^n(1+i)^n}{n!} \langle \xi, \eta \rangle^n,$$

$$e^{-\frac{1}{2}\langle \eta, \eta \rangle} = \sum_{n=0}^{\infty} \frac{(-1)^n}{2^n n!} \langle \eta, \eta \rangle^n.$$

By multiplying these two series together, we get

$$f(\xi, \eta) = \sum_{n=0}^{\infty} (-1)^n \sum_{k=0}^{n} \frac{(1+i)^k}{2^{n-k} k!(n-k)!} \langle \xi, \eta \rangle^k \langle \eta, \eta \rangle^{n-k}.$$

Now, define a generalized function $G_{2n-k,k} \in (\mathcal{S}_c')^{\otimes 2n}$ by

$$G_{2n-k,k}(s_1, \ldots, s_{2n-k}, t_1, \ldots, t_k)$$

$$= \tau(s_1, t_1) \cdots \tau(s_k, t_k)\tau(s_{k+1}, s_{k+2}) \cdots \tau(s_{2n-k-1}, s_{2n-k}),$$

where τ is the trace operator. It is easy to check that

$$\langle \xi, \eta \rangle^k \langle \eta, \eta \rangle^{n-k} = \langle G_{2n-k,k}, \eta^{\otimes(2n-k)} \otimes \xi^{\otimes k} \rangle.$$

Therefore

$$f(\xi, \eta) = \sum_{n=0}^{\infty} (-1)^n \sum_{k=0}^{n} \frac{(1+i)^k}{2^{n-k} k!(n-k)!} \langle G_{2n-k,k}, \eta^{\otimes(2n-k)} \otimes \xi^{\otimes k} \rangle$$

$$= \sum_{n=0}^{\infty} \sum_{k=0}^{n} \left\langle (-1)^n \frac{(1+i)^k}{2^{n-k} k!(n-k)!} G_{2n-k,k}, \eta^{\otimes(2n-k)} \otimes \xi^{\otimes k} \right\rangle.$$

Hence by equation (10.21) and Theorem 10.28 we obtain the decomposition of the Fourier transform $\mathcal{F}$ stated in the next theorem.

Theorem 11.23. *The Fourier transform $\mathcal{F}$ can be decomposed as*

$$\mathcal{F} = \sum_{n=0}^{\infty} \sum_{k=0}^{n} \Xi_{2n-k,k}(\theta_{2n-k,k}),$$

where the kernel function $\theta_{2n-k,k}$ is given by

$$\theta_{2n-k,k}(s_1, \ldots, s_{2n-k}, t_1, \ldots, t_k) = (-1)^n \frac{(1+i)^k}{2^{n-k}k!(n-k)!}$$

$$\times \tau(s_1, t_1) \cdots \tau(s_k, t_k) \tau(s_{k+1}, s_{k+2}) \cdots \tau(s_{2n-m-1}, s_{2n-m}).$$

We make a small remark about this decomposition. It follows from Lemma 10.15 that $\theta_{2n-k,k} \in (\mathcal{S}_c')^{\otimes(2n-k)} \otimes \mathcal{S}_c^{\otimes k}$. Hence in view of Theorem 10.6 each integral kernel operator $\Xi_{2n-k,k}(\theta_{2n-k,k})$ extends by continuity to a continuous linear operator from $(\mathcal{S})_\beta^*$ into itself. This is consistent with the fact that the Fourier transform is continuous from $(\mathcal{S})_\beta^*$ into itself.

11.5 Fourier-Gauss transforms

For the finite dimensional Fourier transform we have $\langle \mathcal{F}f, g \rangle = \langle f, \mathcal{F}g \rangle$ for all test functions f and g in $\mathcal{S}(\mathbb{R}^r)$. Thus the extension $\widetilde{\mathcal{F}}$ of the Fourier transform $\mathcal{F}$ to the space $\mathcal{S}'(\mathbb{R}^r)$ of generalized functions is nothing but its adjoint, i.e., $\widetilde{\mathcal{F}} = \mathcal{F}^*$. In white noise distribution theory the Fourier transform $\mathcal{F}$ is a continuous linear operator from $(\mathcal{E})_\beta^*$ into itself. We will show that $\mathcal{F}$ is the adjoint of a continuous linear operator from $(\mathcal{E})_\beta$ into itself. Thus the Fourier transform on $(\mathcal{E})_\beta^*$ resembles the finite dimensional Fourier transform in the sense that it is the adjoint of a continuous linear operator from the space of test functions into itself.

The fact that $\mathcal{F}$ is the adjoint of a continuous linear operator from $(\mathcal{E})$ into itself has been proved in Hida et al. [57], the book by Hida et al. [58], and Y.-J. Lee [122]. We will consider the general case $(\mathcal{S})_\beta^*$ by studying the Fourier-Gauss transform introduced in Y.-J. Lee [120] [121].

Definition 11.24. Let $a, b \in \mathbb{C}$. The *Fourier-Gauss transform* $\mathcal{G}_{a,b}\varphi$ of a function φ in $(\mathcal{E})_\beta$ is defined to be the function

$$\mathcal{G}_{a,b}\varphi(y) = \int_{\mathcal{E}'} \varphi(ax + by) \, d\mu(x).$$

We will show that $\mathcal{G}_{a,b}$ is a continuous linear operator from $(\mathcal{E})_\beta$ into itself. But first we need to prepare some lemmas about the Hermite polynomial $:x^n:_{\sigma^2}$ and the Wick tensor $:x^{\otimes n}:$ defined in §5.1.

Lemma 11.25. *Let μ_{σ^2} be the Gaussian measure on $\mathbb{R}$ with mean 0 and variance σ^2. Then for any $a \in \mathbb{C}$ and any $n \geq 0$,*

$$\int_{\mathbb{R}} :(ax)^{2n}:_{\sigma^2} d\mu_{\sigma^2}(x) = (2n-1)!!(a^2-1)^n\sigma^{2n},$$

$$\int_{\mathbb{R}} :(ax)^{2n+1}:_{\sigma^2} d\mu_{\sigma^2}(x) = 0.$$

Proof. From the generating function in equation (5.2),

$$\sum_{n=0}^{\infty} \frac{t^n}{n!} :(ax)^n:_{\sigma^2} = e^{atx-\frac{1}{2}\sigma^2 t^2}.$$

We can integrate both sides with respect to μ_{σ^2}, we get

$$\sum_{n=0}^{\infty} \frac{t^n}{n!} \int_{\mathbb{R}} :(ax)^n:_{\sigma^2} d\mu_{\sigma^2}(x) = e^{\frac{1}{2}(a^2-1)\sigma^2 t^2}.$$

This identity yields the equality in the lemma. ☐

Lemma 11.26. *For any $a \in \mathbb{C}, y \in \mathbb{R}$, and $n \geq 0$,*

$$\int_{\mathbb{R}} :(ax+y)^n:_{\sigma^2} d\mu_{\sigma^2}(x) = \sum_{k=0}^{[n/2]} \binom{n}{2k}(2k-1)!!(a^2-1)^k\sigma^{2k}y^{n-2k}.$$

Proof. From Lemma 7.12 we get

$$:(ax+y)^n:_{\sigma^2} = \sum_{k=0}^{n} \binom{n}{k} :(ax)^k:_{\sigma^2} y^{n-k}.$$

By integrating both sides with respect to the Gaussian measure μ_{σ^2} and using Lemma 11.25, we get the equality in the lemma. ☐

Note that Lemma 11.26 is for Hermite polynomials with a parameter. The corresponding result for Wick tensors is given in the next lemma.

Lemma 11.27. *For any $a \in \mathbb{C}, y \in \mathcal{E}'$, and $n \geq 0$,*

$$\int_{\mathcal{E}'} :(ax+y)^{\otimes n}: d\mu(x) = \sum_{k=0}^{[n/2]} \binom{n}{2k}(2k-1)!!(a^2-1)^k y^{\otimes(n-2k)}\widehat{\otimes}\tau^{\otimes k}.$$

Proof. For any $\xi \in \mathcal{E}$, by Lemma 5.2,

$$\left\langle\!\left\langle \int_{\mathcal{E}'} :(ax+y)^{\otimes n}: d\mu(x), \xi^{\otimes n} \right\rangle\!\right\rangle = \int_{\mathcal{E}'} \left\langle\!\left\langle :(ax+y)^{\otimes n}:, \xi^{\otimes n}\right\rangle\!\right\rangle d\mu(x)$$

$$= \int_{\mathcal{E}'} :\langle ax+y, \xi\rangle^n:_{|\xi|_0^2} d\mu(x)$$

$$= \int_{\mathbb{R}} :(au + \langle y, \xi\rangle)^n:_{|\xi|_0^2} d\mu_{|\xi|_0^2}(u).$$

Then we apply Lemma 11.26 to get

$$\left\langle\!\left\langle \int_{\mathcal{E}'} :(ax+y)^{\otimes n}: d\mu(x), \xi^{\otimes n}\right\rangle\!\right\rangle$$

$$= \sum_{k=0}^{[n/2]} \binom{n}{2k}(2k-1)!!(a^2-1)^k |\xi|_0^{2k} \langle y, \xi\rangle^{n-2k}$$

$$= \sum_{k=0}^{[n/2]} \binom{n}{2k}(2k-1)!!(a^2-1)^k \langle y^{\otimes(n-2k)}\widehat{\otimes}\tau^{\otimes k}, \xi^{\otimes n}\rangle$$

$$= \left\langle\!\left\langle \sum_{k=0}^{[n/2]} \binom{n}{2k}(2k-1)!!(a^2-1)^k y^{\otimes(n-2k)}\widehat{\otimes}\tau^{\otimes k}, \xi^{\otimes n}\right\rangle\!\right\rangle.$$

The above equalities hold for all $\xi^{\otimes n}$ with $\xi \in \mathcal{E}_c$. Thus by the polarization identity (see Appendix B), the equality in the lemma holds. $\qquad\square$

Now we can use Lemma 11.27 to represent the Fourier-Gauss transform $\mathcal{G}_{a,b}$ in terms of the Wiener-Itô decomposition.

Theorem 11.28. *Let $\varphi = \sum_{n=0}^{\infty}\langle :\cdot^{\otimes n}:, f_n\rangle \in (\mathcal{E})_\beta$. Then for any $a, b \in \mathbb{C}$,*

$$\mathcal{G}_{a,b}\varphi = \sum_{n=0}^{\infty}\langle\cdot^{\otimes n}, g_n\rangle = \sum_{n=0}^{\infty}\langle :\cdot^{\otimes n}:, h_n\rangle,$$

where g_n and h_n are given by

$$g_n = b^n \sum_{k=0}^{\infty} \binom{n+2k}{2k}(2k-1)!!(a^2-1)^k\langle\tau^{\otimes k}, f_{n+2k}\rangle, \qquad (11.18)$$

$$h_n = b^n \sum_{k=0}^{\infty} \binom{n+2k}{2k}(2k-1)!!(a^2+b^2-1)^k\langle\tau^{\otimes k}, f_{n+2k}\rangle. \qquad (11.19)$$

Remarks. (1) By comparing equation (11.19) with the function g_n for Θ in equation (6.3), and by comparing equation (11.18) with equation (8.23) for $\check{S}$, we see easily that

$$\Theta = \mathcal{G}_{i,1}, \quad \check{S} = \mathcal{G}_{1,1}.$$

(2) Note that from equation (11.18) $\mathcal{G}_{a,b} = \mathcal{G}_{c,d}$ if and only if $a = \pm c$ and $b = d$. This is also obvious from the definition 11.24 since the Gaussian measure μ is symmetric, i.e., $d\mu(-x) = d\mu(x)$.

Proof. Recall that $(E, \mathcal{E}'_r)$ is an abstract Wiener space if $r \geq \frac{\alpha}{2}$. Take and fix a particular $r \geq \frac{\alpha}{2}$. For any $0 \leq \beta < 1$, Fact 2.3 implies that there exists some $c \geq 0$ such that

$$\int_{\mathcal{E}'_r} \exp\left[c|x|_{-r}^{\frac{2}{1+\beta}}\right] d\mu(x) < \infty. \tag{11.20}$$

Let $\varphi_m = \sum_{n=0}^m \langle : \cdot^{\otimes n} :, f_n \rangle$. For any $p > r$, by Theorem 6.8 and the continuity of Θ, there exists $q \geq 0$ such that for all $m \geq 0$,

$$\begin{aligned}
|\varphi_m(x)| &\leq \|\Theta\varphi_m\|_{p,\beta} \exp\left[\frac{1}{2}(1+\beta)|x|_{-p}^{\frac{2}{1+\beta}}\right] \\
&\leq C\|\varphi_m\|_{q,\beta} \exp\left[\frac{1}{2}(1+\beta)\lambda_1^{-(p-r)\frac{2}{1+\beta}}|x|_{-r}^{\frac{2}{1+\beta}}\right] \\
&\leq C\|\varphi\|_{q,\beta} \exp\left[\frac{1}{2}(1+\beta)\lambda_1^{-(p-r)\frac{2}{1+\beta}}|x|_{-r}^{\frac{2}{1+\beta}}\right] \\
&\leq C\|\varphi\|_{q,\beta} \exp\left[c|x|_{-r}^{\frac{2}{1+\beta}}\right],
\end{aligned}$$

where we have chosen large p such that $\frac{1}{2}(1+\beta)\lambda_1^{-(p-r)\frac{2}{1+\beta}} < c$. Hence, in view of (11.20), we can apply the Lebesgue dominated convergence theorem to get

$$\mathcal{G}_{a,b}\varphi(y) = \sum_{n=0}^\infty \int_{\mathcal{E}'} \langle : (ax+by)^{\otimes n} :, f_n \rangle \, d\mu(x).$$

Then by Lemma 11.27,

$$\mathcal{G}_{a,b}\varphi(y) = \sum_{n=0}^\infty \sum_{k=0}^{[n/2]} \binom{n}{2k}(2k-1)!!(a^2-1)^k b^{n-2k}\langle y^{\otimes(n-2k)}\widehat{\otimes}\tau^{\otimes k}, f_n \rangle.$$

By changing the order of summation (which can be justified easily), we get

$$\begin{aligned}
\mathcal{G}_{a,b}\varphi(y) &= \sum_{k=0}^\infty \sum_{n=2k}^\infty \binom{n}{2k}(2k-1)!!(a^2-1)^k b^{n-2k}\langle y^{\otimes(n-2k)}\widehat{\otimes}\tau^{\otimes k}, f_n \rangle \\
&= \sum_{k=0}^\infty \sum_{m=0}^\infty \binom{m+2k}{2k}(2k-1)!!(a^2-1)^k b^m \langle y^{\otimes m}\widehat{\otimes}\tau^{\otimes k}, f_{m+2k} \rangle \\
&= \sum_{n=0}^\infty \left\langle y^{\otimes n}, b^n \sum_{k=0}^\infty \binom{n+2k}{2k}(2k-1)!!(a^2-1)^k \langle \tau^{\otimes k}, f_{n+2k} \rangle \right\rangle.
\end{aligned}$$

This proves the first equality in the lemma. To prove the other equality, suppose

$$\sum_{n=0}^{\infty} \langle \cdot^{\otimes n}, g_n \rangle = \sum_{n=0}^{\infty} \langle :\cdot^{\otimes n}:, h_n \rangle.$$

Then by equation (8.23) h_n is given by

$$h_n = \sum_{k=0}^{\infty} \binom{n+2k}{2k} (2k-1)!! \langle \tau^{\otimes k}, g_{n+2k} \rangle.$$

By using equation (11.18) we can derive

$$h_n = b^n \sum_{k=0}^{\infty} \sum_{j=0}^{\infty} \frac{(n+2k+2j)!}{n!2^{k+j}k!j!} b^{2k}(a^2-1)^j \langle \tau^{\otimes(k+j)}, f_{n+2k+2j} \rangle.$$

Finally, sum over $k+j = m$ for $m \geq 0$ to get

$$h_n = b^n \sum_{m=0}^{\infty} \sum_{k=0}^{m} \frac{(n+2m)!}{n!2^m k!(m-k)!} b^{2k}(a^2-1)^{m-k} \langle \tau^{\otimes m}, f_{n+2m} \rangle$$

$$= b^n \sum_{m=0}^{\infty} \frac{(n+2m)!}{n!2^m m!} \Big(\sum_{k=0}^{m} \frac{m!}{k!(m-k)!} b^{2k}(a^2-1)^{m-k} \Big) \langle \tau^{\otimes m}, f_{n+2m} \rangle$$

$$= b^n \sum_{m=0}^{\infty} \frac{(n+2m)!}{n!2^m m!} (a^2+b^2-1)^m \langle \tau^{\otimes m}, f_{n+2m} \rangle$$

$$= b^n \sum_{m=0}^{\infty} \binom{n+2m}{2m} (2m-1)!!(a^2+b^2-1)^m \langle \tau^{\otimes m}, f_{n+2m} \rangle. \qquad \square$$

Now we use the same idea as in the proof of Theorem 6.2 (for the operator Θ) to show that $\mathcal{G}_{a,b}$ is a continuous linear operator from $(\mathcal{E})_\beta$ into itself.

Theorem 11.29. *For any $a, b \in \mathbb{C}$, the linear operator $\mathcal{G}_{a,b}$ is continuous from $(\mathcal{E})_\beta$ into itself. In fact, for any $p \geq 0$, if $q > p$ satisfies the condition*

$$\lambda_1^{2(q-p)} > \max\{2^{1-\beta}|a^2+b^2-1| \, |\tau|_{-p}, 2^{1-\beta}|b|^2\},$$

then the following inequality holds for all $\varphi \in (\mathcal{E})_\beta$,

$$\|\mathcal{G}_{a,b}\varphi\|_{p,\beta} \leq C_{p,q,\beta} \|\varphi\|_{q,\beta},$$

where $C_{p,q,\beta}$ is the constant given by

$$C_{p,q,\beta} = \big[1 - (2^{1-\beta}\lambda_1^{-2(q-p)}|a^2+b^2-1| \, |\tau|_{-p})^2\big]^{-1/2}$$
$$\times \big(1 - 2^{1-\beta}\lambda_1^{-2(q-p)}|b|^2\big)^{-1/2}.$$

Proof. We use the same argument as in the proof of Theorem 6.2. Let $p \geq 0$ and $q > p$. Similarly to equation (6.4) for $|g_n|_p$, we have

$$|h_n|_p \leq \frac{1}{n!}|b|^n \lambda_1^{-(q-p)n} \sum_{k=0}^{\infty} \frac{(n+2k)!}{2^k k!}$$

$$\times \left[|a^2 + b^2 - 1||\tau|_{-p}\right]^k \lambda_1^{-2k(q-p)} |f_{n+2k}|_q.$$

Then by the same calculation as in the proof of Theorem 6.2,

$$|h_n|_p^2 \leq \|\varphi\|_{q,\beta}^2 \frac{1}{(n!)^{1+\beta}} |b|^{2n} \lambda_1^{-2(q-p)n}$$

$$\times 2^{(1-\beta)n} \left[1 - (2^{1-\beta}\lambda_1^{-2(q-p)}|a^2 + b^2 - 1||\tau|_{-p})^2\right]^{-1},$$

provided that $2^{1-\beta}\lambda_1^{-2(q-p)}|a^2 + b^2 + 1||\tau|_{-p} < 1$. Similarly to equation (6.5), we use the condition $\lambda_1^{2(q-p)} > 2^{1-\beta}|b|^2$ to derive

$$\|\mathcal{G}_{a,b}\varphi\|_{p,\beta}^2 = \sum_{n=0}^{\infty} (n!)^{1+\beta} |h_n|_p^2$$

$$\leq \|\varphi\|_{q,\beta}^2 \left[1 - (2^{1-\beta}\lambda_1^{-2(q-p)}|a^2 + b^2 - 1||\tau|_{-p})^2\right]^{-1}$$

$$\times \left(1 - 2^{1-\beta}\lambda_1^{-2(q-p)}|b|^2\right)^{-1}. \qquad \square$$

Observe that when $b = 0$ the operator $\mathcal{G}_{a,0}$ is not injective. However, when $b \neq 0$ the operator $\mathcal{G}_{a,b}$ is invertible.

Theorem 11.30. *For any $a, b, s, t \in \mathbb{C}$,*

$$\mathcal{G}_{s,t}\,\mathcal{G}_{a,b} = \mathcal{G}_{\pm\sqrt{a^2+b^2s^2},\,bt}.$$

In particular, if $b \neq 0$, then the operator $\mathcal{G}_{a,b}$ is invertible and

$$\mathcal{G}_{a,b}^{-1} = \mathcal{G}_{\pm ia/b, 1/b}.$$

Remark. Note that by the second remark of Theorem 11.28 we have $\mathcal{G}_{a,b} = \mathcal{G}_{-a,b}$ for any $a, b \in \mathbb{C}$. Also note that from the first remark of Theorem 11.28 $\Theta = \mathcal{G}_{i,1}$ and $\check{S} = \mathcal{G}_{1,1}$. Hence we see that $\Theta^{-1} = \check{S}$ (cf. Theorem 8.16).

Proof. Let $\varphi_\xi = :e^{\langle \cdot, \xi \rangle}:, \xi \in \mathcal{E}_c$. Then

$$\varphi_\xi = \sum_{n=0}^{\infty} \frac{1}{n!}\langle :\cdot^{\otimes n}:, \xi^{\otimes n}\rangle.$$

By Theorem 11.28 we have

$$\mathcal{G}_{a,b}\,\varphi_\xi = \sum_{n=0}^{\infty}\langle :\cdot^{\otimes n}:, h_n\rangle,$$

where h_n is given by

$$h_n = b^n \sum_{k=0}^{\infty} \binom{n+2k}{2k} (2k-1)!!(a^2+b^2-1)^k \Big\langle \tau^{\otimes k}, \frac{1}{(n+2k)!}\xi^{\otimes(n+2k)}\Big\rangle$$

$$= \frac{1}{n!}(b\xi)^{\otimes n} \sum_{k=0}^{\infty} \frac{1}{2^k k!}(a^2+b^2-1)^k \langle \xi,\xi\rangle^k$$

$$= \frac{1}{n!}(b\xi)^{\otimes n} \exp\Big[\frac{1}{2}(a^2+b^2-1)\langle\xi,\xi\rangle\Big].$$

Therefore

$$\mathcal{G}_{a,b}\,\varphi_\xi = \Big(\exp\Big[\frac{1}{2}(a^2+b^2-1)\langle\xi,\xi\rangle\Big]\Big) \sum_{n=0}^{\infty} \frac{1}{n!}\langle :\cdot^{\otimes n}:, (b\xi)^{\otimes n}\rangle \quad (11.21)$$

$$= \Big(\exp\Big[\frac{1}{2}(a^2+b^2-1)\langle\xi,\xi\rangle\Big]\Big)\varphi_{b\xi}.$$

Hence for any $a, b, s,$ and t,

$$\mathcal{G}_{s,t}\,\mathcal{G}_{a,b}\,\varphi_\xi = \Big(\exp\Big[\frac{1}{2}(a^2+b^2s^2+b^2t^2-1)\langle\xi,\xi\rangle\Big]\Big)\varphi_{bt\xi}.$$

This implies that

$$\mathcal{G}_{s,t}\,\mathcal{G}_{a,b}\,\varphi_\xi = \mathcal{G}_{\pm\sqrt{a^2+b^2s^2},\,bt}\,\varphi_\xi, \quad \forall \xi \in \mathcal{E}.$$

But the linear span of the set $\{\varphi_\xi; \xi \in \mathcal{E}_c\}$ is dense in $(\mathcal{E})_\beta$. Moreover, by Theorem 11.29 the operators $\mathcal{G}_{s,t}\,\mathcal{G}_{a,b}$ and $\mathcal{G}_{\pm\sqrt{a^2+b^2s^2},\,bt}$ are continuous. Hence

$$\mathcal{G}_{s,t}\,\mathcal{G}_{a,b} = \mathcal{G}_{\pm\sqrt{a^2+b^2s^2},\,bt}.$$

By using the fact that $\mathcal{G}_{0,1} = I$, we can easily check that $\mathcal{G}_{a,b}^{-1} = \mathcal{G}_{\pm ia/b,1/b}$ if $b \neq 0$. $\qquad\square$

Theorem 11.31. *The symbol of the operator $\mathcal{G}_{a,b}$ is given by*

$$F(\xi,\eta) = \exp\Big[\frac{1}{2}(a^2+b^2-1)\langle\xi,\xi\rangle + b\langle\xi,\eta\rangle\Big], \quad \xi,\eta \in \mathcal{E}_c.$$

Proof. Let $\xi, \eta \in \mathcal{E}_c$. By equation (11.21),

$$\mathcal{G}_{a,b}\,\varphi_\xi = \left(\exp\left[\frac{1}{2}(a^2 + b^2 - 1)\langle \xi, \xi \rangle \right] \right) \sum_{n=0}^{\infty} \frac{1}{n!} \langle :\cdot^{\otimes n}:, (b\xi)^{\otimes n} \rangle.$$

On the other hand,

$$\varphi_\eta = \sum_{n=0}^{\infty} \frac{1}{n!} \langle :\cdot^{\otimes n}:, \eta^{\otimes n} \rangle.$$

Thus the symbol of $\mathcal{G}_{a,b}$ is given by

$$F(\xi, \eta) = \left(\exp\left[\frac{1}{2}(a^2 + b^2 - 1)\langle \xi, \xi \rangle \right] \right) \sum_{n=0}^{\infty} n! \langle \frac{1}{n!}(b\xi)^{\otimes n}, \frac{1}{n!}\eta^{\otimes n} \rangle$$

$$= \left(\exp\left[\frac{1}{2}(a^2 + b^2 - 1)\langle \xi, \xi \rangle \right] \right) \sum_{n=0}^{\infty} \frac{1}{n!}(b\langle \xi, \eta \rangle)^n$$

$$= \exp\left[\frac{1}{2}(a^2 + b^2 - 1)\langle \xi, \xi \rangle + b\langle \xi, \eta \rangle \right]. \qquad \square$$

It has been proved in Hida et al. [57], the book by Hida et al. [58], and Y.-J. Lee [122] that the Fourier transform is the adjoint of a continuous linear operator from the space of test functions into itself. We will prove this fact for the Fourier transform on $(\mathcal{E})_\beta^*$.

Theorem 11.32. *The Fourier transform $\mathcal{F}$ is the adjoint operator of the Fourier-Gauss transform $\mathcal{G}_{1,-i}$, i.e.,*

$$\mathcal{F} = \mathcal{G}_{1,-i}^*.$$

Remark. Suppose $\mathcal{G}$ is a continuous linear operator from $(\mathcal{E})_\beta$ into itself such that $\mathcal{F} = \mathcal{G}^*$. Then for any $\Phi \in (\mathcal{E})_\beta^*$ and $\varphi \in (\mathcal{E})_\beta$,

$$\langle\!\langle \mathcal{F}\Phi, \varphi \rangle\!\rangle = \langle\!\langle \Phi, \mathcal{G}\varphi \rangle\!\rangle.$$

In particular, take $\Phi = \widetilde{\delta}_y$. Then by the fact $\mathcal{F}\widetilde{\delta}_y =\, :e^{-i\langle\cdot,y\rangle}:$ (see Example 11.2), we have

$$\langle\!\langle :e^{-i\langle\cdot,y\rangle}:, \varphi \rangle\!\rangle = \langle\!\langle \widetilde{\delta}_y, \mathcal{G}\varphi \rangle\!\rangle,$$

that is,

$$(\mathcal{G}\varphi)(y) = \langle\!\langle :e^{-i\langle\cdot,y\rangle}:, \varphi \rangle\!\rangle.$$

It is shown in the book by Hida et al. [58] that the adjoint of $\mathcal{G}$ is the Fourier transform on $(\mathcal{S})^*$ (except for the minus sign due to the sesquilinear convention used there). On the other hand, it is easy to verify directly that the operator $\mathcal{G}_{1,-i}$ is given by

$$\mathcal{G}_{1,-i}\,\varphi(y) = \langle\!\langle :e^{-i\langle\cdot,y\rangle}:, \varphi\rangle\!\rangle.$$

This implies that $\mathcal{G} = \mathcal{G}_{1,-i}$ and so $\mathcal{F} = \mathcal{G}_{1,-i}^*$.

Proof. It suffices to show that $\mathcal{F}$ and $\mathcal{G}_{1,-i}^*$ have the same symbol. By Lemma 11.22, the symbol F of $\mathcal{F}$ is given by

$$F(\xi,\eta) = e^{-i\langle\xi,\eta\rangle - \frac{1}{2}\langle\eta,\eta\rangle}, \quad \xi,\eta \in \mathcal{E}_c.$$

Let $\varphi_\xi =: e^{\langle\cdot,\xi\rangle}:$. We can apply Theorem 11.31 to get the symbol G of $\mathcal{G}_{1,-i}^*$:

$$\begin{aligned}
G(\xi,\eta) &= \langle\!\langle \mathcal{G}_{1,-i}^*\,\varphi_\xi, \varphi_\eta\rangle\!\rangle \\
&= \langle\!\langle \varphi_\xi, \mathcal{G}_{1,-i}\,\varphi_\eta\rangle\!\rangle \\
&= \langle\!\langle \mathcal{G}_{1,-i}\,\varphi_\eta, \varphi_\xi\rangle\!\rangle \\
&= e^{-i\langle\xi,\eta\rangle - \frac{1}{2}\langle\eta,\eta\rangle}.
\end{aligned}$$

Hence $F = G$ and the theorem is proved.   □

From now on we will use $\mathcal{G}$ to denote $\mathcal{G}_{1,-i}$, i.e., it is the continuous linear operator from $(\mathcal{E})_\beta$ into itself such that $\mathcal{G}^* = \mathcal{F}$. By using Theorems 11.7 and 11.11 we can derive easily the corresponding properties for the operator $\mathcal{G}$ in the next theorem.

Theorem 11.33. *The operator $\mathcal{G}$ has the following properties:*

$$\begin{aligned}
\mathcal{G}^4 &= I, \\
\mathcal{G}Q_\eta &= -iD_\eta^*\mathcal{G}, \quad \eta \in \mathcal{E}, \\
\mathcal{G}D_\eta^* &= -iQ_\eta\mathcal{G}, \quad \eta \in \mathcal{E}, \\
\mathcal{G}D_x &= iD_x\mathcal{G}, \quad x \in \mathcal{E}'.
\end{aligned}$$

Remark. For the general Fourier-Gauss transform $\mathcal{G}_{a,b}$ we have the formulas:

$$\begin{aligned}
D_\eta\mathcal{G}_{a,b} &= b\mathcal{G}_{a,b}D_\eta, \\
\mathcal{G}_{a,b}Q_\eta &= a^2\mathcal{G}_{a,b}D_\eta + bQ_\eta\mathcal{G}_{a,b}.
\end{aligned}$$

The first formula is obvious from the definition of $\mathcal{G}_{a,b}$. The second one can be derived easily from the integration by parts formula in Kuo [97]:

$$\int_{\mathcal{E}'} (D_\eta\varphi)(x)\,d\mu(x) = \int_{\mathcal{E}'} \varphi(x)\langle x,\eta\rangle\,d\mu(x), \quad \varphi \in (\mathcal{E})_\beta.$$

At the end of this section we make some remarks about the operator $\mathcal{G}_{a,b}$. Suppose $a^2 + b^2 = 1$. Then by equation (11.19) we have $h_n = b^n f_n$, i.e.,

$$\mathcal{G}_{a,b}\left(\sum_{n=0}^{\infty}\langle:\cdot^{\otimes n}:, f_n\rangle\right) = \sum_{n=0}^{\infty}\langle:\cdot^{\otimes n}:, b^n f_n\rangle.$$

Hence, if $a^2 + b^2 = 1$, then $\mathcal{G}_{a,b} = \Gamma(bI)$ (the second quantization operator of bI). If, in addition, $|b| = 1$, then $\|\mathcal{G}_{a,b}\varphi\|_{p,\beta} = \|\varphi\|_{p,\beta}$ for all $\varphi \in (\mathcal{E})_\beta$. Thus for such a and b, the operator $\mathcal{G}_{a,b}$ extends to a unitary operator of $(\mathcal{E}_p)_\beta$ for any $p \geq 0$ and $0 \leq \beta < 1$. In particular, $\mathcal{G}_{a,b}$ extends to a unitary operator of (L^2).

Theorem 11.34. *Suppose* $a^2 + b^2 = 1$ *and* $|b| = 1$. *Then the operator* $\mathcal{G}_{a,b}$ *extends to a unitary operator of* $(\mathcal{E}_p)_\beta$ *for any* $p \geq 0$. *Conversely, if* $\mathcal{G}_{a,b}$ *extends to a unitary operator of* $(\mathcal{E}_p)_\beta$ *for some* $p \geq 0$, *then* $a^2 + b^2 = 1$ *and* $|b| = 1$.

Remark. The theorem is also valid if $(\mathcal{E}_p)_\beta$ is replaced with $(\mathcal{E}_p)_\beta^*$.

Proof. The first assertion has already been shown above. Suppose $\mathcal{G}_{a,b}$ extends to a unitary operator of $(\mathcal{E}_p)_\beta$ for some $p \geq 0$. Then

$$\|\mathcal{G}_{a,b}\varphi\|_{p,\beta} = \|\varphi\|_{p,\beta}, \quad \forall \varphi \in (\mathcal{E})_\beta. \tag{11.22}$$

In particular, take $\varphi = \langle\cdot, \xi\rangle$, $\xi \in \mathcal{E}$. Then by the definition of $\mathcal{G}_{a,b}$, $\mathcal{G}_{a,b}\varphi = b\langle\cdot, \xi\rangle$. Thus by equation (11.22), $|b| = 1$. Next, take $\varphi_\xi =\, :e^{\langle\cdot, \xi\rangle}:$. We have

$$\|\varphi_\xi\|_{p,\beta}^2 = \sum_{n=0}^{\infty}\frac{1}{(n!)^{1-\beta}}|\xi|_p^{2n}. \tag{11.23}$$

On the other hand, $\mathcal{G}_{a,b}\varphi_\xi$ is given by equation (11.21) (with $|b| = 1$) and for $\xi \in \mathcal{E}$,

$$\|\mathcal{G}_{a,b}\varphi_\xi\|_{p,\beta}^2 = \left(\exp\left[|\xi|_0^2\,\mathrm{Re}(a^2 + b^2 - 1)\right]\right)\sum_{n=0}^{\infty}\frac{1}{(n!)^{1-\beta}}|\xi|_p^{2n}. \tag{11.24}$$

Hence by equations (11.23) and (11.24),

$$\mathrm{Re}(a^2 + b^2 - 1) = 0.$$

On the other hand, if we take $\varphi_\xi =\, :e^{\langle\cdot, \xi\rangle}:$ with $\xi = (1 + i)\eta$, $\eta \in \mathcal{E}$, then by the same argument we get

$$\mathrm{Im}(a^2 + b^2 - 1) = 0.$$

It follows that $a^2 + b^2 - 1 = 0$. $\qquad\qquad\square$

11.6 Characterization of the Fourier transform

There is a well-known theorem on the characterization of the finite dimensional Fourier transform in terms of coordinate differentiation and multiplication (see Corwin and Greenleaf [26]). Consider the one-dimensional Fourier transform on $S(\mathbb{R})$. Let D be the differentiation operator and Q the multiplication operator by x. Then the Fourier transform $\mathcal{F}$ is the unique continuous linear operator (up to a constant) from $S(\mathbb{R})$ into itself such that $\mathcal{F}D = iQ\mathcal{F}$ and $\mathcal{F}Q = iD\mathcal{F}$.

We will show that this characterization is also true for the Fourier transform $\mathcal{F}$ on $(\mathcal{E})_\beta^*$. In fact, the operator $\mathcal{G}$ can also be characterized in terms of differentiation and multiplication operators.

The key tool for the characterization of $\mathcal{F}$ and $\mathcal{G}$ is the following lemma.

Lemma 11.35. *Suppose L is a continuous linear operator from $(\mathcal{E})_\beta$ into itself such that*

$$LD_\xi = D_\xi L, \quad LQ_\xi = Q_\xi L, \quad \forall \xi \in \mathcal{E}.$$

Then there is a constant c such that $L = cI$.

Proof. Let $\varphi_0 \equiv L1$. Consider $\varphi = \langle \cdot, \xi \rangle^n$ for $\xi \in \mathcal{E}$ and $n \geq 0$. Then $\varphi = Q_\xi^n 1$. Since $LQ_\xi = Q_\xi L$ by assumption, we see that

$$L\varphi = Q_\xi^n L1 = Q_\xi^n \varphi_0 = \varphi_0 \varphi. \tag{11.25}$$

Note that the linear span of the set $\{\langle \cdot, \xi \rangle^n; \xi \in \mathcal{E}, n \geq 0\}$ is dense in $(\mathcal{E})_\beta$. Moreover, the operator L and the multiplication operator by φ_0 are continuous (by assumption and Theorem 8.18, respectively). Thus equation (11.25) implies that

$$L\varphi = \varphi_0 \varphi, \quad \forall \varphi \in (\mathcal{E})_\beta. \tag{11.26}$$

Next we use the assumption $LD_\xi = D_\xi L$ for all $\xi \in \mathcal{E}$ to show that φ_0 is a constant. By Theorem 10.23,

$$T_\xi \varphi = \sum_{n=0}^{\infty} \frac{1}{n!} D_\xi^n \varphi,$$

where T_ξ is the translation by ξ. Hence for any $\varphi \in (\mathcal{E})_\beta$,

$$LT_\xi \varphi = \sum_{n=0}^{\infty} \frac{1}{n!} LD_\xi^n \varphi$$

$$= \sum_{n=0}^{\infty} \frac{1}{n!} D_\xi^n L\varphi$$

$$= T_\xi L\varphi.$$

In particular, for $\varphi = 1$, we have $L1 = \varphi_0$ and so

$$T_\xi \varphi_0 = \varphi_0, \quad \forall \xi \in \mathcal{E}.$$

Thus for $\xi \in \mathcal{E}$,

$$\int_{\mathcal{E}'} \varphi_0(x + \xi)\, d\mu(x) = \int_{\mathcal{E}'} \varphi_0(x)\, d\mu(x). \tag{11.27}$$

Obviously this equation is also true for $\xi \in \mathcal{E}_c$ by taking the analytic extension of φ_0. Note that the S-transform of φ_0 can be expressed as (see §5.2):

$$S\varphi_0(\xi) = \int_{\mathcal{E}'} \varphi_0(x + \xi)\, d\mu(x), \quad \xi \in \mathcal{E}_c.$$

Hence equation (11.27) implies that $S\varphi_0(\xi) = c$ with the constant $c = \int_{\mathcal{E}'} \varphi_0\, d\mu$. Finally, by Proposition 5.10, $\varphi_0 = c$. Thus, in view of equation (11.26), $L\varphi = c\varphi$ for all $\varphi \in (\mathcal{E})_\beta$, i.e., $L = cI$. $\qquad\square$

Theorem 11.36. *The operator $\mathcal{G}$ is the unique (up to a constant) continuous linear operator T from $(\mathcal{E})_\beta$ into itself such that for all $\xi \in \mathcal{E}$,*

$$TQ_\xi = -iD_\xi^* T, \quad TD_\xi^* = -iQ_\xi T. \tag{11.28}$$

Proof. We have already shown in Theorem 11.33 that $\mathcal{G}$ is a continuous linear operator from $(\mathcal{E})_\beta$ into itself satisfying the equalities in (11.28). To show the uniqueness (up to a constant), let T be such an operator. First note that $Q_\xi = D_\xi + D_\xi^*$. Hence the equalities in (11.28) yield that

$$TD_\xi = iD_\xi T. \tag{11.29}$$

Now, observe that $\mathcal{G}$ is invertible ($\mathcal{G}^{-1} = \mathcal{G}^3$ by Theorem 11.33). Let $L \equiv T\mathcal{G}^{-1}$. Then L is a continuous linear operator from $(\mathcal{E})_\beta$ into itself. By using equations (11.28) and (11.29) we can easily show that

$$LD_\xi = D_\xi L, \quad LQ_\xi = Q_\xi L, \quad \forall \xi \in \mathcal{E}.$$

Thus by Lemma 11.35, $L = cI$ for some constant, i.e., $T = c\mathcal{G}$. $\qquad\square$

For the characterization of the Fourier transform, we need the next lemma.

Lemma 11.37. *Suppose L is a continuous linear operator from $(\mathcal{E})_\beta^*$ into itself such that*

$$LD_\xi^* = D_\xi^* L, \quad L\tilde{Q}_\xi = \tilde{Q}_\xi L, \quad \forall \xi \in \mathcal{E}.$$

Then there is a constant c such that $L = cI$.

Proof. Let $\Phi_0 \equiv L1$. It follows from Corollary 9.14 that for any $\Phi \in (\mathcal{E})^*_\beta$, $n \geq 1$,

$$(D^*_\xi)^n \Phi = \langle :\cdot^{\otimes n}:, \xi^{\otimes n} \rangle \diamond \Phi.$$

Hence by using the assumption $LD^*_\xi = D^*_\xi L$, we can derive that for all $\xi \in \mathcal{E}$,

$$L(\langle :\cdot^{\otimes n}:, \xi^{\otimes n} \rangle) = \langle :\cdot^{\otimes n}:, \xi^{\otimes n} \rangle \diamond \Phi_0.$$

But the linear span of the set $\{\langle :\cdot^{\otimes n}:, \xi^{\otimes n} \rangle; \xi \in \mathcal{E}, n \geq 0\}$ is dense in $(\mathcal{E})^*_\beta$. Therefore

$$L\Phi = \Phi_0 \diamond \Phi, \quad \forall \Phi \in (\mathcal{E})^*_\beta.$$

Now, recall that $\widetilde{Q}_\xi = \widetilde{D}_\xi + D^*_\xi$. Thus the assumption implies that

$$L\widetilde{D}_\xi = \widetilde{D}_\xi L, \quad \forall \xi \in \mathcal{E}. \tag{11.30}$$

Note that by Theorems 10.22 and 10.23 the extension $\widetilde{T}_\xi$ of the translation operator T_ξ to $(\mathcal{E})^*_\beta$ is given by

$$\widetilde{T}_\xi \Phi = \sum_{n=0}^\infty \frac{1}{n!} \widetilde{D}^n_\xi \Phi.$$

Apply the operator L to both sides and then use equation (11.30) to show that

$$L\widetilde{T}_\xi \Phi = \widetilde{T}_\xi L\Phi, \quad \forall \Phi \in (\mathcal{E})^*_\beta.$$

In particular, take $\Phi = 1$ to get

$$\widetilde{T}_\xi \Phi_0 = L(\widetilde{T}_\xi 1) = L1 = \Phi_0.$$

This implies, as in the proof of Lemma 11.35, that $S\Phi(\xi) = c$ and so $\Phi_0 = c$. Hence we conclude that $L\Phi = \Phi_0 \diamond \Phi = c\Phi$ for all $\Phi \in (\mathcal{E})^*_\beta$. That is, $L = cI$ and the lemma is proved. $\square$

Theorem 11.38. *The Fourier transform $\mathcal{F}$ is the unique (up to a constant) continuous linear operator T from $(\mathcal{E})^*_\beta$ into itself such that for all $\xi \in \mathcal{E}$,*

$$T\widetilde{D}_\xi = i\widetilde{Q}_\xi T, \quad T\widetilde{Q}_\xi = i\widetilde{D}_\xi T. \tag{11.31}$$

Proof. By Theorems 11.6 and 11.11 the Fourier transform $\mathcal{F}$ is a continuous linear operator from $(\mathcal{E})^*_\beta$ into itself satisfying the equalities in (11.31). To show the uniqueness (up to a constant) of $\mathcal{F}$, let T be such an operator. Since $Q_\xi = D_\xi + D^*_\xi$, the equalities in (11.28) imply that

$$TD^*_\xi = -iD^*_\xi T, \quad \forall \xi \in \mathcal{E}. \tag{11.32}$$

Now, let $L = T\mathcal{F}^{-1}$. Then L is a continuous linear operator from $(\mathcal{E})_\beta^*$ into itself. Since both L and $\mathcal{F}$ satisfy equations (11.31) and (11.32), we can easily check that

$$LD_\xi^* = D_\xi^* L, \quad L\tilde{Q}_\xi = \tilde{Q}_\xi L, \quad \forall \xi \in \mathcal{E}.$$

Thus by Lemma 11.37, $L = cI$ for some constant c, i.e., $T = c\mathcal{F}$. □

11.7 Fourier-Mehler transforms

In this section we study Fourier-Mehler transforms in Hida et al. [57], the book by Hida et al. [58], Kuo [105] [109], and Obata [142]. Fourier-Mehler transforms form a one-parameter group with an interesting infinitesimal generator.

The Fourier-Mehler transform for functions on $\mathbb{R}$ as defined in Hida [47] can be easily extended to functions on $\mathbb{R}^r$. Namely, the Fourier-Mehler transform $\mathcal{F}_\theta f$, $\theta \in \mathbb{R}$, of a function $f \in \mathcal{S}(\mathbb{R}^r)$ is defined by

$$\mathcal{F}_\theta f(y) = \int_{\mathbb{R}^r} K(x,y) f(x) \, dx,$$

where the function K is given by

$$K(x,y) = \left(\pi(1 - e^{2i\theta})\right)^{-r/2} \exp\left[\frac{i}{\sin\theta} \langle x,y \rangle - \frac{i}{2\tan\theta} (|x|^2 + |y|^2)\right].$$

As in the case of Fourier transform, we can rewrite $\mathcal{F}_\theta f$ in a dimension free form by using the renormalization of the exponential function in K. The finite dimensional version of Example 8.7 gives the following renormalization

$$\mathcal{N}_y \exp\left[\frac{i}{\sin\theta} \langle x,y \rangle - \frac{i}{2\tan\theta} (|x|^2 + |y|^2)\right]$$

$$= \left(2^{-1}(1 - e^{2i\theta})\right)^{-r/2} \exp\left[\frac{i}{\sin\theta} \langle x,y \rangle - \frac{i}{2\tan\theta} (|x|^2 + |y|^2) + \frac{1}{2}|x|^2\right],$$

where $\mathcal{N}_y$ denotes the renormalization with respect to the y variable. Then we can rewrite the Fourier-Mehler transform as follows:

$$\mathcal{F}_\theta f(y) = \int_{\mathbb{R}^r} f(x)\mathcal{N}_y \exp\left[\frac{i}{\sin\theta} \langle x,y \rangle - \frac{i}{2\tan\theta} (|x|^2 + |y|^2)\right] d\mu_r(x),$$

where μ_r is the standard Gaussian measure on $\mathbb{R}^r$. Obviously the white noise analogue of this expression is

$$\mathcal{F}_\theta \Phi(y) = \int_{\mathcal{E}'} \Phi(x)\mathcal{N}_y \exp\left[\frac{i}{\sin\theta} \langle x,y \rangle - \frac{i}{2\tan\theta} (|x|_0^2 + |y|_0^2)\right] d\mu(x),$$

where the renormalization $\mathcal{N}_y \exp[\cdots]$ is given in Example 8.7 with the S-transform

$$S\mathcal{N}_y \exp\left[\frac{i}{\sin\theta}\langle x, y\rangle - \frac{i}{2\tan\theta}\left(|x|_0^2 + |y|_0^2\right)\right](\xi)$$
$$= \exp\left[e^{i\theta}\langle x, \xi\rangle - \frac{1}{2}e^{i\theta}\cos\theta\,\langle\xi, \xi\rangle\right], \quad \xi \in \mathcal{E}_c.$$

Thus we define the *Fourier-Mehler transform* $\mathcal{F}_\theta\Phi$ of a generalized function $\Phi \in (\mathcal{E})_\beta^*$ to be the unique element in $(\mathcal{E})_\beta^*$ with the S-transform given by

$$S(\mathcal{F}_\theta\Phi)(\xi) = \left\langle\!\left\langle \Phi, \exp\left[e^{i\theta}\langle\cdot, \xi\rangle - \frac{1}{2}e^{i\theta}\cos\theta\,\langle\xi, \xi\rangle\right]\right\rangle\!\right\rangle, \quad \xi \in \mathcal{E}_c. \quad (11.33)$$

Note that $\mathcal{F}_\theta\Phi$ is well-defined by Theorems 5.7 and 8.2. Moreover, it is easy to check that this definition is equivalent to

$$S(\mathcal{F}_\theta\Phi)(\xi) = (S\Phi)(e^{i\theta}\xi)\exp\left[\frac{i}{2}e^{i\theta}\sin\theta\,\langle\xi, \xi\rangle\right], \quad \xi \in \mathcal{E}_c. \quad (11.34)$$

Obviously we have $\mathcal{F}_0 = I$, $\mathcal{F}_{-\pi/2} = \mathcal{F}$ (the Fourier transform), and $\mathcal{F}_{2\pi+\theta} = \mathcal{F}_\theta$. Moreover, we can use equation (11.34) to check directly that for any $\theta, \tau \in \mathbb{R}$,

$$S(\mathcal{F}_\theta\mathcal{F}_\tau\Phi)(\xi) = (S\Phi)(e^{i(\theta+\tau)}\xi)\exp\left[\frac{i}{2}e^{i(\theta+\tau)}\sin(\theta+\tau)\,\langle\xi, \xi\rangle\right], \quad \xi \in \mathcal{E}_c.$$

This shows that for any $\theta, \tau \in \mathbb{R}$,

$$\mathcal{F}_\theta\mathcal{F}_\tau = \mathcal{F}_{\theta+\tau}.$$

Thus the family $\{\mathcal{F}_\theta;\ \theta \in \mathbb{R}\}$ forms a one-parameter group of linear operators from $(\mathcal{E})_\beta^*$ into itself. In particular, we have $\mathcal{F}_\theta\mathcal{F}_{-\theta} = I$ and so for any $\theta \in \mathbb{R}$,

$$\mathcal{F}_\theta^{-1} = \mathcal{F}_{-\theta}.$$

Hence the inverse Fourier transform is given by $\mathcal{F}^{-1} = \mathcal{F}_{\pi/2}$ and

$$(S\mathcal{F}^{-1}\Phi)(\xi) = \left\langle\!\left\langle \Phi, e^{i\langle\cdot, \xi\rangle}\right\rangle\!\right\rangle = (S\Phi)(i\xi)e^{-\frac{1}{2}\langle\xi, \xi\rangle}, \quad \xi \in \mathcal{E}_c.$$

The Fourier-Mehler transform has many properties similar to those of the Fourier transform. They can be established by suitable modifications of the arguments used for the Fourier transform. Hence we just state these properties without proofs.

1. For any $y \in \mathcal{E}'_c$, the Fourier-Mehler transforms of $\tilde{\delta}_x$ and $: e^{\langle \cdot, x \rangle} :$ are given by

$$\mathcal{F}_\theta(\tilde{\delta}_x) = g_{\sec \theta \, x, -i \tan \theta},$$

$$\mathcal{F}_\theta : e^{\langle \cdot, x \rangle} : = g_{i \csc \theta \, x, i \cot \theta},$$

where $g_{y,c}$ is the Gaussian white noise function given in Example 8.3 and we use the convention $g_{y,0} = \tilde{\delta}_y$.

2. If $\Phi = \sum_{n=0}^{\infty} \langle : \cdot^{\otimes n} :, F_n \rangle \in (\mathcal{E})^*_\beta$, then

$$\mathcal{F}_\theta \Phi = \sum_{n=0}^{\infty} \left\langle : \cdot^{\otimes n} :, \; e^{in\theta} \sum_{k=0}^{[n/2]} \frac{(ie^{-i\theta} \sin \theta)^k}{2^k k!} F_{n-2k} \widehat{\otimes} \tau^{\otimes k} \right\rangle.$$

3. The Fourier-Mehler transform $\mathcal{F}_\theta : (\mathcal{E})^*_\beta \to (\mathcal{E})^*_\beta$ is continuous. In fact, for any $p \geq 0$ and $q > p$ satisfying

$$2^{2-\beta} \lambda_1^{-2(q-p)} \leq 1, \quad 2^{2-\beta} |\tau|_{-q} |\sin \theta| < 1,$$

the following inequality holds for all $\Phi \in (\mathcal{E}_p)^*_\beta$,

$$\|\mathcal{F}_\theta \Phi\|_{-q, -\beta} \leq \left(1 - 2^{2(2-\beta)} |\tau|^2_{-q} \sin^2 \theta\right)^{-1/2} \|\Phi\|_{-p, -\beta}.$$

4. The Fourier-Mehler transform $\mathcal{F}_\theta : (\mathcal{E})^*_\beta \to (\mathcal{E})^*_\beta$ is injective and onto with inverse $\mathcal{F}_\theta^{-1} = \mathcal{F}_{-\theta}$.

5. For any $\eta \in \mathcal{E}$ and $x \in \mathcal{E}'$, the following equalities hold:

$$\mathcal{F}_\theta \tilde{D}_\eta = \cos \theta \, \tilde{D}_\eta \mathcal{F}_\theta - i \sin \theta \, \tilde{Q}_\eta \mathcal{F}_\theta,$$

$$\mathcal{F}_\theta D_x^* = e^{i\theta} D_x^* \mathcal{F}_\theta,$$

$$\mathcal{F}_\theta \tilde{Q}_\eta = -i \sin \theta \, \tilde{D}_\eta \mathcal{F}_\theta + \cos \theta \, \tilde{Q}_\eta \mathcal{F}_\theta.$$

6. For any $\eta \in \mathcal{E}$, $x \in \mathcal{E}'$, and $\Phi \in (\mathcal{E})^*_\beta$, the equalities hold:

$$\mathcal{F}_\theta \tilde{T}_\eta \Phi = \exp \left[i \sin \theta \, \langle \cdot, \eta \rangle - \frac{i}{2} \sin \theta \cos \theta \, \langle \eta, \eta \rangle \right] \tilde{T}_{\cos \theta \, \eta} \mathcal{F}_\theta \Phi,$$

$$\tilde{T}_\eta \mathcal{F}_\theta \Phi = \mathcal{F}_\theta \left(\left(e^{e^{i\theta} \langle \cdot, \eta \rangle} \Phi \right) \diamond \left(e^{-\frac{1}{2} e^{i\theta} \cos \theta \, \langle \eta, \eta \rangle} : e^{-\cos \theta \, \langle \cdot, \eta \rangle} : \right) \right),$$

$$\mathcal{F}_\theta T_x^* \Phi = : e^{e^{i\theta} \langle \cdot, x \rangle} : \diamond (\mathcal{F}_\theta \Phi),$$

$$T_x^* \mathcal{F}_\theta \Phi = \mathcal{F}_\theta \left(: e^{e^{-i\theta} \langle \cdot, x \rangle} : \diamond \Phi \right).$$

7. For any Φ and Ψ in $(\mathcal{E})_\beta^*$, the equalities hold:

$$\mathcal{F}_\theta(\Phi \diamond \Psi) = \big((\mathcal{F}_\theta\Phi) * (\mathcal{F}_\theta\Psi)\big) \diamond g_{-i\tan\theta},$$

$$\mathcal{F}_\theta(\Phi * \Psi) = \big((\mathcal{F}_\theta\Phi) \diamond (\mathcal{F}_\theta\Psi)\big) \diamond g_{i\tan\theta-1},$$

where by convention $g_0 = \widetilde{\delta}_0$ and $g_\infty = 1$.

8. The following equalities hold for any $\Phi \in (\mathcal{E})_\beta^*$,

$$\mathcal{F}_\theta\Phi = \big(\Gamma(e^{i\theta}I)\Phi\big) \diamond g_{i\cot\theta},$$

$$(\mathcal{F}_\theta\Phi) \diamond g_{-i\cot\theta-2} = \Gamma(e^{i\theta}I)\Phi.$$

9. If $\varphi \in (\mathcal{E})_\beta$, then $(\mathcal{F}_\theta\varphi) \diamond g_{-i\cot\theta-2} \in (\mathcal{E})_\beta$ and for any $p \geq 0$,

$$\|(\mathcal{F}_\theta\varphi) \diamond g_{-i\cot\theta-2}\|_{p,\beta} = \|\varphi\|_{p,\beta}.$$

Conversely, if $(\mathcal{F}_\theta\varphi) \diamond g_{-i\cot\theta-2} \in (\mathcal{E})_\beta$, then $\varphi \in (\mathcal{E})_\beta$ and the above equality holds.

10. There exist positive constants C and r depending only on $\lambda_j, j \geq 1, \alpha$ (see §4.2) such that for any $f \in L^p(\mu), 1 < p \leq \infty$,

$$\|(\mathcal{F}_\theta\Phi_f) \diamond g_{-i\cot\theta-2}\|_{-(n-1)r} \leq C^{1-\frac{1}{n}}\|f\|_{L^p(\mu)},$$

where n is any integer such that $n \geq \frac{p}{p-1}$.

11. The symbol of the Fourier-Mehler transform is given by the function

$$F_\theta(\xi,\eta) = \exp\left[e^{i\theta}\langle\xi,\eta\rangle + \frac{i}{2}e^{i\theta}\sin\theta\,\langle\eta,\eta\rangle\right], \quad \xi,\eta \in \mathcal{E}_c.$$

12. The Fourier-Mehler transform $\mathcal{F}_\theta$ can be decomposed as

$$\mathcal{F}_\theta = \sum_{n=0}^{\infty}\sum_{k=0}^{n} \Xi_{2n-k,k}(\rho_{2n-k,k}),$$

where the kernel function $\rho_{2n-k,k}$ is given by

$$\rho_{2n-k,k}(s_1,\ldots,s_{2n-k},t_1,\ldots,t_k) = \frac{a^k b^{n-k}}{2^{n-k}k!(n-k)!}$$

$$\times \tau(s_1,t_1)\cdots\tau(s_k,t_k)\tau(s_{k+1},s_{k+2})\cdots\tau(s_{2n-m-1},s_{2n-m})$$

with $a = e^{i\theta} - 1$ and $b = -ie^{i\theta}\sin\theta$.

13. There exists a unique continuous linear operator $\mathcal{G}_\theta$ from $(\mathcal{E})_\beta$ into itself such that
$$\mathcal{F}_\theta = \mathcal{G}_\theta^*.$$
In fact, $\mathcal{G}_\theta = \mathcal{G}_{a,b}$ with $a = \pm(1 - e^{i\theta}\cos\theta)^{1/2}$ and $b = e^{i\theta}$. (Note that $\mathcal{G}_{a,b} = \mathcal{G}_{-a,b}$.) Moreover, the family $\{\mathcal{G}_\theta; \theta \in \mathbb{R}\}$ forms a one-parameter group of continuous linear operators from $(\mathcal{E})_\beta$ into itself.

14. The operator $\mathcal{G}_\theta$ is the unique (up to a constant) continuous linear operator T from $(\mathcal{E})_\beta$ into itself such that for all $\xi \in \mathcal{E}$,

$$TQ_\xi = \cos\theta\, Q_\xi T + i\sin\theta\, D_\xi^* T,$$

$$TD_\xi^* = i\sin\theta\, Q_\xi T + \cos\theta\, D_\xi^* T.$$

15. The Fourier-Mehler transform $\mathcal{F}_\theta$ is the unique (up to a constant) continuous linear operator T from $(\mathcal{E})_\beta^*$ into itself such that for all $\xi \in \mathcal{E}$,
$$T\widetilde{D}_\xi = \cos\theta\, \widetilde{D}_\xi T - i\sin\theta\, \widetilde{Q}_\xi T,$$

$$T\widetilde{Q}_\xi = -i\sin\theta\, \widetilde{D}_\xi T + \cos\theta\, \widetilde{Q}_\xi T.$$

16. Let g_1 be the Gaussian white noise function with parameter $c = 1$. Then for any $\rho \in (\mathcal{S}_c')^{\otimes n}$,

$$\mathcal{F}_\theta\big(\Xi_{n,0}(\rho)g_1\big) = e^{in\theta}\Xi_{n,0}(\rho)g_1.$$

In particular, when $\rho = (\delta_t)^{\otimes n}$, we have

$$\mathcal{F}_\theta\big((\partial_t^*)^n g_1\big) = e^{in\theta}(\partial_t^*)^n g_1.$$

The generalized function $(\partial_t^*)^n g_1$ is the white noise analogue of the Hermite function of order n.

Now, we study the one-parameter group $\{\mathcal{F}_\theta;\ \theta \in \mathbb{R}\}$ of Fourier-Mehler transforms and the one-parameter group $\{\mathcal{G}_\theta;\ \theta \in \mathbb{R}\}$ of continuous linear operators from $(\mathcal{E})_\beta$ into itself.

Theorem 11.39. *The one-parameter group $\{\mathcal{G}_\theta;\ \theta \in \mathbb{R}\}$ is strongly continuous on $(\mathcal{E})_\beta$ with infinitesimal generator $iN + \frac{i}{2}\Delta_G$.*

Proof. Let $\varphi = \sum_{n=0}^\infty \langle :\cdot^{\otimes n}:, f_n\rangle \in (\mathcal{E})_\beta$. Then by property 13 above and Theorem 11.28,

$$\mathcal{G}_\theta\varphi = \sum_{n=0}^\infty \langle :\cdot^{\otimes n}:, h_n(\theta)\rangle,$$

where $h_n(\theta)$ is given by

$$h_n(\theta) = e^{in\theta} \sum_{k=0}^{\infty} \binom{n+2k}{2k} (2k-1)!! (ie^{i\theta}\sin\theta)^k \langle \tau^{\otimes k}, f_{n+2k} \rangle.$$

Define $J_\theta \varphi$ by

$$J_\theta \varphi = \sum_{n=0}^{\infty} \langle :\cdot^{\otimes n}:, g_n(\theta) \rangle,$$

where $g_n(\theta)$ is given by

$$g_n(\theta) = \sum_{k=0}^{\infty} \binom{n+2k}{2k} (2k-1)!! (ie^{i\theta}\sin\theta)^k \langle \tau^{\otimes k}, f_{n+2k} \rangle.$$

First we show that $J_\theta \varphi$ converges to φ in $(\mathcal{E})_\beta$ as $\theta \to 0$, i.e., for any $p \geq 0$,

$$\lim_{\theta \to 0} \| J_\theta \varphi - \varphi \|_{p,\beta} = 0. \tag{11.35}$$

Note that

$$J_\theta \varphi - \varphi = \sum_{n=0}^{\infty} \Big\langle :\cdot^{\otimes n}:, \sum_{k=1}^{\infty} \frac{(n+2k)!}{n! 2^k k!} (ie^{i\theta}\sin\theta)^k \langle \tau^{\otimes k}, f_{n+2k} \rangle \Big\rangle.$$

It is easy to check that for any $q > p$,

$$\| J_\theta \varphi - \varphi \|_{p,\beta}^2 \leq \| \varphi \|_{q,\beta}^2 \sum_{n=0}^{\infty} \frac{1}{(n!)^{1-\beta}} \sum_{k=1}^{\infty} \frac{((n+2k)!)^{1-\beta}}{2^{2k}(k!)^2}$$
$$\times |\sin\theta|^{2k} |\tau|_{-p}^{2k} \lambda_1^{-2(q-p)(n+2k)}.$$

We can use the fact $(n+2k)! \leq 2^{n+2k} n! (2k)! \leq 2^{n+2k} n! 2^{2k}(k!)^2$ to derive

$$\| J_\theta \varphi - \varphi \|_{p,\beta}^2 \leq \| \varphi \|_{q,\beta}^2 \sum_{n=0}^{\infty} 2^{n(1-\beta)} \lambda_1^{-2(q-p)n}$$
$$\times \sum_{k=1}^{\infty} 2^{2k(1-\beta)} |\sin\theta|^{2k} |\tau|_{-p}^{2k} \lambda_1^{-4(q-p)k}.$$

Now, for any $p \geq 0$, choose q such that

$$\lambda_1^{2(q-p)} > \max\{2^{1-\beta}, 2^{1-\beta}|\tau|_{-p}\}.$$

Then we have

$$\| J_\theta \varphi - \varphi \|_{p,\beta} \leq C_{p,q,\beta}^{(1)} |\sin\theta| \, \| \varphi \|_{q,\beta},$$

where $C^{(1)}_{p,q,\beta}$ is the constant given by

$$C^{(1)}_{p,q,\beta} = 2^{1-\beta}|\tau|_{-p}\lambda_1^{-2(q-p)}\left(1 - 2^{1-\beta}\lambda_1^{-2(q-p)}\right)^{-1/2}$$
$$\times \left(1 - 2^{2(1-\beta)}|\tau|^2_{-p}\lambda_1^{-4(q-p)}\right)^{-1/2}.$$

This yields equation (11.35) immediately.

Next we show that $\mathcal{G}_\theta\varphi - J_\theta\varphi$ converges to 0 in $(\mathcal{E})_\beta$ as $\theta \to 0$, i.e., for any $p \geq 0$,

$$\lim_{\theta \to 0} \|\mathcal{G}_\theta\varphi - J_\theta\varphi\|_{p,\beta} = 0. \qquad (11.36)$$

Note that $|e^{in\theta} - 1| \leq n|\theta| \leq 2^n|\theta|$. By the same argument as above,

$$\|\mathcal{G}_\theta\varphi - J_\theta\varphi\|^2_{p,\beta} \leq \|\varphi\|^2_{q,\beta} \sum_{n=1}^{\infty} 2^{n(2-\beta)}\lambda_1^{-2(q-p)n}|\theta|$$
$$\times \sum_{k=0}^{\infty} 2^{2k(1-\beta)}|\tau|^{2k}_{-p}\lambda_1^{-4(q-p)k}.$$

Choose $q > p$ such that $\lambda_1^{2(q-p)} > \max\{2^{2-\beta}, 2^{1-\beta}|\tau|_{-p}\}$. Then

$$\|\mathcal{G}_\theta\varphi - J_\theta\varphi\|_{p,\beta} \leq C^{(2)}_{p,q,\beta}|\theta|^{1/2}\|\varphi\|_{q,\beta},$$

where $C^{(2)}_{p,q,\beta}$ is the constant given by

$$C^{(2)}_{p,q,\beta} = 2^{(2-\beta)/2}\lambda_1^{-(q-p)}\left(1 - 2^{2-\beta}\lambda_1^{-2(q-p)}\right)^{-1/2}$$
$$\times \left(1 - 2^{2(1-\beta)}|\tau|^2_{-p}\lambda_1^{-4(q-p)}\right)^{-1/2}.$$

This yields equation (11.36). It follows from equations (11.35) and (11.36) that $\mathcal{G}_\theta \to I$ strongly on $(\mathcal{E})_\beta$ as $\theta \to 0$.

Finally we find the infinitesimal generator of $\{\mathcal{G}_\theta; \theta \in \mathbb{R}\}$. Changing the order of limit and summation can be justified to get

$$\lim_{\theta \to 0} \frac{\mathcal{G}_\theta\varphi - \varphi}{\theta} = \sum_{n=0}^{\infty} \langle :\cdot^{\otimes n}:, h'_n(0)\rangle.$$

But it is straightforward to check that

$$h'_n(0) = (in)f_n + \frac{i}{2}(n+2)(n+1)\langle \tau, f_{n+2}\rangle.$$

Therefore

$$\lim_{\theta \to 0} \frac{\mathcal{G}_\theta\varphi - \varphi}{\theta} = i\sum_{n=1}^{\infty} n\langle :\cdot^{\otimes n}:, f_n\rangle + \frac{i}{2}\sum_{n=0}^{\infty}(n+2)(n+1)\langle :\cdot^{\otimes n}:, \langle \tau, f_{n+2}\rangle\rangle.$$

This shows that $\lim_{\theta \to 0} \theta^{-1}(\mathcal{G}_\theta \varphi - \varphi) = iN\varphi + \frac{i}{2}\Delta_G\varphi$ since by Definition 9.22 and Theorem 10.11

$$N\varphi = \sum_{n=1}^{\infty} n \langle :\cdot^{\otimes n} :, f_n \rangle$$

$$\Delta_G\varphi = \sum_{n=0}^{\infty} (n+2)(n+1) \langle :\cdot^{\otimes n} :, \langle \tau, f_{n+2} \rangle \rangle. \qquad \square$$

Theorem 11.40. *The one-parameter group $\{\mathcal{F}_\theta; \theta \in \mathbb{R}\}$ of Fourier-Mehler transforms is strongly continuous on $(\mathcal{E})_\beta^*$. Its infinitesimal generator is given by $iN + \frac{i}{2}\Delta_G^*$.*

Proof. Note that by property 13 stated above, we have $\mathcal{F}_\theta = \mathcal{G}_\theta^*$. Hence this theorem follows from Theorem 11.39. $\qquad \square$

11.8 Initial value problems

In this section we will give some examples of partial differential equations which can be solved by applying the Fourier transform method.

Example 11.41. Consider the initial value problem:

$$\frac{\partial u}{\partial t} = D_\xi u, \ t \in \mathbb{R}, \quad u(0) = \varphi, \tag{11.37}$$

where nonzero $\xi \in \mathcal{E}$ is fixed and $\varphi \in (\mathcal{E})_\beta$. Since the differential operator D_ξ is continuous from $(\mathcal{E})_\beta$ into itself, we expect that $u(t) \in (\mathcal{E})_\beta$. By Theorem 10.23 we have

$$T_\xi = \sum_{n=0}^{\infty} \frac{1}{n!} D_\xi^n.$$

Thus if we think of e^{tD_ξ} as the translation operator $T_{t\xi}$, then the solution u of equation (11.37) is given by

$$u(t, x) = T_{t\xi}\varphi(x) = \varphi(x + t\xi).$$

On the other hand, we can use the Fourier transform to derive the solution as follows. Note that the space $(\mathcal{E})_\beta$ is not invariant under the Fourier transform. Hence we should regard the initial value problem (11.37) as taking values in $(\mathcal{E})_\beta^*$:

$$\frac{\partial u}{\partial t} = \tilde{D}_\xi u, \quad u(0) = \Phi, \tag{11.38}$$

where $\xi \in \mathcal{E}$ is fixed and $\Phi \in (\mathcal{E})_\beta^*$ is the initial condition.

Let $v(t) = \widehat{u}(t)$ be the Fourier transform of $u(t)$. Then by Theorem 11.11, $v(t)$ satisfies the following equation:

$$\frac{\partial v}{\partial t} = i\langle \cdot, \xi \rangle v, \quad v(0) = \widehat{\Phi}.$$

Hence $v(t)$ is given by

$$v(t) = e^{it\langle \cdot, \xi \rangle}\, \widehat{\Phi}.$$

But by Theorem 11.13

$$e^{it\langle \cdot, \xi \rangle}\, \widehat{\Phi} = \mathcal{F}\big(\widetilde{T}_{t\xi}\Phi\big).$$

Therefore

$$v(t) = \mathcal{F}\big(\widetilde{T}_{t\xi}\Phi\big).$$

Upon taking the inverse Fourier transform we get

$$u(t, \cdot) = \widetilde{T}_{t\xi}\Phi = \Phi(\cdot + t\xi).$$

We can check directly that this function $u(t)$ is indeed a solution of equation (11.38). Thus we have proved the existence and uniqueness of a solution of equation (11.38) in the space $(\mathcal{E})_\beta^*$.

We point out two facts:

(1) If the initial condition $\Phi \in (\mathcal{E})_\beta$, then the solution $u(t) \in (\mathcal{E})_\beta$ for all $t \in \mathbb{R}$.

(2) If $\xi \notin E$, then the Fourier transform method can not be applied directly. This is because the function $e^{it\langle \cdot, \xi \rangle}$ has no meaning when $\xi \notin E$.

Example 11.42. Consider the initial value problem:

$$\frac{\partial u}{\partial t} = c\widetilde{D}_\xi^2 u, \; t \geq 0, \quad u(0) = \varphi, \; (c > 0), \tag{11.39}$$

where $\xi \in \mathcal{E}, |\xi|_0 = 1$, is fixed and $\varphi \in (\mathcal{E})_\beta$. Informally, the solution is given by

$$u(t) = e^{ct\widetilde{D}_\xi^2}\varphi.$$

As in the previous example, let $v(t) = \widehat{u}(t)$ be the Fourier transform of $u(t)$. Then $v(t)$ satisfies the equation:

$$\frac{\partial v}{\partial t} = -c\langle \cdot, \xi \rangle^2 v, \quad v(0) = \widehat{\varphi}.$$

Obviously the solution of this equation is given by

$$v(t) = e^{-ct\langle \cdot, \xi \rangle^2}\widehat{\varphi}.$$

This implies that

$$v(t) = \sum_{n=0}^{\infty} (-1)^n \frac{(ct)^n}{n!} \langle \cdot, \xi \rangle^{2n} \widehat{\varphi},$$

where the series converges in $(\mathcal{E})_{\beta}^*$.

Now, recall from Theorem 11.11 that $\langle \cdot, \xi \rangle \widehat{\varphi} = -i \mathcal{F} D_{\xi} \varphi$. Hence by the continuity of the Fourier transform,

$$v(t) = \sum_{n=0}^{\infty} \frac{(ct)^n}{n!} \mathcal{F} D_{\xi}^{2n} \varphi$$

$$= \mathcal{F} \Big(\sum_{n=0}^{\infty} \frac{(ct)^n}{n!} D_{\xi}^{2n} \varphi \Big).$$

Thus the solution of equation (11.39) is given by

$$u(t) = \sum_{n=0}^{\infty} \frac{(ct)^n}{n!} D_{\xi}^{2n} \varphi. \tag{11.40}$$

Note that $D_{\xi}^{2n} = \Xi_{0,2n}(\xi^{\otimes 2n})$. Hence by the same argument as in the proof of Theorem 10.23, the series

$$\sum_{n=0}^{\infty} \frac{(ct)^n}{n!} D_{\xi}^{2n}$$

converges in the sense as stated in Theorem 10.23, i.e., for any $p \geq 0$, there exists $q \geq 0$ such that the series converges in the operator norm $\| \cdot \|_{\{q,p\}}$.

By the analyticity of φ we have

$$\varphi \big(\cdot + \sqrt{2ct} \, \langle x, \xi \rangle \xi \big) = \sum_{n=0}^{\infty} \frac{1}{n!} \big(\sqrt{2ct} \, \langle x, \xi \rangle \big)^n D_{\xi}^n \varphi.$$

Observe that for $|\xi|_0 = 1$,

$$\int_{\mathcal{E}'} \langle x, \xi \rangle^n \, d\mu(x) = \begin{cases} 0, & \text{if } n \text{ is odd}; \\ (n-1)!!, & \text{if } n \text{ is even}. \end{cases}$$

Hence we obtain the equality

$$\int_{\mathcal{E}'} \varphi \big(\cdot + \sqrt{2ct} \, \langle x, \xi \rangle \xi \big) \, d\mu(x) = \sum_{n=0}^{\infty} \frac{(ct)^n}{n!} D_{\xi}^{2n} \varphi.$$

Thus the solution of equation (11.39) is given by

$$u(t) = \int_{\mathcal{E}'} \varphi \big(\cdot + \sqrt{2ct} \, \langle x, \xi \rangle \xi \big) \, d\mu(x). \tag{11.41}$$

It is easy to check that this function $u(t)$ is a solution of equation (11.39). Thus we have shown the existence and uniqueness of a solution of equation (11.39) in the space $(\mathcal{E})^*_\beta$. The solution turns out to be in $(\mathcal{E})_\beta$ if the initial condition $\varphi \in (\mathcal{E})_\beta$.

We make two remarks:

(1) If the initial condition Φ is in $(\mathcal{E})^*_\beta$, then the solution $u(t) \in (\mathcal{E})^*_\beta$. However, in this case the integral for the solution $u(t)$ in equation (11.41) is in the weak sense.

(2) When $\xi \notin E$ and $\varphi \in (\mathcal{E})_\beta$, the solution $u(t)$ can only be expressed as in equation (11.40), while the integration in equation (11.41) has no meaning.

Example 11.43. Consider the initial value problem:

$$\frac{\partial u}{\partial t} = \left(c_1 \widetilde{D}^2_{\xi_1} + c_2 \widetilde{D}^2_{\xi_2} \right) u, \ t \geq 0, \quad u(0) = \varphi, \quad (c_1, c_2 > 0), \tag{11.42}$$

where $\xi_1, \xi_2 \in \mathcal{E}$ are fixed with $|\xi_1|_0 = |\xi_2|_0 = 1, \langle \xi_1, \xi_2 \rangle = 0$, and $\varphi \in (\mathcal{E})_\beta$.

Let $v(t) = \widehat{u}(t)$. By the same argument as in the previous example we can derive that

$$v(t) = \left(\exp \left[- t(c_1 \langle \cdot, \xi_1 \rangle^2 + c_1 \langle \cdot, \xi_1 \rangle^2) \right] \right) \widehat{\varphi}.$$

Similarly to the equation (11.40), we have

$$u(t) = \sum_{j,k=0}^{\infty} \frac{(c_1 t)^j}{j!} \frac{(c_2 t)kj}{k!} D^{2j}_{\xi_1} D^{2k}_{\xi_2} \varphi.$$

Moreover, just as equation (11.41), $u(t)$ can be rewritten as an integral

$$u(t) = \int_{\mathcal{E}'} \varphi \left(\cdot + \sqrt{2c_1 t} \, \langle x, \xi_1 \rangle \xi_1 + \sqrt{2c_2 t} \, \langle x, \xi_2 \rangle \xi_2 \right) d\mu(x).$$

Obviously the remarks at the end of Example 11.42 are also valid for this example.

More generally, consider the initial value problem

$$\frac{\partial u}{\partial t} = \left(\sum_{k=1}^{n} c_k \widetilde{D}^2_{\xi_k} \right) u, \ t \geq 0, \quad u(0) = \varphi, \quad (c_k > 0), \tag{11.43}$$

where $\{\xi_k; 1 \leq k \leq n\} \subset \mathcal{E}$ is orthonormal in E. The solution $u(t)$ is easily seen to be given by

$$u(t) = \left(\sum_{j_1,\ldots,j_n} \frac{(c_{j_1} t)^{j_1}}{j_1!} \cdots \frac{(c_{j_n} t)^{j_n}}{j_n!} D^{2j_1}_{\xi_1} \cdots D^{2j_n}_{\xi_n} \right) \varphi$$

$$= \int_{\mathcal{E}'} \varphi \left(\cdot + \sum_{k=1}^{n} \sqrt{2c_k t} \, \langle x, \xi_k \rangle \xi_k \right) d\mu(x). \tag{11.44}$$

This result holds for more general integral kernel operators and initial conditions; see Dôku et al. [33] for details.

Now, let $\{e_k; k \geq 1\} \subset \mathcal{E}$ be an orthonormal basis for E. The Gross Laplacian Δ_G can be expressed as $\Delta_G = \sum_{k=1}^{\infty} D_{e_k}^2$. In equation (11.43), take $c_k = \frac{1}{2}$ for all k and let $n \to \infty$. Then we get the heat equation for the Gross Laplacian:

$$\frac{\partial u}{\partial t} = \frac{1}{2} \Delta_G u, \quad t \geq 0, \quad u(0) = \varphi, \tag{11.45}$$

where the initial condition $\varphi \in (\mathcal{E})_\beta$. In view of equation (11.44), it is reasonable to expect that the solution of this heat equation is given by

$$u(t) = \int_{\mathcal{E}'} \varphi(\cdot + \sqrt{t}\, x)\, d\mu(x).$$

In the next chapter we will show that this is indeed the case. On the other hand, recall that $(E, \mathcal{E}'_p)$ is an abstract Wiener space for any $p \geq \frac{\alpha}{2}$ (see §4.2). Hence if φ is a bounded Lip-1 function on $\mathcal{E}'_p$ for some $p \geq \frac{\alpha}{2}$, then by Gross [42] the function

$$u(t, x) = \int_{\mathcal{E}'} \varphi(x + \sqrt{t}\, y)\, d\mu(y), \quad x \in \mathcal{E}'_p,$$

is a solution of the equation (11.45).

12

Laplacian Operators

There are four Laplacian operators in white noise distribution theory. We have already studied the Gross Laplacian Δ_G and the number operator N in Chapter 10. We will study the semigroups associated with them in this chapter. The other two Laplacian operators are the Lévy Laplacian Δ_L and the Volterra Laplacian Δ_V. They are defined only for certain generalized functions and are often studied in terms of the S-transforms of these generalized functions.

12.1 Semigroup for the Gross Laplacian

Define a family $\{\mu_t(x, \cdot); t \geq 0, x \in \mathcal{E}'\}$ of probability measures on $\mathcal{E}'$ by $\mu_t(x, \cdot) = \mu((\cdot - x)/\sqrt{t})$. Here μ is the standard Gaussian measure on $\mathcal{E}'$ and $\mu_0(x, \cdot)$ is understood to be $\delta_x(\cdot)$. For each $t \geq 0$ and $\varphi \in (\mathcal{E})_\beta$, define

$$P_t\varphi(x) = \int_{\mathcal{E}'} \varphi(y)\, \mu_t(x, dy) = \int_{\mathcal{E}'} \varphi(x + \sqrt{t}\, y)\, d\mu(y).$$

Obviously $P_t\varphi = \mathcal{G}_{\sqrt{t}, 1}\varphi$ (see the Fourier-Gauss transform in §11.5). Hence by Theorem 11.29, the operator P_t is a continuous linear operator from $(\mathcal{E})_\beta$ into itself. Moreover, we can use Theorem 11.30 to check that $P_tP_s = P_{t+s}$ for all $s, t \geq 0$. This fact can also be checked from the following equality for any $\varphi \in (\mathcal{E})_\beta$:

$$\int_{\mathcal{E}'}\int_{\mathcal{E}'} \varphi(ax + by)\, d\mu(x)d\mu(y) = \int_{\mathcal{E}'} \varphi(\sqrt{a^2 + b^2}\, z)\, d\mu(z).$$

Thus $\{P_t; t \geq 0\}$ is a semigroup of continuous linear operators from $(\mathcal{E})_\beta$ into itself.

Theorem 12.1. *The family* $\{P_t; t \geq 0\}$ *is a strongly continuous semigroup of continuous linear operators from* $(\mathcal{E})_\beta$ *into itself with the infinitesimal generator* $\frac{1}{2}\Delta_G$.

Proof. We need only to show that $P_t\varphi \to \varphi$ as $t \to 0$ for each $\varphi \in (\mathcal{E})_\beta$ and that $\frac{1}{2}\Delta_G$ is the infinitesimal generator of $\{P_t; t \geq 0\}$. Suppose $\varphi \in (\mathcal{E})_\beta$ is represented by

$$\varphi = \sum_{n=0}^{\infty} \langle :\cdot^{\otimes n}:, f_n \rangle.$$

Recall that $P_t = \mathcal{G}_{\sqrt{t},1}$. Hence by Theorem 11.28

$$P_t\varphi = \sum_{n=0}^{\infty} \langle :\cdot^{\otimes n}:, h_n \rangle,$$

where h_n is given by equation (11.19) with $a^2 = t$ and $b = 1$, i.e.,

$$h_n = \sum_{k=0}^{\infty} \binom{n+2k}{2k} (2k-1)!! t^k \langle \tau^{\otimes k}, f_{n+2k} \rangle$$

$$= f_n + t \sum_{k=1}^{\infty} \binom{n+2k}{2k} (2k-1)!! t^{k-1} \langle \tau^{\otimes k}, f_{n+2k} \rangle.$$

For simplicity, let

$$F_n(t) = \sum_{k=1}^{\infty} \binom{n+2k}{2k} (2k-1)!! t^{k-1} \langle \tau^{\otimes k}, f_{n+2k} \rangle.$$

Then we have

$$P_t\varphi - \varphi = t \sum_{n=0}^{\infty} \langle :\cdot^{\otimes n}:, F_n(t) \rangle. \tag{12.1}$$

Suppose $t \leq 1$. For any $p \geq 0$, if q satisfies the condition $\lambda_1^{2(q-p)} > 2^{1-\beta}|\tau|_{-p}$, then by the same calculation as in the proof of Theorem 6.2 we can derive that

$$|F_n(t)|_p^2 \leq C_{p,q,\beta} \|\varphi\|_{q,\beta}^2 \frac{1}{(n!)^{1+\beta}} \lambda_1^{-2(q-p)n} 2^{(1-\beta)n},$$

where $C_{p,q,\beta} = \left(1 - 2^{2(1-\beta)}|\tau|_{-p}^2 \lambda_1^{-4(q-p)}\right)^{-1}$. Therefore, if q also satisfies the condition $\lambda_1^{2(q-p)} > 2^{1-\beta}$, then

$$\|P_t\varphi - \varphi\|_{p,\beta}^2 = t^2 \sum_{n=0}^{\infty} (n!)^{1+\beta} |F_n(t)|_p^2$$

$$\leq t^2 K_{p,q,\beta} \|\varphi\|_{q,\beta}^2,$$

where $K_{p,q,\beta} = C_{p,q,\beta}\left(1 - 2^{1-\beta}\lambda_1^{-2(q-p)}\right)^{-1}$. This shows that for any $p \geq 0$,

$$\lim_{t \to 0} \|P_t\varphi - \varphi\|_{p,\beta} = 0.$$

Hence $P_t\varphi \to \varphi$ in $(\mathcal{E})_\beta$ as $t \to 0$, i.e., the semigroup $\{P_t; t \geq 0\}$ is strongly continuous.

Now, by Theorem 10.11 and equation (12.1), we have

$$\frac{P_t\varphi - \varphi}{t} - \frac{1}{2}\Delta_G\varphi = \sum_{n=0}^{\infty}\langle :\cdot^{\otimes n}:, G_n(t)\rangle,$$

where $G_n(t)$ is given by

$$G_n(t) = \sum_{k=2}^{\infty}\binom{n+2k}{2k}(2k-1)!!t^{k-1}\langle \tau^{\otimes k}, f_{n+2k}\rangle.$$

Thus by the same estimates as above, we see that for any $p \geq 0$,

$$\lim_{t \to 0}\left\|\frac{P_t\varphi - \varphi}{t} - \frac{1}{2}\Delta_G\varphi\right\|_{p,\beta} = 0.$$

Hence $t^{-1}(P_t\varphi - \varphi) \to \frac{1}{2}\Delta_G\varphi$ in $(\mathcal{E})_\beta$ as $t \to 0$. This shows that $\frac{1}{2}\Delta_G$ is the infinitesimal generator of $\{P_t; t \geq 0\}$. □

The heat equation associated with the Gross Laplacian has been studied in Gross [42] (see also Kuo [98]) for an abstract Wiener space. The next theorem is a similar result for the white noise space $\mathcal{E}'$.

Theorem 12.2. Let $\varphi \in (\mathcal{E})_\beta$. Then the heat equation

$$\frac{\partial u}{\partial t} = \frac{1}{2}\Delta_G u, \quad t \geq 0, \quad u(0) = \varphi,$$

has a unique solution in $(\mathcal{E})_\beta$ given by

$$u(t,x) = \int_{\mathcal{E}'}\varphi(x + \sqrt{t}\,y)\,d\mu(y). \tag{12.2}$$

Proof. By Theorem 12.1 the function $u(t,x)$ given in equation (12.2) is a solution of the above heat equation. To show the uniqueness of a solution, let P_t be defined as before. Then $P_t = \mathcal{G}_{\sqrt{t},1}$ and so by Theorem 11.30 P_t is invertible. Let $Q_t \equiv P_t^{-1}$. It is easy to check that $\{Q_t; t \geq 0\}$ is a strongly continuous semigroup of continuous linear operators from $(\mathcal{E})_\beta$ into itself with infinitesimal generator given by $-\frac{1}{2}\Delta_G$.

Now, suppose $u(t,x)$ is a solution of the heat equation with initial condition $\varphi \in (\mathcal{E})_\beta$. Consider $v(t,x) \equiv (Q_t u(t,\cdot))(x)$. Note that

$$\frac{\partial v}{\partial t} = \frac{\partial Q_t}{\partial t} u + Q_t \frac{\partial u}{\partial t}$$

$$= Q_t\left(-\frac{1}{2}\Delta_G\right)u + Q_t\left(\frac{1}{2}\Delta_G\right)u$$

$$= 0.$$

Thus $v(t) = \varphi$ for all t. Hence $Q_t u(t) = \varphi$ or $u(t) = P_t \varphi$. This shows that a solution $u(t,x)$ must be given by equation (12.2). $\square$

We can use the family $\{\mu_t(x,\cdot); t \geq 0, x \in \mathcal{E}'\}$ of probability measures to construct an $\mathcal{E}'$-valued stochastic process. It is called a standard $\mathcal{E}'$-*valued Wiener process* W_t. More precisely, W_t is an $\mathcal{E}'$-valued stochastic process satisfying the conditions:

(1) W_t has independent increments.
(2) For any $t > s \geq 0$ and any $x \in \mathcal{E}'$, $P\{W_t \in \cdot \mid W_s = x\} = \mu_{t-s}(x,\cdot)$.
(3) $P\{\omega; W_{(\cdot)}(\omega)$ is continuous$\} = 1$.

We point out that the σ-field for $\mathcal{E}'$ is understood to be the Borel field of $\mathcal{E}'$. Recall from §2.3 that the three topologies (weak, strong, and inductive limit) on $\mathcal{E}'$ generate the same σ-field, which is called the Borel field of $\mathcal{E}'$.

For any $f \in E$ with $|f|_0 = 1$, the stochastic process $\langle W_t, f \rangle$ is a one-dimensional Brownian motion. The Itô formula holds for any $\varphi \in (\mathcal{E})_\beta$:

$$\varphi(W_t) = \varphi(W_0) + \sum_{n=0}^{\infty} \int_0^t (D_{e_n}\varphi)(W_s)\, d\langle W_s, e_n \rangle$$

$$+ \frac{1}{2}\int_0^t (\Delta_G\varphi)(W_s)\, ds, \tag{12.3}$$

where $\{e_n; n \geq 1\}$ is an orthonormal basis for E. It follows from equation (12.3) that for any $\varphi \in (\mathcal{E})_\beta$

$$\Delta_G\varphi(x) = 2\lim_{t\to 0} \frac{E\big[\varphi(W_t) \mid W_0 = x\big] - \varphi(x)}{t}.$$

Recall that $(E, \mathcal{E}'_p)$ is an abstract Wiener space for any $p \geq \alpha/2$. Take any such p and let $\tau_x^{(\epsilon)}$ be the first exit time of $W_t, W_0 = x$, from the ball $\{y \in \mathcal{E}'_p; |y - x|_{-p} < \epsilon\}$. Then for any $\varphi \in (\mathcal{E})_\beta$, we can apply the above Itô formula for $\tau_x^{(\epsilon)}$ to get

$$\Delta_G\varphi(x) = 2\lim_{\epsilon\to 0} \frac{E\big[\varphi(W_{\tau_x^{(\epsilon)}}) \mid W_0 = x\big] - \varphi(x)}{E\big[\tau_x^{(\epsilon)}\big]}.$$

Next we consider the adjoint operator Δ_G^*. The family $\{P_t^*; t \geq 0\}$ is a strongly continuous semigroup of continuous linear operators from $(\mathcal{E})_\beta^*$ into itself. The infinitesimal generator is $\frac{1}{2}\Delta_G^*$, i.e., for all $\Phi \in (\mathcal{E})_\beta^*$,

$$\lim_{t \to 0} \frac{P_t^*\Phi - \Phi}{t} = \frac{1}{2}\Delta_G^*\Phi.$$

There is a nice expression for the operator P_t^* in terms of the Wick product. Recall from Example 8.4 that for the generalized function $\tilde{\mu}_t$ induced by the Gaussian measure $\mu_t \equiv \mu(\cdot/\sqrt{t})$, we have

$$\langle\!\langle \tilde{\mu}_t, \varphi \rangle\!\rangle = \int_{\mathcal{E}'} \varphi(x)\, d\mu_t(x), \quad \varphi \in (\mathcal{E})_\beta.$$

The S-transform of $\tilde{\mu}_t$ is given by

$$S\tilde{\mu}_t(\xi) = \exp\left[-\frac{1}{2}(1-t)\langle \xi, \xi \rangle\right], \quad \xi \in \mathcal{E}_c. \tag{12.4}$$

Proposition 12.3. *The equality* $P_t^*\Phi = \tilde{\mu}_{1+t} \diamond \Phi$ *holds for all* $\Phi \in (\mathcal{E})_\beta^*$.

Proof. Let $\varphi_\xi =: e^{\langle \cdot, \xi \rangle}:, \ \xi \in \mathcal{E}_c$. Then

$$P_t\varphi_\xi(x) = e^{-\frac{1}{2}\langle \xi, \xi \rangle} \int_{\mathcal{E}'} e^{\langle x + \sqrt{t}\, y, \xi \rangle}\, d\mu(y)$$

$$= \varphi_\xi(x) \int_{\mathcal{E}'} e^{\sqrt{t}\,\langle y, \xi \rangle}\, d\mu(y)$$

$$= \varphi_\xi\, e^{\frac{1}{2}t\langle \xi, \xi \rangle}.$$

Thus for any $\Phi \in (\mathcal{E})_\beta^*$,

$$(SP_t^*\Phi)(\xi) = \langle\!\langle P_t^*\Phi, \varphi_\xi \rangle\!\rangle$$

$$= \langle\!\langle \Phi, P_t\varphi_\xi \rangle\!\rangle$$

$$= e^{\frac{1}{2}t\langle \xi, \xi \rangle} \langle\!\langle \Phi, \varphi_\xi \rangle\!\rangle$$

$$= e^{\frac{1}{2}t\langle \xi, \xi \rangle} S\Phi(\xi).$$

But by equation (12.4) we have

$$S\tilde{\mu}_{1+t}(\xi) = e^{\frac{1}{2}t\langle \xi, \xi \rangle}.$$

Hence $(SP_t^*\Phi)(\xi) = S\tilde{\mu}_{1+t}(\xi)\, S\Phi(\xi)$ for all $\xi \in \mathcal{E}_c$. Obviously this implies that $P_t^*\Phi = \tilde{\mu}_{1+t} \diamond \Phi$. $\qquad\square$

Theorem 12.4. *For any* $\Phi \in (\mathcal{E})_{\beta}^{*}$, *the equality holds:*

$$\Delta_{G}^{*}\Phi = \langle :\cdot^{\otimes 2}:, \tau \rangle \diamond \Phi,$$

where τ *is the trace operator.*

Proof. Let $\varphi_{\xi} = :e^{\langle \cdot, \xi \rangle}:,\ \xi \in \mathcal{E}_{c}$. From the proof of Theorem 10.13, $\Delta_{G}\varphi_{\xi} = \langle \xi, \xi \rangle \varphi_{\xi}$. Hence for any $\Phi \in (\mathcal{E})_{\beta}^{*}$

$$(S\Delta_{G}^{*}\Phi)(\xi) = \langle\!\langle \Delta_{G}^{*}\Phi, \varphi_{\xi} \rangle\!\rangle$$

$$= \langle\!\langle \Phi, \Delta_{G}\varphi_{\xi} \rangle\!\rangle$$

$$= \langle \xi, \xi \rangle \langle\!\langle \Phi, \varphi_{\xi} \rangle\!\rangle$$

$$= \langle \xi, \xi \rangle S\Phi(\xi).$$

Note that $S(\langle :\cdot^{\otimes 2}:, \tau \rangle)(\xi) = \langle \xi, \xi \rangle$. Thus for all $\xi \in \mathcal{E}_{c}$,

$$(S\Delta_{G}^{*}\Phi)(\xi) = S(\langle :\cdot^{\otimes 2}:, \tau \rangle)(\xi)S\Phi(\xi).$$

This shows that $\Delta_{G}^{*}\Phi = \langle :\cdot^{\otimes 2}:, \tau \rangle \diamond \Phi$.   □

12.2 Semigroup for the number operator

Recall that the number operator N is defined in §9.4 by

$$N\Big(\sum_{n=0}^{\infty} \langle :\cdot^{\otimes n}:, f_{n} \rangle \Big) = \sum_{n=1}^{\infty} n \langle :\cdot^{\otimes n}:, f_{n} \rangle.$$

It is a continuous linear operator from $(\mathcal{E})_{\beta}$ into itself and from $(\mathcal{E})_{\beta}^{*}$ into itself.

First we will construct a group $\{O_{t}; t \in \mathbb{R}\}$ with infinitesimal generator $-N$. Observe that symbolically O_{t} is given by $O_{t} = e^{-tN}$. Thus we can define O_{t} as follows. For $\varphi = \sum_{n=0}^{\infty} \langle :\cdot^{\otimes n}:, f_{n} \rangle$, define

$$O_{t}\varphi = \sum_{n=0}^{\infty} e^{-tn} \langle :\cdot^{\otimes n}:, f_{n} \rangle.$$

Lemma 12.5. *For any* $t \in \mathbb{R}$, *the linear operator* O_{t} *is continuous from* $(\mathcal{E})_{\beta}$ *into itself.*

Proof. Let $\varphi \in (\mathcal{E})_\beta$ be represented by

$$\varphi = \sum_{n=0}^{\infty} \langle :\cdot^{\otimes n}:, f_n \rangle.$$

For any $t \geq 0$, we have

$$\|O_t\varphi\|_{p,\beta}^2 = \sum_{n=0}^{\infty} (n!)^{1+\beta} e^{-2tn} |f_n|_p^2$$

$$\leq \sum_{n=0}^{\infty} (n!)^{1+\beta} |f_n|_p^2$$

$$= \|\varphi\|_{p,\beta}^2.$$

On the other hand, if $t < 0$, then for $q > p$,

$$\|O_t\varphi\|_{p,\beta}^2 = \sum_{n=0}^{\infty} (n!)^{1+\beta} e^{-2tn} |f_n|_p^2$$

$$\leq \sum_{n=0}^{\infty} (n!)^{1+\beta} e^{-2tn} \lambda_1^{-2(q-p)n} |f_n|_q^2.$$

We can choose $q > p$ such that $\lambda_1^{q-p} \geq e^{-t}$. Then we get immediately that

$$\|O_t\|_{p,\beta} \leq \|\varphi\|_{q,\beta}.$$

Thus for any $t \in \mathbb{R}$, the operator O_t is continuous from $(\mathcal{E})_\beta$ into itself. $\square$

Theorem 12.6. *The family $\{O_t; t \in \mathbb{R}\}$ is a strongly continuous group of continuous linear operators from $(\mathcal{E})_\beta$ into itself with the infinitesimal generator $-N$.*

Proof. It is obvious that $O_t O_s = O_{t+s}$ for any $t, s \in \mathbb{R}$. Thus $\{O_t; t \in \mathbb{R}\}$ is a group. To show the strong continuity of the group, suppose $|t| \leq 1$. Then we can use the inequality $|e^x - 1| \leq |x|e^{|x|}$, $x \in \mathbb{R}$, to obtain

$$\|O_t\varphi - \varphi\|_{p,\beta}^2 = \sum_{n=1}^{\infty} (n!)^{1+\beta} (e^{-tn} - 1)^2 |f_n|_p^2$$

$$\leq \sum_{n=1}^{\infty} (n!)^{1+\beta} (tn)^2 e^{2|t|n} |f_n|_p^2$$

$$\leq t^2 \sum_{n=1}^{\infty} (n!)^{1+\beta} n^2 e^{2n} \lambda_1^{-2(q-p)n} |f_n|_q^2$$

$$\leq t^2 \sum_{n=1}^{\infty} (n!)^{1+\beta} \left(2e\lambda_1^{-(q-p)}\right)^{2n} |f_n|_q^2.$$

Choose $q > p$ such that $\lambda_1^{q-p} \geq 2e$. Then

$$\|O_t\varphi - \varphi\|_{p,\beta} \leq |t| \|\varphi\|_{q,\beta}.$$

This implies the strong continuity of $\{O_t; t \in \mathbb{R}\}$. To check that $-N$ is the infinitesimal generator of $\{O_t; t \in \mathbb{R}\}$, let $\varphi = \sum_{n=0}^{\infty} \langle : \cdot^{\otimes n} :, f_n \rangle \in (\mathcal{E})_\beta$. Then

$$\frac{O_t\varphi - \varphi}{t} + N\varphi = \sum_{n=1}^{\infty} \left(\frac{e^{-tn} - 1}{t} + n\right) \langle : \cdot^{\otimes n} :, f_n \rangle.$$

Hence for any $p \geq 0$,

$$\left\|\frac{O_t\varphi - \varphi}{t} + N\varphi\right\|_{p,\beta}^2 = \sum_{n=1}^{\infty} (n!)^{1+\beta} \frac{1}{t^2} \left|e^{-tn} - 1 + tn\right|^2 |f_n|_p^2.$$

Note that $|e^x - 1 - x| \leq x^2 e^{|x|}$ for all $x \in \mathbb{R}$. Thus for any $q > p$,

$$\left\|\frac{O_t\varphi - \varphi}{t} + N\varphi\right\|_{p,\beta}^2 \leq t^2 \sum_{n=1}^{\infty} (n!)^{1+\beta} n^4 e^{2|t|n} |f_n|_p^2$$

$$\leq t^2 \sum_{n=1}^{\infty} (n!)^{1+\beta} n^4 e^{2|t|n} \lambda_1^{-2(q-p)n} |f_n|_q^2.$$

Now, suppose $|t| \leq 1$. Since $n^4 \leq 3^{2n}$, we have

$$\left\|\frac{O_t\varphi - \varphi}{t} + N\varphi\right\|_{p,\beta}^2 \leq t^2 \sum_{n=1}^{\infty} (n!)^{1+\beta} \left(3e\lambda_1^{-(q-p)}\right)^{2n} |f_n|_q^2.$$

Choose $q > p$ such that $\lambda_1^{q-p} \geq 3e$. Then

$$\left\|\frac{O_t\varphi - \varphi}{t} + N\varphi\right\|_{p,\beta} \leq |t| \|\varphi\|_{q,\beta}.$$

This implies that $t^{-1}(O_t\varphi - \varphi) \to -N\varphi$ in $(\mathcal{E})_\beta$, i.e., $-N$ is the infinitesimal generator of $\{O_t; t \in \mathbb{R}\}$.   □

Proposition 12.7. *If $t \geq 0$, then for any $\varphi \in (\mathcal{E})_\beta$,*

$$O_t\varphi(x) = \int_{\mathcal{E}'} \varphi\left(e^{-t}x + \sqrt{1 - e^{-2t}}\, y\right) d\mu(y).$$

Proof. Define an operator A_t on $(\mathcal{E})_\beta$ by

$$A_t\varphi(x) = \int_{\mathcal{E}'} \varphi\left(e^{-t}x + \sqrt{1 - e^{-2t}}\, y\right) d\mu(y).$$

For simplicity, let $a = \sqrt{1 - e^{-2t}}$ and $b = e^{-t}$. Then $A_t = \mathcal{G}_{a,b}$ and so by Theorem 11.29 the operator A_t is continuous from $(\mathcal{E})_\beta$ into itself. Let $\varphi_\xi =: e^{\langle \cdot, \xi \rangle} :, \xi \in \mathcal{E}_c$. Then

$$O_t \varphi_\xi = \sum_{n=0}^{\infty} e^{-tn} \frac{1}{n!} \langle :e^{\otimes n}:, \xi^{\otimes n} \rangle$$

$$= \varphi_{e^{-t}\xi}. \tag{12.5}$$

On the other hand, we have

$$A_t \varphi_\xi(x) = e^{-\frac{1}{2}\langle \xi, \xi \rangle} \int_{\mathcal{E}'} e^{\langle bx + ay, \xi \rangle} \, d\mu(y)$$

$$= e^{-\frac{1}{2}\langle \xi, \xi \rangle + b\langle x, \xi \rangle + \frac{1}{2}a^2 \langle \xi, \xi \rangle}$$

$$=: e^{e^{-t}\langle x, \xi \rangle}:$$

$$= \varphi_{e^{-t}\xi}(x). \tag{12.6}$$

It follows from equations (12.5) and (12.6) that $O_t \varphi_\xi = A_t \varphi_\xi$ for all $\xi \in \mathcal{E}_c$. This implies that $O_t = A_t$ as operators on $(\mathcal{E})_\beta$. $\qquad \square$

The heat equation associated with the number operator has been studied in Kuo [99] for an abstract Wiener space and in Kang [78] for a white noise space.

Theorem 12.8. *Let $\varphi \in (\mathcal{E})_\beta$. Then the heat equation*

$$\frac{\partial u}{\partial t} = -N u, \quad t \geq 0, \quad u(0) = \varphi,$$

has a unique solution in $(\mathcal{E})_\beta$ given by

$$u(t, x) = \int_{\mathcal{E}'} \varphi\left(e^{-t}x + \sqrt{1 - e^{-2t}}\, y\right) d\mu(y). \tag{12.7}$$

Proof. By Theorem 12.6 and Proposition 12.7 the function $u(t, x)$ given in equation (12.7) is a solution of the above heat equation. To show the uniqueness of a solution, note that $O_{-t} = O_t^{-1}$ and the semigroup $\{O_{-t}; t \geq 0\}$ is strongly continuous with infinitesimal generator N. Suppose $u(t, x)$ is a solution of the above heat equation with initial condition $\varphi \in (\mathcal{E})_\beta$. Let $v(t, x) \equiv ((O_{-t}u(t, \cdot))(x)$. Then

$$\frac{\partial v}{\partial t} = \frac{\partial O_{-t}}{\partial t} u + O_{-t} \frac{\partial u}{\partial t}$$

$$= O_{-t} N u + O_{-t}(-N u)$$

$$= 0.$$

Hence $v(t) = \varphi$ for all $t \geq 0$, i.e., $O_{-t}u(t) = \varphi$ or $u(t) = O_t\varphi$. This shows that $u(t, x)$ must be given by equation (12.7). □

Next consider the $\mathcal{E}'$-valued stochastic integral equation:

$$U_t = x + \sqrt{2} \int_0^t dW_s - \int_0^t U_s\, ds,$$

where W_t is a standard $\mathcal{E}'$-valued Wiener process starting at 0. This stochastic integral equation has a unique solution. To find the solution, rewrite this equation as

$$dU_t = \sqrt{2}\, dW_t - U_t\, dt.$$

By using the Itô formula in equation (12.3) we get

$$d(e^t U_t) = \sqrt{2}\, e^t\, dW_t.$$

By integrating both sides of this equation we get easily

$$U_t = e^{-t}x + \sqrt{2} \int_0^t e^{-(t-s)}\, dW_s.$$

The diffusion process U_t is called an $\mathcal{E}'$-*valued Ornstein-Uhlenbeck process*. To find the transition probabilities of the process U_t, observe that the $\mathcal{E}'$-valued random variable

$$\sqrt{2} \int_0^t e^{-(t-s)}\, dW_s$$

has a Gaussian distribution with mean zero and variance

$$2 \int_0^t e^{-2(t-s)}\, ds = 1 - e^{-2t}.$$

Hence we get the transition probabilities $\nu_t(x, \cdot)$ of the process U_t:

$$\nu_t(x, \cdot) = P\{U_t \in \cdot \,|\, U_0 = x\} = \mu_{\sqrt{1-e^{-2t}}}(e^{-t}x, \cdot).$$

Note that for any $x \in \mathcal{E}'$, we have

$$\lim_{t \to \infty} \mu_{\sqrt{1-e^{-2t}}}(e^{-t}x, \cdot) = \mu(\cdot).$$

This implies that the Gaussian measure μ on $\mathcal{E}'$ is an invariant measure for the Ornstein-Uhlenbeck process U_t, i.e., for any Borel subset C of $\mathcal{E}'$ and any $t \geq 0$,

$$\int_{\mathcal{E}'} \nu_t(x, C)\, d\mu(x) = \mu(C).$$

Next we show that $-N$ is the infinitesimal generator of U_t. The Itô formula holds for U_t and $\varphi \in (\mathcal{E})_\beta$:

$$\varphi(U_t) = \varphi(U_0) + \sqrt{2} \sum_{n=0}^{\infty} \int_0^t (D_{e_n}\varphi)(U_s) \, d\langle U_s, e_n \rangle$$

$$+ \int_0^t \left[(\Delta_G\varphi)(U_s) - (\Lambda\varphi)(U_s) \right] ds, \qquad (12.8)$$

where $\{e_n; n \geq 1\}$ is an orthonormal basis for E and Λ is the lambda operator defined in §10.4 and $\Lambda\varphi(x) = N\varphi(x) + \Delta_G\varphi(x)$. By using the Itô formula in equation (12.8) we can show that $-N$ is the infinitesimal generator of U_t, i.e., for any $\varphi \in (\mathcal{E})_\beta$

$$\lim_{t \to 0} \frac{E[\varphi(U_t) \,|\, U_0 = x] - \varphi(x)}{t} = -N\varphi(x).$$

Finally, we make some remarks about the number operator acting on the space $(\mathcal{E})_\beta^*$. Obviously the adjoint operator N^* is the extension of N to $(\mathcal{E})_\beta^*$ (we have used the same notation N for the extension). Thus Lemma 12.5 and Theorem 12.6 remain valid when $(\mathcal{E})_\beta$ is replaced with $(\mathcal{E})_\beta^*$. For $\Phi \in (\mathcal{E})_\beta^*$, the generalized function $O_t\Phi$ can not be expressed as an integral as in Proposition 12.7. Theorem 12.8 is true when the initial condition $\Phi \in (\mathcal{E})_\beta^*$, but the solution is given by $u(t) = O_t\Phi$ which can not be expressed as an integral as in equation (12.7).

12.3 Lévy Laplacian

The Gross Laplacian is a natural generalization of the finite dimensional Laplacian to infinite dimensional spaces such as abstract Wiener spaces or white noise spaces. There is another generalization of the finite dimensional Laplacian to a white noise space with a different motivation. Consider the following function F on $\mathbb{R}^n$ and its Laplacian

$$F(x_1, \ldots, x_n) = \sum_{k=1}^n a_k x_k^2, \qquad \Delta F = 2 \sum_{k=1}^n a_k.$$

Note that in the white noise space $(\mathcal{S}'(\mathbb{R}), \mu)$, we take $\{\dot{B}(t); t \in \mathbb{R}\}$ as a coordinate system. Hence a white noise analogue of the above function F with its Laplacian is given by

$$\Phi = \int_0^1 f(t) :\dot{B}(t)^2: dt, \qquad \Delta\Phi = 2 \int_0^1 f(t) \, dt. \qquad (12.9)$$

But what is this Laplacian? What is its relationship to the Gross Laplacian? What kind of properties does this Laplacian have?

Notation: Let $\eta \in \mathcal{E}$. By Theorem 9.10 the differential operator D_η from $(\mathcal{E})_\beta$ into itself has a unique extension by continuity to a continuous linear operator $\widetilde{D}_\eta$ from $(\mathcal{E})_\beta^*$ into itself. For simplicity, we will also use D_η to denote its extension $\widetilde{D}_\eta$.

Note that if $\Psi = \sum_{n=0}^{\infty} \langle :\cdot^{\otimes n}:, f_n \rangle \in (\mathcal{E})_\beta^*$, then

$$D_\eta \Psi = \sum_{n=1}^{\infty} n \langle :\cdot^{\otimes (n-1)}:, \langle f_n, \eta \rangle \rangle. \tag{12.10}$$

Recall that $:\dot{B}(t)^2: := \langle :\cdot^{\otimes 2}:, \delta_t^{\otimes 2} \rangle$. Hence the generalized function Φ in (12.9) can be rewritten as

$$\Phi = \int_0^1 f(t) :\dot{B}(t)^2: \, dt = \langle :\cdot^{\otimes 2}:, \tau_{(1_{[0,1]}f)} \rangle,$$

where for $g \in L^1(\mathbb{R})$, $\tau_g \in (\mathcal{S}_c')^{\widehat{\otimes} 2}$ is defined by

$$\langle \tau_g, \xi \otimes \eta \rangle = \int_{\mathbb{R}} g(t)\xi(t)\eta(t) \, dt.$$

Let $\{e_k; k \geq 1\} \subset \mathcal{S}(\mathbb{R})$ be an orthonormal basis for $L^2[0,1]$. Suppose this orthonormal basis satisfies the following condition

$$\lim_{m \to \infty} \frac{1}{m} \sum_{k=1}^{m} e_k(t)^2 \longrightarrow 1, \quad \text{in } L^2[0,1]. \tag{12.11}$$

Then by using equation (12.10) we get

$$\lim_{m \to \infty} \frac{1}{m} \sum_{k=1}^{m} D_{e_k}^2 \Phi = \lim_{m \to \infty} \frac{1}{m} \sum_{k=1}^{m} 2\langle f, e_k \otimes e_k \rangle$$

$$= 2 \lim_{m \to \infty} \frac{1}{m} \sum_{k=1}^{m} \int_0^1 f(t) e_k(t)^2 \, dt$$

$$= 2 \int_0^1 f(t) \, dt.$$

Thus, in view of (12.9), this new Laplacian can be defined as the limit of the arithmetic mean of the second derivative with respect to an orthonormal basis for $L^2[0,1]$. This Laplacian was introduced by Lévy in [125], [127], and [128].

We will define the Lévy Laplacian for certain generalized functions in $(\mathcal{E})^*_\beta$ by using the above idea. Some properties will be proved. In the next section we will use the S-transform to study the Lévy Laplacian.

Definition 12.9. Let $\{e_k; k \geq 1\} \subset \mathcal{E}$ be an orthonormal basis for E. The *Lévy Laplacian* $\Delta_L \Phi$ of $\Phi \in (\mathcal{E})^*_\beta$ with respect to $\{e_k; k \geq 1\}$ is defined by

$$\Delta_L \Phi = \lim_{m \to \infty} \frac{1}{m} \sum_{k=1}^m D^2_{e_k} \Phi,$$

provided that the limit exists in $(\mathcal{E})^*_\beta$. (The convergence is strong or weak convergence since they are equivalent for sequences; see §2.2.)

We will use $\mathrm{Dom}(\Delta_L)$ to denote the domain of Δ_L, i.e., the subspace of $(\mathcal{E})^*_\beta$ consisting of all Φ such that $\Delta_L \Phi$ exists with respect to a fixed orthonormal basis $\{e_k; k \geq 1\} \subset \mathcal{E}$ for E. (The property in equation (12.11) is not assumed here. It will be assumed in the next section when we use the S-transform to study the Lévy Laplacian.)

Note that if $\varphi \in (\mathcal{E})_\beta$, then its Gross Laplacian is given by

$$\Delta_G \varphi = \sum_{k=1}^\infty D^2_{e_k} \varphi,$$

where $\{e_n; n \geq 1\}$ is any orthonormal basis for E. Obviously this implies that $\Delta_L \varphi = 0$ for all $\varphi \in (\mathcal{E})_\beta$. Observe that the Gross Laplacian acts on the space of test functions, while the Lévy Laplacian is defined for certain generalized functions in $(\mathcal{E})^*_\beta$.

Example 12.10. Consider the generalized function $\Phi = \; : e^{\langle \cdot, x \rangle} :$ with $x \in \mathcal{E}'$. Recall from Definition 5.12 that

$$\Phi = \sum_{n=0}^\infty \frac{1}{n!} \langle :\cdot^{\otimes n}:, x^{\otimes n} \rangle.$$

Let $\{e_k; k \geq 1\} \subset \mathcal{E}$ be an orthonormal basis for E. Then

$$D^2_{e_k} \Phi = \sum_{n=2}^\infty \frac{1}{n!} n(n-1) \langle :\cdot^{\otimes(n-2)}:, \langle x^{\otimes n}, e_k \otimes e_k \rangle \rangle$$

$$= \sum_{n=2}^\infty \frac{1}{(n-2)!} \langle x, e_k \rangle^2 \langle :\cdot^{\otimes(n-2)}:, x^{\otimes(n-2)} \rangle$$

$$= \langle x, e_k \rangle^2 \Phi.$$

Hence the Lévy Laplacian $\Delta_L \Phi$ is given by

$$\Delta_L \Phi = \Phi \lim_{m \to \infty} \frac{1}{m} \sum_{k=1}^{m} \langle x, e_k \rangle^2,$$

provided that the limit exists. For example, if $x \in E$, then $\sum_{k=1}^{\infty} \langle x, e_k \rangle^2 = |x|_0^2 < \infty$, which implies that $\Delta_L \Phi = 0$. We give three examples with $x \notin E$ for the white noise space $\mathcal{S}'(\mathbb{R})$ (A is the operator $A = -D_x^2 + x^2 + 1$ with eigenfunctions given by Hermite functions $e_k, k \geq 0$; see §3.2):

(1) Let $x = a \sum_{k=0}^{\infty} e_k, a \in \mathbb{R}$. Obviously $x \in \mathcal{S}'(\mathbb{R})$ and $\Delta_L \Phi = a^2 \Phi$. Hence Φ is an eigenvector of the Lévy Laplacian.

(2) Let $x = \sum_{k=1}^{\infty} k e_k$. It is easy to check that $x \in \mathcal{S}'(\mathbb{R})$. However, $\Delta_L \Phi$ does not exist since $\lim_{m \to \infty} \frac{1}{m} \sum_{k=1}^{m} k = \infty$.

(3) Let $x = \delta_t$. Recall from §3.4 that $\sup_{t \in \mathbb{R}} |e_n(t)| = O(n^{-1/12})$. Therefore

$$\sum_{k=1}^{m} \langle x, e_k \rangle^2 \approx \sum_{k=1}^{m} n^{-1/6} = O(m^{5/6}).$$

This implies that $\lim_{m \to \infty} \frac{1}{m} \sum_{k=1}^{m} \langle x, e_k \rangle^2 = 0$. Hence $\Delta_L \Phi = 0$ for $x = \delta_t$.

The Lévy Laplacian has some interesting and peculiar properties. For instance, we have $\Delta_G \varphi = 0$ for all $\varphi \in (\mathcal{E})_\beta$ as mentioned above. In fact, this is true for all $\varphi \in (L^2)$ as stated in the next theorem. Thus the Lévy Laplacian acts non-trivially only outside the space (L^2).

Theorem 12.11. *The Lévy Laplacian $\Delta_L \varphi = 0$ for all $\varphi \in (L^2)$.*

Proof. Let $\varphi = \sum_{n=0}^{\infty} \langle :\cdot^{\otimes n}:, f_n \rangle \in (L^2)$. Then by using equation (12.10) we can derive that

$$D_{e_k}^2 \varphi = \sum_{n=0}^{\infty} (n+2)(n+1) \langle :\cdot^{\otimes n}:, \langle f_{n+2}, e_k \otimes e_k \rangle \rangle.$$

Note that $|\langle f_{n+2}, \langle e_k \otimes e_k \rangle \rangle|_{-p} \leq \lambda_1^{-pn} |f_{n+2}|_0$ for $p \geq 0$. Hence

$$\|D_{e_k}^2 \varphi\|_{-p,-\beta}^2 = \sum_{n=0}^{\infty} (n!)^{1-\beta} (n+2)^2 (n+1)^2 |\langle f_{n+2}, e_k \otimes e_k \rangle|_{-p}^2$$

$$\leq \sum_{n=0}^{\infty} (n!)^{1-\beta} (n+2)^2 (n+1)^2 \lambda_1^{-2pn} |f_{n+2}|_0^2.$$

But $(n!)^{-\beta} \leq 1$ and $(n+2)(n+1) \leq 2^{2n+1}$. Hence

$$\|D_{e_k}^2 \varphi\|_{-p,-\beta}^2 \leq 2 \sum_{n=0}^{\infty} (2\lambda_1^{-p})^{2n} (n+2)! |f_{n+2}|_0^2.$$

Thus if we choose $p > 0$ such that $2\lambda_1^{-p} \leq 1$, then

$$\|D_{e_k}^2 \varphi\|_{-p,-\beta}^2 = \sum_{n=0}^{\infty} (n!)^{1-\beta}(n+2)^2(n+1)^2 |\langle f_{n+2}, e_k \otimes e_k\rangle|_{-p}^2$$

$$\leq \sum_{n=0}^{\infty} (n+2)! |f_{n+2}|_0^2.$$

Observe that $\sum_{n=0}^{\infty}(n+2)!|f_{n+2}|_0^2 = \|\varphi\|_0^2 < \infty$ and $\lim_{k\to\infty} |\langle f_{n+2}, e_k \otimes e_k\rangle|_{-p} = 0$. Thus by the dominated convergence theorem (for the summation), we conclude that

$$\lim_{k\to\infty} \|D_{e_k}^2 \varphi\|_{-p,-\beta} = 0.$$

This implies that

$$\lim_{m\to\infty} \frac{1}{m} \sum_{k=1}^{m} \|D_{e_k}^2 \varphi\|_{-p,-\beta} = 0.$$

Therefore

$$\lim_{m\to\infty} \frac{1}{m} \sum_{k=1}^{m} D_{e_k}^2 \varphi = 0, \quad \text{in } (\mathcal{E})_\beta^*.$$

This shows that $\Delta_L \varphi = 0$ with respect to $\{e_k; k \geq 1\}$. $\qquad\square$

Lemma 12.12. For any $\eta \in \mathcal{E}$ and $\Phi \in (\mathcal{E})_\beta^*$, the equality holds:

$$D_\eta \Delta_G^* \Phi = 2D_\eta^* \Phi + \Delta_G^* D_\eta \Phi.$$

Proof. Let $\eta \in \mathcal{E}$. It is straightforward to check that for any $\varphi \in (\mathcal{E})_\beta$,

$$\Delta_G(\langle \cdot, \eta\rangle \varphi) = 2D_\eta \varphi + \langle \cdot, \eta\rangle \Delta_G \varphi.$$

Thus the following equality holds as continuous linear operators on $(\mathcal{E})_\beta$

$$\Delta_G Q_\eta = 2D_\eta + Q_\eta \Delta_G.$$

Recall from Theorem 9.18 that $Q_\eta = D_\eta + D_\eta^*$. Hence

$$\Delta_G(D_\eta + D_\eta^*) = 2D_\eta + (D_\eta + D_\eta^*)\Delta_G.$$

Since $\Delta_G D_\eta = D_\eta \Delta_G$, we get

$$\Delta_G D_\eta^* = 2D_\eta + D_\eta^* \Delta_G.$$

By taking the adjoint operators, we get the equality in the lemma. $\qquad\square$

Theorem 12.13. *The domain* $\mathrm{Dom}(\Delta_L)$ *of the Lévy Laplacian is invariant under* Δ_G^*, *i.e., if* $\Phi \in \mathrm{Dom}(\Delta_L)$, *then* $\Delta_G^* \Phi \in \mathrm{Dom}(\Delta_L)$. *Moreover,*

$$\Delta_L(\Delta_G^* \Phi) = 2\Phi + \Delta_G^*(\Delta_L \Phi), \quad \forall \Phi \in \mathrm{Dom}(\Delta_L).$$

Proof. It follows from Lemma 12.12 that for any $\eta \in \mathcal{E}$ and $\Phi \in (\mathcal{E})_\beta^*$,

$$D_\eta^2 \Delta_G^* \Phi = 2(D_\eta D_\eta^* + D_\eta^* D_\eta)\Phi + \Delta_G^* D_\eta^2 \Phi.$$

But by Theorem 9.15 (d), $D_\eta D_\eta^* = D_\eta^* D_\eta + |\eta|_0^2 I$. Hence

$$D_\eta^2 \Delta_G^* \Phi = 2|\eta|_0^2 \Phi + 4 D_\eta^* D_\eta \Phi + \Delta_G^* D_\eta^2 \Phi. \tag{12.12}$$

Suppose $\Phi \in \mathrm{Dom}(\Delta_L)$ with respect to $\{e_k; k \geq 1\}$. Then by equation (12.12),

$$\sum_{k=1}^m D_{e_k}^2 (\Delta_G^* \Phi) = 2m\Phi + 4 \sum_{k=1}^m D_{e_k}^* D_{e_k} \Phi + \Delta_G^* \left(\sum_{k=1}^m D_{e_k}^2 \Phi \right).$$

It is easy to check that for any $\Phi \in (\mathcal{E})_\beta^*$

$$N\Phi = \sum_{k=1}^\infty D_{e_k}^* D_{e_k} \Phi.$$

Hence

$$\lim_{m \to \infty} \frac{1}{m} \sum_{k=1}^m D_{e_k}^* D_{e_k} \Phi = 0.$$

On the other hand, since $\Phi \in \mathrm{Dom}(\Delta_L)$ and Δ_G^* is a continuous linear operator from $(\mathcal{E})_\beta^*$ into itself,

$$\lim_{m \to \infty} \Delta_G^* \left(\frac{1}{m} \sum_{k=1}^m D_{e_k}^2 \Phi \right) = \Delta_G^*(\Delta_L \Phi).$$

Thus we get

$$\lim_{m \to \infty} \frac{1}{m} \sum_{k=1}^m D_{e_k}^2 (\Delta_G^* \Phi) = 2\Phi + \Delta_G^*(\Delta_L \Phi).$$

This shows that $\Delta_G^* \Phi \in \mathrm{Dom}(\Delta_L)$ and $\Delta_L(\Delta_G^* \Phi) = 2\Phi + \Delta_G^*(\Delta_L \Phi)$. $\quad \square$

Lemma 12.14. *For any* $\eta \in \mathcal{E}$ *and* $\varphi, \psi \in (\mathcal{E})_\beta$,

$$D_\eta^*(\varphi\psi) = (D_\eta^* \varphi)\psi - \varphi(D_\eta \psi).$$

Proof. Note that $Q_\eta = D_\eta + D_\eta^*$ by Theorem 9.18(a). Hence

$$D_\eta^*(\varphi\psi) = (Q_\eta - D_\eta)(\varphi\psi)$$
$$= Q_\eta(\varphi\psi) - ((D_\eta\varphi)\psi + \varphi(D_\eta\psi))$$
$$= (Q_\eta\varphi - D_\eta\varphi)\psi - \varphi(D_\eta\psi)$$
$$= (D_\eta^*\varphi)\psi - \varphi(D_\eta\psi). \qquad \square$$

Theorem 12.15. *Let* $\Phi \in \text{Dom}(\Delta_L)$. *Then* $\varphi\Phi \in \text{Dom}(\Delta_L)$ *for all* $\varphi \in (\mathcal{E})_\beta$ *and*

$$\Delta_L(\varphi\Phi) = \varphi(\Delta_L\Phi).$$

Proof. Let $\Phi \in \text{Dom}(\Delta_L)$ and $\varphi \in (\mathcal{E})_\beta$. Obviously we have

$$D_{e_k}^2(\varphi\Phi) = (D_{e_k}^2\varphi)\Phi + 2(D_{e_k}\varphi)(D_{e_k}\Phi) + \varphi D_{e_k}^2\Phi.$$

Note that

$$\lim_{m\to\infty} \frac{1}{m}\sum_{k=1}^m D_{e_k}^2\varphi = \Delta_L\varphi = 0.$$

Since $\Phi \in \text{Dom}(\Delta_L)$,

$$\lim_{m\to\infty} \frac{1}{m}\sum_{k=1}^m D_{e_k}^2\Phi = \Delta_L\Phi.$$

Thus to finish the proof, we need only to show that

$$\lim_{m\to\infty} \frac{1}{m}\sum_{k=1}^m (D_{e_k}\varphi)(D_{e_k}\Phi) = 0. \qquad (12.13)$$

For any $\psi \in (\mathcal{E})_\beta$, we can use Lemma 12.14 to get

$$\langle\!\langle (D_{e_k}\varphi)(D_{e_k}\Phi), \psi \rangle\!\rangle = \langle\!\langle \Phi, D_{e_k}^*((D_{e_k}\varphi)\psi) \rangle\!\rangle$$
$$= \langle\!\langle \Phi, (D_{e_k}^* D_{e_k}\varphi)\psi \rangle\!\rangle - \langle\!\langle \Phi, (D_{e_k}\varphi)(D_{e_k}\psi) \rangle\!\rangle.$$

By the same argument as in the proof of Theorem 12.11,

$$\lim_{k\to\infty} D_{e_k}\varphi = 0.$$

Moreover, we have $N\varphi = \sum_{k=1}^\infty D_{e_k}^* D_{e_k}\varphi$. Hence

$$\lim_{k\to\infty} \langle\!\langle (D_{e_k}\varphi)(D_{e_k}\Phi), \psi \rangle\!\rangle = 0, \quad \forall\psi \in (\mathcal{E})_\beta.$$

Thus $(D_{e_k}\varphi)(D_{e_k}\Phi)$ converges to 0 weakly (hence strongly) in $(\mathcal{E})^*_\beta$. This implies equation (12.13) and the proof is completed. $\qquad\qquad\Box$

12.4 Lévy Laplacian by the S-transform

We can study Laplacian operators in terms of the S-transform. Suppose $\Phi \in (\mathcal{E})^*_\beta$ and $F = S\Phi$. Then we know from Theorem 8.2 that for any $\xi, \eta \in \mathcal{E}_c$ the function $F(\xi + z\eta)$ is an entire function of $z \in \mathbb{C}$. Hence we have the series expansion

$$F(\xi + z\eta) = F(\xi) + zF'(\xi)(\eta) + \frac{z^2}{2!}F''(\xi)(\eta, \eta)$$

$$+ \cdots + \frac{z^n}{n!}F^{(n)}(\xi)(\eta, \ldots, \eta) + \cdots,$$

where $F^{(n)}(\xi) \colon \mathcal{E}_c \times \cdots \times \mathcal{E}_c \to \mathbb{C}$ is a continuous n-linear functional. Note that by using the Cauchy formula as in the proof of Theorem 8.2 we get a growth condition for the derivative $F^{(n)}(\xi)$. This growth condition on $F^{(n)}(\xi)$ together with the analyticity of F can be used to show that certain functions of $\xi \in \mathcal{E}_c$ are S-transforms of generalized functions. Thus it is often useful to study Laplacian operators in terms of the S-transform. This is more so for the Lévy Laplacian. See Hida [47], the book by Hida et al. [58], Hida and Saitô [62], Kuo [107], Kuo et al. [113], Obata [136] [137] [138], and Saitô [157] [158] [159].

For the Gross Laplacian and number operator, we have the following facts which can be checked easily. Suppose $\varphi \in (\mathcal{E})_\beta$ and $F = S\varphi$. Then

$$S(\Delta_G \varphi)(\xi) = \sum_{k=1}^\infty F''(\xi)(e_k, e_k), \quad \xi \in \mathcal{E}_c,$$

where $\{e_k; k \geq 1\} \subset \mathcal{E}$ is an orthonormal basis for E. For the number operator N, suppose $\Phi \in (\mathcal{E})^*_\beta$ and $F = S\Phi$. Then

$$S(N\Phi)(\xi) = F'(\xi)(\xi), \quad \xi \in \mathcal{E}_c.$$

As for the Lévy Laplacian, let $\Phi \in \mathrm{Dom}(\Delta_L)$ and $F = S\Phi$. Then we have

$$S(\Delta_L \Phi)(\xi) = \lim_{m \to \infty} \frac{1}{m} \sum_{k=1}^m F''(\xi)(e_k, e_k), \quad \xi \in \mathcal{E}_c. \qquad (12.14)$$

In this section we will define and study the Lévy Laplacian in terms of the S-transform. Let $\mathcal{U}_\beta$ denote the set of all S-transforms of generalized functions in $(\mathcal{E})^*_\beta$, i.e.,

$$\mathcal{U}_\beta \equiv \{F; \ F = S\Phi, \Phi \in (\mathcal{E})^*_\beta\}.$$

Equivalently, $\mathcal{U}_\beta$ consists of all functions F on $\mathcal{E}_c$ satisfying the conditions (a) and (b) in Theorem 8.2.

In view of equation (12.14) we would define the Lévy Laplacian $\Delta_L F$ of $F \in \mathcal{U}_\beta$ by

$$\Delta_L F(\xi) = \lim_{m \to \infty} \frac{1}{m} \sum_{k=1}^{m} F''(\xi)(e_k, e_k), \quad \xi \in \mathcal{E}_c, \tag{12.15}$$

provided that the limit exists and $\Delta_L F \in \mathcal{U}_\beta$. However, we will define $\Delta_L F$ below by an integral instead of taking the limit. This is often more convenient for studying certain properties of the Lévy Laplacian.

For the rest of this section we will use the Gel'fand triples from Chapter 3, i.e.,

$$\mathcal{S}(\mathbb{R}) \subset L^2(\mathbb{R}) \subset \mathcal{S}'(\mathbb{R}), \quad (\mathcal{S})_\beta \subset (L^2) \subset (\mathcal{S})^*_\beta.$$

Definition 12.16. A function $F \in \mathcal{U}_\beta$ is called an *L-functional* if its derivatives F' and F'' are given by

$$F'(\xi)(\eta) = \int_{\mathbb{R}} F'(\xi; t)\eta(t)\, dt,$$

$$F''(\xi)(\eta, \zeta) = \int_{\mathbb{R}} F''_L(\xi; t)\eta(t)\zeta(t)\, dt + \int_{\mathbb{R}^2} F''_r(\xi; s, t)\eta(s)\zeta(t)\, ds dt,$$

where the functional derivatives F', F''_L, and F''_r satisfy the following conditions:

(1) $F'(\xi; \cdot) \in L^1_{\text{loc}}(\mathbb{R})$, $F''_L(\xi; \cdot) \in L^1_{\text{loc}}(\mathbb{R})$, and $F''_r(\xi; \cdot, \cdot) \in L^1_{\text{loc}}(\mathbb{R}^2)$ for each $\xi \in \mathcal{S}_c(\mathbb{R})$.

(2) $\int_Z F''_L(\cdot; t)\, dt \in \mathcal{U}_\beta$ for any finite interval $Z \subset \mathbb{R}$.

Notation. Let $\mathcal{D}_L$ denote the set of *L*-functionals. We will also use the same notation $\mathcal{D}_L$ to denote the set of generalized functions $\Phi \in (\mathcal{S})^*_\beta$ such that $S\Phi$ is an *L*-functional. There will be no confusion about this notation since the context is clear.

Remarks. The above integrals for F', F''_L, and F''_r are in the distribution sense, i.e., they are interpreted as the bilinear pairing of a generalized function and a test function. The condition (1) indicates that the generalized functions $F'(\xi; \cdot), F''_L(\xi; \cdot)$, and $F''_r(\xi; \cdot, \cdot)$ are represented by locally integrable functions.

It is easy to see that the decomposition of $F''(\xi)$ in the above definition is unique. We will call F''_L and F''_r the *Lévy and regular parts* of F, respectively. Note that $F''_L(\xi; t) = F''_L(\xi; t)\tau(s, t)$. In general the functional derivative $F''(\xi; s, t)$ of F in $\mathcal{U}_\beta$ is a distribution in $\mathcal{S}'_c(\mathbb{R}^2)$. Thus the functional derivative $F''(\xi; \cdot, \cdot)$ of an *L*-functional F is the sum of a regular function and a distribution concentrated on the diagonal of $\mathbb{R}^2$.

Definition 12.17. (a) Let F be an L-functional. Its *Lévy Laplacian* $\Delta_L^Z F$ on a finite interval $Z \subset \mathbb{R}$ is defined by

$$\Delta_L^Z F(\xi) = \frac{1}{|Z|} \int_Z F_L''(\xi; t) \, dt, \quad \xi \in \mathcal{E}_c.$$

(b) Let $\Phi \in (\mathcal{S})_\beta^*$ be such that $F = S\Phi$ is an L-functional. The *Lévy Laplacian* $\Delta_L^Z \Phi$ of Φ on a finite interval Z is defined by

$$\Delta_L^Z \Phi = S^{-1}(\Delta_L^Z F) = S^{-1}\left(\frac{1}{|Z|} \int_Z (S\Phi)_L''(\cdot; t) \, dt\right).$$

First we mention the relationship between $\mathrm{Dom}(\Delta_L)$ and $\mathcal{D}_L$. We will show below (Theorem 12.25) that $\mathcal{D}_L \subset \mathrm{Dom}(\Delta_L)$ and $\Delta_L^Z(S\Phi) = S(\Delta_L \Phi)$ if $\Phi \in \mathcal{D}_L$ and Φ depends only on Z (defined below). The next example shows that $\mathcal{D}_L \neq \mathrm{Dom}(\Delta_L)$.

Example 12.18. Let $\Phi =: e^{\langle \cdot, x \rangle} :$ be the renormalized exponential function considered in Example 12.10. Its S-transform is $F(\xi) = e^{\langle x, \xi \rangle}$. The derivatives F' and F'' of F are given by

$$F'(\xi)(\eta) = F(\xi) \int_{\mathbb{R}} x(t)\eta(t) \, dt,$$

$$F''(\xi)(\eta, \zeta) = F(\xi) \int_{\mathbb{R}^2} x(s)x(t)\eta(s)\zeta(t) \, dsdt.$$

Thus $F \in \mathcal{D}_L$ if and only if $x \in L_{\mathrm{loc}}^1(\mathbb{R})$. In that case, we have $F_L'' = 0$ and so $\Delta_L F = 0$. On the other hand, as shown in Example 12.10 case (3), $\Phi \in \mathrm{Dom}(\Delta_L)$ for $x = \delta_t$, which is not in $L_{\mathrm{loc}}^1(\mathbb{R})$. This example shows that $\mathcal{D}_L \neq \mathrm{Dom}(\Delta_L)$.

Example 12.19. Let $g_{x,c}$ be the Gaussian white noise function in Example 8.3 and let $F \equiv G_{x,c} = Sg_{x,c}$. From equation (8.10)

$$G_{x,c}(\xi) = \exp\left[\frac{1}{1+c}\langle x, \xi \rangle - \frac{1}{2(1+c)}\langle \xi, \xi \rangle\right], \quad \xi \in \mathcal{E}_c.$$

The derivatives F' and F'' are given by

$$F'(\xi)(\eta) = \frac{1}{1+c} F(\xi)(\langle x, \eta \rangle - \langle \eta, \xi \rangle),$$

$$F''(\xi)(\eta, \zeta) = \frac{1}{1+c} F(\xi)\Big(\frac{1}{1+c}[\langle x, \eta \rangle\langle x, \zeta \rangle - \langle x, \eta \rangle\langle \xi, \zeta \rangle$$

$$- \langle x, \zeta \rangle\langle \xi, \eta \rangle + \langle \xi, \eta \rangle\langle \xi, \zeta \rangle] - \langle \eta, \zeta \rangle\Big).$$

Hence $G_{x,c} \in \mathcal{D}(\Delta_L)$ if and only if $x \in L^1_{\text{loc}}(\mathbb{R})$. In that case, $F''_L(\xi;t) = -\frac{1}{1+c}F(\xi)$ and so we have

$$\Delta^Z_L G_{x,c} = -\frac{1}{1+c}G_{x,c}.$$

This shows that $G_{x,c}$ is an eigenfunction of the Lévy Laplacian with eigenvalue $-\frac{1}{1+c}$ for any $x \in L^1_{\text{loc}}(\mathbb{R})$.

Example 12.20. A *normal functional* is defined to be a finite linear combination of functions of the form

$$\int_{\mathbb{R}^n} f(t_1,\dots,t_n)\xi(t_1)^{p_1}\cdots\xi(t_n)^{p_n}\,dt_1\cdots dt_n,$$

where $f \in L^1_{\text{loc}}(\mathbb{R}^n)$ and the p_j's are nonnegative integers. It is easy to check that all normal functionals are L-functionals. For instance, consider the S-transform F of the generalized function $\int_{\mathbb{R}} f(t) :\dot{B}(t)^2: dt$ with $f \in L^1_{\text{loc}}(\mathbb{R})$ (cf. equation (12.9)):

$$F(\xi) = \int_{\mathbb{R}} f(t)\xi(t)^2\,dt.$$

Its derivatives F' and F'' are given by

$$F'(\xi)(\eta) = 2\int_{\mathbb{R}} f(t)\xi(t)\eta(t)\,dt,$$

$$F''(\xi)(\eta,\zeta) = 2\int_{\mathbb{R}} f(t)\eta(t)\zeta(t)\,dt.$$

Hence F is an L-functional and $\Delta^Z_L = \frac{2}{|Z|}\int_Z f(t)\,dt$.

In Theorem 12.11 we showed that $\Delta_L\varphi = 0$ for all $\varphi \in (L^2)$ by using Definition 12.9. The next theorem is for the same result by using Definition 12.17. We sketch the proof, just for comparison.

Theorem 12.21. *Let $\varphi \in (L^2)$ and $F = S\varphi$. Then $F \in \mathcal{D}_L$ and $\Delta^Z_L F = 0$ for any finite interval $Z \subset \mathbb{R}$.*

Proof. Let $\varphi = \sum_{n=0}^{\infty}\langle :\cdot^{\otimes n}:, f_n\rangle \in (L^2)$ and $F = S\varphi$. Then

$$F(\xi) = \sum_{n=0}^{\infty}\langle f_n, \xi^{\otimes n}\rangle.$$

The derivatives F' and F'' are given by

$$F'(\xi)(\eta) = \sum_{n=1}^{\infty} n\langle f_n, \xi^{\otimes(n-1)}\widehat{\otimes}\eta\rangle,$$

$$F''(\xi)(\eta,\zeta) = \sum_{n=2}^{\infty} n(n-1)\langle f_n, \xi^{\otimes(n-2)}\widehat{\otimes}\eta\widehat{\otimes}\zeta\rangle.$$

Thus we get the functional derivatives of F:

$$F'(\xi; t) = \sum_{n=1}^{\infty} n \langle f_n(t, \cdot), \xi^{\otimes(n-1)} \rangle,$$

$$F_L''(\xi; t) = 0,$$

$$F_r''(\xi; s, t) = \sum_{n=2}^{\infty} n(n-1) \langle f_n(s, t, \cdot), \xi^{\otimes(n-2)} \rangle.$$

It is straightforward to check that $F'(\xi; \cdot) \in L^2(\mathbb{R})$, $F_r''(\xi; \cdot, \cdot) \in L^2(\mathbb{R}^2)$ for each $\xi \in \mathcal{S}_c(\mathbb{R})$. Hence $F'(\xi; \cdot) \in L^1_{\text{loc}}(\mathbb{R})$, $F_r''(\xi; \cdot, \cdot) \in L^1_{\text{loc}}(\mathbb{R}^2)$. Thus F is an L-functional and $\Delta_L^Z F = 0$ for any finite interval Z. $\qquad \square$

Theorem 12.22. *Let* $F, G \in \mathcal{D}_L$. *Then* $FG \in \mathcal{D}_L$ *and the equality holds:*

$$\Delta_L^Z(FG) = (\Delta_L^Z F)G + F(\Delta_L^Z G).$$

Remark. The corresponding result for generalized functions says that if $S\Phi$ and $S\Psi$ are L-functionals, then $S(\Phi \diamond \Psi)$ is also an L-functional and

$$\Delta_L^Z(\Phi \diamond \Psi) = (\Delta_L^Z \Phi) \diamond \Psi + \Phi \diamond (\Delta_L^Z \Psi). \qquad (12.16)$$

This equality seems to indicate that the Lévy Laplacian is a derivation, i.e., it is like a first derivative. However, this is not an intrinsic property of the Lévy Laplacian. It is so because we choose the domain $\mathcal{D}_L$ for Δ_L^Z. On the other hand, if we choose $\text{Dom}(\Delta_L)$, then there is no such equality as in equation (12.16). For example, let $\Phi = :e^{\langle \cdot, x \rangle}:$ with $x = \sum_{n=0}^{\infty} e_k$ as in Example 12.10 (also cf. Example 12.18). It is easy to see that

$$\Phi \diamond \Phi = :e^{2\langle \cdot, x \rangle}:.$$

Hence by Example 12.10, we have

$$\Delta_L \Phi = \Phi, \quad \Delta_L(\Phi \diamond \Phi) = 4(\Phi \diamond \Phi).$$

Thus $\Delta_L(\Phi \diamond \Phi) \neq (\Delta_L \Phi) \diamond \Phi + \Phi \diamond (\Delta_L \Phi)$. This is due to the fact that for this particular generalized function Φ in $\text{Dom}(\Delta_L)$, we have

$$\lim_{m \to \infty} \frac{1}{m} \sum_{k=1}^{m} (D_{e_k} \Phi) \diamond (D_{e_k} \Phi) = \lim_{m \to \infty} \frac{1}{m} \sum_{k=1}^{m} \langle x, e_k \rangle^2 (\Phi \diamond \Phi)$$

$$= (\Phi \diamond \Phi).$$

Proof. Let $H = FG$. Then

$$H'(\xi)(\eta) = F'(\xi)(\eta)G(\xi) + F(\xi)G'(\xi)(\eta),$$

$$H''(\xi)(\eta, \zeta) = F''(\xi)(\eta, \zeta)G(\xi) + F'(\xi)(\eta)G'(\xi)(\zeta)$$
$$+ F'(\xi)(\zeta)G'(\xi)(\eta) + F(\xi)G''(\xi)(\eta, \zeta).$$

Hence we have the following functional derivatives

$$H'(\xi; t) = F'(\xi; t)G(\xi) + F(\xi)G'(\xi; t),$$

$$H''_L(\xi; t) = F''_L(\xi; t)G(\xi) + F(\xi)G''_L(\xi; t),$$

$$H''_r(\xi; s, t) = F''_r(\xi; s, t)G(\xi) + F(\xi)G''_r(\xi; s, t)$$
$$+ F'(\xi; s)G'(\xi; t) + F'(\xi; t)G'(\xi; s).$$

Obviously this implies that H is an L-functional and $\Delta_L^Z H = (\Delta_L^Z F)G + F(\Delta_L^Z G)$. □

Now we consider the relationship between the Lévy Laplacian $\Delta_L^Z F$ as defined in Definition 12.17 and the Lévy Laplacian $\Delta_L F$ as given in equation (12.15). This requires some conditions on the orthonormal basis $\{e_k; k \geq 1\}$.

Definition 12.23. Suppose Z is a finite interval in $\mathbb{R}$ with $|Z| > 0$. An orthonormal basis $\{e_k; k \geq 1\}$ for $L^2(Z)$ is called *equally dense* if

$$\frac{1}{m} \sum_{k=1}^{m} e_k(t)^2 \longrightarrow \frac{1}{|Z|}, \quad \text{in } L^2(Z).$$

It is called *uniformly bounded* if $\sup_{k \geq 1} \sup_{t \in Z} |e_k(t)| < \infty$.

Remarks. The orthonormal basis

$$\left\{ 1, \sqrt{2} \sin 2k\pi t, \sqrt{2} \cos 2k\pi t; \ k \geq 1 \right\}$$

for $L^2(0, 1)$ is equally dense and uniformly bounded. The Walsh system $\{f_{m,k}; 1 \leq k \leq 2^m, m \geq 1\}$ is an equally dense orthonormal basis for $L^2(0, 1)$, but is not uniformly bounded. Here the function $f_{m,k}$ is defined by

$$f_{m,k}(x) = \begin{cases} \sqrt{2^m}, & (k-1)/2^m < x < (k-1/2)/2^m; \\ -\sqrt{2^m}, & (k-1/2)/2^m < x < k/2^m; \\ 0, & \text{elsewhere.} \end{cases}$$

These two examples are taken from Hida [47].

Lemma 12.24. If $\{e_k; k \geq 1\}$ is an equally dense and uniformly bounded orthonormal basis for $L^2(Z)$, then for any $f \in L^1(Z)$,

$$\lim_{m \to \infty} \frac{1}{m} \sum_{k=1}^{m} \int_Z f(t)e_k(t)^2 \, dt = \frac{1}{|Z|} \int_Z f(t) \, dt.$$

Proof. Let $\theta_m(t) = \frac{1}{m} \sum_{k=1}^{m} e_k(t)^2$. Since $\{e_k; k \geq 1\}$ is equally dense, $\theta_m \to \frac{1}{|Z|}$ in $L^2(Z)$. This implies that for any $f \in L^2(Z)$,

$$\int_Z f(t)\theta_m(t) \, dt \longrightarrow \frac{1}{|Z|} \int_Z f(t) \, dt. \tag{12.17}$$

Let $g \in L^1(Z)$. For any $\epsilon > 0$, choose $f \in L^2(Z)$ such that

$$\|g - f\|_1 < \frac{1}{3} \min\{\epsilon|Z|, \, \epsilon c^{-2}\},$$

where $c = \sup_{k \geq 1} \sup_{t \in Z} |e_k(t)|$. Note that

$$\int_Z g(t)\theta_m(t) \, dt - \frac{1}{|Z|} \int_Z g(t) \, dt = \int_Z (g(t) - f(t))\theta_m \, dt$$
$$+ \left(\int_Z f(t)\theta_m(t) \, dt - \frac{1}{|Z|} \int_Z f(t) \, dt \right) + \frac{1}{|Z|} \int_Z (f(t) - g(t)) \, dt.$$

We apply (12.17) to the function f to get a positive integer M such that

$$\left| \int_Z f(t)\theta_m(t) \, dt - \frac{1}{|Z|} \int_Z f(t) \, dt \right| < \epsilon/3, \quad \forall m \geq M.$$

Then we have

$$\left| \int_Z g(t)\theta_m(t) \, dt - \frac{1}{|Z|} \int_Z g(t) \, dt \right| < \epsilon, \quad \forall m \geq M. \qquad \square$$

A function $F \in \mathcal{U}_\beta$ is said to *depend only on* an interval Z if the value of $F(\xi)$ depends only on the values of $\xi(t), t \in Z$. For instance, the following functions depend only on the interval $(0, 1)$:

$$F(\xi) = \int_0^1 f(t)\xi(t)^2 \, dt, \quad G(\xi) = \exp\left[\int_0^1 \xi(t)^2 \, dt \right].$$

Theorem 12.25. Suppose $F \in \mathcal{D}_L$ depends only on a finite interval Z with $|Z| > 0$. Then for any equally dense and uniformly bounded orthonormal basis $\{e_k; k \geq 1\}$ for $L^2(Z)$,

$$\Delta_L^Z F(\xi) = \lim_{m \to \infty} \frac{1}{m} \sum_{k=1}^{m} F''(\xi)(e_k, e_k).$$

Proof. Since F is an L-functional depending only on the interval Z, its second derivative F'' is given by

$$F''(\xi)(\eta, \zeta) = \int_Z F''_L(\xi; t)\eta(t)\zeta(t)\, dt + \int_{Z^2} F_r(\xi; s, t)\eta(s)\zeta(t)\, dsdt. \quad (12.18)$$

By Lemma 12.24, we have

$$\lim_{m \to \infty} \frac{1}{m} \sum_{k=1}^{m} \int_Z F''_L(\xi; t)e_k(t)^2\, dt = \frac{1}{|Z|} \int_Z F''_L(\xi; t)\, dt. \quad (12.19)$$

On the other hand, note that $\{e_j \otimes e_k;\ j, k \geq 1\}$ is an orthonormal basis for $L^2(Z)$. Hence for any $f \in L^2(Z^2)$,

$$\lim_{j,k \to \infty} \int_{Z^2} f(s, t)e_j(s)e_k(t)\, dsdt = 0. \quad (12.20)$$

By using the same argument as in the proof of Lemma 12.24, we can show easily that (12.20) is also true for $f \in L^1(Z^2)$. In particular, we have

$$\lim_{k \to \infty} \int_{Z^2} F''_r(\xi; s, t)e_k(s)e_k(t)\, dsdt = 0.$$

This implies that

$$\lim_{m \to \infty} \frac{1}{m} \sum_{k=1}^{m} \int_{Z^2} F''_r(\xi; s, t)e_k(s)e_k(t)\, dsdt = 0. \quad (12.21)$$

Obviously, equations (12.18), (12.19), and (12.21) yield the conclusion of the theorem. $\qquad\square$

12.5 Spherical mean and the Lévy Laplacian

In this section we will study the mean value property and the probabilistic aspect of the Lévy Laplacian. As in the previous section $\mathcal{U}_\beta$ denotes the set of S-transforms of generalized functions in $(\mathcal{E})^*_\beta$. In view of equation (12.15) and Theorem 12.25, we will regard the Lévy Laplacian $\Delta_L F$ of $F \in \mathcal{U}_\beta$ as

$$\Delta_L F(\xi) = \lim_{m \to \infty} \frac{1}{m} \sum_{k=1}^{m} F''(\xi)(e_k, e_k),$$

provided that the limit exists. Here $\{e_k;\ k \geq 1\} \subset \mathcal{E}$ is a fixed orthonormal basis for E. Note that we now regard the Lévy Laplacian as being defined by the same limit as in §12.3 except that here the S-transform is used. The derivative $D_\eta D_\zeta \Phi$ of Φ and the functional derivative F'' of $F = S\Phi$ is related by the S-transform as follows:

$$S(D_\eta D_\zeta \Phi)(\xi) = F''(\xi)(\eta, \zeta), \quad \xi, \eta, \zeta \in \mathcal{E}_c.$$

First we consider the mean value property of the Lévy Laplacian from Kuo et al. [113]. For each positive integer m, let $S^m = \{(x_1, \ldots, x_m); x_1^2 + \cdots + x_m^2 = 1\}$ be the unit sphere in $\mathbb{R}^m$. We can embed S^m into $\mathcal{E}$ by the mapping $\vartheta_m : S^m \to \mathcal{E}$ defined by

$$\vartheta_m(x_1, \ldots, x_m) = x_1 e_1 + \cdots + x_m e_m.$$

Let $d\sigma_m$ denote the uniform probability measure on S^m.

Definition 12.26. The *spherical mean* of $F \in \mathcal{U}_\beta$ at $\xi \in \mathcal{E}_c$ over the sphere of radius $\rho > 0$ is defined by

$$MF(\xi, \rho) = \lim_{m \to \infty} \int_{S^m} F(\xi + \rho \vartheta_m(x)) \, d\sigma_m(x),$$

provided that the limit exists.

Theorem 12.27. *Assume that $F \in \mathcal{U}_\beta$ satisfies the conditions:*
(a) $F(\xi + \eta) = F(\xi) + F'(\xi)(\eta) + \frac{1}{2}F''(\xi)(\eta, \eta) + o(|\eta|_0^2).$
(b) *F has the spherical mean $MF(\xi, \rho)$ for all small $\rho > 0$.*
Then

$$\Delta_L F(\xi) = 2 \lim_{\rho \to 0} \frac{MF(\xi, \rho) - F(\xi)}{\rho^2},$$

in the sense that if one side exists, so does the other and the equality holds.

Proof. By assumption (a) we have

$$F(\xi + \rho \vartheta_m(x)) = F(\xi) + \rho F'(\xi)(\vartheta_m(x))$$

$$+ \frac{\rho^2}{2} F''(\xi)(\vartheta_m(x), \vartheta_m(x)) + R(\rho \vartheta_m(x)),$$

where $R(\rho \vartheta_m(x)) = o(\rho^2 |\vartheta_m(x)|_0^2)$. By the symmetry of $d\sigma_m$, we get

$$\int_{S^m} F'(\xi)(\vartheta_m(x)) \, d\sigma_m(x) = 0.$$

For the term with the second derivative $F''(\xi)$, note that

$$\int_{S^m} F''(\xi)(\vartheta_m(x), \vartheta_m(x)) \, d\sigma_m(x) = \sum_{j,k=1}^m F''(\xi)(e_j, e_k) \int_{S^m} x_j x_k \, d\sigma_m(x).$$

The symmetry of $d\sigma_m$ implies that if $j \neq k$, then

$$\int_{S^m} x_j x_k \, d\sigma_m(x) = 0.$$

On the other hand, it is obvious that

$$\int_{S^m} x_1^2 \, d\sigma_m(x) = \cdots = \int_{S^m} x_m^2 \, d\sigma_m(x),$$

$$\int_{S^m} (x_1^2 + \cdots + x_m^2) \, d\sigma_m(x) = \int_{S^m} d\sigma_m(x) = 1.$$

These equalities imply that

$$\int_{S^m} x_k^2 \, d\sigma_m(x) = \frac{1}{m}, \quad 1 \le k \le m.$$

Hence for the second derivative $F''(\xi)$, we obtain

$$\int_{S^m} F''(\xi)(\vartheta_m(x), \vartheta_m(x)) \, d\sigma_m(x) = \frac{1}{m} \sum_{k=1}^{m} F''(\xi)(e_k, e_k).$$

Therefore

$$\frac{2}{\rho^2} \left(\int_{S^m} F(\xi + \rho \vartheta_m(x)) \, d\sigma_m(x) - F(\xi) \right)$$

$$= \frac{1}{m} \sum_{k=1}^{m} F''(\xi)(e_k, e_k) + \frac{2}{\rho^2} \int_{S^m} R(\rho \vartheta_m(x)) \, d\sigma_m(x).$$

Now, note that

$$\int_{S^m} |\vartheta_m(x)|_0^2 \, d\sigma_m(x) = \int_{S^m} (x_1^2 + \cdots + x_m^2) \, d\sigma_m(x) = 1.$$

This fact together with the assumption $R(\rho \vartheta_m(x)) = o(\rho^2 |\vartheta_m(x)|_0^2)$ implies the following limit

$$\lim_{\rho \to 0} \frac{1}{\rho^2} \int_{S^m} |R(\rho \vartheta_m(x))| \, d\sigma_m(x) = 0, \quad \text{uniformly in } m.$$

Therefore

$$\lim_{\rho \to 0} \limsup_{m \to \infty} \frac{1}{\rho^2} \int_{S^m} |R(\rho \vartheta_m(x))| \, d\sigma_m(x) = 0.$$

By assumption (b), $MF(\xi,\rho)$ exists for all small $\rho > 0$. Thus if $\Delta_L F$ exists, then

$$\lim_{\rho \to 0} \frac{2}{\rho^2}\left(MF(\xi,\rho) - F(\xi)\right) = \Delta_L F(\xi).$$

On the other hand, it is easy to check that if the limit on the left hand side exists, then $\Delta_L F$ exists and the above equality holds.   □

We make some remarks about the mean value property and Δ_L-harmonic functions. A function $F \in \mathcal{U}_\beta$ is said to satisfy the *mean value property* if $MF(\xi,\rho) = F(\xi)$ for any $\xi \in \mathcal{E}_c$ and $\rho > 0$. Suppose $F \in \mathcal{U}_\beta$ satisfies the mean value property and condition (a) of Theorem 12.27. Then by Theorem 12.27 the function F is Δ_L-harmonic, i.e., $\Delta_L F = 0$.

Next, we study the probabilistic aspect of the Lévy Laplacian. Let W_t be a standard $\mathcal{E}'$-valued Wiener process starting at 0 as defined in §12.1. This Wiener process can be expressed as a series

$$W_t = \sum_{k=1}^{\infty} \langle W_t, e_k \rangle e_k,$$

where $\{e_k; k \geq 1\} \subset \mathcal{E}$ is an orthonormal basis for E. Note that for each fixed m,

$$W_t^{(m)} = \sum_{k=1}^{m} \langle W_t, e_k \rangle e_k$$

is a standard Brownian motion on the linear space spanned by $\{e_1, \ldots, e_m\}$. For each $t > 0$, we have

$$E\left(|W_t^{(m)}|_0^2\right) = mt \to \infty, \quad \text{as } m \to \infty.$$

This reflects the fact that the Wiener process W_t is $\mathcal{E}'$-valued rather than E-valued. On the contrary, consider the weighted Brownian motion

$$X_t^{(m)} = \frac{1}{\sqrt{m}} \sum_{k=1}^{m} \langle W_t, e_k \rangle e_k.$$

This weighted Brownian motion is related to the Lévy Laplacian.

Theorem 12.28. *Suppose $F \in \mathcal{U}_\beta$ satisfies condition (a) in Theorem 12.27. Then*

$$\Delta_L F(\xi) = 2 \lim_{m \to \infty} \lim_{t \to 0} \frac{E\left(F(\xi + X_t^{(m)})\right) - F(\xi)}{t}$$

in the sense that if one side exists, so does the other and the equality holds.

Proof. By assumption (a) in Theorem 12.27, we have

$$F(\xi + X_t^{(m)}) = F(\xi) + F'(\xi)(X_t^{(m)}) + \frac{1}{2}F''(\xi)(X_t^{(m)}, X_t^{(m)}) + R(X_t^{(m)}),$$

where $R(X_t^{(m)}) = o(|X_t^{(m)}|_0^2)$. It is easy to see that

$$E\big(F'(\xi)(X_t^{(m)})\big) = 0,$$

$$E\big(F''(\xi)(X_t^{(m)}, X_t^{(m)})\big) = \frac{t}{m}\sum_{k=1}^{m}F''(\xi)(e_k, e_k).$$

Therefore

$$\frac{E\big(F(\xi + X_t^{(m)})\big) - F(\xi)}{t} = \frac{1}{2m}\sum_{k=1}^{m}F''(\xi)(e_k, e_k) + ER(X_t^{(m)}).$$

Note that $E\big(|X_t^{(m)}|_0^2\big) = t$. Hence by the same argument as in the proof of Theorem 12.27,

$$\lim_{t \to 0}\frac{1}{t}E|R(X_t^{(m)})| = 0, \quad \text{uniformly in } m.$$

This implies that

$$\lim_{m \to \infty}\lim_{t \to 0}\frac{E\big(F(\xi + X_t^{(m)})\big) - F(\xi)}{t} = \frac{1}{2}\Delta_L F(\xi).$$

From the above argument we see that if one side of the last equality exists, then the other side also exists and the equality holds. $\quad\square$

We now make two comments on the weighted Brownian motion $X_t^{(m)}$

$$X_t^{(m)} = \frac{1}{\sqrt{m}}\sum_{k=1}^{m}\langle W_t, e_k\rangle e_k.$$

A. $X_t^{(m)}$ as E-valued stochastic processes

By the strong law of large numbers we have

$$|X_t^{(m)}|_0^2 = \frac{1}{m}\sum_{k=1}^{m}\langle W_t, e_k\rangle^2 \longrightarrow t, \quad \text{almost surely.}$$

Thus for fixed t, $|X_t^{(m)}|_0 \to \sqrt{t}$ as $m \to \infty$. However, by the strong law of large numbers again

$$|X_t^{(2n)} - X_t^{(n)}|_0^2 = \frac{1}{2n} \sum_{k=n+1}^{2n} \langle W_t, e_k \rangle^2 + \left(\frac{1}{\sqrt{2n}} - \frac{1}{\sqrt{n}} \right)^2 \sum_{k=1}^{n} \langle W_t, e_k \rangle^2$$

$$\longrightarrow (2 - \sqrt{2})t, \quad \text{almost surely.}$$

Hence $X_t^{(m)}$ does not converge almost surely in E as $m \to \infty$.

B. $X_t^{(m)}$ as $\mathcal{E}_p'$-valued stochastic processes

Take the orthonormal basis $\{\zeta_k; k \geq 1\}$ for E in §4.2 and let $X_t^{(m)}$ be defined from this orthonormal basis. Then

$$|X_t^{(m)}|_{-p}^2 = \frac{1}{m} \sum_{k=1}^{m} \frac{1}{\lambda_k^{2p}} \langle W_t, \zeta_k \rangle^2.$$

Note that $\sum_{k=1}^{\infty} \lambda_k^{-2p} < \infty$ for any $p \geq \alpha/2$ (see §4.2). This implies that

$$\lim_{m \to \infty} \frac{1}{m} \sum_{k=1}^{m} \lambda_k^{-2p} = 0.$$

Hence by the strong law of large numbers,

$$|X_t^{(m)}|_{-p}^2 = \frac{1}{m} \sum_{k=1}^{m} \lambda_k^{-2p} \langle W_t, \zeta_k \rangle^2 \longrightarrow 0, \quad \text{almost surely.}$$

Thus $X_t^{(m)}$ converges to 0 almost surely in $\mathcal{E}_p'$.

12.6 Relationship between Gross and Lévy Laplacians

The Gross Laplacian is defined on the space of test functions and their S-transforms. On the other hand, the Lévy Laplacian is defined only on certain generalized functions and their S-transforms. A relationship between these two Laplacians has been established in Kuo et al. [113] (see also the book by Hida et al. [58]). In this section we will use a slightly different idea to derive a more general result.

Question: Is it possible to find linear operators J_ϵ of $L^2(Z)$ and positive numbers $c(\epsilon)$ such that for any $F \in \mathcal{D}_L$ depending only on a finite interval Z (see §12.4)

$$\Delta_L^Z F = \lim_{\epsilon \to 0} c(\epsilon) \Delta_G (F \circ J_\epsilon)?$$

If we can find J_ϵ and $c(\epsilon)$ to answer the above question, then we have a very nice relationship between the Lévy Laplacian Δ_L and the Gross Laplacian Δ_G.

To motivate such a family of operators, consider a simple example $F \in \mathcal{D}_L$ given by

$$F(\xi) = \langle \tau_f + g, \xi^{\otimes 2} \rangle,$$

where $f \in L^1(Z), g \in L^1(Z^2)$, and $\tau_f \in \mathcal{S}'_c(\mathbb{R}^2)$ is defined as in §12.3 by

$$\langle \tau_f, \eta \otimes \zeta \rangle = \int_Z f(t)\eta(t)\zeta(t)\, dt.$$

Suppose J_ϵ is such a family of operators. Then we have

$$F(J_\epsilon \xi) = \langle (J_\epsilon^*)^{\otimes 2} \tau_f, \xi^{\otimes 2} \rangle + \langle (J_\epsilon^*)^{\otimes 2} g, \xi^{\otimes 2} \rangle. \tag{12.22}$$

Let F_1 and F_2 denote the first and second terms, respectively, on the right hand side of equation (12.22). Take an orthonormal basis $\{e_k; k \geq 1\} \subset \mathcal{S}(\mathbb{R})$ for $L^2(Z)$. Then $\Delta_G F_1$ is given by

$$\Delta_G F_1 = 2 \sum_{k=1}^\infty \langle (J_\epsilon^*)^{\otimes 2} \tau_f, e_k \otimes e_k \rangle$$

$$= 2 \sum_{k=1}^\infty \langle \tau_f, J_\epsilon^{\otimes 2}(e_k \otimes e_k) \rangle$$

$$= 2 \sum_{k=1}^\infty \int_Z f(t) |(J_\epsilon e_k)(t)|^2 \, dt. \tag{12.23}$$

For the second term F_2, we have

$$\Delta_G F_2 = 2 \sum_{k=1}^\infty \langle (J_\epsilon^*)^{\otimes 2} g, e_k \otimes e_k \rangle$$

$$= 2 \sum_{k=1}^\infty \langle g, J_\epsilon^{\otimes 2}(e_k \otimes e_k) \rangle$$

$$= 2 \sum_{k=1}^\infty \int_{Z^2} g(s,t)(J_\epsilon e_k)(s)(J_\epsilon e_k)(t) \, ds dt. \tag{12.24}$$

It follows from equations (12.22)-(12.24) that

$$\Delta_G(F \circ J_\epsilon) = 2 \sum_{k=1}^\infty \int_Z f(t) |(J_\epsilon e_k)(t)|^2 \, dt$$

$$+ 2 \sum_{k=1}^\infty \int_{Z^2} g(s,t)(J_\epsilon e_k)(s)(J_\epsilon e_k)(t) \, ds dt.$$

On the other hand, we can check easily that $F_L''(\xi; t) = 2f$ and so

$$\Delta_L^Z F = \frac{2}{|Z|} \int_Z f(t)\, dt.$$

Hence in order to have $\Delta_L^Z F = \lim_{\epsilon \to 0} c(\epsilon)\Delta_G(F \circ J_\epsilon)$, we must impose the following conditions on J_ϵ and $c(\epsilon)$:

$$\lim_{\epsilon \to 0} c(\epsilon) \sum_{k=1}^{\infty} \int_Z f(t)|(J_\epsilon e_k)(t)|^2\, dt = \frac{1}{|Z|} \int_Z f(t)\, dt, \qquad (12.25)$$

$$\lim_{\epsilon \to 0} c(\epsilon) \sum_{k=1}^{\infty} \int_{Z^2} g(s,t)(J_\epsilon e_k)(s)(J_\epsilon e_k)(t)\, ds dt = 0. \qquad (12.26)$$

By Lemma 12.24 and its proof, we see that equation (12.25) is valid if $\{e_k; k \geq 1\}$ is uniformly bounded and J_ϵ satisfies the condition:

$$c(\epsilon) \sum_{k=1}^{\infty} (J_\epsilon e_k)(t)^2 \longrightarrow \frac{1}{|Z|}, \quad \text{in } L^2(Z).$$

Note that this condition is similar to the equal denseness of an orthonormal basis. By integrating over Z, we get

$$c(\epsilon) \sum_{k=1}^{\infty} |J_\epsilon e_k|_0^2 \longrightarrow 1.$$

Note that $\|J_\epsilon\|_{HS}^2 = \sum_{k=1}^{\infty} |J_\epsilon e_k|_0^2$. Thus the above condition implies that

$$\lim_{\epsilon \to 0} c(\epsilon)\|J_\epsilon\|_{HS}^2 = 1.$$

The simplest choice of $c(\epsilon)$ satisfying this condition is

$$c(\epsilon) = \frac{1}{\|J_\epsilon\|_{HS}^2}.$$

As for the condition in equation (12.26), it is valid if

$$K_\epsilon(s,t) \equiv c(\epsilon) \sum_{k=1}^{\infty} (J_\epsilon e_k)(s)(J_\epsilon e_k)(t) \longrightarrow 0, \quad \text{in } L^2(Z^2). \qquad (12.27)$$

But we have

$$\int_{Z^2} |K_\epsilon(s,t)|^2\, ds dt = c(\epsilon)^2 \sum_{j,k=1}^{\infty} \langle J_\epsilon e_j, J_\epsilon e_k \rangle^2$$

$$= c(\epsilon)^2 \sum_{j,k=1}^{\infty} \langle J_\epsilon^* J_\epsilon e_j, e_k \rangle^2$$

$$= c(\epsilon)^2 \|J_\epsilon^* J_\epsilon\|_{HS}^2.$$

Hence the condition in equation (12.27) is equivalent to

$$\lim_{\epsilon \to 0} c(\epsilon)\|J_\epsilon^* J_\epsilon\|_{HS} = 0,$$

or with the above choice of $c(\epsilon)$,

$$\lim_{\epsilon \to 0} \frac{1}{\|J_\epsilon\|_{HS}^2}\|J_\epsilon^* J_\epsilon\|_{HS} = 0.$$

To sum up the above discussion, if

$$\Delta_L^Z F = \lim_{\epsilon \to 0} c(\epsilon)\Delta_G(F \circ J_\epsilon)$$

holds for the function $F = \langle \tau_f + g, \xi^{\otimes 2} \rangle$ and if $c(\epsilon) = \|J_\epsilon\|_{HS}^{-2}$, then J_ϵ must satisfy the following conditions:

(1) $\lim_{\epsilon \to 0} \|J_\epsilon\|_{HS}^{-2} \sum_{k=1}^\infty (J_\epsilon e_k)(t)^2 = \frac{1}{|Z|}$ in $L^2(Z)$.

(2) $\lim_{\epsilon \to 0} \|J_\epsilon\|_{HS}^{-2}\|J_\epsilon^* J_\epsilon\|_{HS} = 0$.

It turns out that these conditions together with the fact that $J_\epsilon \to I$ strongly are also sufficient for the equality

$$\Delta_L^Z F = \lim_{\epsilon \to 0} \|J_\epsilon\|_{HS}^{-2} \Delta_G(F \circ J_\epsilon)$$

to hold for all $F \in \mathcal{D}_L$ depending only on Z.

Theorem 12.29. *Suppose $\{J_\epsilon; \epsilon > 0\}$ is a family of continuous linear operators from $\mathcal{S}'(\mathbb{R})$ into $\mathcal{S}(\mathbb{R})$ satisfying the following conditions:*

(a) *$J_\epsilon^* \to I$ strongly on $L^2(\mathbb{R})$ as $\epsilon \to 0$.*

(b) *$\lim_{\epsilon \to 0} \|J_\epsilon\|_{HS}^{-2}\|J_\epsilon^* J_\epsilon\|_{HS} = 0$.*

(c) *There exists a uniformly bounded orthonormal basis $\{e_k; k \geq 1\}$ for $L^2(Z)$ such that as $\epsilon \to 0$,*

$$\|J_\epsilon\|_{HS}^{-2} \sum_{k=1}^\infty (J_\epsilon e_k)(t)^2 \longrightarrow \frac{1}{|Z|}, \quad \text{in } L^2(Z).$$

Then for any $F \in \mathcal{D}_L$ depending only on Z,

$$\Delta_L^Z F = \lim_{\epsilon \to 0} \|J_\epsilon\|_{HS}^{-2} \Delta_G(F \circ J_\epsilon).$$

Remark. We give a simple example to compare this theorem with Theorem 12.25. Let $\{e_k; k \geq 1\} \subset \mathcal{S}(\mathbb{R})$ be a uniformly bounded and equally dense orthonormal basis for $L^2(Z)$. Take $\epsilon = 1/m$ and let J_m be the orthogonal projection onto the linear span of $\{e_1, \ldots, e_m\}$. Then $\|J_m\|_{HS} =$

$\|J_m^* J_m\|_{HS} = \sqrt{m}$. Obviously conditions (a) and (b) hold. Condition (c) is just the equal denseness of $\{e_k; k \geq 1\}$. It is easy to derive that $\Delta_G(F \circ J_m)(\xi) = \sum_{k=1}^m F''(J_m\xi)(e_k, e_k)$. Hence by Theorem 12.29 we have

$$\Delta_L^Z F(\xi) = \lim_{m\to\infty} \frac{1}{m} \sum_{k=1}^m F''(J_m\xi)(e_k, e_k).$$

This equality is slightly different from the one in Theorem 12.25.

Proof. The idea of the proof has already been given in the above derivation of conditions (b) and (c) for J_ϵ. First note that by condition (c) and the same argument as in the proof of Lemma 12.24, we can check easily that for any $f \in L^1(Z)$,

$$\lim_{\epsilon\to 0} \int_Z f(t)\Big(\|J_\epsilon\|_{HS}^{-2} \sum_{k=1}^\infty (J_\epsilon e_k)(t)^2\Big)\, dt = \frac{1}{|Z|} \int_Z f(t)\, dt. \qquad (12.28)$$

On the other hand, by condition (b), equation (12.27), and the same argument as in the proof of Lemma 12.24, we can show that for any $g \in L^1(Z^2)$,

$$\lim_{\epsilon\to 0} \int_{Z^2} g(s,t)\Big(\|J_\epsilon\|_{HS}^{-2} \sum_{k=1}^\infty (J_\epsilon e_k)(s)(J_\epsilon e_k)(t)\Big)\, dsdt = 0. \qquad (12.29)$$

Now, a function F in $\mathcal{D}_L$ depending only on Z can be represented by

$$F(\xi) = \sum_{n=0}^\infty \langle F_n, \xi^{\otimes n}\rangle$$

$$= \sum_{n=0}^\infty \langle (F_n)_L + (F_n)_r, \xi^{\otimes n}\rangle,$$

where $(F_n)_L$ and $(F_n)_r$ are the Lévy and regular parts of F_n, respectively, i.e., the second derivative F'' is given by

$$F''(\xi)(\eta, \zeta) = \sum_{n=2}^\infty n(n-1)\Big\langle \int_Z (F_n)_L(t, \cdot)\eta(t)\zeta(t)\, dt,\ \xi^{\otimes(n-2)}\Big\rangle$$

$$+ \sum_{n=2}^\infty n(n-1)\Big\langle \int_{Z^2} (F_n)_r(s,t,\cdot)\eta(s)\zeta(t)\, dsdt,\ \xi^{\otimes(n-2)}\Big\rangle. \qquad (12.30)$$

Here the integrals are taken only over Z and Z^2 since F depends only on Z. Thus we get the Lévy Laplacian $\Delta_L^Z F$ of F:

$$\Delta_L^Z F(\xi) = \sum_{n=2}^\infty n(n-1)\Big\langle \frac{1}{|Z|} \int_Z (F_n)_L(t, \cdot)\, dt,\ \xi^{\otimes(n-2)}\Big\rangle. \qquad (12.31)$$

Note that
$$(F \circ J_\epsilon)''(\xi)(\eta, \zeta) = F''(J_\epsilon \xi)(J_\epsilon \eta, J_\epsilon \zeta).$$

Hence from equation (12.30) we get
$$\Delta_G(F \circ J_\epsilon)(\xi) \equiv L_\epsilon(\xi) + R_\epsilon(\xi),$$

where L_ϵ and R_ϵ are given by

$$L_\epsilon(\xi) = \sum_{n=2}^\infty n(n-1) \Big\langle \sum_{k=1}^\infty \langle (F_n)_L, J_\epsilon^{\otimes 2}(e_k \otimes e_k)\rangle, \; J_\epsilon^{\otimes(n-2)} \xi^{\otimes(n-2)} \Big\rangle$$

$$R_\epsilon(\xi) = \sum_{n=2}^\infty n(n-1) \Big\langle \sum_{k=1}^\infty \langle (F_n)_r, J_\epsilon^{\otimes 2}(e_k \otimes e_k)\rangle, \; J_\epsilon^{\otimes(n-2)} \xi^{\otimes(n-2)} \Big\rangle.$$

By using condition (a) and equation (12.28) we can verify that

$$\|J_\epsilon\|_{HS}^{-2} L_\epsilon(\xi) = \sum_{n=2}^\infty n(n-1) \Big\langle \int_Z [(J_\epsilon^*)^{\otimes(n-2)}(F_n)_L(t, \cdot)]$$

$$\times \Big(\|J_\epsilon\|_{HS}^{-2} \sum_{k=1}^\infty (J_\epsilon e_k)(t)^2 \Big) dt, \; \xi^{\otimes(n-2)} \Big\rangle$$

$$\longrightarrow \sum_{n=2}^\infty n(n-1) \Big\langle \frac{1}{|Z|} \int_Z (F_n)_L(t, \cdot)\, dt, \; \xi^{\otimes(n-2)} \Big\rangle.$$

Similarly, by using condition (a) and equation (12.29) we can verify that

$$\|J_\epsilon\|_{HS}^{-2} R_\epsilon(\xi) = \sum_{n=2}^\infty n(n-1) \Big\langle \int_{Z^2} [(J_\epsilon^*)^{\otimes(n-2)}(F_n)_r(s, t, \cdot)]$$

$$\times \Big(\|J_\epsilon\|_{HS}^{-2} \sum_{k=1}^\infty (J_\epsilon e_k)(s)(J_\epsilon e_k)(t) \Big) ds\, dt, \; \xi^{\otimes(n-2)} \Big\rangle$$

$$\longrightarrow 0.$$

Thus we have shown that

$$\lim_{\epsilon \to 0} \|J_\epsilon\|_{HS}^{-2} \Delta_G(F \circ J_\epsilon)(\xi)$$

$$= \sum_{n=2}^\infty n(n-1) \Big\langle \frac{1}{|Z|} \int_Z (F_n)_L(t, \cdot)\, dt, \; \xi^{\otimes(n-2)} \Big\rangle. \qquad (12.32)$$

Thus the theorem follows from equations (12.31) and (12.32).   □

12.7 Volterra Laplacian

There is another Laplacian operator in white noise distribution theory, namely, the Volterra Laplacian. It is defined in terms of the S-transforms since it can not be expressed directly as an operator acting on generalized functions. Recall that in Definition 12.16 the Lévy Laplacian picks up the singular part of F on the diagonal and leaves the regular part alone. This regular part gives the Volterra Laplacian of F.

A bilinear functional $B \colon L^2(\mathbb{R}) \times L^2(\mathbb{R}) \to \mathbb{C}$ is said to be of *trace class* if the associated operator is a trace class operator of $L^2(\mathbb{R})$. If B is a trace class bilinear functional, then the *trace* of B is defined to be $\operatorname{tr} B = \sum_{k=1}^{\infty} B(e_k, e_k)$, which can be shown to be independent of the orthonormal basis $\{e_k; k \geq 1\}$ for $L^2(\mathbb{R})$. Moreover, if B is an integral operator with kernel function κ, then its trace is given by

$$\operatorname{tr} B = \sum_{k=1}^{\infty} \int_{\mathbb{R}^2} \kappa(s,t) e_k(s) e_k(t) \, ds dt.$$

Definition 12.30. A function $F \in \mathcal{U}_\beta$ is called a V-*functional* if its derivatives F' and F'' are given by

$$F'(\xi)(\eta) = \int_{\mathbb{R}} F'(\xi; t) \eta(t) \, dt,$$

$$F''(\xi)(\eta, \zeta) = \int_{\mathbb{R}} F_L''(\xi; t) \eta(t) \zeta(t) \, dt + \int_{\mathbb{R}^2} F_V''(\xi; s, t) \eta(s) \zeta(t) \, ds dt,$$

where the functional derivatives F', F_L'', and F_V'' satisfy the following conditions:

(1) $F'(\xi; \cdot) \in L^1_{\text{loc}}(\mathbb{R})$, $F_L''(\xi; \cdot) \in L^1_{\text{loc}}(\mathbb{R})$, and $F_V''(\xi; \cdot, \cdot) \in L^2(\mathbb{R}^2)$ for each $\xi \in \mathcal{S}_c(\mathbb{R})$.

(2) $F_V''(\xi)$ is a trace class bilinear functional for each ξ and $\operatorname{tr} F_V''(\cdot) \in \mathcal{U}_\beta$.

Notation. Let $\mathcal{D}_V$ denote the set of V-functionals. We will also use the same notation $\mathcal{D}_V$ to denote the set of generalized functions $\Phi \in (\mathcal{S})_\beta^*$ such that $S\Phi$ is a V-functional. Like $\mathcal{D}_L$ for the Lévy Laplacian, there will be no confusion about the notation $\mathcal{D}_V$.

It is easy to check that the decomposition of F'' into the Lévy part F_L'' and the Volterra part F_V'' is unique. Moreover, we have $\mathcal{D}_V \subset \mathcal{D}_L$.

Definition 12.31. (a) Let F be a V-functional. Its *Volterra Laplacian* $\Delta_V F$ is defined by

$$\Delta_V F(\xi) = \operatorname{tr} F_V''(\xi), \quad \xi \in \mathcal{E}_c.$$

(b) Let $\Phi \in (\mathcal{S})^*_\beta$ be such that $F = S\Phi$ is a V-functional. Its *Volterra Laplacian* $\Delta_V \Phi$ is defined by

$$\Delta_V \Phi = S^{-1}(\Delta_V F) = S^{-1}\big(\operatorname{tr} F''_V(\cdot)\big).$$

Example 12.32. Consider the function $F(\xi) = e^{\langle x, \xi \rangle}$, $x \in \mathcal{S}'(\mathbb{R})$. We showed in Example 12.18 that $F \in \mathcal{D}_L$ if and only if $x \in L^1_{\mathrm{loc}}(\mathbb{R})$ and in that case $\Delta_L F = 0$. On the other hand, in view of F' and F'' given in Example 12.18, we see that $F \in \mathcal{D}_V$ if and only if $x \in L^2(\mathbb{R})$ and in that case $\Delta_V F = |x|^2_0 F$. Thus F is an eigenvector of the Volterra Laplacian.

Example 12.33. Consider the function $F = G_{x,c}$ defined by

$$G_{x,c}(\xi) = \exp\left[\frac{1}{1+c}\langle x, \xi \rangle - \frac{1}{2(1+c)}\langle \xi, \xi \rangle\right], \quad \xi \in \mathcal{E}_c.$$

We showed in Example 12.19 that $G_{x,c} \in \mathcal{D}_L$ if and only if $x \in L^1_{\mathrm{loc}}(\mathbb{R})$. On the other hand, we get the Volterra part $F''_V(\xi; s, t)$ from Example 12.19

$$F''_V(\xi; s, t) = \frac{1}{(1+c)^2} F(\xi)\big[x(s)x(t) - x(s)\xi(t) - x(t)\xi(s) + \xi(s)\xi(t)\big].$$

Thus $F \in \mathcal{D}_V$ if and only if $x \in L^2(\mathbb{R})$ and in that case we have

$$\Delta_V F(\xi) = \frac{1}{(1+c)^2} \langle x - \xi, x - \xi \rangle F(\xi).$$

Example 12.34. Take a symmetric function $g \in L^2(\mathbb{R}^2)$ which does not define a trace class bilinear functional. Then $\varphi = \langle :\cdot^{\otimes 2}:, g \rangle \in (L^2)$, but its S-transform $F(\xi) = \langle g, \xi^{\otimes 2} \rangle$ is not a V-functional.

Recall that the Gross Laplacian is a continuous linear operator from $(\mathcal{S})_\beta$ into itself. We will show below that the Volterra Laplacian is an extension of the Gross Laplacian to the space $\mathcal{D}_V$ of V-functionals. But first we need a lemma on integral operators. For a function θ in $L^2(\mathbb{R})$, we use K_θ to denote the integral operator with kernel function θ, i.e.,

$$K_\theta f(s) = \int_{\mathbb{R}} \theta(s, t) f(t)\, dt.$$

Then K_θ is a Hilbert-Schmidt operator of $L^2(\mathbb{R})$ (for the proof, see Kuo [98, p.6]). Its Hilbert-Schmidt operator norm is given by

$$\|K_\theta\|_{HS} = |\theta|_0, \tag{12.33}$$

where $|\cdot|_0$ is the $L^2(\mathbb{R}^2)$-norm.

For the rest of this section we use the following Gel'fand triple from Chapter 3:

$$S(\mathbb{R}) \subset L^2(\mathbb{R}) \subset S'(\mathbb{R}).$$

Recall that the associated operator $A = -D_x^2 + x^2 + 1$ has eigenvalues $2j, j \geq 1$. Hence A^{-p} is a Hilbert-Schmidt operator for any $p > 1/2$ and we have

$$\|A^{-p}\|_{HS} = \Big(\sum_{j=1}^{\infty} \frac{1}{(2j)^{2p}} \Big)^{1/2}. \tag{12.34}$$

Lemma 12.35. *If θ is a symmetric function in $S(\mathbb{R}^2)$, then the integral operator K_θ is a trace class operator of $L^2(\mathbb{R})$. For any $p > 1/2$,*

$$\|K_\theta\|_{\mathrm{tr}} \leq 2^{-p} |\theta|_p \Big(\sum_{j=1}^{\infty} \frac{1}{(2j)^{2p}} \Big)^{1/2}$$

where $\|\cdot\|_{\mathrm{tr}}$ denotes the trace class operator norm. Moreover, the trace of K_θ is given by

$$\mathrm{tr}\, K_\theta = \langle \tau, \theta \rangle.$$

Remark. In fact, the proof below shows that if θ is a symmetric function in $S_p(\mathbb{R}^2)$ with $p > 1/2$, then the integral operator K_θ is a trace class operator of $L^2(\mathbb{R})$ and the above inequality for $\|K_\theta\|_{\mathrm{tr}}$ holds.

Proof. Recall that we use $\langle \cdot, \cdot \rangle$ to denote the bilinear pairing of $S'(\mathbb{R})$ and $S(\mathbb{R})$. For any $f, g \in L^2(\mathbb{R})$, we have

$$\langle K_\theta f, g \rangle = \int_{\mathbb{R}^2} \theta(s,t) f(s) g(t) \, ds dt$$

$$= \int_{\mathbb{R}^2} \big((A^{-p})^{\otimes 2} (A^p)^{\otimes 2} \theta \big)(s,t) f(s) g(t) \, ds dt$$

$$= \langle K_{(A^p)^{\otimes 2}\theta} A^{-p} f, A^{-p} g \rangle$$

$$= \langle A^{-p} K_{(A^p)^{\otimes 2}\theta} A^{-p} f, g \rangle.$$

Therefore

$$K_\theta = A^{-p} K_{(A^p)^{\otimes 2}\theta} A^{-p}.$$

Note that for any two Hilbert-Schmidt operators S and T, the product ST is a trace class operator and $\|ST\|_{\mathrm{tr}} \leq \|S\|_{HS} \|T\|_{HS}$ (see Kuo [98, p.10]). Hence by equations (12.33) and (12.34), K_θ is a trace class operator if $p > 1/2$ and

$$\|K_\theta\|_{tr} \le \|A^{-p}\| \|K_{(A^p) \otimes_2 \theta} A^{-p}\|_{tr}$$

$$\le \|A^{-p}\| \|K_{(A^p) \otimes_2 \theta}\|_{HS} \|A^{-p}\|_{HS}$$

$$= 2^{-p} |\theta|_p \Big(\sum_{j=1}^{\infty} \frac{1}{(2j)^{2p}} \Big)^{1/2}.$$

To find the trace of K_θ, let $\{e_k; k \ge 1\}$ be an orthonormal basis for $L^2(\mathbb{R})$. Then we have

$$\operatorname{tr} K_\theta = \sum_{k=1}^{\infty} \int_{\mathbb{R}} \theta(s, t) e_k(s) e_k(t) \, ds dt.$$

On the other hand, recall that $\tau = \sum_{k=1}^{\infty} e_k \otimes e_k$ (see §5.1). Hence $\operatorname{tr} K_\theta = \langle \tau, \theta \rangle$. □

Theorem 12.36. *Suppose $\varphi \in (S)_\beta$ and let $F = S\varphi$. Then F is a V-functional, $\Delta_V F = \Delta_G F$, and $\Delta_V \varphi = \Delta_G \varphi$.*

Remarks. (a) This theorem shows that the Volterra Laplacian is an extension of the Gross Laplacian, i.e., $\Delta_V \varphi = \Delta_G \varphi$ for all $\varphi \in (S)_\beta$. This result should be compared with the fact that $\Delta_L \varphi = 0$ for all $\varphi \in (L^2)$ from Theorem 12.11. Thus we have the following relationships:

$$(S)_\beta \subset (L^2) \subset \mathcal{D}_L \subset (S)_\beta^*,$$

$$(S)_\beta \subset \mathcal{D}_V \subset \mathcal{D}_L \subset (S)_\beta^*.$$

(b) The proof below shows that $(S_p) \subset \mathcal{D}_V$ for all $p > 1/2$. Since $(S_p)_\beta \subset (S_p)$ for any $0 \le \beta < 1$, we also have $(S_p)_\beta \subset \mathcal{D}_V$ for any $p > 1/2$ and $0 \le \beta < 1$.

Proof. Note that $(S)_\beta \subset (S)$ for all $0 \le \beta < 1$. Hence it suffices to prove the theorem for $\varphi \in (S)$. Let $\varphi \in (S)$ be represented by

$$\varphi = \sum_{n=0}^{\infty} \langle : \cdot^{\otimes n} :, f_n \rangle.$$

Its S-transform F is given by

$$F(\xi) = \sum_{n=0}^{\infty} \langle f_n, \xi^{\otimes n} \rangle.$$

Hence we have the second derivative $F''(\xi)$ of F:

$$F''(\xi)(\eta, \zeta) = \sum_{n=0}^{\infty} (n+2)(n+1) \langle f_{n+2}, \xi^{\otimes n} \widehat{\otimes} \eta \widehat{\otimes} \zeta \rangle.$$

Thus $F_L''(\xi) = 0$ and

$$F_V''(\xi; s, t) = \sum_{n=0}^{\infty} (n+2)(n+1)\langle f_{n+2}(s, t, \cdot), \xi^{\otimes n}\rangle. \tag{12.35}$$

By Lemma 12.35 each term in equation (12.35) defines a trace class operator. We show that this series converges in the trace class operator norm. Let $C_p = 2^{-p}\left(\sum_{j=1}^{\infty}(2j)^{-2p}\right)^{1/2}$. By Lemma 12.35

$$\|F_V''(\xi)\|_{\mathrm{tr}} \le C_p \sum_{n=0}^{\infty} (n+2)(n+1)|f_{n+2}|_p\, |\xi|_{-p}^n$$

$$= C_p \sum_{n=0}^{\infty} \left(\sqrt{(n+2)!}\,|f_{n+2}|_p\right)\left(\sqrt{(n+2)(n+1)}\,\frac{|\xi|_{-p}^n}{\sqrt{n!}}\right).$$

Then we use the inequality $(n+2)(n+1) \le 2^{2n+1}$ and the Schwarz inequality to check that the series in equation (12.35) converges in the trace class operator norm and

$$\|F_V''(\xi)\|_{\mathrm{tr}} \le \sqrt{2}\,C_p\|\varphi\|_p\, e^{2|\xi|_{-p}^2}.$$

Thus the Volterra part $F_V''(\xi; s, t)$ of F defines a trace class operator of $L^2(\mathbb{R})$. Hence F is a V-functional. By Lemma 12.35 again,

$$\Delta_V F = \mathrm{tr}F_V''(\xi)$$

$$= \sum_{n=0}^{\infty} (n+2)(n+1)\langle\langle\tau, f_{n+2}\rangle, \xi^{\otimes n}\rangle.$$

On the other hand, by Theorem 10.11,

$$S(\Delta_G\varphi)(\xi) = \sum_{n=0}^{\infty} (n+2)(n+1)\langle\langle\tau, f_{n+2}\rangle, \xi^{\otimes n}\rangle.$$

Hence we have shown that $S(\Delta_G\varphi) = \Delta_V F$. This implies the equalities $\Delta_G\varphi = \Delta_V\varphi$ and $\Delta_G F = \Delta_V F$. $\qquad\square$

Suppose F and G are two V-functionals and let $H = FG$. From the calculation in the proof of Theorem 12.22, we see immediately that the Volterra part of H is given by

$$H_V''(\xi; s, t) = F_V''(\xi; s, t)G(\xi) + F(\xi)G_V''(\xi; s, t)$$

$$+ F'(\xi; s)G'(\xi; t) + F'(\xi; t)G'(\xi; s).$$

Note that $F'(\xi; s)G'(\xi; t)$ and $F'(\xi; t)G'(\xi; s)$ are kernel functions of trace class operators of $L^2(\mathbb{R})$ with the same trace

$$\int_{\mathbb{R}} F'(\xi; t)G'(\xi; t)\, dt.$$

Moreover, $\int_{\mathbb{R}} F'(\cdot; t)G'(\cdot; t)\, dt \in \mathcal{U}_\beta$. Hence H is a V-functional and

$$\Delta_V H = (\Delta_V F)G + F(\Delta_V G) + 2\int_{\mathbb{R}} F'(\xi; t)G'(\xi; t)\, dt.$$

Thus we have proved the next theorem which gives a relationship between the Volterra Laplacian and the Wick product of two generalized functions.

Theorem 12.37. (a) *Let $F, G \in \mathcal{D}_V$. Then $FG \in \mathcal{D}_V$ and the equality holds:*

$$\Delta_V(FG) = (\Delta_V F)G + F(\Delta_V G) + 2\int_{\mathbb{R}} F'(\cdot; t)G'(\cdot; t)\, dt.$$

(b) *If $S\Phi$ and $S\Psi$ are V-functionals, then $S(\Phi \diamond \Psi)$ is also a V-functional and*

$$\Delta_V(\Phi \diamond \Psi) = (\Delta_V \Phi) \diamond \Psi + \Phi \diamond (\Delta_V \Psi) + 2\int_{\mathbb{R}} (\partial_t \Phi) \diamond (\partial_t \Psi)\, dt.$$

As for pointwise product, it is obvious that for any $\varphi, \psi \in (\mathcal{S})_\beta$,

$$\Delta_G(\varphi\psi) = (\Delta_G\varphi)\psi + \varphi(\Delta_G\psi) + \int_{\mathbb{R}} (\partial_t\varphi)(\partial_t\psi)\, dt.$$

Note that the Gross Laplacian is the restriction of the Volterra Laplacian to the space $(\mathcal{S})_\beta$. Thus we can expect this equality to hold for the Volterra Laplacian. Recall that the product $\varphi\Phi$ of $\varphi \in (\mathcal{S})_\beta$ and $\Phi \in (\mathcal{S})_\beta^*$ is a generalized function in $(\mathcal{S})_\beta^*$.

Theorem 12.38. *If $\varphi \in (\mathcal{S})_\beta$ and $\Phi \in \mathcal{D}_V$, then $\varphi\Phi \in \mathcal{D}_V$ and the equality holds:*

$$\Delta_V(\varphi\Phi) = (\Delta_G\varphi)\Phi + \varphi(\Delta_V\Phi) + \int_{\mathbb{R}} (\partial_t\varphi)(\partial_t\Phi)\, dt,$$

where the integral is interpreted as the bilinear pairing of $\mathcal{S}'(\mathbb{R})$ and $\mathcal{S}(\mathbb{R})$.

This theorem can be proved directly by using the fact $S(\partial_t\varphi)(\xi) = (S\varphi)'(\xi; t)$ from Theorem 9.7 and $S(\partial_s\partial_t\varphi)(\xi) = (S\varphi)''(\xi; s, t)$. Note that since $\Phi \in \mathcal{D}_V$, the function $\partial_{(\cdot)}\Phi \in \mathcal{S}'(\mathbb{R})$ (by condition (1) of V-functional in Definition 12.30). On the other hand, it follows from Lemma 10.1 that

$\partial_{(\cdot)}\varphi \in \mathcal{S}(\mathbb{R})$. Thus the integral $\int_{\mathbb{R}}(\partial_t\varphi)(\partial_t\Phi)dt$ can be interpreted as the bilinear pairing of $\mathcal{S}'(\mathbb{R})$ and $\mathcal{S}(\mathbb{R})$.

12.8 Relationships with the Fourier transform

The relationships between the Fourier transform $\mathcal{F}$ and the Laplacian operators are given in the next theorem.

Theorem 12.39. *The following equalities holds:*

$$\mathcal{F}\Delta_G^* = -\Delta_G^*\mathcal{F}, \quad on \ (\mathcal{E})_\beta^*, \tag{12.36}$$

$$\mathcal{F}(N + \tfrac{1}{2}\Delta_G^*) = (N + \tfrac{1}{2}\Delta_G^*)\mathcal{F}, \quad on \ (\mathcal{E})_\beta^*, \tag{12.37}$$

$$\mathcal{F}(\Delta_L^Z + \tfrac{1}{2}) = -(\Delta_L^Z + \tfrac{1}{2})\mathcal{F}, \quad on \ \mathcal{D}_L, \tag{12.38}$$

$$\mathcal{F}(\Delta_V + N) = -(\Delta_V + N)\mathcal{F}, \quad on \ \mathcal{D}_V. \tag{12.39}$$

Remark. Equations (12.38) and (12.39) state implicitly that $\mathcal{F}\Phi \in \mathcal{D}_L$ for all $\Phi \in \mathcal{D}_L$ and $\mathcal{F}\Phi \in \mathcal{D}_V$ for all $\Phi \in \mathcal{D}_V$. That is, both $\mathcal{D}_L$ and $\mathcal{D}_V$ are invariant under the Fourier transform.

Proof. Equation (12.36) follows easily from the definition of the Fourier transform and the fact that for any $\Phi \in (\mathcal{E})_\beta^*$

$$S(\Delta_G^*\Phi)(\xi) = \langle \xi, \xi \rangle S\Phi(\xi), \quad \xi \in \mathcal{E}_c.$$

To prove equation (12.37), note that by Theorem 11.40 the operator $iN + \tfrac{i}{2}\Delta_G^*$ is the infinitesimal generator of the Fourier-Mehler transform $\mathcal{F}_\theta$. Hence we have

$$\mathcal{F}_\theta(N + \tfrac{1}{2}\Delta_G^*) = (N + \tfrac{1}{2}\Delta_G^*)\mathcal{F}_\theta.$$

In particular, since $\mathcal{F}_{-\pi/2} = \mathcal{F}$, we get

$$\mathcal{F}(N + \tfrac{1}{2}\Delta_G^*) = (N + \tfrac{1}{2}\Delta_G^*)\mathcal{F}.$$

Equation (12.38) for the Lévy Laplacian is equivalent to

$$\Delta_L^Z\widehat{\Phi} = -(\Delta_L^Z\Phi)\widehat{\ } - \widehat{\Phi}, \quad \forall \Phi \in \mathcal{D}_L. \tag{12.40}$$

This equality can be shown by checking the S-transforms of both sides. Alternatively, we can prove this equality as follows. By Theorem 11.10

$$\widehat{\Phi} = (\Gamma(-iI)\Phi) \diamond \widetilde{\delta_0}.$$

Note that $S(\Gamma(-iI)\Phi)(\xi) = S\Phi(-i\xi)$. Hence $\Gamma(-iI)\Phi \in \mathcal{D}_L$ and we have

$$\Delta_L^Z(\Gamma(-iI)\Phi) = -\Gamma(-iI)\Delta_L^Z\Phi.$$

On the other hand, by Example 12.19 with $x = 0$ and $c = 0$, $\tilde{\delta}_0 \in \mathcal{D}_L$ and $\Delta_L^Z\tilde{\delta}_0 = -\tilde{\delta}_0$. Thus by Theorem 12.22, $\widehat{\Phi} \in \mathcal{D}_L$ and

$$\Delta_L^Z\widehat{\Phi} = \left(\Delta_L^Z(\Gamma(-iI)\Phi)\right) \diamond \tilde{\delta}_0 + (\Gamma(-iI)\Phi) \diamond (\Delta_L^Z\tilde{\delta}_0)$$

$$= -\left(\Gamma(-iI)\Delta_L^Z\Phi\right) \diamond \tilde{\delta}_0 - (\Gamma(-iI)\Phi) \diamond \tilde{\delta}_0$$

$$= -(\Delta_L^Z\Phi)\widehat{} - \widehat{\Phi}.$$

This gives the equality in (12.40). To prove equation (12.39) for the Volterra Laplacian, observe that by equations (12.36) and (12.37) we have

$$\mathcal{F}N = N\mathcal{F} + \Delta_G^*\mathcal{F}. \tag{12.41}$$

This equality implies that equation (12.39) is equivalent to

$$\Delta_V\widehat{\Phi} = -(\Delta_V\Phi)\widehat{} - 2N\widehat{\Phi} - \Delta_G^*\widehat{\Phi}, \quad \forall\Phi \in \mathcal{D}_V. \tag{12.42}$$

We can show this equality by checking the S-transforms of both sides. Alternatively, we can prove this equality by the same argument as above for the Lévy Laplacian except that Example 12.33 and Theorem 12.37 should be used. □

We can derive some additional properties of the Fourier transform from the last theorem and its proof. Recall that the lambda operator Λ and the multiplication operator Q_τ are defined in §10.4. In particular, the lambda operator Λ is related to the Gross Laplacian Δ_G and the number operator N by $\Lambda = N + \Delta_G$ (see Theorem 10.18). For the multiplication operator Q_τ, we have $Q_\tau = 2N + \Delta_G + \Delta_G^*$ (Corollary 10.20).

(1) $\widehat{\varphi} \in \mathcal{D}_L$ for all $\varphi \in (L^2)$ and $\Delta_L^Z\widehat{\varphi} = -\widehat{\varphi}$.

This follows from equation (12.40) by putting $\Phi = \varphi \in (L^2)$. Recall from Theorem 12.21 that $\Delta_L^Z\varphi = 0$ for all $\varphi \in (L^2)$.

(2) $\widehat{\varphi} \in \mathcal{D}_V$ for all $\varphi \in (\mathcal{S})_\beta$ and $\Delta_V\widehat{\varphi} = -(\Delta_G\varphi)\widehat{} - 2N\widehat{\varphi} - \Delta_G^*\widehat{\varphi}$.

This follows from equation (12.42) with $\Phi = \varphi \in (\mathcal{S})_\beta$. Note that $\Delta_V\varphi = \Delta_G\varphi$ if $\varphi \in (\mathcal{S})_\beta$.

(3) $\mathcal{F}N = \Lambda^*\mathcal{F}$ on $(\mathcal{E})_\beta^*$.

This follows from equations (12.37) since $N + \Delta_G^* = \Lambda^*$ (see Theorem 10.18).

(4) $\mathcal{F}Q_\tau = -\Delta_V \mathcal{F}$ on $(\mathcal{S})_\beta$.

To check this equality, note that $Q_\tau = 2N + \Delta_G + \Delta_G^*$ (Corollary 10.20). Then apply equations (12.36) and (12.41), and comment (2) above.

12.9 Two-dimensional rotations

As in the finite dimensional case, the Gross Laplacian, its adjoint, and the number operator can be characterized in terms of two-dimensional rotations. These results are due to Hida, Obata, and Saitô [60]. We will state their theorems under simpler assumptions and provide elementary proofs.

Consider the one-parameter group of rotations on $\mathbb{R}^2$ given by

$$(R_\theta f)(x, y) = f(x \cos \theta - y \sin \theta, x \sin \theta + y \cos \theta).$$

Its infinitesimal generator is given by the operator

$$x \frac{\partial}{\partial y} - y \frac{\partial}{\partial x}.$$

Thus for the space $\mathbb{R}^n$ we get the following operators as the infinitesimal generators of two-dimensional rotations

$$x_j \frac{\partial}{\partial x_k} - x_k \frac{\partial}{\partial x_j}, \quad 1 \leq j, k \leq n.$$

Recall that the set $\{\dot{B}(t); t \in T\}$ is regarded as a coordinate system in white noise distribution theory. The coordinate multiplication operator is $\dot{B}(t) = \partial_t + \partial_t^*$. Thus the white noise analogues of the above operators are

$$\dot{B}(s)\partial_t - \dot{B}(t)\partial_s = \partial_s^* \partial_t - \partial_t^* \partial_s, \quad s, t \in T.$$

These operators are continuous from $(\mathcal{S})_\beta$ into $(\mathcal{S})_\beta^*$. But for the characterization theorems below, we need continuous linear operators of this type from $(\mathcal{S})_\beta$ into itself. Hence we introduce the following operators

$$R_{\xi,\eta} \equiv D_\xi^* D_\eta - D_\eta^* D_\xi, \quad \xi, \eta \in \mathcal{E}.$$

For any $\xi, \eta \in \mathcal{E}$, the operator $R_{\xi,\eta}$ is a continuous linear operator from $(\mathcal{S})_\beta$ into itself (see Theorem 9.1 and Corollary 9.14).

Recall that the Gross Laplacian and the number operator are integral kernel operators given by

$$\Delta_G = \Xi_{0,2}(\tau) = \int_{T^2} \tau(s, t) \partial_s \partial_t \, ds dt = \int_T \partial_t^2 \, dt,$$

$$N = \Xi_{1,1}(\tau) = \int_{T^2} \tau(s,t)\partial_s^* \partial_t \, dsdt = \int_T \partial_t^* \partial_t \, dt.$$

By Theorem 9.15 (e), we have $[\partial_t, D_\xi^*] = \xi(t)I$, i.e., $\partial_t D_\xi^* = D_\xi^* \partial_t + \xi(t)$, for all $t \in T$ and $\xi \in \mathcal{E}$. Hence for any $\xi, \eta \in \mathcal{E}$,

$$\Delta_G R_{\xi,\eta} = \int_T \partial_t^2 (D_\xi^* D_\eta - D_\eta^* D_\xi) \, dt$$

$$= \int_T \left(\partial_t [D_\xi^* \partial_t + \xi(t)]D_\eta - \partial_t[D_\eta^* \partial_t + \eta(t)]D_\xi \right) dt.$$

Apply the formula $\partial_t D_\xi^* = D_\xi^* \partial_t + \xi(t)$ again to the last equality to get

$$\Delta_G R_{\xi,\eta} = R_{\xi,\eta}\Delta_G + 2\int_T \left(\xi(t)\partial_t D_\eta - \eta(t)\partial_t D_\xi \right) dt.$$

But $\int_T \xi(t)\partial_t \, dt = D_\xi$ for any $\xi \in \mathcal{E}$ by Theorem 9.27 (the measure ν is suppressed as in Chapter 10). Note that $[D_\xi, D_\eta] = 0$ (Theorem 9.15). Therefore

$$\Delta_G R_{\xi,\eta} = R_{\xi,\eta}\Delta_G + 2(D_\xi D_\eta - D_\eta D_\xi) = R_{\xi,\eta}\Delta_G.$$

Thus we have shown that $\Delta_G R_{\xi,\eta} = R_{\xi,\eta}\Delta_G$, i.e., the Gross Laplacian commutes with all operators $R_{\xi,\eta}$, $\xi, \eta \in \mathcal{E}$.

Recall that by Theorem 10.5 the integral kernel operator

$$\Xi_{0,2}(\theta) = \int_{T^2} \theta(s,t)\partial_s \partial_t \, dsdt$$

is a continuous linear operator from $(\mathcal{S})_\beta$ into itself for any $\theta \in (\mathcal{S}_c')^{\otimes 2}$. The next theorem asserts that the Gross Laplacian is the only such operator (up to a constant) that commutes with all operators $R_{\xi,\eta}$.

Theorem 12.40. *Let $\theta \in (\mathcal{S}_c')^{\otimes 2}$ and let*

$$\Xi_{0,2}(\theta) = \int_{T^2} \theta(s,t)\partial_s \partial_t \, dsdt.$$

Suppose $\Xi_{0,2}(\theta)R_{\xi,\eta} = R_{\xi,\eta}\Xi_{0,2}(\theta)$ for all $\xi, \eta \in \mathcal{E}$. Then there exists a constant c such that $\Xi_{0,2}(\theta) = c\Delta_G$.

Proof. Let $\widehat{\theta}$ be the symmetrization of θ. It follows from the remark of Theorem 10.2 that $\Xi_{0,2}(\theta) = \Xi_{0,2}(\widehat{\theta})$. Thus we may assume that θ is symmetric.

Let $\xi, \eta \in \mathcal{E}$. We can use the formula $\partial_t D_\xi^* = D_\xi^* \partial_t + \xi(t)$ and carry out the same calculation as above to get

$$\Xi_{0,2} R_{\xi,\eta} = R_{\xi,\eta} \Xi_{0,2} + \int_{T^2} \theta(s,t) \Big(\xi(s) D_\eta \partial_t $$

$$+ \xi(t) D_\eta \partial_s - \eta(s) D_\xi \partial_t - \eta(t) D_\xi \partial_s \Big) \, ds dt.$$

But $\Xi_{0,2} R_{\xi,\eta} = R_{\xi,\eta} \Xi_{0,2}$ by assumption. Thus for all $\xi, \eta \in \mathcal{E}$,

$$\int_{T^2} \theta(s,t) \Big(\xi(s) D_\eta \partial_t + \xi(t) D_\eta \partial_s $$

$$- \eta(s) D_\xi \partial_t - \eta(t) D_\xi \partial_s \Big) \, ds dt = 0. \qquad (12.43)$$

Let $\varphi = \langle :\cdot^{\otimes 2}:, \eta^{\otimes 2} \rangle \in (\mathcal{S})_\beta$. It is easy to check that

$$D_\eta \partial_t \varphi = 2\eta(t) |\eta|_0^2, \quad D_\xi \partial_t \varphi = 2\eta(t) \langle \xi, \eta \rangle.$$

Suppose $\langle \xi, \eta \rangle = 0$. Then upon applying equation (12.43) to this function φ, we get

$$\int_{T^2} \theta(s,t) \big[\xi(s)\eta(t) + \xi(t)\eta(s) \big] \, ds dt = 0. \qquad (12.44)$$

Since θ is symmetric, we have

$$\int_{T^2} \theta(s,t) \xi(t)\eta(s) \, ds dt = \int_{T^2} \theta(s,t)\xi(s)\eta(t) \, ds dt.$$

Hence equation (12.44) is reduced to

$$\int_{T^2} \theta(s,t)\xi(s)\eta(t) \, ds dt = 0. \qquad (12.45)$$

Thus we have shown that equation (12.45) holds for any $\xi, \eta \in \mathcal{E}$ such that $\langle \xi, \eta \rangle = 0$.

Now, for any $\xi, \eta \in \mathcal{E}$ with $|\xi|_0 = |\eta|_0 = 1$, let $\tilde{\xi} = \xi + \eta$ and $\tilde{\eta} = \xi - \eta$. Then $\langle \tilde{\xi}, \tilde{\eta} \rangle = 0$. Hence we can apply equation (12.45) to $\tilde{\xi}$ and $\tilde{\eta}$ to get

$$\int_{T^2} \theta(s,t) \big[\xi(s) + \eta(s) \big] \big[\xi(t) - \eta(t) \big] \, ds dt = 0.$$

By using the symmetry of θ, we can reduce this equation to

$$\int_{T^2} \theta(s,t) \big[\xi(s)\xi(t) - \eta(s)\eta(t) \big] \, ds dt = 0,$$

or equivalently,

$$\int_{T^2} \theta(s,t)\xi(s)\xi(t)\,dsdt = \int_{T^2} \theta(s,t)\eta(s)\eta(t)\,dsdt. \tag{12.46}$$

Thus we have shown that equation (12.46) holds for any $\xi, \eta \in \mathcal{E}$ with $|\xi|_0 = |\eta|_0 = 1$. This implies that there is a constant c such that

$$\int_{T^2} \theta(s,t)\xi(s)\xi(t)\,dsdt = c, \quad \forall \xi \in \mathcal{E} \text{ with } |\xi|_0 = 1. \tag{12.47}$$

Finally, let $\{\zeta_k; k \geq 1\}$ be the eigenvectors of A (see §4.2). Then by using equations (12.45) and (12.47), we get

$$\theta(s,t) = \sum_{j,k=1}^{\infty} \langle \theta, \zeta_j \otimes \zeta_k \rangle \zeta_j \otimes \zeta_k$$

$$= \sum_{k=1}^{\infty} \langle \theta, \zeta_k \otimes \zeta_k \rangle \zeta_k \otimes \zeta_k$$

$$= c \sum_{k=1}^{\infty} \zeta_k \otimes \zeta_k.$$

Recall from §5.1 that $\tau = \sum_{k=1}^{\infty} \zeta_k \otimes \zeta_k$. Hence $\theta = c\tau$. Thus we conclude that $\Xi_{0,2}(\theta) = c\,\Xi_{0,2}(\tau) = c\,\Delta_G$ and the proof is completed. $\square$

As for the number operator, it can be characterized in the same way as the Gross Laplacian. We show that the number operator N commutes with $R_{\xi,\eta}$ for all $\xi, \eta \in \mathcal{E}$. First we use the formula $\partial_t D_\xi^* = D_\xi^* \partial_t + \xi(t)$ to derive

$$NR_{\xi,\eta} = \int_T \partial_t^* \partial_t (D_\xi^* D_\eta - D_\eta^* D_\xi)\,dt$$

$$= \int_T \left(\partial_t^* [D_\xi^* \partial_t + \xi(t)] D_\eta - \partial_t^* [D_\eta^* \partial_t + \eta(t)] D_\xi \right) dt.$$

Next we apply the formula $\partial_t^* D_\xi = \tilde{D}_\xi \partial_t^* - \xi(t)$ (from Theorem 9.15 (d)) to the last equation to get

$$NR_{\xi,\eta} = R_{\xi,\eta}N - \int_T D_\xi^* \eta(t)\partial_t\,dt + \int_T \xi(t)\tilde{D}_\eta \partial_t^*\,dt$$

$$+ \int_T D_\eta^* \xi(t)\partial_t\,dt - \int_T \eta(t)\tilde{D}_\xi \partial_t^*\,dt.$$

But by Theorem 9.15(d),

$$\int_T \xi(t)\tilde{D}_\eta \partial_t^* \, dt - \int_T D_\xi^* \eta(t) \partial_t \, dt = \tilde{D}_\eta D_\xi^* - D_\xi^* D_\eta$$

$$= [\tilde{D}_\eta, D_\xi^*]$$

$$= \langle \xi, \eta \rangle,$$

$$\int_T \eta(t)\tilde{D}_\xi \partial_t^* \, dt - \int_T D_\eta^* \xi(t) \partial_t \, dt = \tilde{D}_\xi D_\eta^* - D_\eta^* D_\xi$$

$$= [\tilde{D}_\xi, D_\eta^*]$$

$$= \langle \xi, \eta \rangle.$$

Hence $NR_{\xi,\eta} = R_{\xi,\eta}N$ for all $\xi, \eta \in \mathcal{E}$, i.e., the number operator commutes with $R_{\xi,\eta}$.

Now, by Theorem 10.5 the integral kernel operator

$$\Xi_{1,1}(\theta) = \int_{T^2} \theta(s,t)\partial_s^* \partial_t \, dsdt$$

is a continuous linear operator from $(\mathcal{S})_\beta$ into $(\mathcal{S})_\beta^*$. Note that if $\xi \in \mathcal{E}$, then the operator D_ξ extends by continuity to a continuous linear operator $\tilde{D}_\xi$ from $(\mathcal{S})_\beta^*$ into itself (see Theorem 9.10). For any $\xi, \eta \in \mathcal{E}$, let

$$\tilde{R}_{\xi,\eta} = D_\xi^* \tilde{D}_\eta - D_\eta^* \tilde{D}_\xi.$$

Then the operator $\tilde{R}_{\xi,\eta}$ is from $(\mathcal{S})_\beta^*$ into itself.

The next theorem characterizes the number operator as the unique integral kernel operator $\Xi_{1,1}(\theta)$ (up to a constant) such that $\Xi_{1,1}(\theta)R_{\xi,\eta} = \tilde{R}_{\xi,\eta}\Xi_{1,1}(\theta)$ for all $\xi, \eta \in \mathcal{E}$.

Theorem 12.41. *Let $\theta \in (\mathcal{S}_c')^{\otimes 2}$ and let*

$$\Xi_{1,1}(\theta) = \int_{T^2} \theta(s,t)\partial_s^* \partial_t \, dsdt.$$

Suppose $\Xi_{1,1}(\theta)R_{\xi,\eta} = \tilde{R}_{\xi,\eta}\Xi_{1,1}(\theta)$ for all $\xi, \eta \in \mathcal{E}$. Then there exists a constant c such that $\Xi_{1,1}(\theta) = cN$.

Proof. We can use the formulas $\partial_t D_\xi^* = D_\xi^* \partial_t + \xi(t)$ and $\partial_t^* D_\xi = \tilde{D}_\xi \partial_t^* - \xi(t)$ to derive that

$$\Xi_{1,1}(\theta)R_{\xi,\eta} = \tilde{R}_{\xi,\eta}\Xi_{1,1}(\theta) + \int_{T^2} \theta(s,t)\Big(-\eta(s)D_\xi^* \partial_t$$

$$+ \xi(t)\tilde{D}_\eta \partial_s^* - \xi(t)\eta(s) + \xi(s)D_\eta^* \partial_t - \eta(t)\tilde{D}_\xi \partial_s^* + \xi(s)\eta(t) \Big) \, dsdt.$$

Hence by the assumption the following equality holds for all $\xi, \eta \in \mathcal{E}$,

$$\int_{T^2} \theta(s,t) \Big(-\eta(s) D_\xi^* \partial_t + \xi(t) \tilde{D}_\eta \partial_s^* - \xi(t)\eta(s)$$

$$+ \xi(s) D_\eta^* \partial_t - \eta(t) \tilde{D}_\xi \partial_s^* + \xi(s)\eta(t) \Big) \, dsdt = 0. \qquad (12.48)$$

Let $\varphi = \langle \cdot, \eta \rangle$. It is easy to check that

$$D_\xi^* \partial_t \varphi = \eta(t) \langle \cdot, \xi \rangle,$$

$$\tilde{D}_\eta \partial_s^* \varphi = \eta(s) \langle \cdot, \eta \rangle + |\eta|_0^2 \langle \cdot, \delta_s \rangle,$$

$$D_\eta^* \partial_t \varphi = \eta(t) \langle \cdot, \eta \rangle,$$

$$\tilde{D}_\xi \partial_s^* \varphi = \xi(s) \langle \cdot, \eta \rangle + \langle \xi, \eta \rangle \langle \cdot, \delta_s \rangle.$$

Upon applying equation (12.48) to this function $\varphi = \langle \cdot, \eta \rangle$, we get the following equality that is satisfied for all $\xi, \eta \in \mathcal{E}$:

$$\int_{T^2} \theta(s,t) \Big(-\eta(s)\eta(t)\langle \cdot, \xi \rangle + \xi(t)\eta(s)\langle \cdot, \eta \rangle + \xi(t)|\eta|_0^2 \langle \cdot, \delta_s \rangle$$

$$- \xi(t)\eta(s) - \langle \eta, \xi \rangle \eta(t) \langle \cdot, \delta_s \rangle + \xi(s)\eta(t) \Big) \, dsdt = 0.$$

Hence for any $\xi, \eta \in \mathcal{E}$,

$$\int_{T^2} \theta(s,t) \big[\xi(s)\eta(t) - \xi(t)\eta(s) \big] \, dsdt = 0, \qquad (12.49)$$

$$\int_{T^2} \theta(s,t) \Big(-\eta(s)\eta(t)\langle \cdot, \xi \rangle + \xi(t)\eta(s)\langle \cdot, \eta \rangle$$

$$+ \xi(t)|\eta|_0^2 \langle \cdot, \delta_s \rangle - \langle \eta, \xi \rangle \eta(t) \langle \cdot, \delta_s \rangle \Big) \, dsdt = 0. \qquad (12.50)$$

From equation (12.49) we get

$$\int_{T^2} \big[\theta(s,t) - \theta(t,s) \big] \xi(s)\eta(t) \, dsdt = 0, \qquad \forall \xi, \eta \in \mathcal{E}.$$

This implies that θ is symmetric. On the other hand, by equation (5.8),

$$\langle\!\langle \langle \cdot, F \rangle, \langle \cdot, f \rangle \rangle\!\rangle = \langle F, f \rangle, \qquad F \in \mathcal{S}_c', \ f \in \mathcal{S}_c.$$

Hence by applying the generalized function in equation (12.50) to the test function $\varphi = \langle \cdot, \eta \rangle$, we get

$$\int_{T^2} \theta(s,t) \Big(-\eta(s)\eta(t)\langle \xi, \eta \rangle + \xi(t)\eta(s)|\eta|_0^2 \Big) \, dsdt = 0.$$

In particular, when $\langle \xi, \eta \rangle = 0$, we have

$$\int_{T^2} \theta(s,t)\xi(t)\eta(s)\,dsdt = 0. \tag{12.51}$$

Thus we have shown that θ is symmetric and equation (12.51) holds for any $\xi, \eta \in \mathcal{E}$ with $\langle \xi, \eta \rangle = 0$. Then by the proof of Theorem 12.40 (the part after equation (12.45)) we conclude that there exists a constant c such that $\theta = c\tau$. Hence $\Xi_{1,1}(\theta) = c\Xi_{1,1}(\tau) = cN$ and the proof is completed. $\qquad\square$

Finally we consider the adjoint operator Δ_G^*. It is easy to see that $R_{\xi,\eta}^* = \widetilde{R}_{\eta,\xi}$. Thus from the fact that Δ_G commutes with $R_{\xi,\eta}$ for any $\xi, \eta \in \mathcal{E}$, we get $\Delta_G^* R_{\xi,\eta} = \widetilde{R}_{\xi,\eta}\Delta_G^*$ for any $\xi, \eta \in \mathcal{E}$.

The next theorem for the characterization of the adjoint operator Δ_G^* can be proved by the same arguments used in the proofs of Theorems 12.40 and 12.41.

Theorem 12.42. *Let $\theta \in (S_c')^{\otimes 2}$ and let*

$$\Xi_{2,0}(\theta) = \int_{T^2} \theta(s,t)\partial_s^*\partial_t^*\,dsdt.$$

Suppose $\Xi_{2,0}(\theta)R_{\xi,\eta} = \widetilde{R}_{\xi,\eta}\Xi_{2,0}(\theta)$ for all $\xi, \eta \in \mathcal{E}$. Then there exists a constant c such that $\Xi_{2,0}(\theta) = c\Delta_G^*$.

13

White Noise Integration

13.1 Informal motivation

In Chapter 1 we came across the following integrals (see Examples 1.6, 1.7, and 1.8)

$$\int_0^t e^{-c(t-s)}\dot{B}(s)\,ds, \quad \int_0^t e^{-c(t-s)}B(s)\dot{B}(s)\,ds, \quad \int_0^t e^{-c(t-s)} :\dot{B}(s)^2:\,ds.$$

Now that we know both $\dot{B}(s)$ and $:\dot{B}(s)^2:$ are generalized functions, let us reconsider these integrals from the white noise viewpoint.

A. *Wiener integral*

In Example 1.6 the integral $\int_0^t e^{-c(t-s)}\dot{B}(s)\,ds$ is interpreted as a Wiener integral

$$\int_0^t e^{-c(t-s)}\dot{B}(s)\,ds = \int_0^t e^{-c(t-s)}\,dB(s).$$

This Wiener integral is a Gaussian random variable with mean 0 and variance $\frac{1}{2c}(1 - e^{-2ct})$.

From the white noise viewpoint we know that for each s, $\dot{B}(s)$ is a generalized function represented by $\dot{B}(s) = \langle\cdot,\delta_s\rangle$. Hence we get informally

$$\int_0^t e^{-c(t-s)}\dot{B}(s)\,ds = \int_0^t e^{-c(t-s)}\langle\cdot,\delta_s\rangle\,ds$$

$$= \Big\langle\cdot, \int_0^t e^{-c(t-s)}\delta_s\,ds\Big\rangle.$$

Still informally we have

$$\int_0^t e^{-c(t-s)}\delta_s(u)\,ds = 1_{[0,t]}\,e^{-c(t-u)}.$$

Therefore

$$\int_0^t e^{-c(t-s)} \dot{B}(s)\, ds = \langle \cdot, 1_{[0,t]}\, e^{-c(t-\cdot)} \rangle.$$

Note that this equation gives a concrete representation of the integral $\int_0^t e^{-c(t-s)} \dot{B}(s)\, ds$ as a random variable in (L^2). But how do we justify this representation?

B. *Itô integral*

The integral $\int_0^t e^{-c(t-s)} B(s)\dot{B}(s)\, ds$ is interpreted as an Itô integral in Example 1.7 by writing $\dot{B}(s)\, ds = dB(s)$,

$$\int_0^t e^{-c(t-s)} B(s)\dot{B}(s)\, ds = \int_0^t e^{-c(t-s)} B(s)\, dB(s).$$

Let us consider this integral informally by using the white noise $\dot{B}(s)$. It is easy to see from Definition 5.1 that for any $h, k \in L^2(\mathbb{R})$,

$$\langle \cdot, h \rangle \langle \cdot, k \rangle = \langle :\cdot^{\otimes 2}:, h \widehat{\otimes} k \rangle - \langle h, k \rangle.$$

Note that $B(s) = \langle \cdot, 1_{[0,s]} \rangle$ and $\dot{B}(s) = \langle \cdot, \delta_s \rangle$. Thus informally we have

$$B(s)\dot{B}(s) = \langle :\cdot^{\otimes 2}:, 1_{[0,s]} \widehat{\otimes} \delta_s \rangle - \langle 1_{[0,s]}, \delta_s \rangle.$$

The first term on the right hand side is a generalized function. But the second term is only informally equal to $1_{[0,s]}(s) = 1$. This can be viewed in another way. Recall from equation (9.9) that

$$\dot{B}(s) = \partial_s + \partial_s^*$$

as continuous linear operators from $(\mathcal{S})$ into $(\mathcal{S})^*$ and for any $\xi \in \mathcal{S}_c$,

$$\dot{B}(s)\langle \cdot, \xi \rangle = \xi(s) + \langle :\cdot^{\otimes 2}:, \delta_s \widehat{\otimes} 1_{[0,s]} \rangle.$$

Thus we have the informal derivation (since $B(s) = \langle \cdot, 1_{[0,s]} \rangle \notin (\mathcal{S})$):

$$\dot{B}(s)B(s) = (\partial_s + \partial_s^*)\langle \cdot, 1_{[0,s]} \rangle$$

$$= 1_{[0,s]}(s) + \langle :\cdot^{\otimes 2}:, \delta_s \widehat{\otimes} 1_{[0,s]} \rangle$$

$$= 1 + \langle :\cdot^{\otimes 2}:, \delta_s \widehat{\otimes} 1_{[0,s]} \rangle.$$

Hence by using the white noise $\dot{B}(s)$ directly, we get

$$\int_0^t e^{-c(t-s)} B(s)\dot{B}(s)\, ds = \int_0^t e^{-c(t-s)} \left(1 + \langle :\cdot^{\otimes 2}:, \delta_s \widehat{\otimes} 1_{[0,s]} \rangle \right) ds$$

$$= \frac{1}{c}(1 - e^{-ct}) + \int_0^t e^{-c(t-s)} \langle :\cdot^{\otimes 2}:, \delta_s \widehat{\otimes} 1_{[0,s]} \rangle\, ds.$$

However, $B(s)$ can also be represented as $B(s) = \langle \cdot, 1_{[0,s)} \rangle$. But then we would get

$$\int_0^t e^{-c(t-s)} B(s)\dot{B}(s)\,ds = \int_0^t e^{-c(t-s)} \langle :\cdot^{\otimes 2}:, \delta_s \widehat{\otimes} 1_{[0,s)} \rangle \, ds.$$

Hence this is some kind of inconsistency. How can we resolve this difficulty?

C. White noise integral

In Example 1.8 we have the integral $\int_0^t e^{-c(t-s)} : \dot{B}(s)^2 : ds$. Note that the integrand takes values in the space $(\mathcal{S})^*$ of generalized functions. Hence the integral is a "white noise integral." But what is a white noise integral?

From the above discussion we see that more generally we need to define integrals of the following types:

$$\int_a^b \Phi(t)\,dt, \qquad \int_a^b \partial_t^* \Phi(t)\,dt, \qquad \int_a^b \partial_t \Phi(t)\,dt,$$

where Φ is an $(\mathcal{S})^*$-valued measurable function on $[a, b]$. A special case of the second integral, called the Hitsuda-Skorokhod integral, was already defined in §9.4. The third integral is not well-defined because of the inconsistency mentioned above. To resolve this difficulty we will need to define $\partial_{t+}\Phi(t)$ and $\partial_{t-}\Phi(t)$. These two new objects give rise to the forward and backward integrals introduced in Kuo and Russek [116].

13.2 Pettis and Bochner integrals

In this section we briefly describe Pettis and Bochner integrals from Hille and Phillips [63]. Let $(M, \mathcal{B}, m)$ be a sigma-finite measure space. Let $\mathbb{X}$ be a complex Banach space with norm $\| \cdot \|$. A function $f : M \to \mathbb{X}$ is called *weakly measurable* if $\langle w, f(\cdot) \rangle$ is measurable for each $w \in \mathbb{X}^*$ (here $\langle \cdot, \cdot \rangle$ denotes the bilinear pairing of $\mathbb{X}^*$ and $\mathbb{X}$).

Let f be a weakly measurable $\mathbb{X}$-valued function on M such that $\langle w, f(\cdot) \rangle$ belongs to $L^1(M)$ for any $w \in \mathbb{X}^*$. For each $E \in \mathcal{B}$, define a linear operator T from $\mathbb{X}^*$ into $L^1(M)$ by

$$T(w) = \langle w, f(\cdot) \rangle, \quad w \in \mathbb{X}^*.$$

Then T is a closed operator. To see this, suppose $w_n \to w$ in $\mathbb{X}^*$ and $\langle w_n, f(\cdot) \rangle \to g$ in $L^1(M)$. Then, since $\langle w_n, f(\cdot) \rangle \to g$ in $L^1(M)$, we can choose a subsequence $w_{n'}$ of w_n such that $\langle w_{n'}, f(\cdot) \rangle \to g$, m-a.e. On the other hand, since $w_n \to w$ in $\mathbb{X}^*$, we have $\langle w_n, f(\cdot) \rangle \to \langle w, f(\cdot) \rangle$, m-a.e.

Thus $g = \langle w, f(\cdot)\rangle$, m-a.e. and so the linear operator T is closed. Hence by the closed graph theorem T is a bounded operator. This implies that the linear functional

$$w \longmapsto \int_E \langle w, f(u)\rangle\, dm(u), \quad w \in \mathbf{X}^*,$$

is continuous on $\mathbf{X}^*$. Hence there exists a unique element $J_E \in \mathbf{X}^{**}$ such that

$$J_E(w) = \int_E \langle w, f(u)\rangle\, dm(u), \quad w \in \mathbf{X}^*.$$

We use $\int_E f(u)\, dm(u)$ to denote J_E. In general J_E may not be in $\mathbf{X}$ ($\mathbf{X}$ is imbedded in $\mathbf{X}^{**}$ by the canonical map).

Example 13.1. Let c_0 consist of all sequences of complex numbers converging to zero. It is a Banach space with the supremum norm. Define a function $f: (0, \infty) \to c_0$ by

$$f(u) = (e^{-u}, 2e^{-2u}, \dots, ne^{-nu}, \dots).$$

Note that $c_0^* = \ell_1$ and $c_0^{**} = \ell_\infty$. It is easy to see that f is weakly measurable and $\langle w, f(\cdot)\rangle \in L^1(0, \infty)$ for any $w \in \ell_1$. The integral $\int_E f(u)\, du$ belongs to ℓ_∞ for each $E \in \mathcal{B}$. In general it may not be in c_0. For instance, we have

$$\int_{(a,b)} f(u)\, du = (e^{-a} - e^{-b}, \dots, e^{-na} - e^{-nb}, \dots).$$

Hence $\int_{(a,b)} f(u)\, du \in c_0$ if and only if $a > 0$. In particular,

$$\int_{(0,\infty)} f(u)\, du = (1, \dots, 1, \dots) \notin c_0.$$

We point out that it is possible for $\int_M f(u)\, du$ to be in $\mathbf{X}$, while $\int_E f(u)\, du$ is not in $\mathbf{X}$ for some $E \in \mathcal{B}$. This is shown in the next example.

Example 13.2. Let $M = (-\infty, 0) \cup (0, \infty)$ and define a function $f: M \to c_0$ by

$$f(u) = \begin{cases} (e^{-u}, 2e^{-2u}, \dots, ne^{-nu}, \dots), & \text{if } u > 0; \\ -(e^u, 2e^{2u}, \dots, ne^{nu}, \dots), & \text{if } u < 0. \end{cases}$$

It is easy to check that

$$\int_M f(u)\, du = (0, \dots, 0, \dots) \in c_0,$$

$$\int_{(0,\infty)} f(u)\, du = (1, \dots, 1, \dots) \notin c_0.$$

A weakly measurable $\mathbb{X}$-valued function f on M is called *Pettis integrable* if it satisfies the following conditions:

(1) For any $w \in \mathbb{X}^*$, $\langle w, f(\cdot) \rangle \in L^1(M)$.
(2) For any $E \in \mathcal{B}$, there exists $J_E \in \mathbb{X}$ such that

$$\langle w, J_E \rangle = \int_E \langle w, f(u) \rangle \, dm(u), \quad \forall w \in \mathbb{X}^*.$$

Obviously J_E is unique. It is denoted by $(P) \int_E f(u) \, dm(u)$ and is called the *Pettis integral* of f on E.

In general there is no good criterion for Pettis integrability. However, from the above discussion, we see that if $\mathbb{X}$ is reflexive, then f is Pettis integrable if and only if f is weakly measurable and $\langle w, f(\cdot) \rangle \in L^1(M)$ for each $w \in \mathbb{X}^*$.

Next, we consider the Bochner integral. Let f be a countably-valued function from M into $\mathbb{X}$ given by $f = \sum_{k=1}^{\infty} x_k 1_{E_k}$, $E_k \in \mathcal{B}$ disjoint. It is called *Bochner integrable* if

$$\sum_{k=1}^{\infty} \|x_k\| m(E_k) < \infty.$$

For any $E \in \mathcal{B}$, we define the *Bochner integral* of f on E by

$$(B) \int_E f(u) \, dm(u) = \sum_{k=1}^{\infty} x_k m(E \cap E_k).$$

A function $f \colon M \to \mathbb{X}$ is said to be *almost separably-valued* if there exists a set E_0 such that $m(E_0) = 0$ and the set $f(M \setminus E_0)$ is separable. Suppose $f \colon M \to \mathbb{X}$ is weakly measurable and almost separably-valued. Then the function $\|f(\cdot)\|$ is easily seen to be measurable.

Let $\{f_n\}$ be a sequence of weakly measurable functions such that f_n converges almost everywhere. Then the function $f = \lim_{n \to \infty} f_n$ is easily seen to be also weakly measurable. Thus if $\{f_n\}$ is a sequence of countably-valued weakly measurable functions and $\lim_{n \to \infty} f_n = f$ almost everywhere, then both f and $f - f_n$ are weakly measurable and almost separably-valued. Hence the functions $\|f\|$ and $\|f - f_n\|$ are measurable.

A function $f \colon M \to \mathbb{X}$ is called *Bochner integrable* if there exists a sequence $\{f_n\}$ of countably-valued Bochner integrable functions such that

(1) $\lim_{n \to \infty} f_n = f$ almost everywhere.

(2) $\lim_{n \to \infty} \int_M \|f(u) - f_n(u)\| \, dm(u) = 0$.

For each $E \in \mathcal{B}$, it follows from condition (2) that $\int_E f_n(u) \, dm(u), n \geq 1$, is a Cauchy sequence in $\mathbb{X}$. The *Bochner integral* of f on E is defined to be

$$(B) \int_E f(u) \, dm(u) = \lim_{n \to \infty} \int_E f_n(u) \, dm(u).$$

Obviously the Bochner integral $(B)\int_E f(u)\,dm(u)$ is well-defined. Moreover, condition (2) implies that $\int_E \|f(u)\|\,dm(u) < \infty$.

Note that if f is Bochner integrable, then it is also Pettis integrable and we have

$$(P)\int_E f(u)\,dm(u) = (B)\int_E f(u)\,dm(u), \quad \forall E \in \mathcal{B}.$$

However, the converse is false.

Example 13.3. Let M be the set of positive integers with the counting measure. Define a function $f: M \to c_0$ by

$$f(n) = (0, \dots, 0, \tfrac{1}{n}, 0, \dots),$$

where $\frac{1}{n}$ appears at the n-th entry. Then $\sum_{n=1}^{\infty} \|f(n)\| = \infty$. Hence f is not Bochner integrable. On the other hand, suppose $w = (w_1, \dots, w_n, \dots) \in \ell_1$. Then

$$\sum_{n=1}^{\infty} |\langle w, f(\cdot)\rangle(n)| = \sum_{n=1}^{\infty} \tfrac{1}{n}|w_n| \leq \sum_{n=1}^{\infty} |w_n| < \infty.$$

Moreover, for any $E \in \mathcal{B}$,

$$\sum_{n \in E} f(n) = \sum_{n \in E} (0, \dots, 0, \tfrac{1}{n}, 0, \dots) \in c_0.$$

Hence f is Pettis integrable.

The importance of the Bochner integral lies in the following fact that determines Bochner integrability.

Fact: A function $f : M \to \mathbb{X}$ is Bochner integrable if and only if the following conditions are satisfied:

(1) f is weakly measurable.
(2) f is almost separably-valued.
(3) $\int_M \|f(u)\|\,dm(u) < \infty$.

In particular, if $\mathbb{X}$ is separable, then f is Bochner integrable if and only if it is weakly measurable and $\|f(\cdot)\| \in L^1(M)$.

Let $L^1(M; \mathbb{X})$ denote the set of Bochner integrable functions from M into $\mathbb{X}$. Then $L^1(M; \mathbb{X})$ is a Banach space with the norm given by

$$\|f\| = \int_M \|f(u)\|\,dm(u).$$

For a function $f \in L^1(M; \mathbb{X})$, its Bochner integral on E will be simply written as $\int_E f(u)\,dm(u)$. We have

$$\left\| \int_E f(u)\,dm(u) \right\| \leq \int_E \|f(u)\|\,dm(u).$$

13.3 White noise integrals

Let $(\mathcal{E})_\beta \subset (L^2) \subset (\mathcal{E})_\beta^*$ be a fixed Gel'fand triple. In this section we will study white noise integrals of the following type:

$$\int_E \Phi(u)\,dm(u), \quad E \in \mathcal{B},$$

where Φ is an $(\mathcal{E})_\beta^*$-valued function on a measure space $(M, \mathcal{B}, m)$. Although $(\mathcal{E})_\beta^*$ is not a Banach space, we can use the same ideas from the last section to define these integrals in the Pettis or Bochner sense.

A. *White noise integrals in the Pettis sense*

From the white noise viewpoint, it is natural to "define" the integral $\int_E \Phi(u)\,dm(u)$ in terms of the S-transform. That is, $\int_E \Phi(u)\,dm(u)$ is the generalized function in $(\mathcal{E})_\beta^*$ satisfying

$$S\!\left(\int_E \Phi(u)\,dm(u)\right)(\xi) = \int_E S\big(\Phi(u)\big)(\xi)\,dm(u), \quad \xi \in \mathcal{E}_c.$$

Hence we would require the following conditions for Φ:

(a) $S\big(\Phi(\cdot)\big)(\xi)$ is measurable for any $\xi \in \mathcal{E}_c$.
(b) $S\big(\Phi(\cdot)\big)(\xi) \in L^1(M)$ for any $\xi \in \mathcal{E}_c$.
(c) For any $E \in \mathcal{B}$, the function $\int_E S\big(\Phi(u)\big)(\cdot)\,dm(u)$ is the S-transform of a generalized function in $(\mathcal{E})_\beta^*$.

Note that the last equality can be rewritten as

$$\left\langle\!\left\langle \int_E \Phi(u)\,dm(u), \, {:}e^{\langle \cdot, \xi \rangle}{:} \right\rangle\!\right\rangle = \int_E \left\langle\!\left\langle \Phi(u), {:}e^{\langle \cdot, \xi \rangle}{:} \right\rangle\!\right\rangle dm(u), \quad \xi \in \mathcal{E}_c.$$

This implies that the following equality

$$\left\langle\!\left\langle \int_E \Phi(u)\,dm(u), \varphi \right\rangle\!\right\rangle = \int_E \left\langle\!\left\langle \Phi(u), \varphi \right\rangle\!\right\rangle dm(u) \tag{13.1}$$

holds for any φ in the linear span V of the set $\{{:}e^{\langle \cdot, \xi \rangle}{:}; \xi \in \mathcal{E}_c\}$. It is easy to see that $\left\langle\!\left\langle \Phi(\cdot), \varphi \right\rangle\!\right\rangle$ is measurable for any $\varphi \in (\mathcal{E})_\beta$. Hence Φ is weakly measurable. Moreover, note that the left-hand side of equation (13.1) is a continuous linear functional on $(\mathcal{E})_\beta$. This implies that $1_E\left\langle\!\left\langle \Phi(\cdot), \varphi \right\rangle\!\right\rangle \in L^1(M)$ for any $\varphi \in (\mathcal{E})_\beta$ and $E \in \mathcal{B}$ and that equation (13.1) holds. But this is exactly the Pettis integral.

Conversely, let us start with the idea of the Pettis integral. A function $\Phi\colon M \to (\mathcal{E})_\beta^*$ is Pettis integrable if the following conditions are satisfied:

(1) Φ is weakly measurable.

(2) $\langle\!\langle \Phi(\cdot), \varphi \rangle\!\rangle \in L^1(M)$ for all $\varphi \in (\mathcal{E})_\beta$.

(3) For each $E \in \mathcal{B}$, there exists a generalized function in $(\mathcal{E})_\beta^*$, denoted by $\int_E \Phi(u)\, dm(u)$, such that

$$\left\langle\!\!\left\langle \int_E \Phi(u)\, dm(u), \varphi \right\rangle\!\!\right\rangle = \int_E \langle\!\langle \Phi(u), \varphi \rangle\!\rangle\, dm(u), \quad \forall \varphi \in (\mathcal{E})_\beta.$$

In fact, we do not need to require condition (3) since it is implied by conditions (1) and (2) in the present situation. To see this, simply use the same argument for the operator T in the last section.

Obviously these conditions (1)–(3) (in fact, (1) and (2)) imply the above conditions (a)–(c). By letting $\varphi = :e^{\langle \cdot, \xi \rangle}:$, we get

$$S\left(\int_M \Phi(u)\, dm(u) \right)(\xi) = \int_M S(\Phi(u))(\xi)\, dm(u), \quad \xi \in \mathcal{E}_c.$$

Thus to define the white noise integral via S-transform is equivalent to define the white noise integral in the Pettis sense. For application the next theorem is useful since the conditions are easier to verify than those given above.

Theorem 13.4. *Suppose a function* $\Phi \colon M \to (\mathcal{E})_\beta^*$ *satisfies the conditions:*

(1) $S(\Phi(\cdot))(\xi)$ *is measurable for any* $\xi \in \mathcal{E}_c$.

(2) *There exist nonnegative numbers* $K, a,$ *and* p *such that*

$$\int_M |S(\Phi(u))(\xi)|\, dm(u) \leq K \exp\left[a\, |\xi|_p^{\frac{2}{1-\beta}} \right], \quad \xi \in \mathcal{E}_c.$$

Then Φ *is Pettis integrable and for any* $E \in \mathcal{B}$,

$$S\left(\int_E \Phi(u)\, dm(u) \right)(\xi) = \int_E S(\Phi(u))(\xi)\, dm(u), \quad \xi \in \mathcal{E}_c.$$

Proof. Note that condition (2) implies that $S(\Phi(\cdot))(\xi) \in L^1(M)$ for any $\xi \in \mathcal{E}_c$. Hence we need only to check the condition (c) above. For any $E \in \mathcal{B}$, let

$$F(\xi) = \int_E S(\Phi(u))(\xi)\, dm(u), \quad \xi \in \mathcal{E}_c.$$

We can use Morera's theorem to check that for any $\xi, \eta \in \mathcal{E}_c$, the function $F(z\xi + \eta)$ is an entire function of $z \in \mathbb{C}$. Moreover, for any $\xi \in \mathcal{E}_c$,

$$|F(\xi)| \leq \int_E |S(\Phi(u))(\xi)|\, dm(u)$$

$$\leq \int_M |S(\Phi(u))(\xi)|\, dm(u)$$

$$\leq K \exp\left[a\, |\xi|_p^{\frac{2}{1-\beta}} \right].$$

Thus by Theorem 8.2, there exists a generalized function $\Psi \in (\mathcal{E})^*_\beta$ such that

$$S\Psi(\xi) = F(\xi) = \int_E S\big(\Phi(u)\big)(\xi)\, dm(u).$$

Hence condition (c) is satisfied. Since by definition Ψ is the Pettis integral $\int_E \Phi(u)\, dm(u)$, we have

$$S\Big(\int_E \Phi(u)\, dm(u)\Big)(\xi) = \int_E S(\Phi(u))(\xi)\, dm(u), \quad \xi \in \mathcal{E}_c. \qquad \square$$

B. *White noise integrals in the Bochner sense*

Recall from the last section that a function f from M to a separable Banach space is Bochner integrable if and only if it is weakly measurable and $\|f(\cdot)\| \in L^1(M)$. Since the space $(\mathcal{E})^*_\beta$ is not a Banach space, we can not simply use the Bochner integral from the last section to define the white noise integral $\int_M \Phi(u)\, dm(u)$. However, note that $(\mathcal{E})^*_\beta = \cup_{p\geq 0}(\mathcal{E}_p)^*_\beta$ and each $(\mathcal{E}_p)^*_\beta$ is a separable Hilbert space. Thus it is reasonable to define the white noise integral $\int_M \Phi(u)\, dm(u)$ in the Bochner sense as follows. We say that a function $\Phi : M \to (\mathcal{E})^*_\beta$ is *Bochner integrable* if it satisfies the following conditions:

(1) Φ is weakly measurable.
(2) $\exists p \geq 0$ such that $\Phi(u) \in (\mathcal{E}_p)^*_\beta$ for almost all $u \in M$ and $\|\Phi(\cdot)\|_{-p,-\beta} \in L^1(M)$.

Suppose Φ is Bochner integrable. Then by Theorem 13.4 Φ is also Pettis integrable. Moreover, the white noise integral $\int_M \Phi(u)\, dm(u)$ is a generalized function in $(\mathcal{E}_p)^*_\beta$ and

$$\Big\|\int_M \Phi(u)\, dm(u)\Big\|_{-p,-\beta} \leq \int_M \|\Phi(u)\|_{-p,-\beta}\, dm(u),$$

$$\Big\langle\!\Big\langle\int_M \Phi(u)\, dm(u), \varphi\Big\rangle\!\Big\rangle = \int_M \langle\!\langle\Phi(u), \varphi\rangle\!\rangle\, dm(u).$$

In particular, for $\varphi = :e^{\langle \cdot, \xi\rangle}:$, we have

$$S\Big(\int_M \Phi(u)\, dm(u)\Big)(\xi) = \int_M S\big(\Phi(u)\big)(\xi)\, dm(u), \quad \xi \in \mathcal{E}_c.$$

Note that usually it is easier to find the S-transform of $\Phi(u)$ than to calculate the norm $\|\Phi(u)\|_{-p,-\beta}$. In fact, we sometimes estimate the norm $\|\Phi\|_{-p,-\beta}$ of a generalized function $\Phi \in (\mathcal{E})^*_\beta$ from its S-transform $S\Phi$ (see Theorem 8.2). Hence it is desirable to find conditions for Bochner integrability in terms of the S-transform $S\big(\Phi(u)\big)$.

Theorem 13.5. *Let* $\Phi\colon M \to (\mathcal{E})^*_\beta$ *be a function satisfying the conditions:*
(1) $S(\Phi(\cdot))(\xi)$ *is measurable for any* $\xi \in \mathcal{E}_c$.
(2) *There exist* $p \geq 0$ *and nonnegative functions* $L \in L^1(M)$, $b \in L^\infty(M)$, *and an* m-*null set* E_0 *such that*

$$|S(\Phi(u))(\xi)| \leq L(u)\exp\left[b(u)|\xi|_p^{\frac{2}{1-\beta}}\right], \quad \forall \xi \in \mathcal{E}_c, \ u \in E_0^c.$$

Then Φ *is Bochner integrable and* $\int_M \Phi(u)\,dm(u) \in (\mathcal{E}_q)^*_\beta$ *for any* $q > p$ *such that*

$$e^2\left(\frac{2\|b\|_\infty}{1-\beta}\right)^{1-\beta}\|A^{-(q-p)}\|_{HS}^2 < 1, \tag{13.2}$$

where $\|b\|_\infty$ *is the essential supremum of* b. *In fact, for such* q,

$$\left\|\int_M \Phi(u)\,dm(u)\right\|_{-q,-\beta}$$

$$\leq \|L\|_1\left(1 - e^2\left(\frac{2\|b\|_\infty}{1-\beta}\right)^{1-\beta}\|A^{-(q-p)}\|_{HS}^2\right)^{-1/2}. \tag{13.3}$$

Proof. Obviously condition (1) implies that $\langle\!\langle \Phi(\cdot), \varphi\rangle\!\rangle$ is measurable for any $\varphi \in (\mathcal{E})_\beta$. Suppose $q > p$ satisfies the inequality in (13.2). Then for any $u \in E_0^c$,

$$e^2\left(\frac{2|b(u)|}{1-\beta}\right)^{1-\beta}\|A^{-(q-p)}\|_{HS}^2 < 1.$$

Thus by condition (2) and Theorem 8.2,

$$\|\Phi(u)\|_{-q,-\beta} \leq L(u)\left(1 - e^2\left(\frac{2|b(u)|}{1-\beta}\right)^{1-\beta}\|A^{-(q-p)}\|_{HS}^2\right)^{-1/2}$$

$$\leq L(u)\left(1 - e^2\left(\frac{2\|b\|_\infty}{1-\beta}\right)^{1-\beta}\|A^{-(q-p)}\|_{HS}^2\right)^{-1/2}.$$

Hence $\Phi(u) \in (\mathcal{E}_q)^*_\beta$ for all $u \in E_0^c$ and

$$\int_M \|\Phi(u)\|_{-q,-\beta}\,dm(u) \leq \|L\|_1\left(1 - e^2\left(\frac{2\|b\|_\infty}{1-\beta}\right)^{1-\beta}\|A^{-(q-p)}\|_{HS}^2\right)^{-1/2}.$$

Obviously, this implies that Φ is Bochner integrable and the inequality in (13.3) holds. $\qquad\square$

Example 13.6. Consider the integral $\int_0^t e^{-c(t-s)}\dot{B}(s)\,ds$ from §13.1. Let $M = [0,t]$ and $\Phi(s) = e^{-c(t-s)}\dot{B}(s)$, $0 \leq s \leq t$. The S-transform of $\Phi(s)$ is given by

$$S(\Phi(s))(\xi) = e^{-c(t-s)}\xi(s), \quad \xi \in \mathcal{E}_c.$$

Note that

$$S\langle\cdot, 1_{[0,t]}e^{-c(t-\cdot)}\rangle(\xi) = \langle 1_{[0,t]}e^{-c(t-\cdot)}, \xi\rangle = \int_0^t e^{-c(t-s)}\xi(s)\, ds.$$

Thus Φ is Pettis integrable and

$$\int_0^t e^{-c(t-s)}\dot{B}(s)\, ds = \langle\cdot, 1_{[0,t]}e^{-c(t-\cdot)}\rangle.$$

Recall that if $f \in L^2(\mathbb{R})$, then $\langle\cdot, f\rangle$ is a Gaussian random variable with mean 0 and variance $|f|_0^2$. Hence $\int_0^t e^{-c(t-s)}\dot{B}(s)ds$ is Gaussian with mean 0 and variance $\frac{1}{2c}(1 - e^{-2ct})$. We remark that it is easy to check that $\int_0^t e^{-c(t-s)}\dot{B}(s)ds$ is also Bochner integrable. However, Theorem 13.5 does not yield the fact that $\int_0^t e^{-c(t-s)}\dot{B}(s)\, ds \in (L^2)$.

Example 13.7. Consider the integral $\int_0^t e^{-c(t-s)} : \dot{B}(s)^2 : ds$ from §13.1. Let $M = [0,t]$ and $\Phi(s) = e^{-c(t-s)} : \dot{B}(s)^2 :$, $0 \le s \le t$. The S-transform of $\Phi(s)$ is given by

$$S\big(\Phi(s)\big)(\xi) = e^{-c(t-s)}\xi(s)^2, \quad \xi \in \mathcal{E}_c.$$

Define a distribution $G \in \mathcal{S}'_c(\mathbb{R}^2)$ by

$$G(u,v) = e^{-c(t-u)}\tau_{1_{[0,t]}}(u,v),$$

where τ_f is defined in §12.3. Then

$$S\langle :\cdot^{\otimes 2}:, G\rangle(\xi) = \langle G, \xi^{\otimes 2}\rangle = \int_0^t e^{-c(t-u)}\xi(u)^2\, du.$$

Thus Φ is Pettis integrable and

$$\int_0^t e^{-c(t-s)} : \dot{B}(s)^2 : ds = \langle :\cdot^{\otimes 2}:, e^{-c(t-u)}\tau_{1_{[0,t]}}(u,v)\rangle.$$

We remark that Φ satisfies the conditions in Theorem 13.5 and so it is Bochner integrable.

Example 13.8. In Theorem 7.3 we showed that $F(\langle\cdot, f\rangle)$ is a generalized function in $(\mathcal{S})^*$ for any $F \in \mathcal{S}'(\mathbb{R})$ and $f \in L^2(\mathbb{R})$, $f \ne 0$. In this example we show that if the Fourier transform $\widehat{F} \in L^\infty(\mathbb{R})$, then $F(\langle\cdot, f\rangle)$ can be represented as a white noise integral by

$$F(\langle\cdot, f\rangle) = \frac{1}{\sqrt{2\pi}}\int_{\mathbb{R}} e^{iu\langle\cdot, f\rangle}\widehat{F}(u)\, du. \tag{13.4}$$

To prove this representation, let

$$\Phi(u) = e^{iu\langle \cdot, f \rangle} \widehat{F}(u), \quad u \in \mathbb{R}.$$

Then Φ is an $(\mathcal{E})^*_\beta$-valued function. We check that Φ satisfies the conditions in Theorem 13.4. Note that

$$e^{iu\langle \cdot, f \rangle} = :e^{iu\langle \cdot, f \rangle}: e^{-\frac{1}{2}u^2|f|_0^2}.$$

By using Theorem 5.13 we get

$$S(e^{iu\langle \cdot, f \rangle})(\xi) = \exp\left[iu\langle f, \xi \rangle - \frac{1}{2}u^2|f|_0^2\right].$$

Therefore

$$S(\Phi(u))(\xi) = \widehat{F}(u)\exp\left[iu\langle f, \xi \rangle - \frac{1}{2}u^2|f|_0^2\right], \quad \xi \in \mathcal{E}_c.$$

Obviously condition (1) of Theorem 13.4 is satisfied. To check condition (2), note that

$$|S(\Phi(u))(\xi)| \leq \|\widehat{F}\|_\infty \exp\left[|u||f|_0|\xi|_0 - \frac{1}{2}u^2|f|_0^2\right], \quad \xi \in \mathcal{E}_c. \tag{13.5}$$

Hence for all $\xi \in \mathcal{E}_c$,

$$\int_{-\infty}^\infty |S(\Phi(u))(\xi)|\, du$$

$$\leq \|\widehat{F}\|_\infty \left(\int_{-\infty}^\infty e^{u|f|_0|\xi|_0 - \frac{1}{2}u^2|f|_0^2}\, du + \int_{-\infty}^\infty e^{-u|f|_0|\xi|_0 - \frac{1}{2}u^2|f|_0^2}\, du \right)$$

$$= 2\frac{\sqrt{2\pi}}{|f|_0}\|\widehat{F}\|_\infty\, e^{\frac{1}{2}|\xi|_0^2}.$$

This shows that condition (2) of Theorem 13.4 is also satisfied. Thus by Theorem 13.4 Φ is Pettis integrable and so $\frac{1}{\sqrt{2\pi}}\int_{\mathbb{R}} e^{iu\langle \cdot, f \rangle}\widehat{F}(u)\, du \in (\mathcal{E})^*_\beta$ with S-transform

$$S\left(\frac{1}{\sqrt{2\pi}}\int_{\mathbb{R}} e^{iu\langle \cdot, f \rangle}\widehat{F}(u)\, du\right)(\xi)$$

$$= \frac{1}{\sqrt{2\pi}}\int_{-\infty}^\infty \widehat{F}(u)\exp\left[iu\langle f, \xi \rangle - \frac{1}{2}u^2|f|_0^2\right] du. \tag{13.6}$$

Let $\eta \in \mathcal{E}_c$ denote the function

$$\eta(u) = \exp\left[iu\langle f, \xi \rangle - \frac{1}{2}u^2|f|_0^2\right].$$

The inverse Fourier transform $\check{\eta}$ of η is given by

$$\check{\eta}(y) = \frac{1}{\sqrt{2\pi}} \int_{\mathbb{R}} e^{iyu} \eta(u) \, du$$

$$= \frac{1}{|f|_0} \exp\left[-\frac{1}{2|f|_0^2} (y - \langle f, \xi \rangle)^2 \right].$$

Therefore, from equation (13.6), we get

$$S\left(\frac{1}{\sqrt{2\pi}} \int_{\mathbb{R}} e^{iu\langle \cdot, f \rangle} \widehat{F}(u) \, du \right)(\xi)$$

$$= \frac{1}{\sqrt{2\pi}} \langle \widehat{F}, \eta \rangle$$

$$= \frac{1}{\sqrt{2\pi}} \langle F, \check{\eta} \rangle$$

$$= \frac{1}{\sqrt{2\pi}|f|_0} \int_{\mathbb{R}} F(y) \exp\left[-\frac{1}{2|f|_0^2} (y - \langle f, \xi \rangle)^2 \right],$$

where the integral is understood as the bilinear pairing of $\mathcal{S}'(\mathbb{R})$ and $\mathcal{S}(\mathbb{R})$. Then, by Theorem 7.3, we conclude that

$$F(\langle \cdot, f \rangle) = \frac{1}{\sqrt{2\pi}} \int_{\mathbb{R}} e^{iu\langle \cdot, f \rangle} \widehat{F}(u) \, du.$$

We remark that we can not apply Theorem 13.5 to get this representation because the function $b(u) = |f|_0 |u|$ in equation (13.5) is not in $L^\infty(\mathbb{R})$.

Example 13.9. (*Donsker's delta function*) A special case of Example 13.8 with $F = \delta_a$ gives the white noise integral representation

$$\delta_a(\langle \cdot, f \rangle) = \frac{1}{2\pi} \int_{\mathbb{R}} e^{iu(\langle \cdot, f \rangle - a)} \, du.$$

To check this equality, simply note that

$$\widehat{\delta_a}(u) = \frac{1}{\sqrt{2\pi}} \int_{\mathbb{R}} e^{-iuy} \delta_a(y) \, dy = \frac{1}{\sqrt{2\pi}} e^{-iau}.$$

In particular, when $f = 1_{[0,t]}$, we get the white noise integral representation of Donsker's delta function in Example 7.4:

$$\delta_a(B(t)) = \frac{1}{2\pi} \int_{\mathbb{R}} e^{iu(B(t) - a)} \, du.$$

Example 13.10. (*Intersection local time*) Let r be a fixed positive integer and consider the Gel'fand triple $S(\mathbb{R}^r) \subset L^2(\mathbb{R}^r) \subset S'(\mathbb{R}^r)$. Let $(S) \subset (L^2) \subset (S)^*$ be the associated Gel'fand triple arising from some operator on $L^2(\mathbb{R}^r)$ satisfying the conditions in §4.2. Define

$$\mathbb{B}(t)(x_1, \ldots, x_r) = (\langle x_1, 1_{[0,t]} \rangle, \ldots, \langle x_r, 1_{[0,t]} \rangle), \quad (x_1, \ldots, x_r) \in S'(\mathbb{R}^r).$$

Then $\mathbb{B}(t)$ is an r-dimensional Brownian motion. It is easy to check that $\delta_0(\mathbb{B}(t) - \mathbb{B}(s))$ is a generalized function in $(S)^*$ and has the following white noise integral representation

$$\delta_0(\mathbb{B}(t) - \mathbb{B}(s)) = \frac{1}{(2\pi)^r} \int_{\mathbb{R}^r} e^{i\langle u, \mathbb{B}(t) - \mathbb{B}(s) \rangle} \, du.$$

This generalized function has been used to study the intersection local time. For instance, when $0 < a < b < c < d < \infty$, the intersection local time of $\mathbb{B}(t)$ at 0 during $[a, b] \times [c, d]$ is given by

$$\int_a^b \int_c^d \delta_0(\mathbb{B}(t) - \mathbb{B}(s)) \, dt ds.$$

On the other hand, when $c = a$ and $d = b$, we need a renormalization for the integral. For instance, when $r = 2$ or 3,

$$\int_a^b \int_a^b \left[\delta_0(\mathbb{B}(t) - \mathbb{B}(s)) - (2\pi|s - t|)^{-r/2} \right] ds dt.$$

For details on the intersection local time in white noise formulation, see de Faria et al. [28], de Faria and Streit [30], Kuo and Shieh [117], and H. Watanabe [171] [172].

The white noise integrals can be expressed in terms of the Wiener-Itô decomposition. First note that Pettis and Bochner integrals can be defined for $(\mathcal{E}'_c)^{\otimes n}$-valued functions in the same way as before for $(\mathcal{E})^*_\beta$-valued functions (except for the S-transform criteria such as Theorem 13.4). Let $\Phi: M \to (\mathcal{E})^*_\beta$ be a weakly measurable function represented by

$$\Phi(u) = \sum_{n=0}^{\infty} \langle :\cdot^{\otimes n}:, F_n(u) \rangle,$$

where $F_n: M \to (\mathcal{E}'_c)^{\otimes n}$ is weakly measurable for each n. Recall that Φ is Pettis integrable if and only if $\langle\langle \Phi(\cdot), \varphi \rangle\rangle \in L^1(M)$ for all $\varphi \in (\mathcal{E})_\beta$. Let $\varphi = \sum_{n=0}^{\infty} \langle :\cdot^{\otimes n}:, f_n \rangle \in (\mathcal{E})_\beta$. Then

$$\langle\langle \Phi(u), \varphi \rangle\rangle = \sum_{n=0}^{\infty} n! \langle F_n(u), f_n \rangle.$$

If Φ is Pettis integrable, then F_n is Pettis integrable for each n and

$$\int_M \Phi(u)\, du = \sum_{n=0}^{\infty} \left\langle :\cdot^{\otimes n}:, \int_M F_n(u)\, du \right\rangle. \tag{13.7}$$

Conversely, suppose F_n is Pettis integrable for all n and

$$\sum_{n=0}^{\infty} n! \|\langle F_n(\cdot), f_n \rangle\|_1 < \infty$$

for all $\varphi = \sum_{n=0}^{\infty} \langle :\cdot^{\otimes n}:, f_n \rangle \in (\mathcal{E})_\beta$. Then Φ is Pettis integrable and

$$\|\langle\!\langle \Phi(\cdot), \varphi \rangle\!\rangle\|_1 \le \sum_{n=0}^{\infty} n! \|\langle F_n(\cdot), f_n \rangle\|_1.$$

On the other hand, for the Bochner integral, note that

$$\|\Phi(u)\|_{-p,-\beta}^2 = \sum_{n=0}^{\infty} (n!)^{1-\beta} |F_n(u)|_{-p}^2.$$

Hence if Φ is Bochner integrable, then F_n is Bochner integrable for each n. Conversely, suppose Φ is weakly measurable and there exists $p \ge 0$ such that $F_n(u) \in (\mathcal{E}'_{p,c})^{\widehat{\otimes} n}$ for all n and almost all $u \in M$, and

$$\int_M \left(\sum_{n=0}^{\infty} (n!)^{1-\beta} |F_n(u)|_{-p}^2 \right)^{1/2} dm(u) < \infty.$$

Then Φ is Bochner integrable and

$$\int_M \|\Phi(u)\|_{-p,-\beta}\, dm(u) \le \int_M \left(\sum_{n=0}^{\infty} (n!)^{1-\beta} |F_n(u)|_{-p}^2 \right)^{1/2} dm(u).$$

13.4 An extension of the Itô integral

In this section we use the Gel'fand triple $(\mathcal{S})_\beta \subset (L^2) \subset (\mathcal{S})_\beta^*$ arising from the Gel'fand triple $\mathcal{S} \subset L^2(\mathbb{R}) \subset \mathcal{S}'$. We will study white noise integrals of the form

$$\int_a^b \partial_t^* \Phi(t)\, dt.$$

This integral (with a notation different from ∂_t^*), when it defines a random variable in (L^2), has been introduced by Hitsuda [64] and Skorokhod [167]. If $\Phi(t)$ is nonanticipating, then this integral turns out to be the same as the Itô integral of $\Phi(t)$. This result is due to Kubo and Takenaka [94] (see Theorem 13.12 below).

Recall that for any $t \in \mathbb{R}$, the operator ∂_t^* is a continuous linear operator from $(\mathcal{S})_\beta^*$ into itself with operator norm uniformly bounded in t (Theorems 9.1 and 9.12). Thus by the results in the previous section, we have the following facts:

Fact 1. Suppose $\Phi \colon [a, b] \to (\mathcal{S})_\beta^*$ is Pettis integrable. Then the function $t \mapsto \partial_t^* \Phi(t)$ is also Pettis integrable and

$$S\left(\int_a^b \partial_t^* \Phi(t) \, dt \right)(\xi) = \int_a^b \xi(t) S\big(\Phi(t) \big)(\xi) \, dt, \quad \xi \in \mathcal{S}_c. \tag{13.8}$$

Fact 2. Suppose $\Phi \colon [a, b] \to (\mathcal{S})_\beta^*$ is Bochner integrable. Then the function $t \mapsto \partial_t^* \Phi(t)$ is also Bochner integrable and equation (13.8) holds.

Now, define a Brownian motion $B(t)$ by

$$B(t) = \begin{cases} \langle \cdot, 1_{[0,t]} \rangle, & \text{if } t \geq 0; \\ -\langle \cdot, 1_{[t,0]} \rangle, & \text{if } t < 0. \end{cases}$$

We review the definition of Itô integral $\int_a^b \varphi(t) \, dB(t)$ for nonanticipating $\varphi(t)$ such that $\int_a^b \|\varphi\|_0^2 \, dt < \infty$. Here $[a, b]$ is a finite interval in $\mathbb{R}$ and $\varphi(t)$ is nonanticipating with respect to the Brownian motion $B(t)$ defined above, i.e., for any t, the random variable $\varphi(t)$ is measurable with respect to the σ-field $\mathcal{F}_t \equiv \sigma\{B(s); s \leq t\}$.

Step 1: Suppose $\varphi(t)$ is a step function, i.e.,

$$\varphi(t, x) = \sum_{k=0}^{m-1} \varphi(t_k, x) 1_{[t_k, t_{k+1})}(t),$$

where $\varphi(t_k) \in (L^2)$ is $\mathcal{F}_{t_k}$-measurable for all k, $t_0 = a$, and $t_m = b$. The Itô integral of φ is defined by

$$\int_a^b \varphi(t) \, dB(t) = \sum_{k=0}^{m-1} \varphi(t_k) \big(B(t_{k+1}) - B(t_k) \big). \tag{13.9}$$

It is easy to check that

$$\left\| \int_a^b \varphi(t) \, dB(t) \right\|_0^2 = \int_a^b \|\varphi(t)\|_0^2 \, dt. \tag{13.10}$$

Step 2: For any nonanticipating $\varphi(t)$ such that $\int_a^b \|\varphi(t)\|_0^2 \, dt < \infty$, there exists a sequence $\{\varphi_n\}$ of step functions such that

$$\lim_{n\to\infty} \int_a^b \|\varphi(t) - \varphi_n(t)\|_0^2 \, dt = 0.$$

By using equation (13.10), we see that the sequence $\int_a^b \varphi_n(t) \, dB(t)$, $n \geq 1$, is Cauchy in (L^2). Define the Itô integral $\int_a^b \varphi(t) \, dB(t)$ to be

$$\int_a^b \varphi(t) \, dB(t) = \lim_{n\to\infty} \int_a^b \varphi_n(t) \, dB(t) \quad \text{in } (L^2).$$

This integral is well-defined since the limit is independent of the sequence $\{\varphi_n\}$.

Lemma 13.11. *Suppose* $\varphi : [a, b] \to (L^2)$ *is measurable and let* $\varphi(t)$ *be represented by*

$$\varphi(t) = \sum_{n=0}^{\infty} \langle :\cdot^{\otimes n}:, f_n(t)\rangle.$$

Then $\varphi(t)$ *is nonanticipating if and only if for all* $n \geq 1$ *and* $t \in [a, b]$,

$$f_n(t; u_1, \ldots, u_n) = 0 \quad \text{for almost all } (u_1, \ldots, u_n) \in \left([a, t]^n\right)^c. \tag{13.11}$$

Proof. First we show the sufficiency. Suppose $f_n(t)$ satisfies the condition in equation (13.11) for any $n \geq 1$ and $t \in [a, b]$. Then by assertion part (2) in Theorem 5.4,

$$\langle :\cdot^{\otimes n}:, f_n(t)\rangle = I_n(f_n(t))$$
$$= \int_0^t \cdots \int_0^t f_n(t; u_1, \ldots, u_n) \, dB(u_1) \cdots dB(u_n).$$

Hence $\langle :\cdot^{\otimes n}:, f_n(t)\rangle$ is $\mathcal{F}_t$-measurable for any $n \geq 1$. Obviously this implies that $\varphi(t)$ is $\mathcal{F}_t$-measurable for any $t \in [a, b]$. Hence $\varphi(t)$ is nonanticipating.

Next, we show the necessity. Suppose there exist $n_0 \geq 1$ and t_0 such that

$$f_{n_0}(t_0; u_1, \ldots, u_{n_0}) \neq 0 \quad \text{for almost all } (u_1, \ldots, u_{n_0}) \in \left([a, t]^{n_0}\right)^c.$$

Then there exist disjoint intervals $I_1, \ldots, I_{n_0}$ such that at least one of them is contained in $(t_0, b]$ and

$$\langle f_{n_0}(t_0), 1_{I_1} \widehat{\otimes} \cdots \widehat{\otimes} 1_{I_{n_0}}\rangle \neq 0.$$

Let $\psi = \langle :\cdot^{\otimes n_0}:, 1_{I_1} \widehat{\otimes} \cdots \widehat{\otimes} 1_{I_{n_0}}\rangle$. Then

$$\langle\!\langle \varphi(t_0), \psi \rangle\!\rangle = n_0! \langle f_{n_0}(t_0), 1_{I_1} \widehat{\otimes} \cdots \widehat{\otimes} 1_{I_{n_0}}\rangle \neq 0. \tag{13.12}$$

On the other hand, let E be the expectation with respect to the Gaussian measure μ on $\mathcal{S}'$. Since $\varphi(t_0)$ is $\mathcal{F}_{t_0}$-measurable, we can take the conditional expectation to get

$$\langle\langle\varphi(t_0),\psi\rangle\rangle = E\big(\varphi(t_0)\psi\big) = E\big(\varphi(t_0)E(\psi|\mathcal{F}_{t_0})\big). \qquad (13.13)$$

By Corollary 5.3, ψ can be rewritten as

$$\psi = B(I_1)\cdots B(I_{n_0}),$$

where $B(I) = B(v) - B(u)$ if $I = [u,v)$. Note that $E(X|\mathcal{F}) = X$ if X is $\mathcal{F}$-measurable and $E(X|\mathcal{F}) = EX$ if X is independent of $\mathcal{F}$. Since at least one of the I_j's is contained in $(t_0, b]$ and the I_j's are disjoint, we see easily that $E(\psi|\mathcal{F}_{t_0}) = 0$. Thus by equation (13.13) we have

$$\langle\langle\varphi(t_0),\psi\rangle\rangle = 0.$$

But this contradicts equation (13.12). Thus the necessity is proved. $\qquad\square$

The next theorem shows that for a nonanticipating stochastic process $\varphi(t)$ in $L^2([a,b]\times\mathcal{S}')$, the Itô integral $\int_a^b \varphi(t)\,dB(t)$ can be expressed as a white noise integral in the Pettis sense.

Theorem 13.12. *Let $\varphi(t)$ be nonanticipating and $\int_a^b \|\varphi(t)\|_0^2\,dt < \infty$. Then the function $\partial_t^*\varphi(t)$, $t \in [a,b]$, is Pettis integrable and*

$$\int_a^b \partial_t^*\varphi(t)\,dt = \int_a^b \varphi(t)\,dB(t), \qquad (13.14)$$

where the right hand side is the Itô integral of φ.

Proof. First note that if $f \in L^2(\mathbb{R}^n)$ is symmetric, $g \in L^2(\mathbb{R})$ and $\langle f,g\rangle = 0$ on $\mathbb{R}^{n-1}$, then

$$\begin{aligned}
\langle\cdot,g\rangle\langle:\cdot^{\otimes n}:,f\rangle &= (D_g + D_g^*)\langle:\cdot^{\otimes n}:,f\rangle \\
&= n\big\langle:\cdot^{\otimes(n-1)}:,\langle f,g\rangle\big\rangle + \langle:\cdot^{\otimes(n+1)}:,f\widehat{\otimes}g\rangle \\
&= \langle:\cdot^{\otimes(n+1)}:,f\widehat{\otimes}g\rangle.
\end{aligned}$$

For clarity we divide the proof into several steps.

Step 1: Special case with $\varphi(t,x) = \varphi(t_1,x)1_{[t_1,t_2)}(t)$.

Note that by assumption $\varphi(t_1) \in (L^2)$ is $\mathcal{F}_{t_1}$-measurable. Let $\varphi(t_1)$ be represented by

$$\varphi(t_1) = \sum_{n=0}^{\infty}\langle:\cdot^{\otimes n}:,f_n(t_1)\rangle.$$

By the above remark with $g = 1_{[t_1,t_2)}$ and Lemma 13.11, we get

$$\int_a^b \varphi(t)\, dB(t) = \varphi(t_1)\big(B(t_2) - B(t_1)\big)$$

$$= \sum_{n=0}^{\infty} \langle \cdot, 1_{[t_1,t_2)} \rangle \langle :\cdot^{\otimes n} :, f_n(t_1) \rangle$$

$$= \sum_{n=0}^{\infty} \langle :\cdot^{\otimes(n+1)} :, f_n(t_1) \widehat{\otimes} 1_{[t_1,t_2)} \rangle.$$

Thus the S-transform of $\int_a^b \varphi(t)\, dB(t)$ is given by

$$S\left(\int_a^b \varphi(t)\, dB(t) \right)(\xi) = \sum_{n=0}^{\infty} \langle f_n(t_1) \widehat{\otimes} 1_{[t_1,t_2)}, \xi^{\otimes(n+1)} \rangle$$

$$= \sum_{n=0}^{\infty} \langle f_n(t_1), \xi^{\otimes n} \rangle \int_{t_1}^{t_2} \xi(u)\, du$$

$$= S\varphi(t_1)(\xi) \int_{t_1}^{t_2} \xi(u)\, du. \tag{13.15}$$

On the other hand, it is obvious that $\varphi(t,x) = \varphi(t_1,x) 1_{[t_1,t_2)}(t)$ is Pettis integrable. Hence by Fact 1 stated in the beginning of this section, the function $t \mapsto \partial_t^* \varphi(t)$ is also Pettis integrable and

$$S\left(\int_a^b \partial_t^* \varphi(t)\, dt \right)(\xi) = \int_{t_1}^{t_2} S(\partial_t^* \varphi(t_1))(\xi)\, dt$$

$$= \int_{t_1}^{t_2} \xi(t) S\varphi(t_1)(\xi)\, dt$$

$$= S\varphi(t_1)(\xi) \int_{t_1}^{t_2} \xi(u)\, du. \tag{13.16}$$

By equations (13.15) and (13.16), the equality in (13.14) holds for the function $\varphi(t)$.

Step 2: Let $\varphi(t)$ be a simple function, i.e., a function of the form

$$\varphi(t,x) = \sum_{k=1}^{m} \varphi(t_k,x) 1_{[t_k,t_{k+1})}(t).$$

Obviously $\varphi(t)$ is Pettis integrable. This implies that $\partial_t^* \varphi(t)$ is also Pettis integrable. It follows from case 1 and equation (13.9) that the equality in (13.14) holds for $\varphi(t)$.

Step 3: The general case.

Let $\varphi(t)$ be nonanticipating and $\int_a^b \|\varphi(t)\|_0^2 \, dt < \infty$. Choose a sequence $\{\varphi_n(t)\}$ of nonanticipating step functions such that

$$\lim_{n\to\infty} \int_a^b \|\varphi(t) - \varphi_n(t)\|_0^2 \, dt = 0. \tag{13.17}$$

Then by the definition of Itô's integral

$$\int_a^b \varphi(t) \, dB(t) = \lim_{n\to\infty} \int_a^b \varphi_n(t) \, dB(t) \quad \text{in } (L^2).$$

By case 2 for φ_n, we have

$$\int_a^b \partial_t^* \varphi_n(t) \, dt = \int_a^b \varphi_n(t) \, dB(t).$$

Therefore

$$S\left(\int_a^b \varphi(t) \, dB(t)\right)(\xi) = \lim_{n\to\infty} S\left(\int_a^b \partial_t^* \varphi_n(t) \, dt\right)(\xi)$$

$$= \lim_{n\to\infty} \int_a^b \xi(t) S(\varphi_n(t))(\xi) \, dt.$$

But equation (13.17) implies that $\lim_{n\to\infty} S(\varphi_n(t))(\xi) = S(\varphi(t))(\xi)$ for all $\xi \in \mathcal{S}_c$. Hence we obtain that

$$S\left(\int_a^b \varphi(t) \, dB(t)\right)(\xi) = \int_a^b \xi(t) S(\varphi(t))(\xi) \, dt, \quad \xi \in \mathcal{S}_c. \tag{13.18}$$

On the other hand, note that $S(\varphi(\cdot))(\xi)$ is measurable for any $\xi \in \mathcal{S}_c$ and so condition (1) of Theorem 13.4 is satisfied. To check condition (2), let $\varphi(t) = \sum_{n=0}^{\infty} \langle :\cdot^{\otimes n}:, f_n(t)\rangle$. Then

$$|S(\varphi(t))(\xi)| = \left| \sum_{n=0}^{\infty} \langle f_n(t), \xi^{\otimes n}\rangle \right|$$

$$= \sum_{n=0}^{\infty} (\sqrt{n!} \, |f_n(t)|_0) \left(\frac{1}{\sqrt{n!}} |\xi^{\otimes n}|_0 \right)$$

$$\leq \|\varphi(t)\|_0 \left(\sum_{n=0}^{\infty} \frac{1}{n!} |\xi|_0^{2n} \right)^{1/2}$$

$$= \|\varphi(t)\|_0 \, e^{\frac{1}{2}|\xi|_0^2}.$$

This implies that condition (2) of Theorem 13.4 is satisfied. Hence by Theorem 13.4, $\varphi(t)$ is Pettis integrable. It follows that the function $\partial_t^* \varphi(t)$ is also Pettis integrable and

$$S\left(\int_a^b \partial_t^* \varphi(t)\, dt\right)(\xi) = \int_a^b \xi(t) S(\varphi(t))(\xi)\, dt, \quad \xi \in \mathcal{S}_c. \tag{13.19}$$

Finally, by equations (13.18) and (13.19), we have

$$\int_a^b \partial_t^* \varphi(t)\, dt = \int_a^b \varphi(t)\, dB(t). \qquad \square$$

In view of Theorem 13.12 the white noise integral $\int_a^b \partial_t^* \varphi(t)\, dt$ is an extension of the Itô integral to $\varphi(t)$ which may not be nonanticipating. In fact, this integral occurs as the adjoint operator of the gradient in Theorem 9.27. For an abstract Wiener space, it is the divergence introduced in Goodman [39] (see also Kuo [98]).

Definition 13.13. The white noise integral $\int_a^b \partial_t^* \varphi(t)\, dt$ is called the *Hitsuda-Skorokhod integral* of φ if it is a random variable in (L^2).

Example 13.14. $\int_0^1 \partial_t^* B(1)\, dt = B(1)^2 - 1$.

To check this equality, note that $B(1) = \langle \cdot, 1_{[0,1)} \rangle$ and so $(SB(1))(\xi) = \int_0^1 \xi(s)\, ds$. Hence

$$S\left(\int_0^1 \partial_t^* B(1)\, dt\right)(\xi) = \int_0^1 \xi(t)(SB(1))(\xi)\, dt$$

$$= \int_0^1 \int_0^1 \xi(t)\xi(s)\, dt ds$$

$$= S\langle :\cdot^{\otimes 2}:, 1_{[0,1)}^{\otimes 2} \rangle(\xi).$$

Therefore

$$\int_0^1 \partial_t^* B(1)\, dt = \langle :\cdot^{\otimes 2}:, 1_{[0,1)}^{\otimes 2} \rangle.$$

We can apply Lemma 5.2 to get

$$\int_0^1 \partial_t^* B(1)\, dt = :\langle \cdot, 1_{[0,1)} \rangle^2:_1$$

$$= \langle \cdot, 1_{[0,1)} \rangle^2 - 1$$

$$= B(1)^2 - 1.$$

Observe that in view of Theorem 13.12 we have a corresponding integral $\int_0^1 B(1)\,dB(t)$. But this is not an Itô integral since the integrand $B(1)$ is not nonanticipating. Itô has defined stochastic integrals for integrands which may not be nonanticipating in [71]. In particular, he showed that $\int_0^1 B(1)\,dB(t) = B(1)^2$. Clearly, this is different from the Hitsuda-Skorokhod integral of $B(1)$.

Example 13.15. It is not true that $\int_a^b \partial_t^* \varphi(t)\,dt$ is a Hitsuda-Skorokhod integral for any $\varphi \in L^2([a,b];(L^2))$. Here is a simple example. Let

$$\varphi(t) = \sum_{n=1}^{\infty} \frac{1}{n\sqrt{n!}} \langle :\cdot^{\otimes n}:, 1_{[0,1)}^{\otimes n} \rangle.$$

By the same argument as in Example 13.14 we can check easily that

$$\int_0^1 \partial_t^* \varphi(t)\,dt = \sum_{n=1}^{\infty} \frac{1}{n\sqrt{n!}} \langle :\cdot^{\otimes(n+1)}:, 1_{[0,1)}^{\otimes(n+1)} \rangle.$$

Hence we have

$$\int_0^1 \|\varphi(t)\|_0^2\,dt = \int_0^1 \sum_{n=1}^{\infty} n! \frac{1}{n^2 n!} |1_{[0,1)}|_0^{2n}\,dt = \sum_{n=1}^{\infty} \frac{1}{n^2} < \infty,$$

$$\left\| \int_0^1 \partial_t^* \varphi(t)\,dt \right\|_0^2 = \sum_{n=1}^{\infty} (n+1)! \frac{1}{n^2 n!} |1_{[0,1)}|_0^{2(n+1)} = \sum_{n=1}^{\infty} \frac{n+1}{n^2} = \infty.$$

Thus $\int_0^1 \partial_t^* \varphi(t)\,dt$ is not a Hitsuda-Skorokhod integral even though $\varphi \in L^2([a,b];(L^2))$.

Question: When is $\int_a^b \partial_t^* \varphi(t)\,dt$ a Hitsuda-Skorokhod integral?

Recall from §9.4 that the gradient operator ∇ is a continuous linear operator from $\mathcal{W}^{1/2}$ into $L^2([a,b] \times \mathcal{S}')$ (Here we identify $L^2([a,b];(L^2))$ with $L^2([a,b] \times \mathcal{S}')$). Its adjoint ∇^* is a continuous linear operator from $L^2([a,b] \times \mathcal{S}')$ into $(\mathcal{W}^{1/2})^*$. Since $\mathcal{W}^{1/2} \subset (L^2)$ and we identify (L^2) with its dual, we have $(L^2) \subset (\mathcal{W}^{1/2})^*$. It can be checked easily that in fact

$$(\mathcal{W}^{1/2})^* = \mathcal{W}^{-1/2} \equiv \text{ completion of } ((L^2), |\cdot|_{-1/2}),$$

where $|\cdot|_{-1/2}$ is defined similarly as in §9.4, i.e.,

$$|\varphi|_{-1/2} = \|(N+1)^{-1/2}\varphi\|_0.$$

Let φ be in $L^2([a,b];(L^2))$. Then by Theorem 9.28 $\nabla^*\varphi$ is nothing but the white noise integral $\int_a^b \partial_t^* \varphi(t)\,dt$, which belongs to $\mathcal{W}^{-1/2}$. Thus in order for $\int_a^b \partial_t^* \varphi(t)\, dt$ to be a Hitsuda-Skorokhod integral, we need to require a certain condition on the function $\varphi(t)$. This condition can be determined by the number operator N.

Theorem 13.16. *Let $\varphi \in L^2([a,b]; \mathcal{W}^{1/2})$. Then $\int_a^b \partial_t^* \varphi(t)\, dt$ is a Hitsuda-Skorokhod integral and*

$$\left\| \int_a^b \partial_t^* \varphi(t)\, dt \right\|_0^2 = \int_a^b \|\varphi(t)\|_0^2\, dt + \int_a^b \int_a^b ((\partial_t \varphi(s), \partial_s \varphi(t)))_0\, ds dt,$$
$$(13.20)$$

where $((\cdot, \cdot))_0$ is the inner product on (L^2). Moreover,

$$\left| \int_a^b \int_a^b ((\partial_t \varphi(s), \partial_s \varphi(t)))_0\, ds dt \right| \leq \int_a^b \|N^{1/2} \varphi(t)\|_0^2\, dt. \qquad (13.21)$$

Proof. First we prove the inequality in (13.21). Recall that $N = \int_{\mathbb{R}} \partial_s^* \partial_s\, ds$. Hence

$$\|N^{1/2} \varphi\|_0^2 = ((N\varphi, \varphi))_0$$

$$= \int_{\mathbb{R}} ((\partial_s^* \partial_s \varphi, \varphi))_0\, ds$$

$$= \int_{\mathbb{R}} \|\partial_s \varphi\|_0^2\, ds. \qquad (13.22)$$

Suppose $\varphi \in L^2([a,b]; \mathcal{W}^{1/2})$. Then we have

$$\left| \int_a^b \int_a^b ((\partial_t \varphi(s), \partial_s \varphi(t)))_0\, ds dt \right| \leq \int_a^b \int_a^b \|\partial_t \varphi(s)\|_0 \|\partial_s \varphi(t)\|_0\, ds dt$$

$$\leq \frac{1}{2} \int_a^b \int_a^b \left(\|\partial_t \varphi(s)\|_0^2 + \|\partial_s \varphi(t)\|_0^2 \right) ds dt$$

$$= \int_a^b \int_a^b \|\partial_s \varphi(t)\|_0^2\, ds dt.$$

Hence by using equation (13.22), we get

$$\left| \int_a^b \int_a^b ((\partial_t \varphi(s), \partial_s \varphi(t)))_0\, ds dt \right| \leq \int_a^b \left(\int_{\mathbb{R}} \|\partial_s \varphi(t)\|_0^2\, ds \right) dt$$

$$= \int_a^b \|N^{1/2} \varphi(t)\|_0^2\, dt.$$

Thus we have proved the inequality in (13.21). As for equation (13.20), it can be derived informally by using the formula $\partial_s \partial_t^* = \delta_s(t) I + \partial_t^* \partial_s$ (see equation (9.4)). To provide a rigorous proof, let

$$\varphi(t) = \sum_{n=0}^{\infty} \langle :\cdot^{\otimes n}:, f_n(t) \rangle,$$

where $f_n(t; \cdot)$ is a symmetric function in $L^2(\mathbb{R}^n)$ for each t. By assumption, $\varphi \in L^2([a, b]; \mathcal{W}^{1/2})$. Therefore

$$\sum_{n=0}^{\infty} (n+1)n! |f_n(t)|_0^2 < \infty.$$

Obviously φ is Pettis integrable. Hence $\partial_{(\cdot)}^* \varphi(\cdot)$ is also Pettis integrable and by equation (13.7)

$$\int_a^b \partial_t^* \varphi(t)\, dt = \int_a^b \sum_{n=0}^{\infty} \langle :\cdot^{\otimes(n+1)}:, \delta_t \widehat{\otimes} f_n(t) \rangle$$

$$= \sum_{n=0}^{\infty} \langle :\cdot^{\otimes(n+1)}:, \int_a^b \delta_t \widehat{\otimes} f_n(t)\, dt \rangle.$$

For simplicity, let

$$g_{n+1} \equiv \int_a^b \delta_t \widehat{\otimes} f_n(t)\, dt.$$

Then

$$\left\| \int_a^b \partial_t^* \varphi(t)\, dt \right\|_0^2 = \sum_{n=0}^{\infty} (n+1)! |g_{n+1}|_0^2. \tag{13.23}$$

It is straightforward to check that

$$g_{n+1}(u_1, \ldots, u_{n+1}) = \frac{1}{n+1} \Big(1_{[a,b)}(u_1) f_n(u_1; u_2, \ldots, u_{n+1})$$

$$+ 1_{[a,b)}(u_2) f_n(u_2; u_1, u_3, \ldots, u_{n+1})$$

$$+ \cdots + 1_{[a,b)}(u_{n+1}) f_n(u_{n+1}; u_1, \ldots, u_n) \Big).$$

From this equality, we can derive that

$$|g_{n+1}|_0^2 = \frac{1}{n+1} \Big(\int_a^b |f_n(t)|_0^2\, dt + n \int_a^b \int_a^b \big(f_n(s; t, \cdot), f_n(t; s, \cdot) \big)_0\, ds dt \Big),$$

where $(\cdot, \cdot)_0$ denotes the inner product on $L^2(\mathbb{R}^n)$. Hence by equation (13.23) we get

$$\left\| \int_a^b \partial_t^* \varphi(t)\, dt \right\|_0^2 = \sum_{n=0}^{\infty} n! \Big(\int_a^b |f_n(t)|_0^2\, dt$$

$$+ n \int_a^b \int_a^b \big(f_n(s; t, \cdot), f_n(t; s, \cdot) \big)_0\, ds dt \Big). \tag{13.24}$$

On the other hand, we have

$$\int_a^b \|\varphi(t)\|_0^2 \, dt = \sum_{n=0}^{\infty} n! \int_a^b |f_n(t)|_0^2 \, dt. \tag{13.25}$$

Moreover, note that $\partial_t \varphi(s) = \sum_{n=1}^{\infty} n \langle : \cdot^{\otimes(n-1)} :, f_n(s; t, \cdot) \rangle$ and similarly for $\partial_s \varphi(t)$. Hence

$$\left(\left(\partial_t \varphi(s), \partial_s \varphi(t)\right)\right)_0 = \sum_{n=1}^{\infty} (n-1)! n^2 \left(f_n(s; t, \cdot), f_n(t; s, \cdot)\right)_0$$

$$= \sum_{n=1}^{\infty} n n! \left(f_n(s; t, \cdot), f_n(t; s, \cdot)\right)_0.$$

Thus we get

$$\int_a^b \int_a^b \left(\left(\partial_t \varphi(s), \partial_s \varphi(t)\right)\right)_0 \, ds dt$$

$$= \sum_{n=1}^{\infty} n n! \int_a^b \int_a^b \left(f_n(s; t, \cdot), f_n(t; s, \cdot)\right)_0 \, ds dt. \tag{13.26}$$

Finally, by putting equations (13.24)–(13.26) together, we obtain equation (13.20) in the theorem. □

Corollary 13.17. *Let $\varphi \in L^2([a, b]; (L^2))$. Suppose $\left(\left(\partial_t \varphi(s), \partial_s \varphi(t)\right)\right)_0 = 0$ for almost all $(s, t) \in [a, b]^2$. Then*

$$\left\| \int_a^b \partial_t^* \varphi(t) \, dt \right\|_0^2 = \int_a^b \|\varphi(t)\|_0^2 \, dt. \tag{13.27}$$

Proof. By assumption, $\left(\left(\partial_t \varphi(s), \partial_s \varphi(t)\right)\right)_0 = 0$ for almost all $(s, t) \in [a, b]^2$. If in addition $\varphi \in L^2([a, b]; \mathcal{W}^{1/2})$, then the equality in (13.27) follows from Theorem 13.16. On the other hand, by examining the proof of Theorem 13.16, we see that the condition $\varphi \in L^2([a, b]; \mathcal{W}^{1/2})$ can be weakened to $\varphi \in L^2([a, b]; (L^2))$. □

Suppose $\varphi \in L^2([a, b]; (L^2))$ is nonanticipating. It follows from Lemma 13.11 that $\left(\left(\partial_t \varphi(s), \partial_s \varphi(t)\right)\right)_0 = 0$ for almost all $(s, t) \in [a, b]^2$. Hence by the above corollary

$$\left\| \int_a^b \partial_t^* \varphi(t) \, dt \right\|_0^2 = \int_a^b \|\varphi(t)\|_0^2 \, dt.$$

But by Theorem 13.12 we have $\int_a^b \partial_t^* \varphi(t) \, dt = \int_a^b \varphi(t) \, dB(t)$. Therefore

$$\left\| \int_a^b \varphi(t) \, dB(t) \right\|_0^2 = \int_a^b \|\varphi(t)\|_0^2 \, dt. \tag{13.28}$$

This is a well-known identity for the Itô integral. Thus we can regard equation (13.20) as a generalization of equation (13.28) to Hitsuda-Skorokhod integrals.

Example 13.18. Equation (13.27) may hold even if $\varphi(t)$ is not nonanticipating. Here is a simple example. Let $B(t)$ be the Brownian motion $B(t) = \langle \cdot, 1_{[0,t]} \rangle$, $0 \le t \le 1$. Define

$$\varphi(t) = \begin{cases} B(t) + B(1) - B(1-t), & \text{if } 0 \le t \le \frac{1}{2}; \\ B(1-t) + B(1) - B(t), & \text{if } \frac{1}{2} < t \le 1. \end{cases}$$

It is easy to check that $\partial_s \varphi(t)$, $0 < s, t < 1$, is given by

$$\partial_s \varphi(t) = \begin{cases} 1, & \text{if } (t-s)(1-s-t) > 0; \\ 0, & \text{if } (t-s)(1-s-t) < 0. \end{cases}$$

This implies that $\big(\partial_t \varphi(s) \big) \big(\partial_s \varphi(t) \big) = 0$ for almost all $(s,t) \in [0,1]^2$. Hence by Corollary 13.17 we have

$$\left\| \int_a^b \partial_t^* \varphi(t) \, dt \right\|_0^2 = \int_a^b \|\varphi(t)\|_0^2 \, dt.$$

Observe that equation (13.27) holds for $\varphi(t)$ even though $\varphi(t)$ is obviously not nonanticipating.

13.5 Generalizations of Itô's formula

Let $B(t)$ be a Brownian motion given by $B(t) = \langle \cdot, 1_{[0,t)} \rangle$. Consider a simple case of Itô's formula given by a C^2-function θ:

$$\theta(B(t)) = \theta(B(a)) + \int_a^t \theta'(B(s)) \, dB(s) + \frac{1}{2} \int_a^t \theta''(B(s)) \, ds,$$

where $0 \le a \le t$. Suppose $\theta(B(\cdot))$, $\theta'(B(\cdot))$, $\theta''(B(\cdot)) \in L^2([a,b]; (L^2))$. Then by Theorem 13.12 this equality can be written as

$$\theta(B(t)) = \theta(B(a)) + \int_a^t \partial_s^* \theta'(B(s)) \, ds + \frac{1}{2} \int_a^t \theta''(B(s)) \, ds, \tag{13.29}$$

where $\int_a^t \partial_s^* \theta'(B(s))\,ds$ is a Hitsuda-Skorokhod integral.

We give a proof of equation (13.29) from the white noise viewpoint. It suffices to show that both sides of this equation have the same S-transform. For the left hand side,

$$S\theta(B(t))(\xi) = S\theta(\langle \cdot, 1_{[0,t)}\rangle)(\xi)$$

$$= \int_{S'} \theta(\langle x + \xi, 1_{[0,t)}\rangle)\,d\mu(x)$$

$$= \int_{S'} \theta(\langle x, 1_{[0,t)}\rangle + \langle \xi, 1_{[0,t)}\rangle)\,d\mu(x).$$

But the random variable $\langle \cdot, 1_{[0,t)}\rangle$ is Gaussian with mean zero and variance t. Hence

$$S\theta(B(t))(\xi) = \int_{\mathbb{R}} \theta(u + \langle \xi, 1_{[0,t)}\rangle)g_t(u)\,du, \quad \xi \in \mathcal{S}_c, \tag{13.30}$$

where $g_t(u) = (2\pi t)^{-1/2}\exp[-\frac{u^2}{2t}]$. We can differentiate equation (13.30) to get

$$\frac{d}{dt}S\theta(B(t))(\xi) = \xi(t)\int_{\mathbb{R}} \theta'(u + \langle \xi, 1_{[0,t)}\rangle)g_t(u)\,du$$

$$+ \int_{\mathbb{R}} \theta(u + \langle \xi, 1_{[0,t)}\rangle)\frac{d}{dt}(g_t(u))\,du.$$

But $\frac{d}{dt}(g_t(u)) = \frac{1}{2}\frac{d^2}{du^2}(g_t(u))$. Hence by integration by parts we obtain

$$\frac{d}{dt}S\theta(B(t))(\xi) = \xi(t)\int_{\mathbb{R}} \theta'(u + \langle \xi, 1_{[0,t)}\rangle)g_t(u)\,du$$

$$+ \frac{1}{2}\int_{\mathbb{R}} \theta''(u + \langle \xi, 1_{[0,t)}\rangle)g_t(u)\,du. \tag{13.31}$$

On the other hand, by replacing θ in equation (13.30) with θ' we get

$$S\theta'(B(t))(\xi) = \int_{\mathbb{R}} \theta'(u + \langle \xi, 1_{[0,t)}\rangle)g_t(u)\,du.$$

Therefore

$$S(\partial_t^* \theta'(B(t)))(\xi) = \xi(t)\int_{\mathbb{R}} \theta'(u + \langle \xi, 1_{[0,t)}\rangle)g_t(u)\,du. \tag{13.32}$$

Similarly, for θ'', we have

$$S\theta''(B(t))(\xi) = \int_{\mathbb{R}} \theta''(u + \langle \xi, 1_{[0,t)}\rangle)g_t(u)\,du. \tag{13.33}$$

Equations (13.31)–(13.33) show that both sides of equation (13.29) have the same S-transform. This proves equation (13.29).

Now we consider two generalizations of the white noise version of Itô's formula in equation (13.29):

(a) $\theta(X(t), B(c))$ for a C^2-function θ and a Wiener integral $X(t), t \leq c$.

(b) $\theta(B(t))$ with a generalized function θ in $\mathcal{S}'(\mathbb{R})$.

A. $\theta(X(t), B(c))$ for a C^2-function θ and a Wiener integral $X(t), t \leq c$

Consider a simple example $\theta(x, y) = xy$ with $x = B(t)$ and $y = B(1), 0 \leq t \leq 1$. We can not apply Itô's formula for the function θ since $\int_0^1 B(1) \, dB(t)$ is not an Itô integral. In fact, to get an Itô type formula for functions such as $\theta(B(t), B(c)), t < c$, is a motivation in Hitsuda [64] for introducing the Hitsuda-Skorokhod integral.

Theorem 13.19. *Let $0 \leq a \leq c \leq b$ and let $\theta(x, y)$ be a C^2-function on $\mathbb{R}^2$ such that*

$$\theta(B(\cdot), B(c)), \quad \frac{\partial^2 \theta}{\partial x^2}(B(\cdot), B(c)), \quad \frac{\partial^2 \theta}{\partial x \partial y}(B(\cdot), B(c))$$

are all in $L^2([a, b]; (L^2))$. Then for any $a \leq t \leq b$, the integral

$$\int_a^t \partial_s^* \left(\frac{\partial \theta}{\partial x}(B(s), B(c)) \right) ds$$

is a Hitsuda-Skorokhod integral and the following equalities hold in (L^2):

(1) for $a \leq t \leq c$,

$$\theta(B(t), B(c)) = \theta(B(a), B(c)) + \int_a^t \partial_s^* \left(\frac{\partial \theta}{\partial x}(B(s), B(c)) \right) ds$$

$$+ \int_a^t \left(\frac{1}{2} \frac{\partial^2 \theta}{\partial x^2}(B(s), B(c)) + \frac{\partial^2 \theta}{\partial x \partial y}(B(s), B(c)) \right) ds, \qquad (13.34)$$

(2) for $c < t \leq b$,

$$\theta(B(t), B(c)) = \theta(B(a), B(c)) + \int_a^t \partial_s^* \left(\frac{\partial \theta}{\partial x}(B(s), B(c)) \right) ds$$

$$+ \frac{1}{2} \int_a^t \frac{\partial^2 \theta}{\partial x^2}(B(s), B(c)) \, ds + \int_a^c \frac{\partial^2 \theta}{\partial x \partial y}(B(s), B(c)) \, ds. \qquad (13.35)$$

Remark. Take $\theta(x, y) = xy$. Then for $a = 0, c = 1$, and $t = 1$ in the equation (13.34), we get

$$B(1)^2 = \int_0^1 \partial_s^* B(1) \, ds + \int_0^1 ds.$$

Hence $\int_0^1 \partial_s^* B(1) \, ds = B(1)^2 - 1$. This result agrees with Example 13.14.

Proof. We use the same argument as in the above proof for equation (13.29). First let $a \leq t \leq c$. The S-transform of $\theta(B(t), B(c))$ is given by

$$S\theta(B(t), B(c))(\xi)$$

$$= S\theta(\langle \cdot, 1_{[0,t)} \rangle, \langle \cdot, 1_{[0,c)} \rangle)(\xi)$$

$$= \int_{S'} \theta(\langle x, 1_{[0,t)} \rangle + \langle \xi, 1_{[0,t)} \rangle, \langle x, 1_{[0,c)} \rangle + \langle \xi, 1_{[0,c)} \rangle) \, d\mu(x)$$

$$= \int_{S'} \theta(\langle x, 1_{[0,t)} \rangle + \langle \xi, 1_{[0,t)} \rangle, \langle x, 1_{[0,t)} \rangle + \langle x, 1_{[t,c)} \rangle + \langle \xi, 1_{[0,c)} \rangle) \, d\mu(x).$$

Note that $\langle \cdot, 1_{[0,t)} \rangle$ and $\langle \cdot, 1_{[t,c)} \rangle$ are independent Gaussian random variables with mean zero and variances t and $c - t$, respectively. Therefore

$$S\theta(B(t), B(c))(\xi)$$

$$= \int_{\mathbb{R}^2} \theta(u + \langle \xi, 1_{[0,t)} \rangle, u + v + \langle \xi, 1_{[0,c)} \rangle) g_t(u) g_{c-t}(v) \, du \, dv, \quad (13.36)$$

where $g_s(w)$ is the Gaussian density function with mean zero and variance s, i.e.,

$$g_s(w) = (2\pi s)^{-1/2} \exp\left[-\frac{w^2}{2s} \right].$$

Differentiate both sides of equation (13.36) in the variable t to get

$$\frac{d}{dt} S\theta(B(t), B(c))(\xi)$$

$$= \xi(t) \int_{\mathbb{R}^2} \frac{\partial \theta}{\partial x}(u + \langle \xi, 1_{[0,t)} \rangle, u + v + \langle \xi, 1_{[0,c)} \rangle) g_t(u) g_{c-t}(v) \, du \, dv$$

$$+ \int_{\mathbb{R}^2} \theta(u + \langle \xi, 1_{[0,t)} \rangle, u + v + \langle \xi, 1_{[0,c)} \rangle) \left(\frac{d}{dt} g_t(u) \right) g_{c-t}(v) \, du \, dv$$

$$+ \int_{\mathbb{R}^2} \theta(u + \langle \xi, 1_{[0,t)} \rangle, u + v + \langle \xi, 1_{[0,c)} \rangle) g_t(u) \left(\frac{d}{dt} g_{c-t}(v) \right) \, du \, dv$$

$$\equiv I_1 + I_2 + I_3. \quad (13.37)$$

Now, by replacing θ in equation (13.36) with $\frac{\partial \theta}{\partial x}$, we get

$$I_1 = S\left[\partial_t^* \left(\frac{\partial \theta}{\partial x}(B(t), B(c)) \right) \right](\xi). \quad (13.38)$$

As for the term I_2, we use the fact that $\frac{d}{dt}(g_t(u)) = \frac{1}{2}\frac{d^2}{du^2}(g_t(u))$ and then apply the integration by parts formula to get

$$I_2 = \frac{1}{2}S\Big(\frac{\partial^2\theta}{\partial x^2}(B(t), B(c))\Big)(\xi) + S\Big(\frac{\partial^2\theta}{\partial x\partial y}(B(t), B(c))\Big)(\xi)$$
$$+ \frac{1}{2}S\Big(\frac{\partial^2\theta}{\partial y^2}(B(t), B(c))\Big)(\xi). \tag{13.39}$$

Similarly, for the term I_3, we have

$$I_3 = -\frac{1}{2}S\Big(\frac{\partial^2\theta}{\partial y^2}(B(t), B(c))\Big)(\xi). \tag{13.40}$$

It follows from equations (13.37)–(13.40) that both sides of equation (13.34) have the same S-transform. Hence $\int_a^t \partial_s^*\big(\frac{\partial\theta}{\partial x}(B(s), B(c))\big)ds$ is a Hitsuda-Skorokhod integral and equation (13.34) holds.

Next, consider the case $c < t \leq b$. Calculations similar to the above give

$$\frac{d}{dt}S\theta(B(t), B(c))(\xi)$$
$$= S\Big[\partial_t^*\Big(\frac{\partial\theta}{\partial x}(B(t), B(c))\Big)\Big](\xi) + \frac{1}{2}S\Big(\frac{\partial^2\theta}{\partial x^2}(B(t), B(c))\Big)(\xi).$$

This implies that there is a constant K such that

$$\theta\big(B(t), B(c)\big)$$
$$= K + \int_a^t \partial_s^*\Big(\frac{\partial\theta}{\partial x}(B(s), B(c))\Big)\,ds + \frac{1}{2}\int_a^t \frac{\partial^2\theta}{\partial x^2}(B(s), B(c))\,ds.$$

But $\theta\big(B(t), B(c)\big)$ is continuous in t. Hence K is given by

$$K = \theta\big(B(a), B(c)\big) + \int_a^c \frac{\partial^2\theta}{\partial x\partial y}(B(s), B(c))\,ds.$$

By putting this constant K into the last equation, we obtain equation (13.35) right away.   □

Example 13.20. For $\theta(x, y) = xy^2, a = 0$, and $c = 1$ in Theorem 13.19, we have

$$\int_0^t \partial_s^*(B(1)^2)\,ds = \begin{cases} B(t)B(1)^2 - 2tB(1), & \text{if } 0 \leq t \leq 1; \\ B(t)B(1)^2 - 2B(1), & \text{if } 1 < t. \end{cases}$$

In particular, we have $\int_0^1 \partial_t^*(B(1)^2)\,dt = B(1)^3 - 2B(1)$.

In general, suppose $c_1 \leq \cdots \leq c_{i-1} \leq a \leq b \leq c_{i+1} \leq \cdots \leq c_r$. Let $\theta(x_1, \ldots, x_r)$ be a function on $\mathbb{R}^r$ such that

$$\chi\big(B(c_1), \ldots, B(c_{i-1}), B(\cdot), B(t_{i+1}), \ldots, B(t_r)\big) \in L^2\big([a,b]; (L^2)\big),$$

where $\chi = \theta, \frac{\partial^2 \theta}{\partial x_i^2}$, and $\frac{\partial^2 \theta}{\partial x_i \partial x_j}$ for all $j > i$. Then for any $a \leq t \leq b$,

$$\theta\big(B(c_1), \ldots, B(c_{i-1}), B(t), B(c_{i+1}), \ldots, B(c_r)\big)$$

$$= \theta\big(B(c_1), \ldots, B(c_{i-1}), B(a), B(c_{i+1}), \ldots, B(c_r)\big)$$

$$+ \int_a^t \partial_s^* \Big(\frac{\partial \theta}{\partial x_i} \big(B(c_1), \ldots, B(c_{i-1}), B(s), B(c_{i+1}), \ldots, B(c_r)\big)\Big)\, ds$$

$$+ \int_a^t \Big(\frac{1}{2}\frac{\partial^2 \theta}{\partial x_i^2} + \sum_{j>i} \frac{\partial^2 \theta}{\partial x_i \partial x_j}\Big) \big(B(c_1), \ldots,$$

$$B(c_{i-1}), B(s), B(c_{i+1}), \ldots, B(c_r)\big)\, ds.$$

Observe that the anticipating part of

$$\theta\big(B(c_1), \ldots, B(c_{i-1}), B(t), B(c_{i+1}), \ldots, B(c_r)\big)$$

occurs at those $B(c_j)$ for $j > i$. This part is reflected in the Hitsuda-Skorokhod integral and the summation $\sum_{j>i}$.

Now we consider the generalization to $\theta\big(X(t), B(c)\big)$ for a Wiener integral $X(t)$. The formulas are similar to those given in Theorem 13.19.

Theorem 13.21. *Let $0 \leq a \leq c \leq b$. Let $X(t) = \int_a^t f(s)\, dB(s)$ be a Wiener integral with $f \in L^\infty([a,b])$ and let $\theta(x,y)$ be a C^2-function on $\mathbb{R}^2$ such that*

$$\theta\big(X(\cdot), B(c)\big),\ \frac{\partial^2 \theta}{\partial x^2}\big(X(\cdot), B(c)\big),\ \frac{\partial^2 \theta}{\partial x \partial y}\big(X(\cdot), B(c)\big)$$

are all in $L^2\big([a,b]; (L^2)\big)$. Then for any $a \leq t \leq b$, the integral

$$\int_a^t \partial_s^*\Big(f(s)\frac{\partial \theta}{\partial x}\big(X(s), B(c)\big)\Big)\, ds$$

is a Hitsuda-Skorokhod integral and the following equalities hold in (L^2):
(1) for $a \leq t \leq c$,

$$\theta\big(X(t), B(c)\big) = \theta\big(X(a), B(c)\big) + \int_a^t \partial_s^*\Big(f(s)\frac{\partial \theta}{\partial x}\big(X(s), B(c)\big)\Big)\, ds$$

$$+ \int_a^t \Big(\frac{1}{2}f(s)^2 \frac{\partial^2 \theta}{\partial x^2}\big(X(s), B(c)\big) + f(s)\frac{\partial^2 \theta}{\partial x \partial y}\big(X(s), B(c)\big)\Big)\, ds,$$

(2) for $c < t \le b$,

$$\theta\big(X(t), B(c)\big) = \theta\big(X(a), B(c)\big) + \int_a^t \partial_s^*\left(f(s)\frac{\partial\theta}{\partial x}(X(s), B(c))\right) ds$$

$$+ \frac{1}{2}\int_a^t f(s)^2\frac{\partial^2\theta}{\partial x^2}(X(s), B(c))\, ds + \int_a^c f(s)\frac{\partial^2\theta}{\partial x\partial y}(X(s), B(c))\, ds.$$

Proof. First suppose $f \in L^\infty([a, b])$ is a simple function so that

$$X(t) = \sum_{k=0}^{n-1} f(t_k)\big(B(t_{k+1}) - B(t_k)\big) = \sum_{k=0}^{n-1} f(t_k)\langle\cdot, 1_{[t_k, t_{k+1}]}\rangle,$$

where $t_0 = a$ and $t_n = t$. If $t \le c$, then

$$B(c) = \langle\cdot, 1_{[a,c)}\rangle = \sum_{k=0}^{n-1}\langle\cdot, 1_{[t_k, t_{k+1}]}\rangle + \langle\cdot, 1_{[t,c)}\rangle.$$

Hence the S-transform of $\theta\big(X(t), B(c)\big)$ is given by

$$S\theta\big(X(t), B(c)\big)(\xi)$$

$$= \int_{S'} \theta\Big(\sum_{k=0}^{n-1} f(t_k)\langle x, 1_{[t_k, t_{k+1}]}\rangle + \sum_{k=0}^{n-1} f(t_k)\langle\xi, 1_{[t_k, t_{k+1}]}\rangle,$$

$$\sum_{k=0}^{n-1}\langle x, 1_{[t_k, t_{k+1}]}\rangle + \langle x, 1_{[t,c)}\rangle + \langle\xi, 1_{[a,c)}\rangle\Big)\, d\mu(x)$$

$$= \int_{\mathbb{R}^{n+1}} \theta\Big(\sum_{k=0}^{n-1} f(t_k)u_{k+1} + \sum_{k=0}^{n-1} f(t_k)\langle\xi, 1_{[t_k, t_{k+1}]}\rangle,$$

$$\sum_{k=0}^{n-1} u_{k+1} + v + \langle\xi, 1_{[a,c)}\rangle\Big)$$

$$\times g_{t_1-t_0}(u_1)\cdots g_{t-t_{n-1}}(u_n)g_{c-t}(v)\, du_1\cdots du_n dv.$$

Observe that in the last equation the variable t appears in three places including the one with $t_n = t$. We can use the same argument as in the proof of Theorem 13.19 to check that

$$\frac{d}{dt}S\theta\big(X(t), B(c)\big)(\xi) = S\Big(\partial_t^* f(t_{n-1})\frac{\partial\theta}{\partial x}(X(t), B(c))\Big)(\xi)$$

$$+ \frac{1}{2}S\Big(f(t_{n-1})^2\frac{\partial^2\theta}{\partial x^2}(X(t), B(c))\Big)(\xi) + S\Big(f(t_{n-1})\frac{\partial^2\theta}{\partial x\partial y}(X(t), B(c))\Big)(\xi).$$

Note that $f(t_{n-1}) = f(t)$. By taking the inverse S-transform we obtain the first equality in the theorem for the case when f is a simple function.

Then we use approximation to show that this equality holds for any $f \in L^\infty([a, b])$. The second equality in the theorem can be proved by the similar argument. $\square$

B. $\theta(B(t))$ *with a generalized function* θ *in* $\mathcal{S}'(\mathbb{R})$

The generalization of equation (13.29) to a generalized functions θ in $\mathcal{S}'(\mathbb{R})$ was first obtained by Kubo in [88].

Recall that for $F \in \mathcal{S}'_c(\mathbb{R})$ and nonzero $f \in L^2(\mathbb{R})$, the generalized function $F(\langle \cdot, f \rangle)$ in $(\mathcal{S})^*$ is defined in equation (7.2). Its S-transform is given in Theorem 7.3. In particular, when $f = 1_{[0,t]}, t > 0$, we have $\langle \cdot, 1_{[0,t]} \rangle = B(t)$ and

$$(SF(B(t)))(\xi) = \frac{1}{\sqrt{2\pi t}} \int_{\mathbb{R}} F(y) \exp\left[-\frac{1}{2t}\left(y - \int_0^t \xi(u)\,du\right)^2\right] dy, \quad (13.41)$$

where the integral in the y variable is understood to be the bilinear pairing of $\mathcal{S}'_c(\mathbb{R})$ and $\mathcal{S}_c(\mathbb{R})$.

Lemma 13.22. *Let* $F \in \mathcal{S}'_c(\mathbb{R})$. *Then the function* $F(B(\cdot))$ *is Bochner integrable on any finite interval* $[a, b]$ *with* $a > 0$.

Proof. It follows from equation (13.41) that $F(B(\cdot))$ is weakly measurable. Moreover, from the proof of Theorem 7.3 we have the inequality

$$\|F(B(t))\|_{-p} = \|F(\langle \cdot, 1_{[0,t]} \rangle)\|_{-p} \le \frac{1}{\sqrt{2\pi}} K_p(t) \left(\sum_{n=0}^{\infty} (n+1)^{2p} 2^{-2np}\right)^{1/2},$$

where $K_p(t) = 2^{p+\frac{1}{2}} \sqrt[4]{\pi} |F(\sqrt{2t}\,(\cdot))e^{-\frac{1}{2}(\cdot)^2}|_{-p}$. It can be checked (similar to Theorem 11.18 on the scaling operator) that there exists some $p > 0$ such that $K_p(t) < \infty$ for all $t \in \mathbb{R}$ and $K_p(t)$ is a continuous function of t. Hence $\|F(B(\cdot))\|_{-p} \in L^1([a, b])$ for any finite interval $[a, b]$ with $a > 0$. This shows that $F(B(\cdot))$ is Bochner integrable on $[a, b]$. $\square$

Now, note that if θ is a generalized function in $\mathcal{S}'_c(\mathbb{R})$, then the distribution derivatives θ' and θ'' are also generalized functions in $\mathcal{S}'_c(\mathbb{R})$. Hence $\theta(B(t)), \theta'(B(t))$, and $\theta''(B(t))$ are all Bochner integrable on $[a, b]$ with $a > 0$.

Theorem 13.23. *Let* $\theta \in \mathcal{S}'_c(\mathbb{R})$. *Then for any* $0 < a \le t < \infty$,

$$\theta(B(t)) = \theta(B(a)) + \int_a^t \partial_s^* \theta'(B(s))\,ds + \frac{1}{2} \int_a^t \theta''(B(s))\,ds \quad \text{in } (\mathcal{S})^*,$$

where θ' and θ'' are distribution derivatives of θ, and the integrals are $(\mathcal{S})^*$-valued white noise integrals.

Remarks. (1) Note that if three terms of the equality are in (L^2), then the other term is also in (L^2) and the equality holds almost surely as random variables.

(2) Suppose $\theta \in S'_c(\mathbb{R})$ is a continuous function. Then by letting $a \to 0^+$ we get

$$\theta(B(t)) = \theta(B(0)) + \int_0^t \partial_s^* \theta'(B(s)) \, ds + \frac{1}{2} \int_0^t \theta''(B(s)) \, ds. \qquad (13.42)$$

Observe that if $\theta'(B(\cdot)) \in L^2([0,t]; (L^2))$, then by Theorem 13.12

$$\int_0^t \partial_s^* \theta'(B(s)) \, ds = \int_0^t \theta'(B(s)) \, dB(s).$$

Thus by the first remark the equality in (13.42) holds almost surely as random variables.

Proof. It follows from Lemma 13.22 that the functions $\theta'(B(\cdot))$ and $\theta''(B(\cdot))$ are Bochner integrable on $[a,t]$. Hence by Fact 2 stated in the beginning of §13.4, the function $\partial_{(\cdot)}^* \theta'(B(\cdot))$ is also Bochner integrable. By equation (13.41) with $F = \theta$ we have

$$(S\theta(B(t)))(\xi) = \int_{\mathbb{R}} \theta(y) g_t\left(y - \int_0^t \xi(u) \, du\right) dy, \qquad (13.43)$$

where $g_t(u) = (2\pi t)^{-1/2} \exp[-\frac{u^2}{2t}]$ and the integral is understood to be the bilinear pairing of $S'_c(\mathbb{R})$ and $S_c(\mathbb{R})$. We can differentiate equation (13.43) to get

$$\frac{d}{dt}(S\theta(B(t)))(\xi) = -\xi(t) \int_{\mathbb{R}} \theta(y) g_t'\left(y - \int_0^t \xi(u) \, du\right) dy$$

$$+ \int_{\mathbb{R}} \theta(y) \left(\frac{d}{dt} g_t\right)\left(y - \int_0^t \xi(u) \, du\right) dy. \qquad (13.44)$$

Now, note that $\langle \theta', \xi \rangle = -\langle \theta, \xi' \rangle$ for any $\xi \in S_c$. Hence we have

$$\xi(t) \int_{\mathbb{R}} \theta(y) g_t'\left(y - \int_0^t \xi(u) \, du\right) dy = -\xi(t) \int_{\mathbb{R}} \theta'(y) g_t\left(y - \int_0^t \xi(u) \, du\right) dy.$$

Thus by equation (13.41) with $F = \theta'$

$$\xi(t) \int_{\mathbb{R}} \theta(y) g_t'\left(y - \int_0^t \xi(u) \, du\right) dy = -\xi(t) (S\theta'(B(t)))(\xi). \qquad (13.45)$$

Similarly, by using the fact that $\frac{d}{dt}(g_t(u)) = \frac{1}{2}\frac{d^2}{du^2}(g_t(u))$ and equation (13.41) with $F = \theta''$, we get

$$\int_{\mathbb{R}} \theta(y) \left(\frac{d}{dt} g_t\right)\left(y - \int_0^t \xi(u) \, du\right) dy = \frac{1}{2}(S\theta''(B(t)))(\xi). \qquad (13.46)$$

Obviously equations (13.44)–(13.46) yield the equality in the theorem. □

We point out that the same argument as in the above proof can be used to show that the formulas in Theorem 13.19 remain valid in the space $(\mathcal{S})^*$ for a generalized function $\theta \in \mathcal{S}'_c(\mathbb{R}^2)$.

Example 13.24. (*Tanaka formula*) Take the function $\theta(x) = |x|$. The distribution derivatives θ' and θ'' of θ are given by

$$\theta'(x) = \operatorname{sgn}(x), \quad \theta''(x) = 2\delta_0(x),$$

where $\operatorname{sgn}(x)$ is the signum function, i.e., $\operatorname{sgn}(0) = 0$ and $\operatorname{sgn}(x) = x/|x|$ if $x \neq 0$. Then equation (13.42) for the function $\theta(x) = |x|$ gives the following equality

$$|B(t)| = \int_0^t \partial_s^* \operatorname{sgn}(B(s))\, ds + \int_0^t \delta_0(B(s))\, ds.$$

But by Theorem 13.12 the first integral is actually an Itô integral and we have

$$|B(t)| = \int_0^t \operatorname{sgn}(B(s))\, dB(s) + \int_0^t \delta_0(B(s))\, ds. \qquad (13.47)$$

This equality is related to the local time of Brownian motion as follows. The local time $L(t, x)$ of $B(t)$ at x was introduced in Lévy [126] (see also Ikeda and Watanabe [67]) as the limit:

$$L(t, x) = \lim_{\epsilon \to 0^+} \frac{1}{2\epsilon} m\{0 \leq s < t; |B(s) - x| < \epsilon\},$$

where m denotes the Lebesgue measure on $\mathbb{R}$. The local time $L(t, x)$ is jointly measurable and is uniquely determined by the property that for all $t \geq 0$ and $C \in \mathcal{B}(\mathbb{R})$,

$$\int_C L(t, x)\, dx = \int_0^t 1_C(B(s))\, ds.$$

Upon applying Fubini's theorem informally, we get

$$\int_C \int_0^t \delta_x(B(s))\, ds\, dx = \int_0^t \int_C \delta_x(B(s))\, dx\, ds$$

$$= \int_0^t 1_C(B(s))\, ds.$$

This shows that the local time $L(t, x)$ is given by

$$L(t, x) = \int_0^t \delta_x(B(s))\, ds.$$

Observe that the white noise integral $\int_0^t \delta_x(B(s))\,ds$ is actually a random variable in (L^2). Thus equation (13.47) can be rewritten as

$$|B(t)| = \int_0^t \operatorname{sgn}(B(s))\,dB(s) + L(t,0).$$

This equality is known as Tanaka's formula (see Chung and Williams [24] and McKean [131, p.68]).

13.6 One-sided white noise differentiation

The Hitsuda-Skorokhod integral $\int_a^b \partial_t^* \varphi(t)\,dt$ provides an extension of the Itô integral for integrands that may be anticipating. On the other hand, we can use the symbolic expression $dB(t) = \dot{B}(t)\,dt$ to rewrite an Itô integral $\int_a^b \varphi(t)\,dB(t)$ as

$$\int_a^b \varphi(t)\,dB(t) = \int_a^b \dot{B}(t)\varphi(t)\,dt.$$

Recall that $\dot{B}(t) = \partial_t + \partial_t^*$ by equation (9.9). Thus the integral

$$\int_a^b (\partial_t + \partial_t^*)\varphi(t)\,dt$$

seems to be another extension of the Itô integral. Obviously, in order for this integral to make sense, we need to define the integral

$$\int_a^b \partial_t \varphi(t)\,dt.$$

However, $\partial_t \varphi(t)$ is not well-defined for a stochastic process $\varphi(t)$. For example, let $B(t)$ be a Brownian motion given by $B(t) = \langle \cdot, 1_{[0,t)} \rangle$. Then we have

$$\partial_t B(t) = 1_{[0,t)}(t) = 0.$$

On the other hand, note that $B(t)$ defined by $B(t) = \langle \cdot, 1_{[0,t]} \rangle$ is also a Brownian motion. But for this $B(t)$ we have

$$\partial_t B(t) = 1_{[0,t]}(t) = 1.$$

This shows that $\partial_t B(t)$ is not well-defined. Clearly, we need to consider one-sided limits when white noise differentiation is applied to stochastic processes.

Recall from the proof of Theorem 13.16 that if $\varphi \in L^2([a, b]; \mathcal{W}^{1/2})$ is represented by $\varphi(t) = \sum_{n=0}^{\infty} \langle : \cdot^{\otimes n} :, f_n(t) \rangle$, then $\partial_s \varphi(t)$ is defined for almost all $(s, t) \in \mathbb{R} \times [a, b]$ by

$$\partial_s \varphi(t) = \sum_{n=1}^{\infty} n \langle : \cdot^{\otimes(n-1)} :, f_n(t; s) \rangle. \qquad (13.48)$$

Observe that the restriction of $\partial_s \varphi(t)$ to the diagonal $s = t$ has no meaning. But we can use equation (13.48) as a guideline to define one-sided white noise derivatives.

Definition 13.25. Let $\varphi \in L^2([a, b]; (L^2))$ be represented by

$$\varphi(t) = \sum_{n=0}^{\infty} \langle : \cdot^{\otimes n} :, f_n(t) \rangle.$$

The *right-hand white noise derivative* of the stochastic process $\varphi(t)$ is said to exist if the following conditions are satisfied:

(a) There is a null set $C \subset [a, b]$ such that $f_n(t; t^+)$ exists and belongs to $L^2(\mathbb{R}^{n-1})$ for all $n \geq 1$ and $t \in [a, b] \setminus C$.

(b) $\sum_{n=1}^{\infty} n! n \int_a^b |f_n(t; t^+)|_0^2 \, dt < \infty$. (As before, $|\cdot|_0$ denotes $L^2(\mathbb{R}^k)$-norm for any k.)

Notations: Let $\mathbb{D}^+$ denote the set of all stochastic processes $\varphi(t)$ in $L^2([a, b]; (L^2))$ such that the right-hand white noise derivative of $\varphi(t)$ exists. For $\varphi \in \mathbb{D}^+$, we define the right-hand white noise derivative $\partial_{t+}\varphi(t)$ of φ to be the stochastic process

$$\partial_{t+}\varphi(t) = \sum_{n=1}^{\infty} n \langle : \cdot^{\otimes(n-1)} :, f_n(t; t^+) \rangle.$$

Note that if $\varphi \in \mathbb{D}^+$, then conditions (a) and (b) imply that $\partial_{t+}\varphi(t) \in (L^2)$ for almost all t and the function $\partial_{(\cdot+)}\varphi(\cdot)$ is Bochner integrable from $[a, b]$ into (L^2). Hence $\partial_{(\cdot+)}\varphi(\cdot) \in L^2([a, b]; (L^2))$ and we have

$$\int_a^b \|\partial_{t+}\varphi(t)\|_0^2 \, dt = \sum_{n=1}^{\infty} n! n \int_a^b |f_n(t; t^+)|_0^2 \, dt.$$

By replacing the right-hand limit in the above discussion with the left-hand limit, we can define the *left-hand white noise derivative* $\partial_{t-}\varphi(t)$ of $\varphi(t)$ to be

$$\partial_{t-}\varphi(t) = \sum_{n=1}^{\infty} n \langle : \cdot^{\otimes(n-1)} :, f_n(t; t^-) \rangle.$$

Let $\mathbb{D}^-$ denote the set of all $\varphi(t)$ such that $\partial_{t-}\varphi(t)$ exists. For $\varphi \in \mathbb{D}^-$, we have

$$\int_a^b \|\partial_{t-}\varphi(t)\|_0^2 \, dt = \sum_{n=1}^\infty n!n \int_a^b |f_n(t;t^-)|_0^2 \, dt.$$

Example 13.26. Let $B(t)$ be the Brownian motion $B(t) = \langle\cdot, 1_{[0,t)}\rangle$. It is easy to check that

$$\partial_{t+}B(t) = 0, \quad \partial_{t-}B(t) = 1, \qquad t > 0,$$

$$\partial_{t\pm}B(1) = 1, \qquad\qquad 0 < t < 1,$$

$$\partial_{t\pm}B(1-t) = 1, \qquad 0 < t < 1/2,$$

$$\partial_{t\pm}B(1-t) = 0, \qquad 1/2 < t < 1.$$

These equalities yield the following integrals:

$$\int_0^1 \partial_{t+}B(t) \, dt = 0, \quad \int_0^1 \partial_{t-}B(t) \, dt = 1,$$

$$\int_0^1 \partial_{t+}B(1) \, dt = \int_0^1 \partial_{t-}B(1) \, dt = 1,$$

$$\int_0^1 \partial_{t+}B(1-t) \, dt = \int_0^1 \partial_{t-}B(1-t) \, dt = \frac{1}{2}.$$

Proposition 13.27. *Suppose $\varphi \in L^2([a,b]; (L^2))$ is nonanticipating. Then $\varphi \in \mathbb{D}^+$ and $\partial_{t+}\varphi(t) = 0$ for almost all $t \in [a,b]$.*

Remark. Note that the conclusion implies that $\int_a^b \partial_{t+}\varphi(t) \, dt = 0$.

Proof. The assertion is a simple consequence of Lemma 13.11.   □

For $\varphi(t)$ which may not be nonanticipating, the existence of $\partial_{t+}\varphi(t)$ and $\partial_{t-}\varphi(t)$ can be assured by the trace theorem of Sobolev spaces. Let Ω_n^+ denote the region

$$\Omega_n^+ = \{(t; u_1, \ldots, u_n) \in (a,b) \times \mathbb{R}^n; \, u_1 > t\}.$$

For a real number $\gamma > 0$, let $W^\gamma(\Omega_n^+)$ denote the Sobolev space of order γ (see the book by Adams [3] page 45 if γ is an integer and page 205 if γ is not an integer). The $W^\gamma(\Omega_n^+)$-norm is denoted by $\|\cdot\|_{W^\gamma(\Omega_n^+)}$.

Let $g \in W^\gamma(\Omega_n^+)$. The trace of g is defined to be its restriction to the boundary of Ω_n^+. The trace theorem (see Adams [3, pp.216–217]) says that if $g \in W^\gamma(\Omega_n^+)$, then its trace $g|_{\text{bdry }\Omega_n^+}$ belongs to $W^{\gamma-\frac{1}{2}}(\text{bdry }\Omega_n^+)$ and

$$\|g|_{\text{bdry }\Omega_n^+}\|_{W^{\gamma-\frac{1}{2}}(\text{bdry }\Omega_n^+)} \leq K\|g\|_{W^\gamma(\Omega_n^+)},$$

where K is a constant independent of g and n. In particular, if $\gamma > \frac{1}{2}$ and we take the boundary of Ω_n^+ along $t = u_1$, then we get

$$\int_a^b |g(t; t^+, \cdot)|_0^2 \, dt \le C\|g\|_{W^\gamma(\Omega_n^+)}^2,$$

where $C = K^2$. Thus, in view of condition (b) in Definition 13.25, we have the next theorem for the existence of the stochastic process $\partial_{t+}\varphi(t)$.

Theorem 13.28. *Let* $\varphi(t) = \sum_{n=0}^\infty \langle : \cdot^{\otimes n} :, f_n(t)\rangle \in L^2([a,b]; (L^2))$ *and assume there exists* $\gamma > \frac{1}{2}$ *such that*

(1) $f_n \in W^\gamma(\Omega_n^+)$ *for all* $n \ge 1$.

(2) $\sum_{n=1}^\infty n! n \|f_n\|_{W^\gamma(\Omega_n^+)} < \infty$.

Then $\varphi \in \mathbb{D}^+$, *i.e., the right-hand white noise derivative* $\partial_{t+}\varphi(t)$ *exists. Moreover,*

$$\int_a^b \|\partial_{t+}\varphi(t)\|_0^2 \, dt \le C \sum_{n=1}^\infty n! n \|f_n\|_{W^\gamma(\Omega_n^+)},$$

where C *is a constant independent of* φ.

Remark. Similarly, let Ω_n^- denote the region

$$\Omega_n^- = \{(t; u_1, \ldots, u_n) \in (a,b) \times \mathbb{R}^n; \, u_1 < t\}.$$

Then we get a corresponding theorem for the left-hand white noise derivative when the region Ω_n^+ is replaced by Ω_n^-.

Recall from Theorem 13.16 that if $\varphi \in L^2([a,b]; \mathcal{W}^{1/2})$, then $\int_a^b \partial_t^* \varphi(t) \, dt$ is a random variable in (L^2). If $\varphi \in \mathbb{D}^+$, then $\int_a^b \partial_{t+}\varphi(t) \, dt$ is also a random variable in (L^2). Similarly, $\int_a^b \partial_{t-}\varphi(t) \, dt \in (L^2)$ for all $\varphi \in \mathbb{D}^-$.

Question: How are these integrals related to the Itô integral?

We showed in the beginning of this section that $\dot{B}(t)\varphi(t) = (\partial_t + \partial_t^*)\varphi(t)$ is not well-defined. Clearly, we need to consider multiplication by $\dot{B}(t^+)$ and by $\dot{B}(t^-)$ defined by

$$\dot{B}(t^+)\varphi(t) = (\partial_{t+} + \partial_t^*)\varphi(t),$$

$$\dot{B}(t^-)\varphi(t) = (\partial_{t-} + \partial_t^*)\varphi(t).$$

They induce two different integrals. First we define

$$\int_a^b \varphi(t) \, dB(t^+) \equiv \int_a^b \partial_{t+}\varphi(t) \, dt + \int_a^b \partial_t^*\varphi(t) \, dt, \qquad (13.49)$$

provided that both integrals are random variables in (L^2). Note that this integral is defined when $\varphi \in \mathbb{D}^+ \cap L^2([a,b]; \mathcal{W}^{1/2})$ (by Theorem 13.16). It is also defined when $\varphi \in L^2([a,b]; (L^2))$ is nonanticipating (by Theorem 13.12 and Proposition 13.27) and in this case it coincides with the Itô integral of $\varphi(t)$. Thus the integral $\int_a^b \varphi(t)\, dB(t^+)$ is an extension of the Itô integral. Moreover, by Examples 13.14 and 13.26,

$$\int_0^1 B(1)\, dB(t^+) = B(1)^2.$$

This equality agrees with $\int_0^1 B(t)\, dB(t) = B(1)^2$ as given in Itô [71].

Similarly, we define

$$\int_a^b \varphi(t)\, dB(t^-) \equiv \int_a^b \partial_{t-}\varphi(t)\, dt + \int_a^b \partial_t^* \varphi(t)\, dt, \qquad (13.50)$$

provided that both integrals are random variables in (L^2). By Theorem 13.16 this integral is defined for $\varphi \in \mathbb{D}^- \cap L^2([a,b]; \mathcal{W}^{1/2})$.

The integrals in Equations (13.49) and (13.50) are called the *forward* and *backward integrals* of φ, respectively. The forward integral corresponds to choosing the right endpoint t_k to evaluate φ in equation (13.9), while for the backward integral the function φ is evaluated at the left endpoint. Thus it is natural to define the *Stratonovich integral* of φ by

$$\int_a^b \varphi(t) \circ dB(t) \equiv \frac{1}{2}\left(\int_a^b \varphi(t)\, dB(t^+) + \int_a^b \varphi(t)\, dB(t^-) \right).$$

Example 13.29. For the integrand $\varphi(t) = B(t)$, we have

$$\int_0^t B(t)\, dB(t^+) = \frac{1}{2}(B(1)^2 - 1),$$

$$\int_0^t B(t)\, dB(t^-) = \frac{1}{2}(B(1)^2 + 1),$$

$$\int_0^t B(t) \circ dB(t) = B(1)^2.$$

A generalization of Itô's formula for forward integrals has been obtained in Asch and Potthoff [8]. Let $X(t)$ be a forward integral given by

$$X(t) = X(a) + \int_a^t \varphi(s)\, dB(t^+).$$

Suppose θ is a C^2-function satisfying certain conditions (see the book by Hida et al. [58] for details). Then

$$\theta(X(t)) = \theta(X(a)) + \int_a^t \theta'(X(s))\varphi(s)\,dB(s^+) + \frac{1}{2}\int_a^t \theta''(X(s))\varphi(s)^2\,ds.$$

Hence Itô's formula has the same expression for forward integrals.

13.7 Stochastic integral equations

Let $B(t)$ be the Brownian motion $B(t) = \langle\cdot, 1_{[0,t)}\rangle$ and let $0 \le a < b < \infty$. Consider the stochastic integral equation of Itô type:

$$X(t) = X(a) + \int_a^t f(s, X(s))\,dB(s) + \int_a^t g(s, X(s))\,ds, \qquad (13.51)$$

where $a \le t \le b$, and $X(a)$ is measurable with respect to $\sigma\{B(s); s \le a\}$. Suppose f and g satisfy the following conditions:

(a) (Lipschitz condition) There exists a constant C_1 such that for all $t \in [a, b]$, $x, y \in \mathbb{R}$,

$$|f(t, x) - f(t, y)| + |g(t, x) - g(t, y)| \le C_1|x - y|.$$

(b) (Growth condition) There exists a constant C_2 such that for all $t \in [a, b]$, $x \in \mathbb{R}$,
$$|f(t, x)|^2 + |g(t, x)|^2 \le C_2(1 + x^2).$$

Then equation (13.51) has a unique continuous solution. The solution $X(t)$ is a Markov process on $[a, b]$ (e.g., see Arnold [7]). A standard way to show the existence of a solution $X(t)$ is to apply the iteration method, i.e., define $X_0(t) \equiv X(a)$ and for $n \ge 1$,

$$X_n(t) = X(a) + \int_a^t f(s, X_{n-1}(s))\,dB(s) + \int_a^t g(s, X_{n-1}(s))\,ds. \quad (13.52)$$

Then with probability 1, the sequence $X_n(t)$ converges to the solution $X(t)$ uniformly on the interval $[a, b]$.

Now, suppose we do not assume that the initial condition $X(a)$ is measurable with respect to $\sigma\{B(s); s \le a\}$. Then the process $f(s, X(a))$ may not be nonanticipating and so the integral $\int_a^t f(s, X(a))\,dB(s)$ can not be defined as an Itô integral. Hence $X_1(t)$ in equation (13.52) is not defined. Moreover, observe that we can not interpret equation (13.51) in the Itô sense.

To overcome the above difficulties when the nonanticipating property is not assumed, we may replace equation (13.51) with the following equation

$$X(t) = X(a) + \int_a^t \partial_s^* f(s, X(s)) \, ds + \int_a^t g(s, X(s)) \, ds, \qquad (13.53)$$

where $\int_a^t \partial_s^* f(s, X(s)) \, ds$ is a Hitsuda-Skorokhod integral. Note that equation (13.53) is supposed to hold as random variables in (L^2).

From the white noise viewpoint, it is natural to use the S-transform to study equation (13.53). By taking the S-transform on both sides, we get the ordinary integral equation for each fixed $\xi \in S_c$:

$$SX(t)(\xi) = SX(a)(\xi) + \int_a^t \xi(s) Sf(s, X(s))(\xi) \, ds$$

$$+ \int_a^t Sg(s, X(s))(\xi) \, ds. \qquad (13.54)$$

Suppose we can solve this integral equation for each ξ. Then we simply take the inverse S-transform to get the solution of equation (13.53). However, there are several difficulties. For instance, equation (13.54) is almost impossible to solve. This is due to the fact that $Sf(s, X(s))$ can not be expressed as a simple function of s and $SX(s)$ (in the next section we will find some way to overcome this difficulty). But assuming we can solve equation (13.54) for each $\xi \in S_c$, we still have to make sure that the inverse S-transform is in (L^2).

First we use the S-transform to solve some simple examples of Hitsuda-Skorokhod type equation (13.53). We will see a distinct feature of the solution, i.e., the solution $X(t)$ of an anticipating equation (13.53) may not be continuous in t.

Example 13.30. Consider the following equation given in Buckdahn [17]:

$$X(t) = \text{sgn}\,(B(1)) + \int_0^t \partial_s^* X(s) \, ds, \quad 0 \le t \le 1. \qquad (13.55)$$

Let $F(t)$ and G be the S-transforms of $X(t)$ and $\text{sgn}\,(B(1))$, respectively. Then by taking the S-transform of the above equation we get

$$F(t)(\xi) = G(\xi) + \int_0^t \xi(s) F(s)(\xi) \, ds.$$

Hence for each $\xi \in S_c$, the function $F(t)(\xi)$ satisfies the differential equation:

$$F'(t) = \xi(t) F(t), \ F(0) = G(\xi), \quad 0 \le t \le 1.$$

Obviously the solution is given by

$$F(t)(\xi) = G(\xi) \exp\left[\int_0^t \xi(s)\, ds\right]$$

$$= G(\xi) \exp\left[\langle 1_{[0,t)}, \xi\rangle\right]. \tag{13.56}$$

Recall from Chapter 5 that for $f \in L^2(\mathbb{R})$, we have $:e^{\langle \cdot, f\rangle}: \; = e^{\langle \cdot, f\rangle - \frac{1}{2}|f|_0^2}$ and

$$S(:e^{\langle \cdot, f\rangle}:)(\xi) = e^{\langle f, \xi\rangle}, \quad \xi \in S_c.$$

Therefore

$$F(t)(\xi) = G(\xi) S(:e^{\langle \cdot, 1_{[0,t)}\rangle}:)(\xi).$$

Note that $S(\Phi \diamond \Psi) = (S\Phi)(S\Psi)$ (see the Wick product in Definition 8.11). Thus by taking the inverse S-transform we obtain

$$X(t) = (\operatorname{sgn}(B(1))) \diamond (:e^{\langle \cdot, 1_{[0,t)}\rangle}:).$$

But $:e^{\langle \cdot, 1_{[0,t)}\rangle}: \; = e^{B(t) - \frac{1}{2}t}$. Hence

$$X(t) = (\operatorname{sgn}(B(1))) \diamond (e^{B(t) - \frac{1}{2}t}).$$

It is not clear from this Wick product that $X(t) \in (L^2)$ for each t. In order to see this fact, we need to express the solution $X(t)$ in another form.

Let $\varphi(t) = \operatorname{sgn}(B(1) - t)\, e^{B(t) - \frac{1}{2}t}$. Then $\varphi(t) \in (L^2)$ can be rewritten as

$$\varphi(t) = \operatorname{sgn}(\langle \cdot, 1_{[0,1)}\rangle - t)\, e^{\langle \cdot, 1_{[0,t)}\rangle - \frac{1}{2}t}.$$

From §5.2 the S-transform of $\varphi(t)$ is given by

$$S\varphi(t)(\xi) = \int_{S'} \varphi(t)(x + \xi)\, d\mu(x)$$

$$= \int_{S'} \operatorname{sgn}(\langle x + \xi, 1_{[0,1)}\rangle - t)\, e^{\langle x + \xi, 1_{[0,t)}\rangle - \frac{1}{2}t}\, d\mu(x).$$

We can make a change of variables $x = y + 1_{[0,t)}$ and apply the translation formula for the Gaussian measure μ on S' in §2.1 to get

$$S\varphi(t)(\xi) = \int_{S'} \operatorname{sgn}(\langle y + \xi, 1_{[0,1)}\rangle) e^{\langle 1_{[0,t)}, \xi\rangle}\, d\mu(y)$$

$$= e^{\langle 1_{[0,t)}, \xi\rangle} \int_{S'} \operatorname{sgn}(\langle y + \xi, 1_{[0,1)}\rangle)\, d\mu(y)$$

$$= e^{\langle 1_{[0,t)}, \xi\rangle} S(\operatorname{sgn}(B(1)))(\xi). \tag{13.57}$$

Thus in view of equation (13.56), $X(t)$ and $\varphi(t)$ have the same S-transform. Hence $X(t) = \varphi(t)$ and so the solution of equation (13.55) is given by

$$X(t) = \operatorname{sgn}\left(B(1) - t\right) e^{B(t) - \frac{1}{2}t}.$$

Observe that $X(t)$ is not continuous in t if $0 < B(1) < 1$. This solution is derived by using a different method in Buckdahn [17].

Example 13.31. The discontinuity of a solution of equation (13.53) can be a consequence of the initial condition as seen in the previous example. It can also be a consequence of the function g. Consider the equation:

$$X(t) = 1 + \int_0^t \partial_s^* X(s)\, ds + \int_0^t \operatorname{sgn}\left(B(1) - s\right) e^{B(s) - \frac{1}{2}s}\, ds, \qquad (13.58).$$

where $0 \le t \le 1$. To solve this equation, let $F(t)$ and G be the S-transforms of $X(t)$ and $\operatorname{sgn}\left(B(1)\right)$, respectively. Then by using equation (13.57), we can check easily that the S-transform of $X(t)$ satisfies the integral equation

$$F(t)(\xi) = 1 + \int_0^t \xi(s)F(s)(\xi)\, ds + \int_0^t G(\xi)\, e^{\int_0^s \xi(\tau)\, d\tau}\, ds.$$

Hence for each $\xi \in \mathcal{S}_c$, $F(t)$ satisfies the differential equation

$$F'(t) = \xi(t)F(t) + G(\xi)\, e^{\int_0^t \xi(s)\, ds}, \quad F(0) = 1, \quad 0 \le t \le 1.$$

The solution of this differential equation is given by

$$F(t) = \exp\left[\int_0^t \xi(s)\, ds\right] + t\, G(\xi)\, \exp\left[\int_0^t \xi(s)\, ds\right].$$

Note that $S e^{B(t) - \frac{1}{2}t}(\xi) = \exp\left[\int_0^t \xi(s)\, ds\right]$ and from the previous example we have

$$S\left(\operatorname{sgn}\left(B(1) - t\right) e^{B(t) - \frac{1}{2}t}\right)(\xi) = G(\xi)\, \exp\left[\int_0^t \xi(s)\, ds\right].$$

Thus the solution of equation (13.58) is given by

$$X(t) = e^{B(t) - \frac{1}{2}t} + t\operatorname{sgn}\left(B(1) - t\right) e^{B(t) - \frac{1}{2}t}$$

$$= e^{B(t) - \frac{1}{2}t}\left(1 + t\operatorname{sgn}\left(B(1) - t\right)\right).$$

As in the previous example, $X(t)$ is not continuous in t if $0 < B(1) < 1$. But in this example the discontinuity comes from the function g in equation (13.53).

Now, let us consider the general linear equation of Hitsuda-Skorokhod type

$$X(t) = \varphi + \int_a^t \partial_s^* \big(f(s)X(s) \big)\, ds + \int_a^t \big(g(s)X(s) + \psi(s) \big)\, ds,$$

where f and g are deterministic functions, φ is a random variable, and $\psi(s)$ is a stochastic process. We will use the S-transform to solve this equation.

Lemma 13.32. *Suppose* $f \in L^2([a,b])$ *and* $\varphi \in L^p(\mathcal{S}')$ *for some* $p > 2$. *Then*

$$\varphi \diamond \exp\left[\int_a^t f(s)\, dB(s) - \frac{1}{2}\int_a^t f(s)^2\, ds \right]$$

$$= (T_{-1_{[a,t)}f}\, \varphi) \exp\left[\int_a^t f(s)\, dB(s) - \frac{1}{2}\int_a^t f(s)^2\, ds \right],$$

where $\diamond$ *denotes the Wick multiplication in §8.4 and* T_h *is the translation operator by* h *in §10.5, i.e.,* $T_h\varphi(x) = \varphi(x + h)$.

Proof. For simplicity, let

$$Z(t) \equiv \exp\left[\int_a^t f(s)\, dB(s) - \frac{1}{2}\int_a^t f(s)^2\, ds \right].$$

We can rewrite $Z(t)$ as

$$Z(t) = \exp\left[\langle \cdot, 1_{[a,t)}f\rangle - \tfrac{1}{2}|1_{[a,t)}f|_0^2 \right] = \,:e^{\langle \cdot, 1_{[a,t)}f\rangle}:\,.$$

Hence by Theorem 5.13

$$SZ(t)(\xi) = e^{\langle 1_{[a,t)}f,\xi\rangle}, \quad \xi \in \mathcal{S}_c. \tag{13.59}$$

We can use the assumption to check that $(T_{-1_{[a,t)}f}\,\varphi)Z(t) \in (L^2)$. Thus its S-transform is given by

$$S\big((T_{-1_{[a,t)}f}\,\varphi)Z(t)\big)(\xi)$$

$$= \int_{\mathcal{S}'} \varphi(x + \xi - 1_{[a,t)}f)\exp\left[\langle x + \xi, 1_{[a,t)}f\rangle - \tfrac{1}{2}|1_{[a,t)}f|_0^2\right] d\mu(x).$$

Make a change of variables $y = x - 1_{[a,t)}f$ and then use the translation formula for μ in §2.1 to get

$$S\big((T_{-1_{[a,t)}f}\,\varphi)Z(t)\big)(\xi) = e^{\langle 1_{[a,t)}f,\xi\rangle}\int_{\mathcal{S}'} \varphi(y + \xi)\, d\mu(y)$$

$$= (S\varphi)(\xi)\, e^{\langle 1_{[a,t)}f,\xi\rangle}. \tag{13.60}$$

From equations (13.59) and (13.60) we get $(T_{-1_{[a,t)}f} \varphi) Z(t) = \varphi \diamond Z(t)$, which is the conclusion of the lemma. $\square$

Theorem 13.33. *Suppose that $f(t)$ and $g(t)$ are deterministic functions, φ is a random variable, and $\psi(t)$ is a stochastic process satisfying the following conditions:*

(1) $f, g \in L^2([a, b])$.
(2) $\varphi \in L^p(\mathcal{S}')$ *for some $p > 2$.*
(3) $\psi \in L^q([a, b] \times \mathcal{S}')$ *for some $q > 2$.*

Then the stochastic integral equation

$$X(t) = \varphi + \int_a^t \partial_s^*\big(f(s)X(s)\big)\, ds + \int_a^t \big(g(s)X(s) + \psi(s)\big)\, ds \qquad (13.61)$$

has a unique solution $X(t)$ in $L^2([a, b]; (L^2))$ given by

$$X(t) = (T_{-1_{[a,t)}f}\, \varphi) \exp\left[\int_a^t f(s)\, dB(s) + \int_a^t \big(g(s) - \tfrac{1}{2}f(s)^2\big)\, ds\right]$$
$$+ \int_a^t (T_{-1_{[s,t)}f}\, \psi(s)) \exp\left(\int_s^t f(\tau)\, dB(\tau) + \int_s^t (g(\tau) - \tfrac{1}{2}f(\tau)^2)\, d\tau\right) ds.$$

Remark. Suppose $X(t)$ is a solution of equation (13.61) in $L^2([a, b]; (L^2))$. Then $X(t)$ is in (L^2) for almost all t in $[a, b]$. Condition (2) implies that φ is in (L^2). Moreover, it follows from conditions (1) and (3) that the last integral in equation (13.61) belongs to (L^2) for all $t \in [a, b]$. Thus $\int_a^t \partial_s^*(f(s)X(s))ds$ is a Hitsuda-Skorokhod integral for almost all t in $[a, b]$.

Proof. First we show the uniqueness of a solution in $L^2([a, b]; (L^2))$. Suppose $Y(t)$ is another such solution and let $Z(t) = X(t) - Y(t)$. Then $Z(t)$ satisfies the equation

$$Z(t) = \int_a^t \partial_s^*\big(f(s)Z(s)\big)\, ds + \int_a^t g(s)Z(s)\, ds, \quad a \le t \le b.$$

Let $G(t) = SZ(t)$. The above equation implies that for almost all $t \in [a, b]$ we have

$$G(t)(\xi) = \int_a^t \big(\xi(s)f(s) + g(s)\big)G(s)(\xi)\, ds, \quad \forall \xi \in \mathcal{S}_c.$$

Note that *a priori* $G(t)(\xi)$ is defined only almost everywhere in t for any fixed ξ. Thus $G'(t)(\xi)$ is meaningless. In order to overcome this difficulty, for each fixed $\xi \in \mathcal{S}_c$, define $H_\xi(t)$ by

$$H_\xi(t) = \int_a^t \big(\xi(s)f(s) + g(s)\big)G(s)(\xi)\, ds, \quad a \le t \le b.$$

Then for almost all $t \in [a, b]$, $H_\xi(t) = G(t)(\xi)$ for all $\xi \in S_c$. Moreover, for each ξ, the function H_ξ is absolutely continuous, $H_\xi(a) = 0$, and

$$H'_\xi(t) = \big(\xi(t)f(t) + g(t)\big) H_\xi(t), \quad t-\text{a.e. on } [a, b].$$

This implies that $H_\xi(t) = 0$ for all t in $[a, b]$. Hence for almost all t in $[a, b]$, we have $G(t)(\xi) = H_\xi(t) = 0$ for all $\xi \in S_c$. But this means that for almost all t in $[a, b]$, $Z(t) = 0$, i.e., $X(t) = Y(t)$. Thus we have proved the uniqueness of a solution of equation (13.61).

Now we use the S-transform method to derive the solution. Let

$$F(t) = SX(t), \quad G = S\varphi, \quad V(t) = S\psi(t).$$

Then by taking the S-transform of equation (13.61) we get

$$F(t)(\xi) = G(\xi) + \int_a^t \big(\xi(s)f(s) + g(s)\big) F(s)(\xi)\, ds + \int_a^t V(s)(\xi)\, ds.$$

This equality holds in the sense that for almost all t in $[a, b]$, it holds for all $\xi \in S_c$. Since $F(t)(\xi)$ is defined only almost everywhere on $[a, b]$, $F'(t)(\xi)$ is meaningless. Thus as in the proof for uniqueness, for each fixed $\xi \in S_c$, we define a function H_ξ on $[a, b]$ by

$$H_\xi(t) = G(\xi) + \int_a^t \big(\xi(s)f(s) + g(s)\big) F(s)(\xi)\, ds + \int_a^t V(s)(\xi)\, ds.$$

Then for almost all $t \in [a, b]$, $H_\xi(t) = F(t)(\xi)$ for all $\xi \in S_c$. Obviously for each fixed ξ, the function H_ξ is absolutely continuous and satisfies the differential equation

$$H'_\xi(t) = \big(\xi(t)f(t) + g(t)\big) H_\xi(t) + V(t)(\xi), \quad t-\text{a.e. on } [a, b],$$

with the initial condition $H_\xi(a) = G(\xi)$. The unique solution of this differential equation is easily seen to be given by

$$H_\xi(t) = G(\xi) \exp\left[\int_a^t \big(\xi(s)f(s) + g(s)\big)\, ds \right]$$

$$+ \int_a^t V(s)(\xi) \exp\left[\int_s^t \big(\xi(\tau)f(\tau) + g(\tau)\big)\, d\tau \right] ds.$$

But recall that for almost all $t \in [a, b]$, $H_\xi(t) = F(t)(\xi)$ for all $\xi \in S_c$. Thus for almost all $t \in [a, b]$, the S-transform $F(t)$ of $X(t)$ is given by

$$F(t)(\xi) = G(\xi) \exp\left[\int_a^t \big(\xi(s)f(s) + g(s)\big)\, ds \right]$$

$$+ \int_a^t V(s)(\xi) \exp\left[\int_s^t \big(\xi(\tau)f(\tau) + g(\tau)\big)\, d\tau \right] ds. \tag{13.62}$$

We need to find the inverse S-transform of $F(t)$. First note that for $f \in L^2([a,b])$,

$$\left(S \exp\left[\int_a^t f(s)\,dB(s) - \frac{1}{2}\int_a^t f(s)^2\,ds\right]\right)(\xi) = \exp\left[\int_a^t \xi(s)f(s)\,ds\right].$$

Hence we have

$$S\left(\varphi \diamond \exp\left[\int_a^t f(s)\,dB(s) - \frac{1}{2}\int_a^t f(s)^2\,ds\right]\right)(\xi)$$

$$= G(\xi)\exp\left[\int_a^t \xi(s)f(s)\,ds\right]. \qquad (13.63)$$

Similarly, for almost all $a \le s \le t$,

$$S\left(\psi(s) \diamond \exp\left[\int_s^t f(\tau)\,dB(\tau) - \frac{1}{2}\int_s^t f(\tau)^2\,d\tau\right]\right)(\xi)$$

$$= V(s)(\xi)\exp\left[\int_s^t \xi(\tau)f(\tau)\,d\tau\right].$$

Then we use conditions (1) and (3) to get

$$S\left(\int_a^t \psi(s) \diamond \exp\left[\int_s^t f(\tau)\,dB(\tau) - \frac{1}{2}\int_s^t f(\tau)^2\,d\tau\right]\,ds\right)(\xi)$$

$$= \int_a^t V(s)(\xi)\exp\left[\int_s^t \xi(\tau)f(\tau)\,d\tau\right]\,ds. \qquad (13.64)$$

It follows from equations (13.62)–(13.64) that for almost all $t \in [a,b]$,

$$X(t) = \varphi \diamond \exp\left[\int_a^t f(s)\,dB(s) + \int_a^t \left(g(s) - \tfrac{1}{2}f(s)^2\right)\,ds\right]$$

$$+ \int_a^t \psi(s) \diamond \exp\left[\int_s^t f(\tau)\,dB(\tau) + \int_s^t \left(g(\tau) - \tfrac{1}{2}f(\tau)^2\right)\,d\tau\right]\,ds.$$

Then we apply Lemma 13.32 to rewrite $X(t)$ in the form as given in the theorem. Note that $X(t) \in (L^2)$ for almost all $t \in [a,b]$. Finally, by going backward in the above argument, we can check that $X(t)$ is a solution of the equation (13.61). □

We have used the S-transform to study stochastic integral equations. However, there are occasions when the S-transform can not be applied. For instance, consider the following example in Buckdahn [16]:

$$X(t) = 1 + \int_0^t \partial_s^*\left(B(1)X(s)\right)\,ds.$$

Here the S-transform method is not practical since the the expression of $S(B(1)X(s))$ in terms of $S(B(1))$ and $S(X(t))$ is rather complicated. However, we can use Itô's formula in Theorem 13.21 to solve this equation.

Theorem 13.34. *Let $0 \le a < b < \infty$ and let $f \in L^\infty([a,b])$. Then the stochastic integral equation*

$$X(t) = x_0 + \int_a^t \partial_s^* (f(s)B(b)X(s)) \, ds, \quad a \le t \le b, \quad (x_0 \in \mathbb{R}), \quad (13.65)$$

has a unique solution in $L^2([a,b];(L^2))$ given by

$$X(t) = x_0 \exp \left[B(b) \int_a^t f(s) e^{-\int_s^t f(\tau) \, d\tau} dB(s) \right.$$
$$\left. - \frac{1}{2} B(b)^2 \int_a^t f(s)^2 e^{-2 \int_s^t f(\tau) \, d\tau} ds - \int_a^t f(s) \, ds \right]. \quad (13.66)$$

Proof. Suppose $Y(t)$ is another such solution and let $Z(t) = X(t) - Y(t)$. Then $Z(t)$ satisfies the equation

$$Z(t) = \int_a^t \partial_s^* (f(s)B(b)Z(s)) \, ds. \quad (13.67)$$

Let $Z(t)$ be represented by

$$Z(t) = \sum_{n=0}^\infty \langle :\cdot^{\otimes n} :, f_n(t) \rangle.$$

Note that the right hand side of equation (13.67) has no constant term. Hence $f_0(t) = 0$ a.e. Assume that $f_k(t) = 0$ a.e. for all $k \le n - 1$. Then it follows from equation (13.67) that for almost all t and u,

$$f_n(t; u) = n \, 1_{[a,t)}(u) \int_a^b f_n(u; \tau) \, d\tau \quad \text{on } \mathbb{R}^{n-1}.$$

Obviously this implies that $f_n(t) = 0$ a.e. on $\mathbb{R}^n$ for almost all t. Hence by mathematical induction, $f_n(t) = 0$ a.e. for all n. This shows that $Z(t) = 0$, i.e., $X(t) = Y(t)$ for almost all t. Thus we have proved the uniqueness of a solution in $L^2([a,b];(L^2))$ for equation (13.65).

To find the solution of equation (13.65), consider a stochastic process given by

$$Q(t) = x_0 \exp \left[B(b)h(t) \int_a^t \frac{f(s)}{h(s)} dB(s) \right.$$
$$\left. - \frac{1}{2} B(b)^2 h(t)^2 \int_a^t \frac{f(s)^2}{h(s)^2} ds - g(t) \right],$$

where the functions $g(t)$ and $h(t)$ are to be chosen so that $Q(t)$ is a solution of equation (13.65). Note that the form of $Q(t)$ is suggested by the corresponding linear equation in the Itô theory. Let

$$\theta(t, x, y) = x_0 \exp\left[h(t)xy - \frac{1}{2}h(t)^2 y^2 \int_a^t \frac{f(s)^2}{h(s)^2}\, ds - g(t)\right].$$

We can apply Theorem 13.21 to the function θ with $X(t) = \int_a^t \frac{f(s)}{h(s)}\, dB(s)$ and $y = B(b)$ (note that the dependence of θ on t requires an extra term given by $\partial \theta/\partial t$). It is easy to check that

$$\frac{d}{dt}\theta(t, X(t), B(b))$$

$$= \theta\left(h'(t)xy - h(t)h'(t)y^2 \int_a^t \frac{f(s)^2}{h(s)^2}\, ds - \frac{1}{2}y^2 f(t)^2 - g'(t)\right) + \partial_t^*\left(f(t)\theta y\right)$$

$$+ \theta\left(\frac{1}{2}y^2 f(t)^2 + f(t) + yf(t)\left[h(t)x - h(t)^2 y \int_a^t \frac{f(s)^2}{h(s)^2}\, ds\right]\right),$$

where we let $x = X(t)$ and $y = B(b)$ for convenience. Choose functions g and h such that $g'(t) = f(t)$ and $h'(t) = -f(t)h(t)$; for instance,

$$g(t) = \int_a^t f(s)\, ds, \quad h(t) = e^{-\int_a^t f(s)\, ds}.$$

With this choice of g and h, we see that $\theta(t, X(t), B(b))$ satisfies the equation

$$\frac{d}{dt}\theta(t, X(t), B(b)) = \partial_t^*\left(f(t)B(b)\theta(t, X(t), B(b))\right),$$

where $X(t) = \int_a^t f(s) \exp\left[\int_a^s f(\tau)\, d\tau\right] dB(s)$. But $Q(t) = \theta(t, X(t), B(b))$. Hence $Q(t)$ satisfies equation (13.65). Moreover, it is easy to check that $Q \in L^2\left([a, b]; (L^2)\right)$. This completes the proof. $\qquad\square$

Example 13.35. Consider the stochastic integral equation

$$X(t) = 1 + \int_0^t \partial_s^*\left(B(1)X(s)\right) ds, \quad 0 \le t \le 1. \tag{13.68}$$

By Theorem 13.34 the solution is given by

$$X(t) = \exp\left[B(1)\int_0^t e^{-(t-s)}\, dB(s) - \frac{1}{4}B(1)^2\left(1 - e^{-2t}\right) - t\right].$$

This solution was derived by using a different method in Buckdahn [16].

Example 13.36. Consider the stochastic integral equation

$$Y(t) = 1 + \int_0^t \partial_s^* \left(B(1) \diamond Y(s) \right) ds, \quad 0 \le t \le 1. \tag{13.69}$$

Note that this equation is similar to equation (13.68) except that we use the Wick product instead of pointwise product. This equation has been studied in Kuo and Potthoff [114].

We can use the S-transform to solve this new equation. Suppose $Y \in L^2\left([0,1]; (L^2)\right)$ is a solution of equation (13.69). It follows from Theorem 13.4 that the function $\partial_t^* \left(B(1) \diamond Y(t) \right)$, $t \in [0,1]$, is Pettis integrable. Hence the S-transform $F(t) = SY(t)$ satisfies the following equation for each fixed $\xi \in \mathcal{S}_c$,

$$F(t) = 1 + \int_0^t \xi(s) \int_0^1 \xi(\tau) \, d\tau \, F(s) \, ds.$$

Here we have used the fact that $SB(1)(\xi) = \int_0^1 \xi(\tau) \, d\tau$. Note that $F(t)$ is defined only for almost all t and so its derivative $F'(t)$ is meaningless. However, we can use the same argument as in the proof of Theorem 13.33 to extend $F(t)$ so that $F(t)$ is absolutely continuous and for almost all t,

$$F'(t) = \xi(t) F(t) \int_0^1 \xi(\tau) \, d\tau, \quad F(0) = 1.$$

Obviously the solution of this differential equation is given by

$$F(t)(\xi) = \exp\left[\int_0^1 \xi(\tau) \, d\tau \int_0^t \xi(s) \, ds \right], \quad \xi \in \mathcal{S}_c.$$

Observe that the above argument (with initial condition $F(0) = 0$) shows the uniqueness of a solution of equation (13.69).

In order to find $Y(t) = S^{-1} F(t)$, we rewrite $F(t)(\xi)$ as follows:

$$F(t)(\xi) = \exp\left[-\frac{1}{2} \langle L\xi, \xi \rangle \right], \tag{13.70}$$

where L is the linear operator of $L^2(\mathbb{R})$ defined by

$$L\xi = -\langle \xi, 1_{[0,1)} \rangle 1_{[0,t)} - \langle \xi, 1_{[0,t)} \rangle 1_{[0,1)}.$$

The operator L is self-adjoint. It is easy to find the eigenvalues of L

$$\alpha_1 = -t - \sqrt{t}, \quad \alpha_2 = -t + \sqrt{t}.$$

The corresponding unit eigenvectors are given respectively by

$$e_1 = \frac{1}{\sqrt{2(1+\sqrt{t})}}\left(1_{[0,1)} + \frac{1}{\sqrt{t}}1_{[0,t)}\right),$$

$$e_2 = \frac{1}{\sqrt{2(1-\sqrt{t})}}\left(1_{[0,1)} - \frac{1}{\sqrt{t}}1_{[0,t)}\right).$$

Thus we can rewrite the function $F(t)(\xi)$ in equation (13.70) as

$$F(t)(\xi) = \exp\left[-\frac{\alpha_1}{2}\langle\xi,e_1\rangle^2 - \frac{\alpha_2}{2}\langle\xi,e_2\rangle^2\right].$$

Now, let e be a unit vector in $L^2(\mathbb{R})$ and $c < -1$ or $c > 0$. From Example 8.7 we have

$$S\left(\sqrt{\frac{1+c}{c}}\exp\left[-\frac{1}{2c}\langle\cdot,e\rangle^2\right]\right)(\xi) = \exp\left[-\frac{1}{2(1+c)}\langle\xi,e\rangle^2\right], \quad \xi\in\mathcal{S}_c.$$

In fact, this equality can be checked by direct calculation. By using this equality we see that the inverse S-transform $Y(t) = S^{-1}F(t)$ is given by

$$Y(t) = \frac{1}{\sqrt{(1-\alpha_1)(1-\alpha_2)}}\exp\left[-\frac{\alpha_1}{2(1-\alpha_1)}\langle\cdot,e_1\rangle^2 - \frac{\alpha_2}{2(1-\alpha_2)}\langle\cdot,e_2\rangle^2\right].$$

Finally note that $\langle\cdot,1_{[0,1)}\rangle = B(1)$ and $\langle\cdot,1_{[0,t)}\rangle = B(t)$. Thus after putting the eigenvalues and eigenvectors into the last equation we get the solution $Y(t)$ of equation (13.69)

$$Y(t) = \frac{1}{\sqrt{1+t+t^2}}\exp\left[-\frac{1}{2(1+t+t^2)}\right.$$
$$\left. \times\left(tB(1)^2 - 2(1+t)B(1)B(t) + B(t)^2\right)\right].$$

Remark. It is interesting to compare the solution $X(t)$ of equation (13.68) and the solution $Y(t)$ of equation (13.69). Note that the randomness of $Y(t)$ comes from only $B(1)$ and $B(t)$, while $X(t)$ depends on $B(1)$ and $B(s)$ for $0 \le s \le t$. Furthermore, let us determine the effect of $B(1)$ on these solutions for small t. In order to do this, we regard $B(t)$ as $B(t) \approx \pm\sqrt{t}$ for small t. Similarly, note that the Wiener integral $\int_0^t e^{-(t-s)}\,dB(s)$ is normally distributed with mean 0 and variance $\frac{1}{2}(1-e^{-2t})$. Hence this random variable can be regarded informally as

$$\int_0^t e^{-(t-s)}\,dB(s) \approx \pm\frac{1}{\sqrt{2}}\left(1-e^{-2t}\right)^{1/2}.$$

With this informal consideration, we can check easily that

$$\ln X(t) \approx \pm B(1)\sqrt{t} - \frac{1}{2}(2 + B(1)^2)t \pm \frac{1}{2}B(1)t^{3/2} + \frac{1}{2}B(1)^2 t^2 + \cdots,$$

$$\ln Y(t) \approx \pm B(1)\sqrt{t} - \frac{1}{2}(2 + B(1)^2)t + \frac{1}{4}t^2 + \cdots.$$

This shows that $B(1)$ has the same effect on $X(t)$ and $Y(t)$ up to order t for small values of t. This justifies in some sense the use of Wick product in equation (13.69).

Next we give simple examples that are similar to Examples 13.35 and 13.36, but the product of $B(1)$ and $X(s)$ appears in the other integral.

Example 13.37. Consider the stochastic integral equations:

$$X(t) = B(t) + \int_0^t B(1)X(s)\,ds, \qquad (13.71)$$

$$Y(t) = B(t) + \int_0^t B(1) \diamond Y(s)\,ds. \qquad (13.72)$$

By iteration we can easily derive the solution of equation (13.71)

$$X(t) = B(t) + B(1)\int_0^t e^{(t-s)B(1)}B(s)\,ds.$$

As for equation (13.72), we can use the S-transform method to find its solution

$$Y(t) = \int_0^t \partial_s^* e^{(t-s)B(1) - \frac{1}{2}(t-s)^2}\,ds.$$

By using the same argument as in the previous example we see the effect of $B(1)$ on $X(t)$ and $Y(t)$ for small values of t:

$$X(t) \approx \pm\sqrt{t} + \frac{2}{3}B(1)t^{3/2} + \cdots,$$

$$Y(t) \approx \pm\sqrt{t} \pm \frac{1}{\sqrt{3}}B(1)t^{3/2} + \cdots.$$

13.8 White noise integral equations

Note that the integral $\int_a^t \partial_s^* f(s, X(s))\,ds$ in equation (13.53) is a Hitsuda-Skorokhod integral, i.e., it is a random variable in (L^2). Suppose we regard equation (13.53) as an $(\mathcal{S})^*_\beta$-valued integral equation, i.e., $X(t) \in (\mathcal{S})^*_\beta$ for each t, then $\int_a^t \partial_s^* f(s, X(s))\, ds \in (\mathcal{S})^*_\beta$ is a white noise integral. Hence if we look for a solution in the space $(\mathcal{S})^*_\beta$, then the two integrals in equation (13.53) should be combined together and we get the equation

$$X(t) = X(a) + \int_a^t h(s, X(s))\, ds, \quad a \le t \le b, \tag{13.73}$$

where $h : [a, b] \times \mathbb{R} \to \mathbb{R}$. But in general $h(s, X(s))$ has no meaning. For instance, when $X(s) = \dot{B}(s)$ and $h(x) = x^2$, the composition $h(X(s)) = \dot{B}(s)^2$ is meaningless. Thus equation (13.73) should be replaced by the following equation

$$X(t) = X(a) + \int_a^t f(s, X(s))\, ds, \quad a \le t \le b, \tag{13.74}$$

where f is a function from $[a, b] \times (\mathcal{S})^*_\beta$ into $(\mathcal{S})^*_\beta$. Obviously, the integral $\int_a^t f(s, X(s))\, ds$ should be interpreted as a white noise integral.

First we give several examples of white noise integral equations. The solutions in these examples are generalized functions induced by Hida measures. Recall from Definition 8.14 that a measure ν on $\mathcal{S}'$ is called a Hida measure if it induces a generalized function $\tilde{\nu} \in (\mathcal{S})^*_\beta$ such that

$$\langle\!\langle \tilde{\nu}, \varphi \rangle\!\rangle = \int_{\mathcal{S}'} \varphi(x)\, d\nu(x), \quad \varphi \in (\mathcal{S})_\beta.$$

In particular, $\mu_{x,t}(\cdot) = \mu(\frac{\cdot - x}{\sqrt{t}})$ is a Hida measure for any $x \in \mathcal{S}'$ and $t > 0$. The induced generalized function $\tilde{\mu}_{x,t}$ has S-transform

$$S\tilde{\mu}_{x,t}(\xi) = \exp\left[\langle x, \xi \rangle - \frac{1}{2}(1 - t)\langle \xi, \xi \rangle\right], \quad \xi \in \mathcal{S}_c.$$

When $x = 0$, $\mu_{x,t}$ and $\tilde{\mu}_{x,t}$ are denoted by μ_t and $\tilde{\mu}_t$, respectively. Note that

$$\tilde{\mu}_{x,s} * \tilde{\mu}_{y,t} = (\mu_{x,x} * \mu_{y,t})^\sim = \tilde{\mu}_{x+y,s+t},$$

where convolution is defined in §8.4.

Example 13.38. Consider the white noise integral equation

$$X(t) = \Phi_0 + \int_0^t \Delta_G^* X(s)\, ds, \quad 0 \le t \le 1. \tag{13.75}$$

We look for a solution $X(t)$ of this equation such that the white noise integral $\int_0^t \Delta_G^* X(s)\, ds$ exists in the Pettis sense. Note that

$$S(\Delta_G^* \Phi)(\xi) = \langle \xi, \xi \rangle S\Phi(\xi).$$

Thus for each fixed $\xi \in \mathcal{S}_c$, the function $F(t)(\xi) = SX(t)(\xi)$ satisfies the following integral equation

$$F(t)(\xi) = S\Phi_0(\xi) + \int_0^t \langle \xi, \xi \rangle F(s)(\xi)\, ds.$$

Obviously the solution of this equation is given by

$$F(t)(\xi) = (S\Phi_0(\xi))e^{t\langle \xi, \xi \rangle}.$$

Hence we get the solution of equation (13.75)

$$X(t) = \Phi_0 \diamond \widetilde{\mu}_{1+2t}.$$

For some initial conditions, the solutions are induced by Hida measures. Here are some examples:

(1) For $\Phi_0 = 1$, we have $X(t) = \widetilde{\mu}_{1+2t}$.
(2) For $\Phi_0 = \widetilde{\delta}_x$, the Kubo-Yokoi delta function at $x \in \mathcal{S}'$, we have $X(t) = \widetilde{\mu}_{x,2t}$.
(3) For $\Phi_0 = :e^{\langle \cdot, x \rangle}:$, $x \in \mathcal{S}'$, we have $X(t) = \widetilde{\mu}_{x,1+2t}$.

In fact, if the initial condition Φ_0 is induced by a Hida measure ν on $\mathcal{S}'$, i.e., $\Phi_0 = \widetilde{\nu}$, then the solution $X(t)$ is also induced by a Hida measure

$$X(t) = (\nu * \mu_{2t})^{\widetilde{\;}},$$

where $\nu_1 * \nu_2$ is the convolution of ν_1 and ν_2 (see §8.4). On the other hand, if $\Phi_0 = -1$, then the solution is not induced by a Hida measure.

Example 13.39. Consider the same equation as given in Example 13.30, but with a different initial condition

$$X(t) = \widetilde{\delta}_0 + \int_0^t \partial_s^* X(s)\, ds.$$

The S-transform $F(t) = SX(t)$ satisfies the equation

$$F(t)(\xi) = e^{-\frac{1}{2}\langle \xi, \xi \rangle} + \int_0^t \xi(s)F(s)(\xi)\, ds.$$

Hence $F(t)(\xi)$ is given by

$$F(t)(\xi) = \exp\left[\langle 1_{[0,t)}, \xi \rangle - \frac{1}{2}\langle \xi, \xi \rangle\right].$$

Therefore we get the solution $X(t) = \widetilde{\delta}_{1_{[0,t)}}$. Note that $X(t)$ is not a random variable. But it is induced by a Hida measure. Moreover, observe that $\int_0^t \partial_s^* X(s)\, ds$ is not a Hitsuda-Skorokhod integral.

Example 13.40. Consider the white noise integral equation

$$X(t) = \Phi_0 + \int_0^t \left(cX(s) + :e^{\dot{B}(s)}: \right) ds, \quad (c \in \mathbb{R}),$$

where $:e^{\dot{B}(s)}:$ is regarded as a noise for this equation. The solution of this equation is given by

$$X(t) = e^{ct}\Phi_0 + \int_0^t e^{c(t-s)} :e^{\dot{B}(s)}: ds.$$

It is easy to check that the noise $:e^{\dot{B}(s)}:$ is induced by the measure $T_{\delta_s}\mu$ (the translation of μ by δ_s), i.e.,

$$:e^{\dot{B}(s)}: = \left(T_{\delta_s}\mu\right)^{\sim}.$$

Hence if the initial condition Φ_0 is induced by a Hida measure, then the solution $X(t)$ is also induced by a Hida measure.

Example 13.41. Let $k \geq 2$ be an integer and consider the white noise integral equation

$$X(t) = 1 + \int_0^t (\Delta_G^*)^k X(s)\, ds. \tag{13.76}$$

The S-transform $F(t)$ of $X(t)$ satisfies

$$F(t)(\xi) = 1 + \int_0^t \langle \xi, \xi \rangle^k F(s)(\xi)\, ds.$$

Hence $F(t)(\xi) = e^{t\langle \xi, \xi \rangle^k}$ and the solution of equation (13.76) is given by

$$X(t) = \sum_{n=0}^{\infty} \frac{t^n}{n!} \left\langle :\cdot^{\otimes 2kn}:, \tau^{\otimes kn} \right\rangle.$$

By Theorem 8.2 we have $X(t) \in (\mathcal{S})^*_{\frac{k-1}{k}}$. Moreover, it can be checked easily that $X(t)$ is not induced by a Hida measure.

Now, we study the white noise integral equation (13.74). We will use the S-transform to establish the existence and uniqueness of a solution of this equation. Recall from §13.3 that the white noise integral in terms of the S-transform is equivalent to the Pettis integral. Hence it is natural to use the white noise integral in the Pettis sense. Of course, we may take the white noise integral in the Bochner sense since it implies the white noise integral in the Pettis sense. However, the Bochner integrability is a rather strong condition. Hence in general we will use the white noise integral in the Pettis sense.

Definition 13.42. A function $X : [a, b] \rightarrow (S)^*_\beta$ is called a *weak solution* of the equation (13.74) on $[a, b]$ if it satisfies the following conditions:
(a) X is weakly measurable.
(b) The function $f(s, X(s)), s \in [a, b]$, is Pettis integrable.
(c) For each $\varphi \in (S)_\beta$, the equality holds for almost all $t \in [a, b]$

$$\langle\!\langle X(t), \varphi \rangle\!\rangle = \langle\!\langle X(a), \varphi \rangle\!\rangle + \int_a^t \langle\!\langle f(s, X(s)), \varphi \rangle\!\rangle \, ds.$$

Theorem 13.43. *Suppose f is a function from $[a, b] \times (S)^*_\beta$ into $(S)^*_\beta$ satisfying the following conditions:*
(1) *(Measurability condition) The function $f(t, X(t)), t \in [a, b]$, is weakly measurable for any weakly measurable function $X : [a, b] \rightarrow (S)^*_\beta$.*
(2) *(Lipschitz condition) For almost all t in $[a, b]$,*

$$|Sf(t, \Phi)(\xi) - Sf(t, \Psi)(\xi)|$$
$$\leq L(t, \xi)|S\Phi(\xi) - S\Psi(\xi)|, \qquad \forall \xi \in \mathcal{S}_c, \ \Phi, \Psi \in (S)^*_\beta,$$

where L is nonnegative and $\int_a^b L(t, \xi) \, dt \leq K\left(1 + |\xi|_p^{\frac{2}{1-\beta}}\right)$ for some constants $K, p \geq 0$.
(3) *(Growth condition) For almost all t in $[a, b]$,*

$$|Sf(t, \Phi)(\xi)| \leq \rho(t, \xi)\left(1 + |S\Phi(\xi)|\right), \quad \forall \xi \in \mathcal{S}_c, \ \Phi \in (S)^*_\beta,$$

where ρ is nonnegative and $\int_a^b \rho(t, \xi) \, dt \leq K \exp\left[c|\xi|_p^{\frac{2}{1-\beta}}\right]$ for some constants K, p (same as above), and $c \geq 0$.
Then for any $X(a) \in (S)^*_\beta$, the equation (13.74) has a unique weak solution X such that

$$\operatorname*{ess\,sup}_{a \leq t \leq b} |SX(t)(\xi)| < \infty, \quad \forall \xi \in \mathcal{S}_c. \tag{13.77}$$

Before proving this theorem we first prepare two lemmas. For the uniqueness part, we need the following version of Bellman-Gronwall inequality (cf. Fleming and Rishel [36, p.198]).

Lemma 13.44. *Let $f \in L^\infty([a, b])$ and $f \geq 0$ a.e. Suppose f satisfies the condition*

$$f(t) \leq \rho(t) + \int_a^t \theta(s) f(s) \, ds, \quad \text{a.e. on } [a, b],$$

where $\rho \in L^\infty([a, b])$ and $\theta \in L^1([a, b])$, $\theta \geq 0$ a.e. Then

$$f(t) \leq \rho(t) + \int_a^t \rho(s)\theta(s) e^{\int_s^t \theta(u) \, du} \, ds, \quad \text{a.e. on } [a, b].$$

Proof. Let $g(t) = \int_a^t \theta(s)f(s)\,ds$. Then we have $f(t) \leq \rho(t) + g(t)$. The function g is absolutely continuous and $g'(t) = \theta(t)f(t)$ a.e. Hence for almost all t on $[a,b]$,

$$g'(t) - \theta(t)g(t) = \theta(t)(f(t) - g(t)) \leq \rho(t)\theta(t).$$

By multiplying the integrating factor $\exp\left[-\int_a^t \theta(u)\,du\right]$ to both sides, we get

$$\frac{d}{dt}\left(g(t)e^{-\int_a^t \theta(u)\,du}\right) \leq \rho(t)\theta(t)e^{-\int_a^t \theta(u)\,du}.$$

This implies that

$$g(t)e^{-\int_a^t \theta(u)\,du} \leq \int_a^t \rho(s)\theta(s)e^{-\int_a^s \theta(u)\,du}\,ds.$$

Hence we obtain

$$g(t) \leq \int_a^t \rho(s)\theta(s)e^{\int_s^t \theta(u)\,du}\,ds.$$

Therefore

$$f(t) \leq \rho(t) + g(t) \leq \rho(t) + \int_a^t \rho(s)\theta(s)e^{\int_s^t \theta(u)\,du}\,ds. \qquad \square$$

Lemma 13.45. *Let $f: [a,b] \times (\mathcal{S})_\beta^* \to (\mathcal{S})_\beta^*$ be a function satisfying the measurability and growth conditions stated in Theorem 13.43. Suppose X is a weakly measurable function from $[a,b]$ into $(\mathcal{S})_\beta^*$ and there exist nonnegative constants K_1, c_1, and p_1 such that*

$$\operatorname*{ess\,sup}_{a \leq t \leq b} |SX(t)(\xi)| \leq K_1 \exp\left[c_1|\xi|_{p_1}^{\frac{2}{1-\beta}}\right], \quad \forall \xi \in \mathcal{S}_c.$$

Then the function $f(t, X(t)), t \in [a,b]$, is Pettis integrable and

$$\int_a^b |Sf(t, X(t))(\xi)|\,dt \leq K_2 \exp\left[c_2|\xi|_{p_2}^{\frac{2}{1-\beta}}\right], \quad \forall \xi \in \mathcal{S}_c, \qquad (13.78)$$

where $K_2 = K(1 + K_1)$, $c_2 = c + c_1$, and $p_2 = p \vee p_1$.

Proof. We check that conditions in Theorem 13.4 are satisfied for the function $f(t, X(t))$. It follows from the measurability condition of f that $Sf(t, X(t))(\xi), t \in [a,b]$, is measurable for each $\xi \in \mathcal{S}_c$. Hence condition (1)

in Theorem 13.4 is satisfied. On the other hand, by the growth condition
of f and the assumption on $X(t)$, we have

$$\int_a^b |Sf(t, X(t))(\xi)|\, dt \le \int_a^b \rho(t, \xi)\big(1 + |SX(t)(\xi)|\big)\, dt$$

$$\le \int_a^b \rho(t, \xi)\Big(1 + K_1 \exp\big[c_1|\xi|_{p_1}^{\frac{2}{1-\beta}}\big]\Big)\, dt$$

$$\le \Big(1 + K_1 \exp\big[c_1|\xi|_{p_1}^{\frac{2}{1-\beta}}\big]\Big)\, K \exp\big[c|\xi|_p^{\frac{2}{1-\beta}}\big]$$

$$\le K(1 + K_1) \exp\big[(c + c_1)|\xi|_{p\vee p_1}^{\frac{2}{1-\beta}}\big].$$

This proves the inequality in (13.78), which is condition (2) of Theorem
13.4. Hence by Theorem 13.4, the function $f(t, X(t)), t \in [a, b]$, is Pettis
integrable. □

Proof of Theorem 13.43.

A. Uniqueness. Suppose $X(t)$ and $Y(t)$ are two weak solutions of equation
(13.74) satisfying the condition in (13.77). Let $F(t) = SX(t)$ and $G(t) = SY(t)$. Then for fixed $\xi \in \mathcal{S}$, it follows from Definition 13.42 and the
Lipschitz condition of f that

$$|F(t)(\xi) - G(t)(\xi)| \le \int_a^t |Sf(s, X(s))(\xi) - Sf(s, Y(s))(\xi)|\, ds$$

$$\le \int_a^t L(s, \xi)|F(s)(\xi) - G(s)(\xi)|\, ds, \quad \text{a.e. on } [a, b].$$

Hence by Lemma 13.44 $F(t)(\xi) = G(t)(\xi)$, t-a.e. on $[a, b]$. For each $\xi \in \mathcal{S}$,
let N_ξ be a null subset of $[a, b]$ such that $F(t)(\xi) = G(t)(\xi)$ if $t \in N_\xi^c$. Since
the space $\mathcal{S}$ is separable, there is a countable dense subset $\{\xi_n; n \ge 1\}$ of
$\mathcal{S}$. Let $N_0 = \cup_{n\ge1} N_{\xi_n}$. Then N_0 is a null subset of $[a, b]$ and we have
$F(t)(\xi_n) = G(t)(\xi_n)$ for all $t \in N_0$ and $n \ge 1$. This implies that if $t \in N_0$,
then $F(t)(\xi) = G(t)(\xi)$ for all $\xi \in \mathcal{S}$ and so for all $\xi \in \mathcal{S}_c$. Hence $X = Y$
and the uniqueness of a weak solution is proved.

B. Existence. To show the existence of a weak solution, we use the
iteration method. Let $G = SX(a)$. Define $X_0(t) \equiv X(a)$. Obviously X_0 is
weakly measurable from $[a, b]$ into $(\mathcal{S})_\beta^*$ and we have

$$\operatorname*{ess\,sup}_{a\le t\le b} |SX_0(t)(\xi)| = |G(\xi)| \le K_0 \exp\big[c_0|\xi|_{p_0}^{\frac{2}{1-\beta}}\big], \tag{13.79}$$

where K_0, c_0, and p_0 are nonnegative constants associated with $X(a)$ as in
Theorem 8.2. Hence by Lemma 13.45 the function $f(s, X_0(s)), s \in [a, b]$, is

Pettis integrable and there are nonnegative constants $K_0', c_0',$ and p_0' such that

$$\int_a^b |Sf(s, X_0(s))(\xi)| \, ds \leq K_0' \exp\left[c_0' |\xi|_{p_0'}^{\frac{2}{1-\beta}}\right]. \tag{13.80}$$

This allows us to define X_1 by

$$X_1(t) = X(a) + \int_a^t f(s, X_0(s)) \, ds, \quad a \leq t \leq b.$$

Then X_1 is weakly measurable and by the inequalities in (13.79) and (13.80) there exist nonnegative constants $K_1, c_1,$ and p_1 such that

$$\operatorname*{ess\,sup}_{a \leq t \leq b} |SX_1(t)(\xi)| \leq K_1 \exp\left[c_1 |\xi|_{p_1}^{\frac{2}{1-\beta}}\right].$$

Thus by Lemma 13.45 $f(s, X_1(s)), s \in [a, b]$, is Pettis integrable and there are nonnegative constants $K_1', c_1',$ and p_1' such that

$$\int_a^b |Sf(s, X_1(s))(\xi)| \, ds \leq K_1' \exp\left[c_1' |\xi|_{p_1'}^{\frac{2}{1-\beta}}\right].$$

Hence we can define X_2 by

$$X_2(t) = X(a) + \int_a^t f(s, X_1(s)) \, ds, \quad a \leq t \leq b.$$

Then repeat the above argument to define X_3. Thus we can define inductively a sequence of weakly measurable functions $X_n : [a, b] \rightarrow (\mathcal{S})_\beta^*$ by

$$X_n(t) = X(a) + \int_a^t f(s, X_{n-1}(s)) \, ds, \quad n \geq 1. \tag{13.81}$$

Let $F_n(t) = SX_n(t), n \geq 0, (F_0(t) = G)$. Note that $F_n(t)(\xi)$ is defined for all $t \in [a, b]$ and $\xi \in \mathcal{S}_c$. By using the growth condition of f, we get

$$|F_1(t)(\xi) - G(\xi)| \leq \int_a^t |Sf(s, X(a))(\xi)| \, ds$$

$$\leq \int_a^t \rho(s, \xi)(1 + |G(\xi)|) \, ds$$

$$\leq H(\xi),$$

where $H(\xi) \equiv K(1 + |G(\xi)|) \exp\left[c|\xi|_p^{\frac{2}{1-\beta}}\right]$. By the Lipschitz condition of f,

$$|F_2(t)(\xi) - F_1(t)(\xi)| \le \int_a^t |Sf(s, X_1(s))(\xi) - Sf(s, X(a))(\xi)|\, ds$$

$$\le \int_a^t L(s, \xi)|F_1(s)(\xi) - G(\xi)|\, ds$$

$$\le H(\xi) \int_a^b L(s, \xi)\, ds.$$

It is easy to check that in general

$$|F_n(t)(\xi) - F_{n-1}(t)(\xi)|$$

$$\le H(\xi) \int_a^t \int_a^{s_{n-1}} \cdots \int_a^{s_2} L(s_1, \xi) \cdots L(s_{n-1}, \xi)\, ds_1 \cdots ds_{n-1}$$

$$= H(\xi) \frac{1}{(n-1)!} \int_a^t \cdots \int_a^t L(s_1, \xi) \cdots L(s_{n-1}, \xi)\, ds_1 \cdots ds_{n-1}$$

$$\le H(\xi) \frac{1}{(n-1)!} \left(\int_a^b L(s, \xi)\, ds \right)^{n-1}. \tag{13.82}$$

This implies that the series

$$G(\xi) + \sum_{n=1}^{\infty} \left(F_n(t)(\xi) - F_{n-1}(t)(\xi) \right)$$

converges absolutely and uniformly for $t \in [a, b]$. Hence $F_n(t)(\xi)$, being a partial sum of this series, converges uniformly on $[a, b]$ for each $\xi \in S_c$ and by the inequality in (13.82),

$$|F_n(t)(\xi)| \le |G(\xi)| + \sum_{k=1}^{n} |F_k(t)(\xi) - F_{k-1}(t)(\xi)|$$

$$\le |G(\xi)| + \sum_{k=1}^{\infty} |F_k(t)(\xi) - F_{k-1}(t)(\xi)|$$

$$\le |G(\xi)| + H(\xi) \exp\left[\int_a^b L(s, \xi)\, ds \right].$$

Thus by the condition on L, there are nonnegative constants $\widetilde{K}, \widetilde{c}$, and $\widetilde{p}$ such that for all n, t, and ξ,

$$|F_n(t)(\xi)| \le \widetilde{K} \exp\left[\widetilde{c}|\xi|_{\widetilde{p}}^{\frac{2}{1-\beta}} \right].$$

Finally, we define

$$F(t)(\xi) = \lim_{n \to \infty} F_n(t)(\xi).$$

Obviously for all $t \in [a, b]$ and $\xi \in \mathcal{S}_c$,

$$|F(t)(\xi)| \leq \tilde{K} \exp\left[\tilde{c}|\xi|_p^{\frac{2}{1-\beta}}\right]. \tag{13.83}$$

By Theorem 8.6 there exists $X(t) \in (\mathcal{S})_\beta^*$ such that $F(t) = SX(t)$. We show below that this function X is a weak solution of equation (13.74).

First of all, $SX(\cdot)(\xi)$ is measurable since it is the limit of measurable functions $F_n(\cdot)(\xi)$. This implies that $\langle\!\langle X(\cdot), \varphi \rangle\!\rangle$ is measurable for any $\varphi \in (\mathcal{S})_\beta$. Hence X is weakly measurable and so condition (a) in Definition 13.42 is satisfied. By the inequality in (13.83) and Lemma 13.45 the function $f(s, X(s)), s \in [a, b]$, is Pettis integrable and so condition (b) in Definition 13.42 is also satisfied.

By taking the S-transform in equation (13.81) we get

$$F_n(t)(\xi) = G(\xi) + \int_a^t Sf(s, X_{n-1}(s))(\xi)\, ds. \tag{13.84}$$

Note that for each $\xi \in \mathcal{S}_c$,

$$\int_a^b |Sf(s, X_n(s))(\xi) - Sf(s, X(s))(\xi)|\, ds$$

$$\leq \int_a^b L(s, \xi)|F_n(s)(\xi) - F(s)(\xi)|\, ds$$

$$\leq \sup_{a \leq s \leq b} |F_n(s)(\xi) - F(s)(\xi)| \int_a^b L(s, \xi)\, ds$$

$$\longrightarrow 0, \quad \text{as } n \to \infty.$$

Thus by letting $n \to \infty$ in equation (13.84), we obtain

$$F(t)(\xi) = G(\xi) + \int_a^t Sf(s, X(s))(\xi)\, ds, \quad \xi \in \mathcal{S}_c,$$

or equivalently, for any $\xi \in \mathcal{S}_c$,

$$\langle\!\langle X(t), \varphi_\xi \rangle\!\rangle = \langle\!\langle X(a), \varphi_\xi \rangle\!\rangle + \int_a^t \langle\!\langle f(s, X(s)), \varphi_\xi \rangle\!\rangle\, ds,$$

where $\varphi_\xi =: \,e^{\langle\cdot,\xi\rangle}:$. But the linear span of the set $\{\varphi_\xi; \xi \in \mathcal{S}_c\}$ is dense in $(\mathcal{S})_\beta$. Hence we conclude that for any $\varphi \in (\mathcal{S})_\beta$,

$$\langle\!\langle X(t), \varphi \rangle\!\rangle = \langle\!\langle X(a), \varphi \rangle\!\rangle + \int_a^t \langle\!\langle f(s, X(s)), \varphi \rangle\!\rangle\, ds.$$

This shows that condition (c) in Definition 13.42 is satisfied. Hence X is a weak solution of equation (13.74). □

Comments: We give several comments on Theorem 13.43, Lemmas 13.44 and 13.45, and their proofs.

(1) The interval $[a, b]$: For the interval $[a, b]$, we did not use the fact that b is a finite number except for the condition in (13.77). Thus the interval $[a, b]$ can be replaced by $[a, \infty)$ and everything remains valid when the condition in (13.77) is replaced by the condition that for all finite number $\tau > a$ and $\xi \in \mathcal{S}_c$,

$$\operatorname*{ess\,sup}_{a \leq t \leq \tau} |SX(t)(\xi)| < \infty.$$

(2) Lipschitz and growth conditions: By carefully examining the proofs of Theorem 13.43 and Lemma 13.45, we see that we have only used the following inequalities:

$$\int_a^t |Sf(s, X(s))(\xi) - Sf(s, Y(s))(\xi)|\, ds$$

$$\leq \int_a^t L(s, \xi)|SX(s)(\xi) - SY(s)(\xi)|\, ds,$$

$$\int_a^t |Sf(s, X(s))(\xi)|\, ds \leq \int_a^t \rho(s, \xi)\big(1 + |SX(s)(\xi)|\big)\, ds.$$

These two inequalities are satisfied under the following modified Lipschitz and growth conditions: For any two weakly measurable functions X and Y, there exists a null set $N_0 \subset [a, b]$ such that for all $t \in N_0^c$ and $\xi \in \mathcal{S}_c$,

$$|Sf(t, X(t))(\xi) - Sf(t, Y(t))(\xi)| \leq L(t, \xi)|SX(t)(\xi) - SY(t)(\xi)|,$$

$$|Sf(t, X(t))(\xi)| \leq \rho(t, \xi)\big(1 + |SX(t)(\xi)|\big),$$

where the functions L and ρ satisfy the conditions as stated in Theorem 13.43. Obviously these modified Lipschitz and growth conditions are weaker than those given in Theorem 13.43. But the proofs of Theorem 13.43 and Lemma 13.45 can be easily modified so that Theorem 13.43 remains valid when the Lipschitz and growth conditions are replaced with the above modified conditions.

(3) Bochner integrability: Suppose the function $\rho(t, \xi)$ in the growth condition of Theorem 13.43 satisfies the condition that for all t and ξ,

$$\rho(t, \xi) \leq \gamma(t) \exp\big[c|\xi|_p^{\frac{2}{1-\beta}}\big],$$

where γ is a nonnegative function such that $\int_a^b \gamma(t)\,dt < \infty$. Then we get a stronger conclusion in Lemma 13.45, i.e., the function $f(\cdot, X(\cdot))$ is Bochner integrable and there exist nonnegative numbers C and q such that

$$\int_a^b \|f(t, X(t))\|_{-q,-\beta}\,dt \leq C \int_a^b \gamma(t)\,dt.$$

This implies that we can replace Pettis integrability by Bochner integrability in the proof of Theorem 13.43. In particular, for the solution X, the integral $\int_a^t f(s, X(s))\,ds$ is a Bochner integral.

Example 13.46. Let $\theta \in (\mathcal{S}_c')^{\otimes k}$, $k \geq 2$, and let $\Xi_{k,0}(\theta)$ be the associated integral kernel operator. By Theorem 10.6 the operator $\Xi_{k,0}(\theta)$ extends by continuity to a continuous linear operator $\widetilde{\Xi}_{k,0}(\theta)$ from $(\mathcal{S})_\beta^*$ into itself for any β. Consider the $(\mathcal{S})_\beta^*$-valued white noise integral equation:

$$X(t) = X(0) + \int_0^t \widetilde{\Xi}_{k,0}(\theta) X(s)\,ds. \tag{13.85}$$

This is a white noise integral equation with $f(t, \Phi) = \widetilde{\Xi}_{k,0}(\theta)\Phi$. Obviously f satisfies the measurability condition in Theorem 13.43. It is easy to see that

$$Sf(t, \Phi)(\xi) = S(\widetilde{\Xi}_{k,0}(\theta)\Phi)(\xi) = \langle \theta, \xi^{\otimes k}\rangle S\Phi(\xi).$$

Hence we can take the following functions L and ρ for the Lipschitz and growth conditions,

$$L(t, \xi) = \rho(t, \xi) = |\langle \theta, \xi^{\otimes k}\rangle|.$$

Note that there exists some $p \geq 0$ such that $\theta \in (\mathcal{S}_{p,c}')^{\otimes k}$. Hence

$$|\langle \theta, \xi^{\otimes k}\rangle| \leq |\theta|_{-p}|\xi|_p^k.$$

Thus the Lipschitz and growth conditions are satisfied when $\beta \geq 1 - \frac{2}{k}$. This shows that for a given $\theta \in (\mathcal{S}_c')^{\otimes k}$, $k \geq 2$, the corresponding equation (13.85) is an $(\mathcal{S})_\beta^*$-valued white noise integral equation if $\beta \geq 1 - \frac{2}{k}$. By Theorem 13.43 this equation has a unique solution X. It is easy to see that the S-transform of $X(t)$ is given by

$$SX(t)(\xi) = (SX(0)(\xi)) \exp\left[t\langle \theta, \xi^{\otimes k}\rangle\right], \quad \xi \in \mathcal{S}_c.$$

In particular, when $\theta = y^{\otimes k}$, $y \in \mathcal{S}_c'$, we have

$$SX(t)(\xi) = (SX(0)(\xi)) \exp\left[t\langle y, \xi\rangle^k\right], \quad \xi \in \mathcal{S}_c.$$

Recall from §5.3 that the additive renormalization $:e^{z\langle \cdot, y\rangle^k}:$, $z \in \mathbb{C}, y \in \mathcal{S}_c'$, is a generalized function in $(\mathcal{S})_{\frac{k-2}{k}}^*$ with S-transform

$$(S :e^{z\langle \cdot, y\rangle^k}:)(\xi) = e^{z\langle y, \xi\rangle^k}, \quad \xi \in \mathcal{S}_c.$$

Hence the unique solution of equation (13.85) with $\theta = y^{\otimes k}$ is given by the following Wick product

$$X(t) = X(0) \diamond {:} e^{t\langle \cdot, y \rangle^k} {:} \, .$$

Example 13.47. Consider the white noise integral equation

$$X(t) = X(0) + \int_0^t (D_{-\delta_s'})^* X(s) \, ds, \qquad (13.86)$$

where δ_s' is the distribution derivative of δ_s and D_y^* denotes an adjoint operator with $y \in \mathcal{S}'$ (see §9.2). By Theorem 13.43 this equation has a unique solution. To find the solution, note that by Theorem 9.13,

$$S((D_{-\delta_s'})^* X(s))(\xi) = \langle -\delta_s', \xi \rangle SX(s)(\xi) = \xi'(s) SX(s)(\xi).$$

With this fact, we can easily find the S-transform of $X(t)$

$$F(t)(\xi) = SX(0)(\xi) \, e^{\xi(t) - \xi(0)}, \quad \xi \in \mathcal{S}_c.$$

But by Theorem 5.13, we have

$$S {:} e^{\langle \cdot, \delta_t - \delta_0 \rangle} {:} (\xi) = e^{\langle \delta_t - \delta_0, \xi \rangle} = e^{\xi(t) - \xi(0)}.$$

Hence the solution of equation (13.86) is given by

$$X(t) = X(0) \diamond {:} e^{\langle \cdot, \delta_t - \delta_0 \rangle} {:} = X(0) \diamond {:} e^{\dot{B}(t) - \dot{B}(0)} {:} \, .$$

More generally, consider the white noise integral equation

$$X(t) = X(0) + \int_0^t \left(D_{(-1)^k \delta_s^{(k)}} \right)^* X(s) \, ds,$$

where $\delta_s^{(k)}$ is the k-th distribution derivative of δ_s. By the same argument as above, we can check that the solution of this equation is given by

$$X(t) = X(0) \diamond \left({:} \exp \left[(-1)^{k-1} \langle \cdot, \delta_t^{(k-1)} - \delta_0^{(k-1)} \rangle \right] {:} \right)$$

$$= X(0) \diamond \left({:} \exp \left[B^{(k)}(t) - B^{(k)}(0) \right] {:} \right).$$

Here we have used the expression $B^{(k)}(t) = (-1)^{k-1} \langle \cdot, \delta_t^{(k-1)} \rangle$ for the k-th derivative of the Brownian motion $B(t) = \langle \cdot, 1_{[0,t)} \rangle$.

Example 13.48. We can use the same methods described in this section to study systems of white noise integral equations. Consider a simple example

$$X(t) = X(0) + \int_0^t (D_{-\delta_s'})^* \begin{bmatrix} 0 & 1 \\ -1 & 0 \end{bmatrix} X(s) \, ds. \qquad (13.87)$$

Let $G = SX(0)$ and $F(t) = SX(t)$. Then $F(t)$ satisfies the equation

$$F(t)(\xi) = G(\xi) + \int_0^t \xi'(s) \begin{bmatrix} 0 & 1 \\ -1 & 0 \end{bmatrix} F(s)(\xi)\, ds.$$

The solution of this equation is given by

$$F(t)(\xi) = \left(\exp\left((\xi(t) - \xi(0)) \begin{bmatrix} 0 & 1 \\ -1 & 0 \end{bmatrix} \right) \right) G(\xi).$$

Note that for any $z \in \mathbb{C}$,

$$\exp\left(z \begin{bmatrix} 0 & 1 \\ -1 & 0 \end{bmatrix} \right) = \cos z \begin{bmatrix} 1 & 0 \\ 0 & 1 \end{bmatrix} + \sin z \begin{bmatrix} 0 & 1 \\ -1 & 0 \end{bmatrix}.$$

Hence $F(t)(\xi)$ can be rewritten as

$$F(t)(\xi) = \begin{bmatrix} \cos\left(\xi(t) - \xi(0)\right) & \sin\left(\xi(t) - \xi(0)\right) \\ -\sin\left(\xi(t) - \xi(0)\right) & \cos\left(\xi(t) - \xi(0)\right) \end{bmatrix} G(\xi).$$

Since $(S\langle \cdot, \delta_t - \delta_0 \rangle)(\xi) = \xi(t) - \xi(0)$, we have

$$S\left(\cos {}^{\diamond}\langle \cdot, \delta_t - \delta_0 \rangle \right)(\xi) = \cos\left(\xi(t) - \xi(0)\right),$$
$$S\left(\sin {}^{\diamond}\langle \cdot, \delta_t - \delta_0 \rangle \right)(\xi) = \sin\left(\xi(t) - \xi(0)\right),$$

where $f {}^{\diamond}\Phi$ denotes the Wick composition of f and Φ (see Definition 8.20). Hence the solution of equation (13.87) is given by

$$X(t) = \begin{bmatrix} \left(\cos {}^{\diamond}\langle \cdot, \delta_t - \delta_0 \rangle \right) \diamond X_1(0) + \left(\sin {}^{\diamond}\langle \cdot, \delta_t - \delta_0 \rangle \right) \diamond X_2(0) \\ -\left(\sin {}^{\diamond}\langle \cdot, \delta_t - \delta_0 \rangle \right) \diamond X_1(0) + \left(\cos {}^{\diamond}\langle \cdot, \delta_t - \delta_0 \rangle \right) \diamond X_2(0) \end{bmatrix},$$

where $X_1(0)$ and $X_2(0)$ are the components of $X(0)$.

14

Feynman Integrals

In this chapter we will describe very briefly an important application of white noise distribution theory to Feynman integrals. This application was first discussed by Hida and Streit in [168]. We will explain the basic ideas of white noise approach to Feynman integrals and carry out explicit calculations for some simple cases. For recent developments, see de Falco and Khandekar [27], de Faria et al. [29], the book by Hida et al. [58], and Khandekar and Streit [79].

14.1 Informal derivation

We begin with an informal derivation of Feynman integrals based on the one from Albeverio and Høegh-Krohn [6]. Consider a nonrelativistic particle of mass m moving in $\mathbb{R}$ under the influence of a conservative force given by a potential V. In quantum mechanics, the state of the particle at time t is described by a function $\psi(t, x)$ satisfying the Schrödinger equation:

$$\begin{cases} i\hbar \, \dfrac{\partial \psi}{\partial t} = -\dfrac{\hbar^2}{2m} \Delta \psi + V(x)\psi, \\ \psi(0, x) = f(x), \end{cases}$$

where $\hbar$ is the Planck constant and $\int_{\mathbb{R}} |f(x)|^2 \, dx = 1$. Let H denote the Hamiltonian of this particle, i.e.,

$$H = -\frac{\hbar^2}{2m} \Delta + V.$$

Then the Schrödinger equation can be rewritten as

$$\begin{cases} \dfrac{\partial \psi}{\partial t} = -\dfrac{i}{\hbar} H\psi, \\ \psi(0, x) = f(x). \end{cases}$$

Hence the solution is given informally by

$$\psi(t,x) = \left(\exp\left[-\frac{i}{\hbar} tH \right] \right) f(x).$$

The question is how to find an explicit expression for $\psi(t,x)$. Such an expression is very useful for studying properties of the solution $\psi(t,x)$.

By the Trotter product formula (see Reed and Simon [155]), we have

$$\exp\left[-\frac{i}{\hbar} tH \right] = \text{strong-}\lim_{n\to\infty} \left(\exp\left[-\frac{i}{\hbar}\frac{t}{n} V \right] \exp\left[\frac{i\hbar}{2m}\frac{t}{n} \Delta \right] \right)^n.$$

Note that for $s > 0$,

$$e^{\frac{1}{2}s\Delta} f(x) = \frac{1}{\sqrt{2\pi s}} \int_{\mathbb{R}} \exp\left[-\frac{1}{2s}(y-x)^2 \right] f(y)\,dy.$$

Apply this equality informally for the complex number $s = \frac{i\hbar t}{mn}$ and iterate the integration n times to get

$$\left(\exp\left[-\frac{i}{\hbar}\frac{t}{n} V \right] \exp\left[\frac{i\hbar}{2m}\frac{t}{n} \Delta \right] \right)^n f(x)$$

$$= \left(2\pi \frac{i\hbar t}{mn} \right)^{-\frac{n}{2}} \int_{\mathbb{R}^n} \exp\left[\frac{i}{\hbar}\frac{t}{n} \sum_{k=1}^{n} \left\{ \frac{m}{2} \left(\frac{x_k - x_{k-1}}{t/n} \right)^2 \right.\right.$$

$$\left.\left. - V(x_k) \right\} \right] f(x_n)\,dx_1 \cdots dx_n,$$

where $x_0 = x$. Hence the solution $\psi(t,x)$ can be expressed as the following limit:

$$\psi(t,x) = \lim_{n\to\infty} \left(2\pi \frac{i\hbar t}{mn} \right)^{-\frac{n}{2}} \int_{\mathbb{R}^n} \exp\left[\frac{i}{\hbar}\frac{t}{n} \sum_{k=1}^{n} \left\{ \frac{m}{2} \left(\frac{x_k - x_{k-1}}{t/n} \right)^2 \right.\right.$$

$$\left.\left. - V(x_k) \right\} \right] f(x_n)\,dx_1 \cdots dx_n.$$

But what is this limit? Consider the space $\mathcal{C}_x$ of all continuous functions y on $[0,\infty)$ with $y(0) = x$. Let $x_k = y(k\frac{t}{n}), 1 \le k \le n$. Then the above summation is a Riemann sum and as $n \to \infty$, we see that $\psi(t,x)$ can be written informally as

$$\psi(t,x) = \mathcal{M} \int_{\mathcal{C}_x} \exp\left[\frac{i}{\hbar} \int_0^t \left(\frac{m}{2}\dot{y}(u)^2 - V(y(u)) \right) du \right] f(y(t))\, \mathcal{D}_t^\infty[y], \quad (14.1)$$

where $\mathcal{M}$ and $\mathcal{D}_t^\infty[y]$ are the following symbolic expressions

$$\mathcal{M} = \lim_{n\to\infty} \left(2\pi\frac{i\hbar t}{mn}\right)^{-\frac{n}{2}}, \quad \mathcal{D}_t^\infty[y] = \prod_{k=1}^\infty dx_k.$$

The integral in equation (14.1), denoted by $\mathbb{I}_{V,f}(t,x)$, is called a Feynman integral. This integral is only an informal expression because of the following reasons:

(1) The limit $\mathcal{M}$ does not exist.
(2) The infinite dimensional Lebesgue measure $\mathcal{D}_t^\infty[y]$ does not exist.
(3) If $\mathcal{D}_t^\infty[y]$ is replaced by a Gaussian measure on $\mathcal{C}_x$, then the integral $\int_0^t \dot{y}(u)^2 du$ exists only on a set of measure zero.

In fact, the Feynman integral in (14.1) is not the integral of an integrable function with respect to a measure. There are several methods to give a rigorous mathematical definition of Feynman integrals (e.g., see Albeverio and Høegh-Krohn [6] and the references in it). White noise distribution theory provides a very useful and natural tool to study Feynman integrals.

14.2 White noise formulation

Recall that in §11.1 we encountered the same three difficulties above when we tried to define the Fourier transform of a generalized function. Thus it is very natural to use white noise distribution theory to define Feynman integrals. We can use the same idea for the Fourier transform to overcome these difficulties. However, the calculation for Feynman integrals is much more involved and rather complicated.

Before we give a white noise formulation of equation (14.1) we prepare a lemma. Take the white noise space $(\mathcal{S}'(\mathbb{R}), \mu)$. An element in $\mathcal{S}'(\mathbb{R})$ will be denoted by $\dot{B}$ when it is more explanatory. For a real number $c < -1$ or $c > 0$, consider

$$\exp\left[-\frac{1}{2c}\int_0^t \dot{B}(u)^2 du\right].$$

Note that this is just an informal expression since the integral does not exist μ-a.e. This informal expression is similar to the one given in Example 8.7 except for the limits of integration. To renormalize this quantity, take an orthonormal basis $\{e_n; n \geq 1\}$ for $L^2([0,t])$ and define

$$\Phi_n(\dot{B}) \equiv \prod_{k=1}^n \left(\frac{1+c}{c}\right)^{1/2} \exp\left[-\frac{1}{2c}\langle\dot{B}, e_k\rangle^2\right].$$

Obviously $\Phi_n \in (L^2)$ for all n. Moreover, by Example 8.7 (see also Example 13.36), the S-transform of Φ_n is given by

$$S\Phi_n(\xi) = \exp\left[-\frac{1}{2(1+c)}\sum_{k=1}^{n}\langle\xi, e_k\rangle^2\right], \quad \xi \in \mathcal{S}_c(\mathbb{R}).$$

Hence for each ξ, we have

$$\lim_{n\to\infty} S\Phi_n(\xi) = \exp\left[-\frac{1}{2(1+c)}\int_0^t \xi(u)^2\,du\right].$$

It is easy to see that for all $n \geq 1$,

$$|S\Phi_n(\xi)| \leq \exp\left[\frac{1}{2|1+c|}|\xi|_0^2\right].$$

Thus by Theorem 8.6, the sequence Φ_n converges strongly in $(\mathcal{S})^*$. Its limit, denoted by $\mathcal{N}\exp\left[-\frac{1}{2c}\int_0^t \dot{B}(u)^2\,du\right]$, is called the renormalization of $\exp\left[-\frac{1}{2c}\int_0^t \dot{B}(u)^2\,du\right]$ and

$$S\left(\mathcal{N}\exp\left[-\frac{1}{2c}\int_0^t \dot{B}(u)^2\,du\right]\right)(\xi)$$

$$= \exp\left[-\frac{1}{2(1+c)}\int_0^t \xi(u)^2\,du\right], \quad \xi \in \mathcal{S}_c(\mathbb{R}). \qquad (14.2)$$

Note that the right hand side of this equality is defined for any complex number $c \neq -1$. By Theorem 8.2 it is the S-transform of some generalized function. Hence for such a complex number we define the renormalization $\mathcal{N}\exp\left[-\frac{1}{2c}\int_0^t \dot{B}(u)^2\,du\right]$ to be the generalized function with S-transform given by equation (14.2).

The above discussion yields the next lemma.

Lemma 14.1. *The renormalization $\mathcal{N}\exp\left[-\frac{1}{2c}\int_0^t \dot{B}(u)^2\,du\right]$, $c \in \mathbb{C}$, $c \neq -1$, is a generalized function in $(\mathcal{S})^*$ with the following S-transform*

$$S\left(\mathcal{N}\exp\left[-\frac{1}{2c}\int_0^t \dot{B}(u)^2\,du\right]\right)(\xi)$$

$$= \exp\left[-\frac{1}{2(1+c)}\int_0^t \xi(u)^2\,du\right], \quad \xi \in \mathcal{S}_c(\mathbb{R}).$$

Remark. From the above discussion we have the informal expression

$$\mathcal{N}\exp\left[-\frac{1}{2c}\int_0^t \dot{B}(u)^2\,du\right]$$

$$= \left(\sqrt{\tfrac{1+c}{c}}\right)^{\infty}\exp\left[-\frac{1}{2c}\int_0^t \dot{B}(u)^2\,du\right]. \qquad (14.3)$$

Now, we are ready to give the white noise formulation of the Feynman integral $\mathbb{I}_{V,f}(t,x)$ in equation (14.1). For clarity, we do this in several steps:

Step 1. *Still an informal expression, but on the white noise space.*

Let $B(t)$ be the Brownian motion $B(t) = \langle \cdot, 1_{[0,t)} \rangle$ and let y in equation (14.1) be given by

$$y(t) = x - B(t), \quad t \geq 0.$$

Here we use $-B(t)$ for convenience in equation (14.6) below. Note that $-B(t)$ is also a Brownian motion. By using this $y(t)$, we can rewrite equation (14.1) as an informal expression,

$$\mathbb{I}_{V,f}(t,x) = \mathcal{M} \int_{\mathcal{S}'(\mathbb{R})} \exp\left[\frac{im}{2\hbar} \int_0^t \dot{B}(u)^2 \, du\right]$$

$$\times \exp\left[-\frac{i}{\hbar} \int_0^t V\big(x - B(u)\big) \, du\right] f\big(x - B(t)\big) \, \mathcal{D}_t^\infty[\dot{B}],$$

where $\dot{B}$ denotes an element in $\mathcal{S}'(\mathbb{R})$ as mentioned before. Note that the integral is over the white noise space $\mathcal{S}'(\mathbb{R})$ because we want to use the Gaussian measure μ.

Step 2. *The trick: Renormalization.*

In the informal expression for $\mathbb{I}_{V,f}(t,x)$ from step 1, each one of the three quantities $\mathcal{M}$, $\mathcal{D}_t^\infty[\dot{B}]$, and $\exp\left[\frac{im}{2\hbar} \int_0^t \dot{B}(u)^2 \, du\right]$ is bad individually. However, their product turns out to be a well-defined quantity through the renormalization of $\exp\left[\frac{im}{2\hbar} \int_0^t \dot{B}(u)^2 \, du\right]$. To see this, note that $\mathcal{D}_t^\infty[\dot{B}]$ and the Gaussian measure μ are related informally by

$$\mathcal{D}_t^\infty[\dot{B}] = \left(\sqrt{2\pi}\right)^\infty \exp\left[\frac{1}{2} \int_0^t \dot{B}(u)^2 \, du\right] d\mu(\dot{B}).$$

Moreover, in view of equation (14.3), it is reasonable to assert that

$$\mathcal{M} \exp\left[\frac{im}{2\hbar} \int_0^t \dot{B}(u)^2 \, du\right]$$

$$= \left(\mathcal{N} \exp\left[\frac{1}{2}\left(\frac{im}{\hbar} + 1\right) \int_0^t \dot{B}(u)^2 \, du\right]\right) \left(\frac{1}{\sqrt{2\pi}}\right)^\infty \exp\left[-\frac{1}{2} \int_0^t \dot{B}(u)^2 \, du\right].$$

Thus we get the product of these three bad quantities,

$$\mathcal{M} \exp\left[\frac{im}{2\hbar} \int_0^t \dot{B}(u)^2 \, du\right] \mathcal{D}_t^\infty[\dot{B}]$$

$$= \left(\mathcal{N} \exp\left[\frac{1}{2}\left(\frac{im}{\hbar} + 1\right) \int_0^t \dot{B}(u)^2 \, du\right]\right) d\mu(\dot{B}).$$

Hence $\mathbb{I}_{V,f}(t, x)$ can be rewritten as follows:

$$\mathbb{I}_{V,f}(t, x) = \int_{S'(\mathbb{R})} \left(\mathcal{N} \exp\left[\frac{1}{2}\left(\frac{im}{\hbar} + 1 \right) \int_0^t \dot{B}(u)^2 \, du \right] \right)$$

$$\times \exp\left[-\frac{i}{\hbar} \int_0^t V(x - B(u)) \, du \right] f(x - B(t)) \, d\mu(\dot{B}). \quad (14.4)$$

Step 3. *Feynman integrand as a generalized function.*

Equation (14.4) shows that $\mathbb{I}_{V,f}(t, x)$ is an "integral" with respect to μ. But what is this integral? In order to answer this question, we need to know what the integrand is. Let $\mathbb{F}_{V,f}(t, x)$ denote the integrand in equation (14.4), i.e.,

$$\mathbb{F}_{V,f}(t, x) \equiv \left(\mathcal{N} \exp\left[\frac{1}{2}\left(\frac{im}{\hbar} + 1 \right) \int_0^t \dot{B}(u)^2 \, du \right] \right)$$

$$\times \exp\left[-\frac{i}{\hbar} \int_0^t V(x - B(u)) \, du \right] f(x - B(t)). \quad (14.5)$$

For convenience, the integrand $\mathbb{F}_{V,f}(t, x)$ is called a *Feynman integrand*. It depends on the potential function V and the initial condition f. Note that there are three factors in the Feynman integrand $\mathbb{F}_{V,f}(t, x)$. By Lemma 14.1 the first factor is a generalized function in $(\mathcal{S})^*$. Hence the integral in equation (14.4) should be interpreted as the bilinear pairing of a generalized function and a test function. Observe that in general the last two factors in equation (14.5) can not be test functions because they are functions of $B(u), 0 \le u \le t$. Hence the only way to interpret the integral in equation (14.4) is the evaluation of the Feynman integrand $\mathbb{F}_{V,f}(t, x)$ at the test function $\varphi \equiv 1$, i.e.,

$$\int_{S'(\mathbb{R})} \mathbb{F}_{V,f}(t, x) \, d\mu = \langle\langle \mathbb{F}_{V,f}(t, x), 1 \rangle\rangle.$$

Thus if the Feynman integrand $\mathbb{F}_{V,f}(t, x)$ is a generalized function in $(\mathcal{S})^*_\beta$ for some β, then the Feynman integral $\mathbb{I}_{V,f}(t, x)$ is given by

$$\mathbb{I}_{V,f}(t, x) = \int_{S'(\mathbb{R})} \mathbb{F}_{V,f}(t, x) \, d\mu = \langle\langle \mathbb{F}_{V,f}(t, x), 1 \rangle\rangle.$$

To summarize the above discussion, we give the definition of Feynman integrals in white noise language as follows.

Definition 14.2. Suppose the Feynman integrand $\mathbb{F}_{V,f}(t, x)$ in equation (14.5) is a generalized function in the space $(\mathcal{S})^*_\beta$ for some β. Then the *Feynman integral* $\mathbb{I}_{V,f}(t, x)$ is defined by

$$\mathbb{I}_{V,f}(t, x) = \int_{S'(\mathbb{R})} \mathbb{F}_{V,f}(t, x) \, d\mu = \langle\langle \mathbb{F}_{V,f}(t, x), 1 \rangle\rangle.$$

Now comes the most difficult part of the white noise approach to Feynman integrals, namely, the following

Question: Is the Feynman integrand $\mathbb{F}_{V,f}(t,x)$ a generalized function in some $(\mathcal{S})^*_\beta$?

Note that the Feynman integrand $\mathbb{F}_{V,f}(t,x)$ in equation (14.5) is the product of three factors. Unlike the Wick product, the pointwise product is not always defined for two generalized functions. Thus we need to check that the product of the three factors in equation (14.5) is well-defined so that $\mathbb{F}_{V,f}(t,x)$ is a generalized function. As a matter of fact, it is sufficient to consider the case $f = \delta_0$, the Dirac delta function at 0. To see this, note that if $\psi_0(t,x)$ is the solution of the Schrödinger equation with initial condition δ_0, then the solution $\psi(t,x)$ with initial condition f is given by

$$\psi(t,x) = (\psi_0(t) * f)(x).$$

Thus we have $\mathbb{I}_{V,f}(t,x) = (\mathbb{I}_{V,\delta_0}(t) * f)(x)$. Hence we need only to check that $\mathbb{F}_{V,\delta_0}(t,x)$ is a generalized function in some space $(\mathcal{S})^*_\beta$. This has been verified in Streit and Hida [168] for the following cases: (1) $V \equiv 0$ (free particle), (2) $V = $ a quadratic polynomial (e.g., harmonic oscillator), and (3) $V = $ the Fourier transform of a finite measure.

14.3 Explicit calculation

We now carry out explicit calculations to show that the Feynman integrand for a free particle ($V \equiv 0$) is a generalized function in $(\mathcal{S})^*$. The basic idea for the calculation is very simple. In fact, the same idea also works for constant external field ($V(x) = -ax$, a: constant force) and harmonic oscillator ($V(x) = \frac{1}{2}gx^2$, g: gravitation constant). However, the calculations for these cases are rather complicated.

For simplicity, let $\mathbb{F}(t,x)$ denote the Feynman integrand in equation (14.5) with $V \equiv 0$ and $f = \delta_0$, i.e.,

$$\mathbb{F}(t,x) = \left(\mathcal{N} \exp\left[\frac{1}{2}\left(\frac{im}{\hbar} + 1\right) \int_0^t \dot{B}(u)^2 \, du\right]\right) \delta_x(B(t)). \tag{14.6}$$

We have rewritten $\delta_0(x - B(t))$ from equation (14.5) with $f = \delta_0$ as $\delta_x(B(t))$. Note that $\delta_x(B(t))$ is Donsker's delta function in Example 7.4.

Here is the basic idea to show that $\mathbb{F}(t,x)$ is a generalized function in $(\mathcal{S})^*$. Recall from Example 13.9 that $\delta_x(B(t))$ can be represented as a white noise integral by

$$\delta_x(B(t)) = \frac{1}{2\pi} \int_{\mathbb{R}} e^{i\tau(B(t) - x)} \, d\tau.$$

In view of this representation we expect that $\mathbb{F}(t, x)$ can be represented by

$$\mathbb{F}(t, x) = \frac{1}{2\pi} \int_{\mathbb{R}} \left(\mathcal{N} \exp\left[\frac{1}{2}\left(\frac{im}{\hbar} + 1 \right) \int_0^t \dot{B}(u)^2 \, du \right] \right) e^{i\tau(B(t)-x)} \, d\tau. \quad (14.7)$$

Thus what we really need to show is that the product

$$\left(\mathcal{N} \exp\left[\frac{1}{2}\left(\frac{im}{\hbar} + 1 \right) \int_0^t \dot{B}(u)^2 \, du \right] \right) e^{i\tau(B(t)-x)} \quad (14.8)$$

is well-defined and gives a generalized function. Note that the first factor $\mathcal{N} \exp[\cdots]$ is now multiplied by $e^{i\tau(B(t)-x)}$ instead of $\delta_x(B(t))$ as before. Thus the problem of multiplication has been simplified.

Now we carry out the calculations. Let c be a real number $c < -1$ or $c > 0$, and let $g \in L_c^2([0, t])$ (sub-c denotes the complexification). Similar to the sequence $\{\Phi_n\}$ in the derivation of Lemma 14.1, we define

$$\Psi_n(\dot{B}) \equiv \left(\prod_{k=1}^n \left(\frac{1+c}{c} \right)^{1/2} \exp\left[-\frac{1}{2c}\langle \dot{B}, e_k \rangle^2 \right] \right) \exp\left[\langle \dot{B}, g \rangle \right],$$

where $\{e_n; n \geq 1\}$ is an orthonormal basis for $L^2([0, t])$. We can write g as

$$g = \sum_{k=1}^n \langle g, e_k \rangle e_k + h,$$

where $h \perp e_k$ for all $1 \leq k \leq n$. Then

$$\Psi_n(\dot{B})$$
$$= \left(\prod_{k=1}^n \left(\frac{1+c}{c} \right)^{1/2} \exp\left[-\frac{1}{2c}\langle \dot{B}, e_k \rangle^2 + \langle g, e_k \rangle \langle \dot{B}, e_k \rangle \right] \right) \exp\left[\langle \dot{B}, h \rangle \right].$$

The S-transform of Ψ_n is easily checked to be

$$S\Psi_n(\xi) = \exp\left[-\frac{1}{2(1+c)} \sum_{k=1}^n \langle g + \xi, e_k \rangle^2 + \frac{1}{2}\langle g, g \rangle + \langle g, \xi \rangle \right].$$

Hence for each $\xi \in \mathcal{S}_c(\mathbb{R})$,

$$\lim_{n \to \infty} S\Psi_n(\xi) = \exp\left[\frac{c}{2(1+c)} \int_0^t g(u)^2 \, du \right.$$

$$\left. + \frac{c}{1+c} \int_0^t g(u)\xi(u) \, du - \frac{1}{2(1+c)} \int_0^t \xi(u)^2 \, du \right]. \quad (14.9)$$

Moreover, there exist positive constants K and a, independent of n, such that

$$|S\Psi_n(\xi)| \le K \exp\left[a|\xi|_0^2\right], \quad \forall \xi \in \mathcal{S}_c(\mathbb{R}).$$

Thus by Theorem 8.6, Ψ_n converges strongly in $(\mathcal{S})^*$. The limit is denoted by

$$\left(\mathcal{N}\exp\left[-\frac{1}{2c}\int_0^t \dot{B}(u)^2\,du\right]\right)\exp\left[\langle\dot{B},g\rangle\right]. \tag{14.10}$$

Its S-transform is given by the right hand side of equation (14.9). But the right hand side of equation (14.9) is meaningful for any complex number $c \ne -1$. For such a complex number, we can apply Theorem 8.2 to conclude that the right hand side of equation (14.9) is the S-transform of some generalized function, which is defined to be the product in (14.10).

We sum up the above discussion as the next lemma.

Lemma 14.3. *Let* $c \in \mathbb{C}$, $c \ne -1$, *and* $g \in L_c^2([0,t])$. *The product*

$$\left(\mathcal{N}\exp\left[-\frac{1}{2c}\int_0^t \dot{B}(u)^2\,du\right]\right)\exp\left[\langle\dot{B},g\rangle\right]$$

is a generalized function in $(\mathcal{S})^*$ *with the following S-transform*

$$S\left(\left(\mathcal{N}\exp\left[-\frac{1}{2c}\int_0^t \dot{B}(u)^2\,du\right]\right)\exp\left[\langle\dot{B},g\rangle\right]\right)(\xi)$$

$$= \exp\left[\frac{c}{2(1+c)}\int_0^t g(u)^2\,du + \frac{c}{1+c}\int_0^t g(u)\xi(u)\,du\right.$$

$$\left. -\frac{1}{2(1+c)}\int_0^t \xi(u)^2\,du\right].$$

Remark. The subscript c in $L_c^2([0,t])$ denotes the complexification.

Now, apply Lemma 14.3 to the case with $c = -\frac{\hbar}{im+\hbar}$ and $g = i\tau 1_{[0,t)}$. Then the product in (14.8) is well-defined and gives a generalized function in $(\mathcal{S})^*$ with S-transform

$$S\left(\left(\mathcal{N}\exp\left[\frac{1}{2}\left(\frac{im}{\hbar}+1\right)\int_0^t \dot{B}(u)^2\,du\right]\right)e^{i\tau(B(t)-x)}\right)(\xi)$$

$$= \exp\left[-i\tau x - \frac{i\hbar t}{2m}\tau^2 - \frac{\hbar}{m}\tau\int_0^t \xi(u)\,du - \frac{1}{2}\left(1-\frac{i\hbar}{m}\right)\int_0^t \xi(u)^2\,du\right].$$

Moreover, the Feynman integrand $\mathbb{F}(t,x)$ in equation (14.6) can be represented by a white noise integral as in equation (14.7), i.e.,

$$\mathbb{F}(t,x) = \frac{1}{2\pi}\int_{-\infty}^{\infty}\left(\mathcal{N}\exp\left[\frac{1}{2}\left(\frac{im}{\hbar}+1\right)\int_0^t \dot{B}(u)^2\,du\right]\right)e^{i\tau(B(t)-x)}\,d\tau.$$

In fact, we can not apply Theorem 13.4 to get this representation since the integrand does not satisfy the second condition in Theorem 13.4. However, we can use contour integration to show that this representation is still valid as an improper integral in the τ variable. To find the S-transform of $\mathbb{F}(t, x)$, we need to use the formula

$$\frac{1}{\sqrt{2\pi c}} \int_{-\infty}^{\infty} e^{-\frac{1}{2c}\tau^2 - ia\tau} \, d\tau = e^{-\frac{c}{2}a^2}. \tag{14.11}$$

This formula is true for any complex number $c \neq 0$ with the left hand side being an improper integral. By using formula (14.11) with $c = \frac{m}{i\hbar t}$ and $a = x - \frac{i\hbar}{m} \int_0^t \xi(u) \, du$, we can easily derive the S-transform of $\mathbb{F}(t, x)$:

$$S\mathbb{F}(t, x)(\xi) = \sqrt{\frac{m}{2\pi i\hbar t}} \, \exp\left[-\frac{1}{2}\left(1 - \frac{i\hbar}{m}\right) \int_0^t \xi(u)^2 \, du\right.$$

$$\left. + \frac{im}{2\hbar t}\left(x - \frac{i\hbar}{m} \int_0^t \xi(u) \, du\right)^2\right]. \tag{14.12}$$

In particular, we get the Feynman integral $\mathbb{I}(t, x)$ for a free particle

$$\mathbb{I}(t, x) = \langle\!\langle \mathbb{F}(t, x), 1 \rangle\!\rangle$$

$$= S\mathbb{F}(t, x)(0)$$

$$= \sqrt{\frac{m}{2\pi i\hbar t}} \, \exp\left[\frac{im}{2\hbar t} x^2\right].$$

We sum up the above discussion as the next theorem.

Theorem 14.4. *The Feynman integrand $\mathbb{F}(t, x)$ for a free particle* ($V \equiv 0$)

$$\mathbb{F}(t, x) = \left(\mathcal{N} \exp\left[\frac{1}{2}\left(\frac{im}{\hbar} + 1\right) \int_0^t \dot{B}(u)^2 \, du\right]\right) \delta_x(B(t))$$

is a generalized function in $(S)^*$ with S-transform as in equation (14.12). Moreover, the corresponding Feynman integral is given by

$$\mathbb{I}(t, x) = \sqrt{\frac{m}{2\pi i\hbar t}} \, \exp\left[\frac{im}{2\hbar t} x^2\right].$$

We now consider a nonzero potential function $V(x)$. The associated Feynman integrand $\mathbb{F}_V(t, x)$ is given by

$$\mathbb{F}_V(t, x) = \left(\mathcal{N} \exp\left[\frac{1}{2}\left(\frac{im}{\hbar} + 1\right) \int_0^t \dot{B}(u)^2 \, du\right]\right)$$

$$\times \exp\left[-\frac{i}{\hbar} \int_0^t V(x - B(u)) \, du\right] \delta_x(B(t)).$$

We expect that $\mathbb{F}_V(t,x)$ can be represented by a formula similar to the one in equation (14.7)

$$\mathbb{F}_V(t,x) = \frac{1}{2\pi}\int_{\mathbb{R}} \left(\mathcal{N}\exp\left[\frac{1}{2}\left(\frac{im}{\hbar}+1\right)\int_0^t \dot{B}(u)^2\,du\right]\right)$$

$$\times \exp\left[-\frac{i}{\hbar}\int_0^t V(x - B(u))\,du\right] e^{i\tau(B(t)-x)}\,d\tau. \quad (14.13)$$

Thus we need to show that the above integrand

$$\left(\mathcal{N}\exp\left[\frac{1}{2}\left(\frac{im}{\hbar}+1\right)\int_0^t \dot{B}(u)^2\,du\right]\right)$$

$$\times \exp\left[-\frac{i}{\hbar}\int_0^t V(x - B(u))\,du\right] e^{i\tau(B(t)-x)}$$

is a generalized function in $(\mathcal{S})^*$ (or in some space $(\mathcal{S})^*_\beta$). By examining the derivation for Lemma 14.3, we see that similar calculations can be carried out for the potential functions $V_1(x) = -ax$ (constant external force) and $V_2(x) = \frac{1}{2}gx^2$ (harmonic oscillator, g : gravitation constant). Hence we conclude that the Feynman integrands $\mathbb{F}_{V_1}(t,x)$ and $\mathbb{F}_{V_2}(t,x)$ (with $f(x) = \delta_0(x)$ in equation (14.5)) are generalized functions in $(\mathcal{S})^*$. The corresponding Feynman integrals are given by

$$\mathbb{I}_{V_1}(t,x) = \sqrt{\frac{m}{2\pi i\hbar t}}\,\exp\left[\frac{imx^2}{2\hbar t} + \frac{iaxt}{2\hbar} - \frac{ia^2t^3}{24\hbar m}\right],$$

$$\mathbb{I}_{V_2}(t,x) = \sqrt{\frac{m\omega}{2\pi i\hbar \sin(\omega t)}}\,\exp\left[\frac{im\omega x^2}{2\hbar \tan(\omega t)}\right],$$

where $\omega = \sqrt{g/m}$.

Finally, consider the case when V is the Fourier transform of a finite measure m,

$$V(x) = \int_{\mathbb{R}} e^{irx}\,dm(r). \quad (14.14)$$

Informally we have the series expansion

$$\exp\left[-\frac{i}{\hbar}\int_0^t V(x - B(u))\,du\right] = \sum_{n=0}^{\infty} \frac{(-i)^n}{n!\hbar^n}\int_{[0,t]^n}\int_{\mathbb{R}^n}$$

$$\exp\left[i\sum_{k=1}^{n} r_k(x - B(u_k))\right] dm(r_1)\cdots dm(r_n)du_1\cdots du_n.$$

Thus we expect that the Feynman integrand in equation (14.13) with V as in equation (14.14) is given by

$$\mathbb{F}_V(t,x) = \frac{1}{2\pi} \sum_{n=0}^{\infty} \frac{(-i)^n}{n!\hbar^n} \int_{[0,t]^n} \int_{\mathbb{R}^n} \int_{\mathbb{R}}$$

$$\left(\mathcal{N} \exp\left[\frac{1}{2}\left(\frac{im}{\hbar}+1\right)\int_0^t \dot{B}(u)^2\, du\right]\right)$$

$$\times \exp\left[i\tau\left(B(t)-x\right) + i\sum_{k=1}^n r_k\left(x - B(u_k)\right)\right]$$

$$\times d\tau dm(r_1)\cdots dm(r_n) du_1 \cdots du_n. \qquad (14.15)$$

This is the Dyson series for the Feynman integrand in the field of a potential V. By Lemma 14.3 the integrand

$$\left(\mathcal{N} \exp\left[\frac{1}{2}\left(\frac{im}{\hbar}+1\right)\int_0^t \dot{B}(u)^2\, du\right]\right)$$

$$\times \exp\left[i\tau\left(B(t)-x\right) + i\sum_{k=1}^n r_k\left(x - B(u_k)\right)\right]$$

is a generalized function in $(\mathcal{S})^*$. Although rather complicated, it can be shown that the series in equation (14.15) converges in $(\mathcal{S})^*$. Hence the Feynman integrand $\mathbb{F}_V(t,x)$ is a generalized function in $(\mathcal{S})^*$. We can also use Lemma 14.3 to find the S-transform of $\mathbb{F}_V(t,x)$. In particular, the Feynman integral is given by

$$\mathbb{I}_V(t,x) = \sqrt{\frac{m}{2\pi i \hbar t}} \exp\left[\frac{im}{2\hbar t} x^2\right] \sum_{n=0}^{\infty} \frac{(-i)^n}{n!\hbar^n} \int_{[0,t]^n} \int_{\mathbb{R}^n}$$

$$\exp\left[-\frac{i\hbar}{2m} \sum_{j,k=1}^n r_j r_k \left(u_j \wedge u_k - \frac{u_j u_k}{t}\right) + \frac{ix}{t} \sum_{k=1}^n r_k u_k\right]$$

$$\times dm(r_1)\cdots dm(r_n) du_1 \cdots du_n.$$

15

Positive Generalized Functions

In finite dimensional distribution theory it is well-known that a positive generalized function (i.e., $\langle F, \xi \rangle \geq 0$ for any nonnegative test function ξ) is represented by a measure. The same fact holds in white noise distribution theory. Moreover, measures inducing generalized functions can be characterized. In order to prove this characterization theorem, we need to introduce the construction of Lee mentioned in §4.1.

15.1 Positive generalized functions

Among the examples of white noise integral equations in §13.8, some equations have solutions that are induced by Hida measures. In general, it is of interest to determine whether the solution of a given white noise integral equation is induced by a Hida measure. This is due to the fact that we can use this measure to study properties of the solution. For example, if X_t is induced by a Hida measure ν_t which has Radon-Nikodym derivative with respect to the Gaussian measure μ and $d\nu_t/d\mu \in (L^2)$, then X_t is a random variable in (L^2).

Recall that Hida complex measures are defined in §8.4. We will be concerned with only Hida measures in this section. A measure ν on $\mathcal{E}'$ is a Hida measure if $(\mathcal{E})_\beta \subset L^1(\nu)$ and the mapping $\varphi \mapsto \int_{\mathcal{E}'} \varphi(x) \, d\nu(x)$ is continuous on $(\mathcal{E})_\beta$. Thus it induces a generalized function $\tilde{\nu}$ in $(\mathcal{E})_\beta^*$ such that

$$\langle\langle \tilde{\nu}, \varphi \rangle\rangle = \int_{\mathcal{E}'} \varphi(x) \, d\nu(x), \quad \varphi \in (\mathcal{E})_\beta. \tag{15.1}$$

We mention that test functions in $(\mathcal{E})_\beta$ are understood to be their continuous versions. Note that equation (15.1) with $\varphi \equiv 1$ gives $\nu(\mathcal{E}') = \langle\langle \tilde{\nu}, 1 \rangle\rangle$. Hence a Hida measure must be a finite measure.

Question: What generalized functions in $(\mathcal{E})^*_\beta$ are induced by Hida measures?

The idea to answer this question is very simple. Suppose $\Phi \in (\mathcal{E})^*_\beta$ is induced by a Hida measure ν on $\mathcal{E}'$. Then

$$\langle\!\langle \Phi, \varphi \rangle\!\rangle = \int_{\mathcal{E}'} \varphi(x) \, d\nu(x), \quad \varphi \in (\mathcal{E})_\beta. \tag{15.2}$$

In particular, let $\varphi = e^{i\langle \cdot, \xi \rangle}$, $\xi \in \mathcal{E}$. Then

$$\langle\!\langle \Phi, e^{i\langle \cdot, \xi \rangle} \rangle\!\rangle = \int_{\mathcal{E}'} e^{i\langle x, \xi \rangle} \, d\nu(x), \quad \forall \xi \in \mathcal{E}.$$

Note that the left hand side is the restriction of the $\mathcal{T}$-transform $\mathcal{T}\Phi$ of Φ (see Definition 8.13) to $\mathcal{E}$ and the right hand side is the characteristic function $\hat{\nu}$ of ν. Hence we get

$$\mathcal{T}\Phi(\xi) = \hat{\nu}(\xi), \quad \forall \xi \in \mathcal{E}, \tag{15.3}$$

that is, the restriction of $\mathcal{T}\Phi$ to $\mathcal{E}$ and the characteristic function $\hat{\nu}$ are equal.

Now, by the Minlos theorem, the existence of ν can be established by checking that the function $\hat{\nu}$, i.e., $\mathcal{T}\Phi$ (by equation (15.3)) satisfies the following two conditions:

(1) $\mathcal{T}\Phi$ is continuous on $\mathcal{E}$.
(2) $\mathcal{T}\Phi$ is positive definite on $\mathcal{E}$.

Note that we do not need the condition $\mathcal{T}\Phi(0) = 1$ since the measure ν inducing Φ may not be a probability measure.

Consider the first condition that $\mathcal{T}\Phi$ is continuous on $\mathcal{E}$. By the next lemma, this condition is automatically satisfied.

Lemma 15.1. *Let* $\Phi \in (\mathcal{E})^*_\beta$ *be fixed. Then the* $\mathcal{T}$*-transform* $\mathcal{T}\Phi$ *of* Φ *is a continuous function from* $\mathcal{E}_c$ *into* $\mathbb{C}$.

Remark. Since $S\Phi(\xi) = \mathcal{T}\Phi(-i\xi) \exp\left[-\frac{1}{2}\langle \xi, \xi \rangle\right]$, the function $S\Phi$ is also continuous from $\mathcal{E}_c$ into $\mathbb{C}$.

Proof. For any $\xi, \eta \in \mathcal{E}_c$, define

$$f(t) = e^{it\langle \cdot, \xi - \eta \rangle + i\langle \cdot, \eta \rangle}.$$

Then $f(1) = e^{i\langle \cdot, \xi \rangle}$, $f(0) = e^{i\langle \cdot, \eta \rangle}$, and

$$f'(t) = i\langle \cdot, \xi - \eta \rangle \, e^{it\langle \cdot, \xi - \eta \rangle + i\langle \cdot, \eta \rangle}.$$

Hence by the fundamental theorem of calculus, we have

$$e^{i\langle\cdot,\xi\rangle} - e^{i\langle\cdot,\eta\rangle} = i\langle\cdot,\xi-\eta\rangle\, e^{i\langle\cdot,\eta\rangle} \int_0^1 e^{it\langle\cdot,\xi-\eta\rangle}\, dt.$$

Here the integral is a white noise integral and defines a test function in $(\mathcal{E})_\beta$. For any $p \geq 0$, by Theorem 8.18, there exist positive constants C and q such that for all $\xi, \eta \in \mathcal{E}_c$,

$$\left\| e^{i\langle\cdot,\xi\rangle} - e^{i\langle\cdot,\eta\rangle} \right\|_{p,\beta} \leq C \|\langle\cdot,\xi-\eta\rangle\|_{q,\beta} \left\| e^{i\langle\cdot,\eta\rangle} \right\|_{q,\beta} \int_0^1 \left\| e^{it\langle\cdot,\xi-\eta\rangle} \right\|_{q,\beta}\, dt.$$

But $\|\langle\cdot,\xi-\eta\rangle\|_{q,\beta} = |\xi-\eta|_q$ and by Theorem 5.7

$$\left\| e^{i\langle\cdot,\eta\rangle} \right\|_{q,\beta} = \left| e^{\frac{1}{2}\langle\eta,\eta\rangle} \right| \left\| :e^{i\langle\cdot,\eta\rangle}: \right\|_{q,\beta}$$

$$\leq e^{\frac{1}{2}|\eta|_0^2} 2^{\beta/2} \exp\left[(1-\beta)2^{\frac{2\beta-1}{1-\beta}}|\eta|_q^{\frac{2}{1-\beta}}\right].$$

Similarly, for any $0 \leq t \leq 1$, we have

$$\left\| e^{it\langle\cdot,\xi-\eta\rangle} \right\|_{q,\beta} \leq e^{\frac{1}{2}|\xi-\eta|_0^2} 2^{\beta/2} \exp\left[(1-\beta)2^{\frac{2\beta-1}{1-\beta}}|\xi-\eta|_q^{\frac{2}{1-\beta}}\right].$$

Thus for any $\xi, \eta \in \mathcal{E}_c$,

$$\left\| e^{i\langle\cdot,\xi\rangle} - e^{i\langle\cdot,\eta\rangle} \right\|_{p,\beta} \leq C|\xi-\eta|_q\, e^{\frac{1}{2}(|\eta|_0^2+|\xi-\eta|_0^2)} 2^\beta$$

$$\times \exp\left[(1-\beta)2^{\frac{2\beta-1}{1-\beta}}\left(|\eta|_q^{\frac{2}{1-\beta}}+|\xi-\eta|_q^{\frac{2}{1-\beta}}\right)\right].$$

Note that $\mathcal{T}\Phi(\xi) = \langle\!\langle \Phi, e^{i\langle\cdot,\xi\rangle}\rangle\!\rangle$, $\xi \in \mathcal{E}_c$. Hence the last inequality implies that for any $\Phi \in (\mathcal{E})_\beta^*$, the function $\mathcal{T}\Phi \colon \mathcal{E}_c \to \mathbb{C}$ is continuous. $\square$

Next, consider the condition that $\mathcal{T}\Phi$ is positive definite on $\mathcal{E}$. Let $\xi_k \in \mathcal{E}$, $z_k \in \mathbb{C}$, $k = 1,\ldots,n$. Then

$$\sum_{j,k=1}^n z_j \mathcal{T}\Phi(\xi_j - \xi_k)\overline{z_k} = \sum_{j,k=1}^n z_j\langle\!\langle \Phi, e^{i\langle\cdot,\xi_j-\xi_k\rangle}\rangle\!\rangle\overline{z_k}$$

$$= \left\langle\!\left\langle \Phi, \left|\sum_{j=1}^n z_j e^{i\langle\cdot,\xi_j\rangle}\right|^2\right\rangle\!\right\rangle. \tag{15.4}$$

Thus in order for $\mathcal{T}\Phi$ to be positive definite on $\mathcal{E}$, Φ must satisfy the condition that for any $\xi_k \in \mathcal{E}$, $z_k \in \mathbb{C}$, $k = 1,\ldots,n$,

$$\left\langle\!\left\langle \Phi, \left|\sum_{j=1}^n z_j e^{i\langle\cdot,\xi_j\rangle}\right|^2\right\rangle\!\right\rangle \geq 0.$$

This motivates the definition of positivity for generalized functions.

Definition 15.2. A generalized function Φ in $(\mathcal{E})_\beta^*$ is called *positive* if $\langle\!\langle \Phi, \varphi \rangle\!\rangle \geq 0$ for all nonnegative test functions φ in $(\mathcal{E})_\beta$.

Suppose $\Phi \in (\mathcal{E})_\beta^*$ is induced by a Hida measure ν on $\mathcal{E}'$. Then for any nonnegative test function $\varphi \in (\mathcal{E})_\beta$,

$$\langle\!\langle \Phi, \varphi \rangle\!\rangle = \int_{\mathcal{E}'} \varphi(x)\, d\nu(x) \geq 0.$$

Hence Φ is a positive generalized function. It turns out (by the next theorem) that the converse is also true. Thus the set of positive generalized functions consists of precisely those generalized functions induced by Hida measures on $\mathcal{E}'$.

Theorem 15.3. *Let* $\Phi \in (\mathcal{E})_\beta^*$. *Then the following are equivalent:*
(a) Φ *is positive.*
(b) $\mathcal{T}\Phi$ *is positive definite on* $\mathcal{E}$.
(c) Φ *is induced by a Hida measure, i.e., there exists a (finite) measure* ν *on* $\mathcal{E}'$ *such that* $(\mathcal{E})_\beta \subset L^1(\nu)$ *and*

$$\langle\!\langle \Phi, \varphi \rangle\!\rangle = \int_{\mathcal{E}'} \varphi(x)\, d\nu(x), \quad \forall \varphi \in (\mathcal{E})_\beta.$$

Remark. For the case $\beta = 0$, the equivalence of (a) and (c) is due to Kondratiev and Samoilenko [82] and Yokoi [179].

Proof. (c) $\Rightarrow$ (a): This is obvious as mentioned above.
(a) $\Rightarrow$ (b): Let $\xi_k \in \mathcal{E}$, $z_k \in \mathbb{C}$, $k = 1, \ldots, n$. By equation (15.4) we have

$$\sum_{j,k=1}^{n} z_j \mathcal{T}\Phi(\xi_j - \xi_k)\overline{z_k} = \left\langle\!\!\left\langle \Phi, \left| \sum_{j=1}^{n} z_j e^{i\langle \cdot, \xi_j \rangle} \right|^2 \right\rangle\!\!\right\rangle.$$

Observe that

$$\left| \sum_{j=1}^{n} z_j e^{i\langle \cdot, \xi_j \rangle} \right|^2 = \sum_{j,k=1}^{n} z_j \overline{z_k}\, e^{i\langle \cdot, \xi_j - \xi_k \rangle}$$

is a nonnegative test function in $(\mathcal{E})_\beta$. Hence by the positivity of Φ,

$$\sum_{j,k=1}^{n} z_j \mathcal{T}\Phi(\xi_j - \xi_k)\overline{z_k} \geq 0.$$

This shows that $\mathcal{T}\Phi$ is positive definite.

(b) $\Rightarrow$ (c): Suppose $\mathcal{T}\Phi$ is positive definite on $\mathcal{E}$. By Lemma 15.1, $\mathcal{T}\Phi$ is continuous on $\mathcal{E}$. Hence by the Minlos theorem there exists a finite measure ν on $\mathcal{E}'$ such that

$$\mathcal{T}\Phi(\xi) = \int_{\mathcal{E}'} e^{i\langle x,\xi\rangle}\, d\nu(x), \quad \forall \xi \in \mathcal{E},$$

or equivalently,

$$\langle\!\langle \Phi, e^{i\langle\cdot,\xi\rangle}\rangle\!\rangle = \int_{\mathcal{E}'} e^{i\langle x,\xi\rangle}\, d\nu(x), \quad \forall \xi \in \mathcal{E}. \tag{15.5}$$

We need to show that ν is a Hida measure inducing Φ, i.e., $(\mathcal{E})_\beta \subset L^1(\nu)$ and

$$\langle\!\langle \Phi, \varphi\rangle\!\rangle = \int_{\mathcal{E}'} \varphi(x)\, d\nu(x), \quad \forall \varphi \in (\mathcal{E})_\beta. \tag{15.6}$$

Let V be the subspace of $(\mathcal{E})_\beta$ spanned by the set $\{e^{i\langle\cdot,\xi\rangle}; \xi \in \mathcal{E}\}$. It follows from equation (15.5) that

$$\langle\!\langle \Phi, \varphi\rangle\!\rangle = \int_{\mathcal{E}'} \varphi(x)\, d\nu(x), \quad \forall \varphi \in V. \tag{15.7}$$

Note that if $\varphi, \psi \in V$, then $\varphi\psi \in V$. In particular, if $\varphi \in V$, then $|\varphi|^2 \in V$ and by equation (15.7) we have

$$\int_{\mathcal{E}'} |\varphi(x)|^2\, d\nu(x) = \langle\!\langle \Phi, |\varphi|^2\rangle\!\rangle < \infty.$$

Hence $\varphi \in L^2(\nu)$. This shows that $V \subset L^2(\nu)$.

Now, let $\varphi \in (\mathcal{E})_\beta$. Since V is dense in $(\mathcal{E})_\beta$, we can choose a sequence $\{\varphi_n\}$ in V such that $\varphi_n \to \varphi$ in $(\mathcal{E})_\beta$ as $n \to \infty$. Then $\varphi_n \in L^2(\nu)$ since $V \subset L^2(\nu)$ as shown above. Note that the pointwise multiplication on $(\mathcal{E})_\beta$ is continuous (by Theorem 8.18). Hence $|\varphi_n - \varphi|^2 \to 0$ in $(\mathcal{E})_\beta$ as $n \to \infty$. This implies that $|\varphi_n - \varphi_m|^2 \to 0$ in $(\mathcal{E})_\beta$ as $n, m \to \infty$ and so

$$\langle\!\langle \Phi, |\varphi_n - \varphi_m|^2\rangle\!\rangle \to 0, \quad \text{as } n, m \to \infty.$$

Thus by equation (15.7),

$$\int_{\mathcal{E}'} |\varphi_n(x) - \varphi_m(x)|^2\, d\nu(x) = \langle\!\langle \Phi, |\varphi_n - \varphi_m|^2\rangle\!\rangle \to 0, \quad \text{as } n, m \to \infty.$$

Hence $\{\varphi_n\}$ is a Cauchy sequence in $L^2(\nu)$. Let

$$\psi \equiv \lim_{n\to\infty} \varphi_n \quad \text{in } L^2(\nu).$$

Choose a subsequence $\{\varphi_{n'}\}$ of $\{\varphi_n\}$ such that $\varphi_{n'} \to \psi$, ν-a.e. On the other hand, since $\tilde{\delta}_x \in (\mathcal{E})^*$ for any $x \in \mathcal{E}'$ (see §7.2), we have

$$\varphi_n(x) - \varphi(x) = \langle\langle \tilde{\delta}_x, \varphi_n - \varphi \rangle\rangle \to 0, \qquad \text{as } n \to \infty.$$

Hence $\varphi_n(x) \to \varphi(x)$ for all $x \in \mathcal{E}'$. In particular, $\varphi_{n'}(x) \to \varphi(x)$ for all $x \in \mathcal{E}'$. But as shown above we also have $\varphi_{n'} \to \psi$, ν-a.e. Thus we conclude that $\varphi = \psi$, ν-a.e. This implies that $\varphi \in L^2(\nu)$. Moreover, we have

$$\langle\langle \Phi, \varphi \rangle\rangle = \lim_{n' \to \infty} \langle\langle \Phi, \varphi_{n'} \rangle\rangle$$

$$= \lim_{n' \to \infty} \int_{\mathcal{E}'} \varphi_{n'}(x) \, d\nu(x)$$

$$= \int_{\mathcal{E}'} \psi(x) \, d\nu(x)$$

$$= \int_{\mathcal{E}'} \varphi(x) \, d\nu(x).$$

Thus we have shown that $(\mathcal{E})_\beta \subset L^2(\nu)$ (so $(\mathcal{E})_\beta \subset L^1(\nu)$) and equation (15.6) holds. This completes the proof of (b) $\Rightarrow$ (c). $\qquad\square$

Corollary 15.4. *Let $\Phi \in (\mathcal{E})^*_\beta$ and let ν be a finite measure on $\mathcal{E}'$ such that*

$$\mathcal{T}\Phi(\xi) = \int_{\mathcal{E}'} e^{i\langle x, \xi \rangle} \, d\nu(x), \quad \forall \xi \in \mathcal{E}.$$

Then Φ is induced by ν.

Proof. Obviously the assumption implies that $\mathcal{T}\Phi$ is positive definite on $\mathcal{E}$. Hence by Theorem 15.3, Φ is induced by some measure m. Then for all $\varphi \in (\mathcal{E})_\beta$,

$$\langle\langle \Phi, \varphi \rangle\rangle = \int_{\mathcal{E}'} \varphi(x) \, dm(x).$$

By putting $\varphi = e^{i\langle x, \xi \rangle}$ in this equation, we see that m and ν have the same characteristic function. Hence $m = \nu$ and Φ is induced by ν. $\qquad\square$

Example 15.5. Let $x \in \mathcal{E}'$ and let e_x be the evaluation at x, i.e.,

$$\langle\langle e_x, \varphi \rangle\rangle = \varphi(x), \quad \varphi \in (\mathcal{E}).$$

Obviously e_x is positive. In fact, e_x is the generalized function induced by δ_x, i.e., $e_x = \tilde{\delta}_x$.

Example 15.6. Let $x \in \mathcal{E}'$. Then the renormalization $:e^{\langle \cdot, x \rangle}:$ is a positive generalized function. To see this, note that from Theorem 5.13 we have

$$(S :e^{\langle \cdot, x \rangle}:)(\xi) = e^{\langle x, \xi \rangle}, \quad \xi \in \mathcal{E}_c.$$

Hence its $\mathcal{T}$-transform is given by

$$(\mathcal{T}:e^{\langle \cdot, x \rangle}:)(\xi) = e^{-\frac{1}{2}\langle \xi, \xi \rangle}(S:e^{\langle \cdot, x \rangle}:)(i\xi)$$

$$= e^{i\langle x, \xi \rangle - \frac{1}{2}\langle \xi, \xi \rangle}, \quad \xi \in \mathcal{E}_c.$$

In particular, we have

$$(\mathcal{T}:e^{\langle \cdot, x \rangle}:)(\xi) = e^{i\langle x, \xi \rangle - \frac{1}{2}|\xi|_0^2}, \quad \xi \in \mathcal{E}.$$

On the other hand, let μ_x be the translation of the Gaussian measure μ by x. Then for any $\xi \in \mathcal{E}$,

$$\int_{\mathcal{E}'} e^{i\langle y, \xi \rangle} \, d\mu_x(y) = \int_{\mathcal{E}'} e^{i\langle y, \xi \rangle} \, d\mu(y - x)$$

$$= \int_{\mathcal{E}'} e^{i\langle x+w, \xi \rangle} \, d\mu(w)$$

$$= e^{i\langle x, \xi \rangle - \frac{1}{2}|\xi|_0^2}.$$

Thus by Corollary 15.4, the generalized function $:e^{\langle \cdot, x \rangle}:$ is induced by μ_x. Hence it is positive. In particular, when $x = \delta_t$, we have the following positive generalized function

$$:e^{\langle \cdot, \delta_t \rangle}: = :e^{\dot{B}(t)}: .$$

This positive generalized function has been used in Lindstrøm et al. [129].

Example 15.7. Let $f \in (L^2)$ and consider the generalized function Φ_f defined by

$$\langle\!\langle \Phi_f, \varphi \rangle\!\rangle = \langle\!\langle f, \varphi \rangle\!\rangle, \quad \varphi \in (\mathcal{E}).$$

Suppose $f \geq 0$, μ-a.e. Then obviously Φ_f is positive. On the other hand, let $d\nu_f = f \, d\mu$. Then ν_f is a finite measure on $\mathcal{E}'$ and for any $\xi \in \mathcal{E}$,

$$\mathcal{T}\Phi_f(\xi) = \langle\!\langle \Phi_f, e^{i\langle \cdot, \xi \rangle} \rangle\!\rangle$$

$$= \int_{\mathcal{E}'} e^{i\langle x, \xi \rangle} f(x) \, d\mu(x)$$

$$= \int_{\mathcal{E}'} e^{i\langle x, \xi \rangle} \, d\nu_f(x).$$

Thus by Corollary 15.4, Φ_f is induced by ν_f.

Example 15.8. The Gaussian white noise function $g_{x,c}$ with $x \in \mathcal{E}'_c$ and $c \in \mathbb{C}$, $c \neq -1$, is defined in Example 8.3 by its S-transform

$$Sg_{x,c}(\xi) = \exp\left[\frac{1}{1+c}\langle x, \xi \rangle - \frac{1}{2(1+c)}\langle \xi, \xi \rangle\right], \quad \xi \in \mathcal{E}_c.$$

We can easily determine when $g_{x,c}$ is positive as follows. Recall that by equation (8.21) $T\Phi(\xi) = \exp\left[-\frac{1}{2}\langle\xi,\xi\rangle\right]S\Phi(i\xi)$. Hence the T-transform of $g_{x,c}$ on $\mathcal{E}$ is given by

$$Tg_{x,c}(\xi) = \exp\left[\frac{i}{1+c}\langle x,\xi\rangle - \frac{c}{2(1+c)}|\xi|_0^2\right], \quad \xi \in \mathcal{E}.$$

If $g_{x,c}$ is positive, then there exists a finite measure ν on $\mathcal{E}'$ such that $Tg_{x,c} = \hat{\nu}$. Hence

$$\exp\left[\frac{i}{1+c}\langle x,\xi\rangle - \frac{c}{2(1+c)}|\xi|_0^2\right] = \int_{\mathcal{E}'} e^{i\langle x,\xi\rangle}\, d\nu(x), \quad \xi \in \mathcal{E}.$$

By replacing ξ with $\lambda\xi$ in the above equality and then letting $\lambda \to \infty$, we would get a contradiction if $\frac{c}{1+c} < 0$, i.e., $-1 < c < 0$. Thus in order for $g_{x,c}$ to be positive, c must be either $c < -1$ or $0 \le c \le \infty$ ($g_{x,0}$ is understood to be the Kubo-Yokoi delta function $\tilde{\delta}_x$ and $g_{x,\infty} = 1$).

Suppose now that $c < -1$ or $0 \le c \le \infty$. Then by Example 8.4 we have

$$g_{x,c} = \left(\delta_{\frac{1}{1+c}x} * \mu_{\frac{c}{1+c}}\right)^{\sim}.$$

This implies that x must be in $\mathcal{E}'$ in order for $g_{x,c}$ to be induced by a measure on $\mathcal{E}'$. But then by Example 8.4 we have

$$g_{x,c} = \tilde{\mu}_{\frac{1}{1+c}x, \frac{c}{1+c}}. \tag{15.8}$$

Hence the Gaussian white noise function $g_{x,c}$ in Example 8.3 is positive if and only if $x \in \mathcal{E}'$ and c satisfies $c < -1$ or $0 \le c \le \infty$. In that case, $g_{x,c}$ is induced by the Gaussian measure given in equation (15.8).

Example 15.9. Consider Donsker's delta function $\delta_a(B(t))$ in Example 7.4. It is a generalized function in $(\mathcal{S})^*$ with S-transform

$$S\delta_a(B(t))(\xi) = \frac{1}{\sqrt{2\pi t}} \exp\left[-\frac{1}{2t}\left(a - \int_0^t \xi(u)\, du\right)^2\right], \quad \xi \in \mathcal{S}_c(\mathbb{R}).$$

Hence its T-transform with $\xi \in \mathcal{S}(\mathbb{R})$ is given by

$$T\delta_a(B(t))(\xi) = \frac{1}{\sqrt{2\pi t}} \exp\left[-\frac{1}{2}|\xi|_0^2 - \frac{1}{2t}\left(a - i\int_0^t \xi(u)\, du\right)^2\right]$$

$$= \frac{1}{\sqrt{2\pi t}} e^{-\frac{1}{2t}a^2} \exp\left[\frac{ia}{t}\int_0^t \xi(u)\, du\right]$$

$$\times \exp\left[-\frac{1}{2}|\xi|_0^2 + \frac{1}{2t}\left(\int_0^t \xi(u)\, du\right)^2\right].$$

For the first exponential factor, note that the Kubo-Yokoi delta function $\tilde{\delta}_x$ at $x \in \mathcal{S}'$ has $\mathcal{T}$-transform

$$\mathcal{T}\tilde{\delta}_x(\xi) = \exp\left[i\langle x, \xi\rangle\right], \quad \xi \in \mathcal{S}(\mathbb{R}).$$

In particular, for $x = \frac{a}{t}1_{[0,t)}$, we have

$$\mathcal{T}\tilde{\delta}_{\frac{a}{t}1_{[0,t)}}(\xi) = \exp\left[\frac{ia}{t}\int_0^t \xi(u)\,du\right], \quad \xi \in \mathcal{S}(\mathbb{R}). \tag{15.9}$$

On the other hand, let K be the orthogonal projection of $L^2(\mathbb{R})$ onto the subspace spanned by $1_{[0,t)}$, i.e.,

$$Kf = \frac{1}{t}\langle f, 1_{[0,t)}\rangle 1_{[0,t)}.$$

Note that K extends to a measurable transformation on $\mathcal{S}'(\mathbb{R})$. Let $T = I - K$. Then $\mu \circ T^{-1}$ is a probability measure on $\mathcal{S}'(\mathbb{R})$ with characteristic function

$$(\mu \circ T^{-1})\hat{\,}(\xi) = \int_{\mathcal{S}'(\mathbb{R})} e^{i\langle x, \xi\rangle}\,d\mu \circ T^{-1}(x)$$

$$= \int_{\mathcal{S}'(\mathbb{R})} e^{i\langle Ty, \xi\rangle}\,d\mu(y)$$

$$= \int_{\mathcal{S}'(\mathbb{R})} e^{i\langle y, T\xi\rangle}\,d\mu(y).$$

But for any $\xi \in \mathcal{S}(\mathbb{R})$,

$$\int_{\mathcal{S}'(\mathbb{R})} e^{i\langle y, \xi\rangle}\,d\mu(y) = \exp\left[-\frac{1}{2}|\xi|_0^2\right].$$

Hence we get

$$(\mu \circ T^{-1})\hat{\,}(\xi) = \exp\left[-\frac{1}{2}|T\xi|_0^2\right]$$

$$= \exp\left[-\frac{1}{2}|\xi|_0^2 + \frac{1}{2t}\left(\int_0^t \xi(u)\,du\right)^2\right]. \tag{15.10}$$

By comparing $\mathcal{T}\delta_a(B(t))(\xi)$ with equations (15.9) and (15.10), we see immediately that

$$\mathcal{T}\delta_a(B(t))(\xi) = \frac{1}{\sqrt{2\pi t}}e^{-\frac{1}{2t}a^2}\left(\tilde{\delta}_{\frac{a}{t}1_{[0,t)}} * (\mu \circ T^{-1})\right)\hat{\,}(\xi), \quad \xi \in \mathcal{S}(\mathbb{R}).$$

Thus by Corollary 15.4, $\delta_a(B(t))$ is induced by the following measure on $\mathcal{S}'(\mathbb{R})$

$$\frac{1}{\sqrt{2\pi t}} e^{-\frac{1}{2t}a^2} \left(\delta_{\frac{a}{t}1_{[0,t)}} * (\mu \circ T^{-1})\right).$$

Hence Donsker's delta function $\delta_a(B(t))$ is positive.

At the end of this section we mention that the positivity is preserved by convolution (see Definition 8.15). To check this fact, let Φ_1 and Φ_2 be positive generalized functions. By Theorem 15.3 there exist Hida measures ν_1 and ν_2 such that $\Phi_1 = \tilde{\nu}_1$ and $\Phi_2 = \tilde{\nu}_2$. Then $\Phi_1 * \Phi_2 = (\nu_1 * \nu_2)\tilde{}$ and so $\Phi_1 * \Phi_2$ is also positive.

However, the positivity is not preserved by the Wick product. For instance, it is easy to check that $g_{1/2} \diamond g_{1/2} = g_{-1/4}$. By Example 15.8, $g_{1/2}$ is positive, but $g_{-1/4}$ is not.

15.2 Construction of Lee

From the last section we know that a generalized function in $(\mathcal{E})^*_\beta$ is positive if and only if it is induced by a Hida measure on $\mathcal{E}'$. But then how can we tell whether a given measure on $\mathcal{E}'$ is a Hida measure? This is the next question.

Question: How to characterize Hida measures on $\mathcal{E}'$?

In order to obtain a characterization theorem for Hida measures we need to introduce the construction of Lee mentioned in §4.1. In [122] Y.-J. Lee constructed the spaces $\mathcal{A}$ and $\mathcal{A}^*$ of test functions and generalized functions, respectively, without using the Wiener-Itô decomposition theorem. In this section we will follow the same idea to construct general spaces $\mathcal{A}_\beta$ and $\mathcal{A}^*_\beta$ with index $0 \le \beta < 1$ ($\beta = 0$ is Lee's case). As an application we will prove a characterization theorem in the next section to answer the above question.

Recall from Theorem 6.13 that every test function $\varphi \in (\mathcal{E})_\beta$ has a unique extension to $\mathcal{E}'_c$ such that φ is analytic on $\mathcal{E}'_{p,c}$ for any $p \ge 0$ and

$$|\varphi(x)| \le C_{p,q,\beta} \|\varphi\|_{p+q,\beta} \exp\left[\frac{1}{2}(1+\beta)|x|_{-p}^{\frac{2}{1+\beta}}\right], \quad x \in \mathcal{E}'_{p,c}, \qquad (15.11)$$

where $C_{p,q,\beta} = \left(1 - \left(2\lambda_1^{-2q}|\tau|_{-p}\right)^2\right)^{-1/2}\left(1 - 2^{1-\beta}\lambda_1^{-2q}\right)^{-1/2}$.

The above fact is the motivation to define test functions in terms of analyticity and a growth condition.

Definition 15.10. Let $\mathcal{A}_{p,\beta}$, $p \geq 1, 0 \leq \beta < 1$, consist of all functions φ defined on $\mathcal{E}'_c$ satisfying the following conditions:
(1) φ is an analytic function on $\mathcal{E}'_{p,c}$.
(2) There exists some constant $C \geq 0$ such that

$$|\varphi(x)| \leq C \exp\left[\frac{1}{2}(1+\beta)|x|^{\frac{2}{1+\beta}}_{-p}\right], \quad \forall x \in \mathcal{E}'_{p,c}.$$

For $\varphi \in \mathcal{A}_{p,\beta}$, define $\|\varphi\|_{\mathcal{A}_{p,\beta}}$ by

$$\|\varphi\|_{\mathcal{A}_{p,\beta}} = \sup_{x \in \mathcal{E}'_{p,c}} |\varphi(x)| \exp\left[-\frac{1}{2}(1+\beta)|x|^{\frac{2}{1+\beta}}_{-p}\right]. \tag{15.12}$$

Convention: An element φ in $\mathcal{A}_{p,\beta}$ is defined on $\mathcal{E}'_c$. Its restriction to $\mathcal{E}'$ will also be denoted by φ. Thus we will regard φ as a function defined on $\mathcal{E}'_c$ and also as a function defined on $\mathcal{E}'$. However, the norm $\|\varphi\|_{\mathcal{A}_{p,\beta}}$ is the supremum taken over x in $\mathcal{E}'_{p,c}$ as in equation (15.12), *not* over x in $\mathcal{E}'_p$.

Obviously $\|\cdot\|_{\mathcal{A}_{p,\beta}}$ is a norm on $\mathcal{A}_{p,\beta}$ and $(\mathcal{A}_{p,\beta}, \|\cdot\|_{\mathcal{A}_{p,\beta}})$ is a Banach space. Moreover, $\mathcal{A}_{p,\beta} \subset \mathcal{A}_{q,\beta}$ for any $p \geq q \geq 1$ and the inclusion is continuous. Define the space $\mathcal{A}_\beta$ of test functions on $\mathcal{E}'$ by

$$\mathcal{A}_\beta \equiv \bigcap_{p \geq 1} \mathcal{A}_{p,\beta}.$$

We can endow $\mathcal{A}_\beta$ with the projective limit topology in the same way as in §2.2 (for a sequence of Hilbert spaces), i.e., the coarsest topology on $\mathcal{A}_\beta$ such that the inclusion from $\mathcal{A}_\beta$ into $\mathcal{A}_{p,\beta}$ is continuous for each p. Then $\mathcal{A}_\beta$ is a Fréchet space, i.e., a complete metrizable locally convex space (see Reed and Simon [155]). Let

$$\mathcal{A}^*_\beta \equiv \text{the dual space of } \mathcal{A}_\beta,$$

$$\mathcal{A}^*_{p,\beta} \equiv \text{the dual space of } \mathcal{A}_{p,\beta}.$$

The elements in $\mathcal{A}^*_\beta$ are called generalized functions on $\mathcal{E}'$.

Facts. The following properties can be checked easily:
(a) $\mathcal{A}_\beta$ is closed under pointwise multiplication, i.e., $\varphi\psi \in \mathcal{A}_\beta$ for any $\varphi, \psi \in \mathcal{A}_\beta$.
(b) $\mathcal{A}^*_\beta = \cup_{p \geq 1} \mathcal{A}^*_{p,\beta}$.
(c) The inductive limit topology on $\mathcal{A}^*_\beta$ (i.e., the finest locally convex topology such that the inclusion from $\mathcal{A}^*_{p,\beta}$ into $\mathcal{A}^*_\beta$ is continuous for each p) coincides with the strong topology on $\mathcal{A}^*_\beta$.

Note that, on the one hand, we have the spaces $(\mathcal{E})_\beta, (\mathcal{E}_p)_\beta, (\mathcal{E}_p)^*_\beta$, and $(\mathcal{E})^*_\beta$. On the other hand, we have the spaces $\mathcal{A}_\beta, \mathcal{A}_{p,\beta}, \mathcal{A}^*_{p,\beta}$, and $\mathcal{A}^*_\beta$. It is natural to ask whether the spaces $(\mathcal{E})_\beta$ and $\mathcal{A}_\beta$ of test functions are equal as vector spaces and, if so, whether they have the same topology.

Lemma 15.11. *For any $p \geq 1$, there exist positive constants C and $q > p$ such that*

$$\|\varphi\|_{A_{p,\beta}} \leq C\|\varphi\|_{q,\beta}, \quad \forall \varphi \in (\mathcal{E}_q)_\beta.$$

Remark. This lemma implies that for any $p \geq 1$, there exists some $q > p$ such that $(\mathcal{E}_q)_\beta \subset \mathcal{A}_{p,\beta}$ and the inclusion is continuous.

Proof. Let $p \geq 1$ be given. It follows from equation (15.11) that there exist positive constants C and $q \geq p$ such that for any $\varphi \in (\mathcal{E}_q)_\beta$

$$|\varphi(x)| \leq C\|\varphi\|_{q,\beta} \, \exp\left[\frac{1}{2}(1+\beta)|x|^{\frac{2}{1+\beta}}_{-p}\right], \quad \forall x \in \mathcal{E}'_{p,c}. \tag{15.13}$$

Hence by equations (15.12) and (15.13)

$$\|\varphi\|_{A_{p,\beta}} = \sup_{x \in \mathcal{E}'_{p,c}} |\varphi(x)| \exp\left[-\frac{1}{2}(1+\beta)|x|^{\frac{2}{1+\beta}}_{-p}\right] \leq C\|\varphi\|_{q,\beta}. \quad \square$$

Now, recall from §4.2 that $A^{-\alpha/2}$ is a Hilbert-Schmidt operator of E. Hence $(E, \mathcal{E}'_{\alpha/2})$ is an abstract Wiener space. Then by the Fernique theorem in Fact 2.3 (see Fernique [35] or Kuo [98]) there exists a constant $\gamma > 0$ such that

$$\int_{\mathcal{E}'} \exp\left[\gamma|x|^2_{-\alpha/2}\right] d\mu(x) < \infty.$$

Lemma 15.12. *Let $q_0 > \alpha/2$ satisfy $(1+\beta)2^{-\frac{2\beta}{1+\beta}}\lambda_1^{-(2q_0-\alpha)} \leq \gamma$. Then for any $q \geq q_0$, the inequality holds:*

$$\int_{\mathcal{E}'} \exp\left[(1+\beta)2^{-\frac{2\beta}{1+\beta}}|x|^{\frac{2}{1+\beta}}_{-q}\right] d\mu(x)$$

$$\leq \exp\left[(1+\beta)2^{-\frac{2\beta}{1+\beta}}\right] \int_{\mathcal{E}'} \exp\left[c|x|^2_{-\alpha/2}\right] d\mu(x).$$

Remark. Here λ_1 is the smallest eigenvalue of the operator A in §4.2.

Proof. Since $\frac{2}{1+\beta} \leq 2$, we have

$$\int_{\mathcal{E}'} \exp\left[(1+\beta)2^{-\frac{2\beta}{1+\beta}}|x|^{\frac{2}{1+\beta}}_{-q}\right] d\mu(x)$$

$$\leq \int_{\mathcal{E}'} \exp\left[(1+\beta)2^{-\frac{2\beta}{1+\beta}}\left(1+|x|^2_{-q}\right)\right] d\mu(x)$$

$$= \exp\left[(1+\beta)2^{-\frac{2\beta}{1+\beta}}\right] \int_{\mathcal{E}'} \exp\left[(1+\beta)2^{-\frac{2\beta}{1+\beta}}|x|^2_{-q}\right] d\mu(x).$$

Note that $|x|_{-q_0} \leq \lambda_1^{-(q_0-\alpha/2)}|x|_{-\alpha/2}$ since $q_0 \geq \alpha/2$. Hence for any $q \geq q_0$,

$$(1+\beta)2^{-\frac{2\beta}{1+\beta}}|x|^2_{-q} \leq (1+\beta)2^{-\frac{2\beta}{1+\beta}}|x|^2_{-q_0}$$

$$\leq (1+\beta)2^{-\frac{2\beta}{1+\beta}}\lambda_1^{-(2q_0-\alpha)}|x|^2_{-\alpha/2}$$

$$\leq c|x|^2_{-\alpha/2}.$$

This yields the inequality in the lemma immediately. □

Lemma 15.13. *For any $p \geq 1$, there exist positive constants C and $q > p$ such that*

$$\|\varphi\|_{p,\beta} \leq C\|\varphi\|_{\mathcal{A}_{q,\beta}}, \quad \forall \varphi \in \mathcal{A}_{q,\beta}.$$

Remark. This lemma implies that for any $p \geq 1$, there exists some $q > p$ such that $\mathcal{A}_{q,\beta} \subset (\mathcal{E}_p)_\beta$ and the inclusion is continuous.

Proof. We need to carry out calculations similar to those in the proofs of Theorems 8.2 and 8.9. Let $\varphi = \sum_{n=0}^{\infty}\langle :\cdot^{\otimes n}:, f_n\rangle$ and let $F = S\varphi$ (the S-transform of φ). Then as in the proof of Theorem 8.9, we have

$$f_n = \sum_{j_1,\ldots,j_n} J_n(A^p\zeta_{j_1},\ldots,A^p\zeta_{j_n})(A^{-p}\zeta_{j_1}) \otimes \cdots \otimes (A^{-p}\zeta_{j_n}), \quad (15.14)$$

where ζ_n's are eigenvectors of the operator A in §4.2 and J_n is the n-linear functional given as in equation (8.8) by

$$J_n(\xi_1,\ldots,\xi_n) = \frac{1}{n!}\frac{\partial}{\partial z_1}\cdots\frac{\partial}{\partial z_n}F(z_1\xi_1+\cdots+z_n\xi_n)\Big|_{z_1=\cdots=z_n=0}.$$

By the Cauchy formula as in the proof of Theorem 8.2,

$$J_n(\xi_1,\ldots,\xi_n) = \frac{1}{n!}\frac{1}{(2\pi i)^n}\int_{|z_1|=r_1}\cdots\int_{|z_n|=r_n}$$

$$\frac{F(z_1\xi_1+\cdots+z_n\xi_n)}{z_1^2\cdots z_n^2}\,dz_1\cdots dz_n. \quad (15.15)$$

Observe that for any $q \geq 1$,

$$|F(\xi)| \leq \int_{\mathcal{E}'}|\varphi(x+\xi)|\,d\mu(x)$$

$$= \int_{\mathcal{E}'}\left(|\varphi(x+\xi)|\exp\left[-\frac{1}{2}(1+\beta)|x+\xi|^{\frac{2}{1+\beta}}_{-q}\right]\right)$$

$$\times \exp\left[\frac{1}{2}(1+\beta)|x+\xi|^{\frac{2}{1+\beta}}_{-q}\right]d\mu(x)$$

$$\leq \|\varphi\|_{\mathcal{A}_{q,\beta}}\int_{\mathcal{E}'}\exp\left[\frac{1}{2}(1+\beta)|x+\xi|^{\frac{2}{1+\beta}}_{-q}\right]d\mu(x).$$

By using the inequality $(a + b)^{\frac{2}{1+\beta}} \leq 2^{\frac{1-\beta}{1+\beta}}\left(a^{\frac{2}{1+\beta}} + b^{\frac{2}{1+\beta}}\right)$ for $a, b \geq 0$, we can easily get

$$|F(\xi)| \leq \|\varphi\|_{A_{q,\beta}} \exp\left[(1 + \beta)2^{-\frac{2\beta}{1+\beta}}|\xi|_{-q}^{\frac{2}{1+\beta}}\right]$$

$$\times \int_{\mathcal{E}'} \exp\left[(1 + \beta)2^{-\frac{2\beta}{1+\beta}}|x|_{-q}^{\frac{2}{1+\beta}}\right] d\mu(x).$$

Let q_0 be chosen as in Lemma 15.12. Then for any $q \geq q_0$, we can use Lemma 15.12 to get

$$|F(\xi)| \leq L\|\varphi\|_{A_{q,\beta}} \exp\left[(1 + \beta)2^{-\frac{2\beta}{1+\beta}}|\xi|_{-q}^{\frac{2}{1+\beta}}\right], \qquad (15.16)$$

where L is the following constant

$$L = \exp\left[(1 + \beta)2^{-\frac{2\beta}{1+\beta}}\right] \int_{\mathcal{E}'} \exp\left[c|x|_{-\alpha/2}^2\right] d\mu(x).$$

Suppose $|\xi_1|_{-q} = \cdots = |\xi_n|_{-q} = 1$. It follows from equations (15.15) and (15.16) that

$$|J_n(\xi_1, \ldots, \xi_n)|$$

$$\leq L\|\varphi\|_{A_{q,\beta}}\frac{1}{n!}(r_1 \cdots r_n)^{-1} \exp\left[(1 + \beta)2^{-\frac{2\beta}{1+\beta}}(r_1 + \cdots + r_n)^{\frac{2}{1+\beta}}\right].$$

From the proof of Theorem 8.9 we see that the minimum of the right hand side occurs at $r_1 = \cdots = r_n = (2n)^{-\frac{1-\beta}{2}}$. Thus we get the estimate

$$|J_n(\xi_1, \ldots, \xi_n)| \leq L\|\varphi\|_{A_{q,\beta}}\frac{1}{n!}(2n)^{\frac{n(1-\beta)}{2}}e^{\frac{n(1+\beta)}{2}}.$$

This implies that for any $\xi_1, \ldots, \xi_n \in \mathcal{E}_c$, we have

$$|J_n(\xi_1, \ldots, \xi_n)| \leq L\|\varphi\|_{A_{q,\beta}}\frac{1}{n!}(2n)^{\frac{n(1-\beta)}{2}}e^{\frac{n(1+\beta)}{2}}|\xi_1|_{-q}\cdots|\xi|_{-q}.$$

Now, for any given $p \geq 1$, choose some q such that $q > p \vee q_0$. Then we use the same argument as in the proof of Theorem 8.2 to estimate f_n in equation (15.14)

$$|f_n|_p^2 = \sum_{j_1,\ldots,j_n} |J_n(A^p\zeta_{j_1}, \ldots, A^p\zeta_{j_n})|^2$$

$$\leq L^2\|\varphi\|_{A_{q,\beta}}^2\frac{1}{(n!)^2}(2n)^{n(1-\beta)}e^{n(1+\beta)}\|A^{-(q-p)}\|_{HS}^{2n}.$$

Therefore

$$\|\varphi\|_{p,\beta}^2 = \sum_{n=0}^{\infty} (n!)^{1+\beta} |f_n|_p^2$$

$$\leq L^2 \|\varphi\|_{\mathcal{A}_{q,\beta}}^2 \sum_{n=0}^{\infty} \frac{1}{(n!)^{1-\beta}} (2n)^{n(1-\beta)} e^{n(1+\beta)} \|A^{-(q-p)}\|_{HS}^{2n}.$$

Hence by using the inequality $n! \geq (n/e)^n$ (see the proof of Theorem 8.2) we can easily derive that

$$\|\varphi\|_{p,\beta}^2 \leq L^2 \|\varphi\|_{\mathcal{A}_{q,\beta}}^2 \sum_{n=0}^{\infty} 2^{n(1-\beta)} e^{2n} \|A^{-(q-p)}\|_{HS}^{2n}.$$

Finally, choose $q > q_0$ such that $2^{\frac{1-\beta}{2}} e \|A^{-(q-p)}\|_{HS} < 1$. Then we have

$$\|\varphi\|_{p,\beta} \leq L\|\varphi\|_{\mathcal{A}_{q,\beta}} \left(1 - 2^{1-\beta} e^2 \|A^{-(q-p)}\|_{HS}^2\right)^{-1/2}.$$

This yields the inequality in the lemma. $\qquad\square$

Theorem 15.14. *For any* $0 \leq \beta < 1$, $\mathcal{A}_\beta = (\mathcal{E})_\beta$ *as vector spaces. Moreover,* $\mathcal{A}_\beta$ *and* $(\mathcal{E})_\beta$ *have the same topology.*

Remarks. (a) The fact that $\mathcal{A}_\beta$ and $(\mathcal{E})_\beta$ have the same topology means that the two families $\{\| \cdot \|_{\mathcal{A}_{p,\beta}}; p \geq 1\}$ and $\{\| \cdot \|_{p,\beta}; p \geq 1\}$ of norms are equivalent.
(b) This theorem implies that $\mathcal{A}_\beta^* = (\mathcal{E})_\beta^*$ and they have the same inductive limit topology (which coincides with the strong topology).

Proof. For any $p \geq 1$, by Lemma 15.11 there exists some q such that $(\mathcal{E}_q)_\beta \subset \mathcal{A}_{p,\beta}$. But $(\mathcal{E})_\beta \subset (\mathcal{E}_q)_\beta$. Hence $(\mathcal{E})_\beta \subset \mathcal{A}_{p,\beta}$ for any $p \geq 1$. This implies that $(\mathcal{E})_\beta \subset \cap_{p \geq 1} \mathcal{A}_{p,\beta}$ and so $(\mathcal{E})_\beta \subset \mathcal{A}_\beta$. Conversely, we can use Lemma 15.13 to show that $\mathcal{A}_\beta \subset (\mathcal{E})_\beta$. Thus $\mathcal{A}_\beta = (\mathcal{E})_\beta$ as vector spaces. As for the topologies, note that by Lemma 15.11 the identity mapping $(\mathcal{E})_\beta \hookrightarrow \mathcal{A}_\beta$ is continuous. On the other hand, Lemma 15.13 implies that the identity mapping $\mathcal{A}_\beta \hookrightarrow (\mathcal{E})_\beta$ is also continuous. Hence $\mathcal{A}_\beta$ and $(\mathcal{E})_\beta$ have the same topology. $\qquad\square$

15.3 Characterization of Hida measures

We will prove a characterization theorem for Hida measures to answer the question raised in the previous section. But first we point out a simple fact about Hida measures.

Theorem 15.15. *Let ν be a Hida measure on $\mathcal{E}'$ inducing a generalized function $\tilde{\nu}$ in $(\mathcal{E})_\beta^*$. Then $(\mathcal{E})_\beta \subset \cap_{1 \le p < \infty} L^p(\nu)$.*

Proof. Let $\varphi \in (\mathcal{E})_\beta$. Since $(\mathcal{E})_\beta$ is closed under conjugation and pointwise multiplication, $|\varphi|^{2q} \in (\mathcal{E})_\beta$ for any integer $q \ge 1$. Hence we have

$$\int_{\mathcal{E}'} |\varphi(x)|^{2q} \, d\nu = \langle\!\langle \tilde{\nu}, |\varphi|^{2q} \rangle\!\rangle < \infty.$$

Thus $\varphi \in L^{2q}(\nu)$ for any integer $q \ge 1$. Obviously this implies that $\varphi \in L^p(\nu)$ for any integer $p \ge 1$.   □

The next lemma is essential for the proof of the characterization theorem for Hida measures.

Lemma 15.16. *For any $r > 0$, the following inequality holds*

$$e^{ru/4} \le e^{3r/2} \sum_{n=0}^{\infty} \left(\frac{u^n}{2^n n!} \right)^r, \quad \forall u \ge 0.$$

Proof. First suppose $u \ge 2$. Let $k \ge 1$ be the integer such that $2k \le u < 2k + 2$. Consider the term with $n = k$ in the series:

$$\left(\frac{u^k}{2^k k!} \right)^r \ge \left(\frac{(2k)^k}{2^k k!} \right)^r = k^{kr} \left(\frac{1}{k!} \right)^r.$$

But $k! \le e \sqrt{k} \left(\frac{k}{e} \right)^k$ and $k \le e^k$ for any $k \ge 1$. Hence

$$\left(\frac{u^k}{2^k k!} \right)^r \ge k^{kr} \left(\frac{1}{e} \frac{1}{\sqrt{k}} \left(\frac{e}{k} \right)^k \right)^r = \frac{1}{e^r} \frac{1}{k^{r/2}} e^{kr}$$

$$\ge \frac{1}{e^r} \frac{1}{e^{kr/2}} e^{kr} = \frac{1}{e^r} e^{kr/2}.$$

Note that $u \le 2k + 2$. Thus we get

$$\left(\frac{u^k}{2^k k!} \right)^r \ge \frac{1}{e^r} e^{\frac{r}{2}\left(\frac{u}{2} - 1 \right)} \ge e^{-3r/2} e^{ru/4}.$$

Obviously this implies that

$$e^{ru/4} \le e^{3r/2} \sum_{n=0}^{\infty} \left(\frac{u^n}{2^n n!} \right)^r.$$

On the other hand, when $0 \le u < 2$, it is obvious that

$$\sum_{n=0}^{\infty} \left(\frac{u^n}{2^n n!} \right)^r \ge 1 \ge e^{ru/4} e^{-3r/2}.$$

Thus we have proved the inequality in the lemma. $\qquad\square$

Theorem 15.17. *A measure ν on $\mathcal{E}'$ is a Hida measure with $\tilde{\nu} \in (\mathcal{E})^*_\beta$ if and only if ν is supported in $\mathcal{E}'_p$ for some $p \geq 1$ and*

$$\int_{\mathcal{E}'_p} \exp\left[\frac{1}{2}(1+\beta)|x|_{-p}^{\frac{2}{1+\beta}}\right] d\nu(x) < \infty.$$

Remark. The case $\beta = 0$ has been proved in Y.-J. Lee [122]. However, the proof for the necessity part in Y.-J. Lee [122] can not be adapted to the case when $\beta \neq 0$. Note that the condition in the theorem is a Fernique type integrability for ν.

Proof. To prove sufficiency, let $\varphi \in \mathcal{A}_\beta$. By assumption we have

$$\int_{\mathcal{E}'_p} |\varphi(x)| \, d\nu(x)$$

$$= \int_{\mathcal{E}'_p} \left(|\varphi(x)| \exp\left[-\frac{1}{2}(1+\beta)|x|_{-p}^{\frac{2}{1+\beta}}\right]\right) \exp\left[\frac{1}{2}(1+\beta)|x|_{-p}^{\frac{2}{1+\beta}}\right] d\nu(x)$$

$$\leq \|\varphi\|_{A_{p,\beta}} \int_{\mathcal{E}'_p} \exp\left[\frac{1}{2}(1+\beta)|x|_{-p}^{\frac{2}{1+\beta}}\right] d\nu(x).$$

This implies that the linear functional Φ_ν

$$\Phi_\nu(\varphi) = \int_{\mathcal{E}'_p} \varphi(x) \, d\nu(x)$$

is defined on $\mathcal{A}_\beta$ and is continuous. Hence by Theorem 15.14, Φ_ν is a continuous linear functional on $(\mathcal{E})_\beta$, i.e., $\Phi_\nu \in (\mathcal{E})^*_\beta$. Thus ν is a Hida measure inducing a generalized function in $(\mathcal{E})^*_\beta$.

To prove necessity, suppose ν is a Hida measure inducing a generalized function $\tilde{\nu} \in (\mathcal{E})^*_\beta$. By Theorem 15.14 we have $(\mathcal{E})^*_\beta = \mathcal{A}^*_\beta = \cup_{q \geq 1} \mathcal{A}^*_{q,\beta}$. Hence there exists some q such that $\tilde{\nu} \in \mathcal{A}^*_{q,\beta}$. Then

$$\tilde{\nu}(\varphi) = \int_{\mathcal{E}'} \varphi(x) \, d\nu(x), \quad \varphi \in \mathcal{A}_{q,\beta}. \tag{15.17}$$

Define a function ψ on $\mathcal{E}'_{q,c}$ by

$$\psi(x) = \sum_{n=0}^{\infty} \frac{1}{(2^n n!)^{1+\beta}} |x|_{-q}^{2n}.$$

Obviously ψ is analytic on $\mathcal{E}'_{q,c}$. Moreover, by Lemma 6.6

$$\psi(x) = \sum_{n=0}^{\infty} \left(\frac{1}{n!}\left[\frac{1}{2}|x|_{-q}^{\frac{2}{1+\beta}}\right]^n\right)^{1+\beta} \leq \exp\left[\frac{1}{2}(1+\beta)|x|_{-q}^{\frac{2}{1+\beta}}\right].$$

Hence $\psi \in \mathcal{A}_{q,\beta}$. Thus by equation (15.17),

$$\tilde{\nu}(\psi) = \int_{\mathcal{E}'} \psi(x)\,d\nu(x). \tag{15.18}$$

On the other hand, by Lemma 15.16 with $r = 1 + \beta$ and $u = |x|_{-q}^{\frac{2}{1+\beta}}$, we have

$$\exp\left[\frac{1+\beta}{4}|x|_{-q}^{\frac{2}{1+\beta}}\right] \leq e^{3(1+\beta)/2}\,\psi(x), \quad x \in \mathcal{E}'_{q,c}. \tag{15.19}$$

It follows from equations (15.18) and (15.19) that

$$\int_{\mathcal{E}'} \exp\left[\frac{1+\beta}{4}|x|_{-q}^{\frac{2}{1+\beta}}\right] d\nu(x) \leq e^{3(1+\beta)/2}\,\tilde{\nu}(\psi). \tag{15.20}$$

Finally, note that $|x|_{-p} \leq \lambda_1^{-(p-q)}|x|_{-q}$ for any $p \geq q$. We can choose large $p > q$ such that $\lambda_1^{-(p-q)\frac{2}{1+\beta}} \leq \frac{1}{2}$. Then

$$\int_{\mathcal{E}'} \exp\left[\frac{1}{2}(1+\beta)|x|_{-p}^{\frac{2}{1+\beta}}\right] d\nu(x) \leq \int_{\mathcal{E}'} \exp\left[\frac{1+\beta}{4}|x|_{-q}^{\frac{2}{1+\beta}}\right] d\nu(x).$$

Thus by equation (15.20) we have

$$\int_{\mathcal{E}'} \exp\left[\frac{1}{2}(1+\beta)|x|_{-p}^{\frac{2}{1+\beta}}\right] d\nu(x) \leq e^{3(1+\beta)/2}\,\tilde{\nu}(\psi) < \infty.$$

Since $|x|_{-p} = \infty$ for any $x \in \mathcal{E}' \setminus \mathcal{E}'_p$, this inequality shows that the measure ν is supported in $\mathcal{E}'_p$ and

$$\int_{\mathcal{E}'_p} \exp\left[\frac{1}{2}(1+\beta)|x|_{-p}^{\frac{2}{1+\beta}}\right] d\nu(x) < \infty. \qquad \square$$

Example 15.18. Recall from Example 8.5 that the grey noise measure ν_λ, $0 < \lambda \leq 1$, is the probability measure on $\mathcal{E}'$ with characteristic function

$$\int_{\mathcal{E}'} e^{i\langle x,\xi\rangle}\,d\nu_\lambda(x) = L_\lambda(|\xi|_0^2), \quad \xi \in \mathcal{E},$$

where L_λ is the Mittag-Leffler function. It is shown in Example 8.5 that ν_λ induces a generalized function $\tilde{\nu}_\lambda$ in $(\mathcal{E})_{1-\lambda}^*$. Hence by Theorem 15.17 the grey noise measure ν_λ is supported in $\mathcal{E}'_p$ for some p and

$$\int_{\mathcal{E}'_p} \exp\left[\frac{1}{2}(2-\lambda)|x|_{-p}^{\frac{2}{2-\lambda}}\right] d\nu_\lambda(x) < \infty.$$

Appendix A: Notes and Comments

Chapter 1

§1.1 For several constructions of a Brownian motion, see Hida [47]. Wiener constructed a Brownian motion in Wiener [176] by introducing a Gaussian measure on the space $C[0,1]$ of continuous functions on $[0,1]$. See Kuo [98] for details. On the other hand, a Gaussian measure can be introduced on the space of tempered distributions by the Minlos theorem. Then a Brownian motion can be defined by equation (3.1) in Chapter 3. For properties of Brownian motion, see McKean [131]. As for generalized Gaussian processes, see Gel'fand and Vilenkin [38].

Chapter 2

§2.2 Countably-Hilbert spaces are used throughout the book. But in §15.2 and §15.3 we also use Fréchet spaces. A Fréchet space is defined to be a complete metrizable locally convex space (see Reed and Simon [155]). Equivalently, a topological vector space V with topology generated by a family $\{|\cdot|_n; n \geq 1\}$ of norms is called a Fréchet space if it is complete with respect to the metric $d(u,v)$ given in §2.2. Suppose $\{(V_n, |\cdot|_n); n \geq 1\}$ is a sequence of Banach spaces such that V_{n+1} is continuously imbedded in V_n for each n. Let $V = \cap_{n=1}^{\infty} V_n$ and endow V with the coarsest topology such that for each n the inclusion from V into V_n is continuous. Then V, called the projective limit of $\{V_n; n \geq 1\}$, is a Fréchet space. The dual space V' of V is the inductive limit of $\{V'_n; n \geq 1\}$ and we have $V' = \cup_{n=1}^{\infty} V'_n$.

§2.3 A Fréchet space V with a family $\{|\cdot|_n; n \geq 1\}$ of norms is called a nuclear space if there exists an equivalent family $\{\|\cdot\|_n; n \geq 1\}$ of inner product norms such that V is a nuclear space with respect to $\{\|\cdot\|_n; n \geq 1\}$ as defined in §2.3.

Chapter 3

This chapter serves as a follow-up to Chapter 1 and provides a smooth transition to the more general constructions of test and generalized functions in Chapter 4.

For another infinite dimensional calculus, the Malliavin calculus, see the recent book by Nualart [134].

§3.3 The decomposition of (L^2) into homogeneous chaos was introduced by Wiener [177]. In the case of Brownian motion, Itô defined multiple Wiener integrals in [68] and proved the decomposition of (L^2) into multiple Wiener integrals. In this section we take for granted multiple Wiener integrals and the Wiener-Itô decomposition theorem. However, in §4.2 we briefly define multiple Wiener integrals. The second quantization operator $\Gamma(A)$ associated with $A = -d^2/dx^2 + x^2 + 1$ was first used in white noise distribution theory by Potthoff and Streit [150].

Chapter 4

In [89] Kubo has given a direct construction of generalized functions by using a weighted number operator. In Kondratiev and Streit [84] a space $\mathcal{M}$ of test functions and its dual $\mathcal{M}^*$ due to Meyer-Yan are discussed. These spaces are related to $(\mathcal{E})_\beta$ and $(\mathcal{E})_\beta^*$ as follows:

$$\mathcal{M} \subset (\mathcal{E})_\beta \subset (L^2) \subset (\mathcal{E})_\beta^* \subset \mathcal{M}^*.$$

However, the space $\mathcal{M}$ is not a nuclear space. Recently, a new space $[\mathcal{E}]$ of test functions has been introduced in Cochran et al. [25]. This space is a nuclear space and its dual $[\mathcal{E}]^*$ contains interesting noises such as the Poisson noise. These spaces are related to the above spaces by:

$$\mathcal{M} \subset [\mathcal{E}] \subset (\mathcal{E})_\beta \subset (L^2) \subset (\mathcal{E})_\beta^* \subset [\mathcal{E}]^* \subset \mathcal{M}^*.$$

Chapter 5

§5.1 Wick tensors are usually defined by induction. However, it is more convenient to define $: x^{\otimes n} :$ directly as in Definition 5.1. For instance, the relationships between Wick tensors and Hermite polynomials with a parameter become apparent and can be checked very easily.

§**5.2**   The S-transform $S\Phi$ of a generalized function Φ has long been regarded as a function defined on the real nuclear space $\mathcal{E}$; e.g., see the book by Hida et al. [58], Kubo and Takenaka [92] [93] [94] [95], and Potthoff and Streit [150]. In this case, the analyticity is expressed in terms of being a ray entire function. Recently, the S-transform $S\Phi$ is regarded as a function defined on the complexification $\mathcal{E}_c$ of $\mathcal{E}$ in Kondratiev and Streit [83] [84] and Obata [142]. This is more convenient for the discussion of analyticity. However, these two concepts are actually equivalent as shown in a recent paper by Kondratiev et al. [81].

In the proof of Theorem 5.7 we can split the n-th term in a different way and then apply the Hölder inequality. This gives a more general estimate as given in Kondratiev and Streit [84]. We have chosen the particular estimate in Theorem 5.7 just for convenience.

§**5.3**   The additive renormalization $:e^{z\langle\cdot,y\rangle^k}:$ is a motivation for introducing the more general space $(\mathcal{E})^*_\beta$ of generalized functions with $\beta \neq 0$. There are other motivations; e.g., the grey noise measure in Example 8.5 and white noise integral equations in Chapter 13.

• **The Segal-Bargmann transform**:   When Hida introduced the theory of white noise in [45], he used the $\mathcal{T}$-transform (see Definition 8.13). The S-transform for white noise was first used in Kubo and Takenaka [92]. However, the S-transform acting on the space (L^2) is known as the Segal-Bargmann transform. We make some comments about this transform based on private conversations with L. Gross and information from Gross and Malliavin [43]. Let $\mathcal{E} \subset E \subset \mathcal{E}'$ be a Gel'fand triple and μ the standard Gaussian measure on $\mathcal{E}'$. When $E = \mathbb{R}^n$, Segal [162] [163] [164] and Bargmann [10] showed that the S-transform is a unitary operator from $L^2(\mathbb{R}^n, \mu)$ onto a Hilbert space H^n of analytic functions on $\mathbb{C}^n$. See also a recent paper by Kubo and Kuo [90]. In fact, H^n is the Hilbert space $\mathcal{H}L^2(\mathbb{C}^n, m)$ consisting of holomorphic L^2-functions with respect to the measure $dm(z) = (\pi)^{-n} \exp[-|z|^2]\, dz$ on $\mathbb{C}^n$ (here $dz = dxdy$) (see Gross and Malliavin [43]). When E is infinite dimensional, it has been shown in Kondratiev [80], Krée [85] [86] [87], and Y.-J. Lee [123] that the S-transform is a unitary operator from (L^2) onto a Hilbert space H of analytic functions on E_c. Here H is the Hilbert space $\mathcal{H}L^2(E_c, m)$ consisting of holomorphic functions f on E_c such that $\|f\|^2 \equiv \sup\{\int_F |f(z)|^2\, dm_F(z)\} < \infty$ with the supremum being taken over finite dimensional subspaces F of E_c and $m_F = m \circ F^{-1}$. The inverse of the Segal-Bargmann transform is related to the Fourier-Wiener transform introduced in Cameron and Martin [19]. Segal gave a generalization of the Fourier-Wiener transform to an arbitrary Hilbert space in [162]. Hitsuda [65] rediscovered the result of Cameron and Martin in terms of conditional expectations of Brownian motion. For a recent exposition of this theory, see Gross and Malliavin [43]. See also notes and comments for §11.5.

Chapter 6

In finite dimensional distribution theory we have the Gel'fand triple
$\mathcal{S}(\mathbb{R}^n) \subset L^2(\mathbb{R}^n) \subset \mathcal{S}'(\mathbb{R}^n)$. A test function f in $\mathcal{S}(\mathbb{R}^n)$ satisfies smoothness and growth conditions, i.e., C^∞ and rapidly decreasing. In white noise distribution theory we use the Gel'fand triple $(\mathcal{E})_\beta \subset (L^2) \subset (\mathcal{S})^*_\beta$. A test function φ in $(\mathcal{E})_\beta$ satisfies analyticity and growth conditions (Theorem 6.13).

Theorem 6.5 (continuous version of a test function) is due to Kubo and Yokoi [96], while Theorem 6.13 (analytic version and growth condition of a test function for the case $\beta = 0$) is due to Y.-J. Lee [122]. Our proof of Theorem 6.5 follows more or less the same idea as in Kubo and Yokoi [96]. However, our proof of Theorem 6.13 is different from Y.-J. Lee [122].

The main idea in Chapter 6 is very simple, i.e., to realize that a test function φ in $(\mathcal{E})_\beta$ can be represented by $\varphi(x) = \langle\!\langle :e^{\langle \cdot, x \rangle}:, \Theta\varphi \rangle\!\rangle$. This idea is essentially due to Kubo and Yokoi [96]. The operator Θ is continuous from $(\mathcal{E})_\beta$ into itself (Theorem 6.2) and the mapping $x \mapsto :e^{\langle \cdot, x \rangle}:$ is continuous from $\mathcal{E}'$ into $(\mathcal{E})^*$ (Theorem 6.4). From the representation $\varphi(x) = \langle\!\langle :e^{\langle \cdot, x \rangle}: , \Theta\varphi \rangle\!\rangle$, all nice properties of φ follow immediately; e.g., continuous version, analytic version, growth condition, and norm estimate.

Chapter 7

§7.1 Let f be a nonzero function in $L^2(\mathbb{R})$. If $F \in \mathcal{S}'_c(\mathbb{R})$, then $F(\langle \cdot, f \rangle) \in (\mathcal{S})^*$ (Theorem 7.3). It is natural to ask whether the converse is also true. On the other hand, let $\xi \in \mathcal{S}(\mathbb{R})$. Suppose g is a function on $\mathbb{R}$ such that its Fourier transform belongs to $L^1(\mathbb{R})$ and has compact support. Then it is shown in Kuo et al. [115] that $g(\langle \cdot, \xi \rangle) \in (\mathcal{S})$. However, $g \in \mathcal{S}(\mathbb{R})$ does not imply that $g(\langle \cdot, \xi \rangle) \in (\mathcal{S})$; e.g., $\exp[-\langle \cdot, \xi \rangle^2] \notin (\mathcal{S})$. It is natural to ask whether it is possible to characterize those functions g so that $g(\langle \cdot, \xi \rangle) \in (\mathcal{S})$.

The finite dimensional analogue of $(\mathcal{E}) \subset (L^2) \subset (\mathcal{E})^*$ is the Gel'fand triple $\mathcal{H}(\mathbb{R}^n) \subset L^2(\mathbb{R}^n, \mu_n) \subset \mathcal{H}^*(\mathbb{R}^n)$ introduced in Kubo and Kuo [90]. Here μ_n is the standard Gaussian measure on $\mathbb{R}^n$. By Kubo and Kuo [90, Theorem 2.3] we have $\mathcal{S}'_c(\mathbb{R}^n) \subset \mathcal{H}^*(\mathbb{R}^n)$ as *sets*. Let $\{f_k; 1 \leq k \leq n\} \subset E$ be orthonormal. Then by Kubo and Kuo [90, Theorems 3.2 and 3.3] $F(\langle \cdot, f_1 \rangle, \ldots, \langle \cdot, f_n \rangle) \in (\mathcal{E})^*$ if and only if $F \in \mathcal{H}^*(\mathbb{R}^n)$. In particular, we see that Theorem 7.3 holds for $F \in \mathcal{H}^*(\mathbb{R})$. On the other hand, let $\{\xi_k; 1 \leq k \leq n\} \subset \mathcal{E}$ be orthonormal in E. Then by Kubo and Kuo [90, Theorems 3.4 and 3.5] $g(\langle \cdot, \xi_1 \rangle, \ldots, \langle \cdot, \xi_n \rangle) \in (\mathcal{E})$ if and only if $g \in \mathcal{H}(\mathbb{R}^n)$. For the generalizations of these results to the spaces $(\mathcal{E})_\beta$ and $(\mathcal{E})^*_\beta$, see K. Sim Lee

[119]. For example, the Gel'fand triple $\mathcal{H}_\beta(\mathbb{R}^n) \subset L^2(\mathbb{R}^n, \mu_n) \subset \mathcal{H}_\beta^*(\mathbb{R}^n)$ can be defined. For nonzero $f \in E$, we have $F(\langle \cdot, f \rangle) \in (\mathcal{E})_\beta^*$ if and only if $F \in \mathcal{H}_\beta^*(\mathbb{R})$.

§7.2 The basic idea in this section is the expansion in Theorem 7.8

$$\widetilde{\delta}_x = \sum_{n=0}^{\infty} \frac{1}{n!} \langle :\cdot^{\otimes n}:, :x^{\otimes n}: \rangle.$$

From this expansion we can derive a variety of results regarding the Kubo-Yokoi delta function $\widetilde{\delta}_x$. The right hand side of the inequality in Theorem 7.9 is due to Kubo and Yokoi [96]. In the proof of Theorem 7.9 we essentially follow their ideas.

§7.3 Observe the similarity of the expansion $\widetilde{\delta}_x = \sum_{n=0}^{\infty} \frac{1}{n!} \langle :\cdot^{\otimes n}:, :x^{\otimes n}: \rangle$ and the expansion of $:e^{\langle \cdot, x \rangle}:$ given in Definition 5.12

$$:e^{\langle \cdot, x \rangle}: = \sum_{n=0}^{\infty} \frac{1}{n!} \langle :\cdot^{\otimes n}:, x^{\otimes n} \rangle.$$

Because of this similarity, the proofs of Theorems 7.18 and 7.20 are similar to those of Theorem 6.4 and Lemma 6.10, respectively. The estimate in Theorem 7.20 is for the norm $\|\widetilde{\delta}_y - \widetilde{\delta}_x\|_{-q}$. However, we can easily modify the argument in the proof to get an estimate for $\|\widetilde{\delta}_y - \widetilde{\delta}_x\|_{-q, -\beta}$. On the other hand, the generalized functions $\widetilde{\delta}_x$ and $:e^{\langle \cdot, x \rangle}:$ are related by $\widetilde{\delta}_x = \widetilde{\delta}_0 \diamond :e^{\langle \cdot, x \rangle}:$. Here $\diamond$ denotes the Wick product (Definition 8.11). Hence the properties of $\widetilde{\delta}_x$ can also be derived by using the continuity of the Wick product (Remark (2) of Theorem 8.12) and the corresponding properties of $:e^{\langle \cdot, x \rangle}:$.

Chapter 8

§8.1 In the original proof of Theorem 8.2 for the space $(\mathcal{S})^*$ in Potthoff and Streit [150], J_n is defined on the diagonal $(\xi, \ldots, \xi)$ and then the polarization identity is used to define f_n. Our idea in the proof is essentially the same as in Kubo and Kuo [91], i.e., $J_n(\xi_1, \ldots, \xi_n)$ and $f_n \in (\mathcal{E}_c')^{\widehat{\otimes} n}$ are defined directly without using the polarization identity. The grey noise measure in Example 8.5 provides an important motivation for introducing the space $(\mathcal{E})_\beta^*$ of generalized functions with $0 \leq \beta < 1$.

§8.2 Theorem 8.6 was often used for renormalization in the early stage of white noise distribution theory. For instance, consider the informal expression $\dot{B}(t)^2$. Intuitively we have $\dot{B}(t)^2 = \lim_{\epsilon \to 0} \epsilon^{-2} \big(B(t + \epsilon) - B(t) \big)^2$.

But this limit does not exist in (L^2), nor in $(\mathcal{S})^*$. However, if we consider $\lim_{\epsilon \to 0} \left[\epsilon^{-2} \left(B(t+\epsilon) - B(t) \right)^2 - \epsilon^{-1} \right]$, then by Theorem 8.6 the limit exists in $(\mathcal{S})^*$. The limit is the renormalization $:\dot{B}(t)^2:$.

§8.4 The Wick product as defined in the form $S(\varphi \diamond \psi) = (S\varphi)(S\psi)$ was introduced in Hida and Ikeda [56]. It has also been discussed in Meyer and Yan [132]. Theorem 8.18 for the case $\beta = 0$ is due to Kubo and Takenaka [93]. Various proofs of this special case have been given in the book by Hida et al. [58], Kuo [110], Y.-J. Lee [122], Obata [142], and Potthoff and Yan [152]. Corollary 8.19 with $\beta = 0$ is due to Obata [142].

§8.5 Theorems 8.21 and 8.23 are due to Obata [142].

• **Hida complex measures:** Let ν be a complex measure and let $|\nu|$ be the total variation of ν. Then there exists a measurable function f with $|f| = 1$ such that $d\nu = f\,d|\nu|$. Suppose $|\nu|$ is a Hida measure inducing a generalized function $\Phi_{|\nu|}$ in $(\mathcal{E})^*_\beta$ and $f \in (\mathcal{E})_\beta$. Then it is obvious that ν is a Hida complex measure inducing a generalized function Φ_ν in $(\mathcal{E})^*_\beta$ and we have $\Phi_\nu = f\,\Phi_{|\nu|}$. In general we can not expect ν to be a Hida complex measure simply because $|\nu|$ is a Hida measure. For example, by the remark preceding Theorem 8.21, an $L^1(\mu)$-function may not define a generalized function. But then is it possible to characterize f so that ν is a Hida complex measure?

• **The generalized Radon-Nikodym derivative:** Consider the translation μ_x of μ by $x \in \mathcal{E}'$, i.e., $\mu_x(\cdot) = \mu(\cdot - x)$. The measure μ_x is absolutely continuous with respect to μ if and only if $x \in E$. In that case, its Radon-Nikodym derivative is given by

$$\frac{d\mu_x}{d\mu} = e^{\langle \cdot, x \rangle - \frac{1}{2}|x|_0^2} = :e^{\langle \cdot, x \rangle}: .$$

Observe that $:e^{\langle \cdot, x \rangle}:$ is a generalized function for any $x \in \mathcal{E}'$. Thus for $x \in \mathcal{E}'$, we may say that the measure μ_x is absolutely continuous with respect to μ in the generalized sense. The generalized function $d\mu_x/d\mu$ in $(\mathcal{E})^*$ is called a *generalized Radon-Nikodym derivative* (see Potthoff and Streit [151] and Potthoff and Yan [152]). In fact, $d\mu_x/d\mu$ is nothing but the generalized function induced by μ_x. Similarly, for any $t > 0$, the measure $\mu_t(\cdot) = \mu(\cdot/\sqrt{t})$ is absolutely continuous with respect to μ in the generalized sense and its generalized Radon-Nikodym derivative $d\mu_t/d\mu$ has S-transform

$$S\frac{d\mu_t}{d\mu}(\xi) = \exp\left[-\frac{1}{2}(1-t)\langle \xi, \xi \rangle \right], \quad \xi \in \mathcal{E}_c.$$

Suppose L is a continuous linear transformation from $\mathcal{E}$ into itself. Then it is easy to see that the measure $\mu \circ (L^*)^{-1}$ is a Hida measure and its generalized Radon-Nikodym derivative has S-transform

$$S\frac{d(\mu \circ (L^*)^{-1})}{d\mu}(\xi) = \exp\left[-\frac{1}{2}\big(\langle\xi,\xi\rangle - \langle L\xi, L\xi\rangle\big)\right], \quad \xi \in \mathcal{E}_c.$$

Now, consider a nonlinear transformation T from $\mathcal{E}'$ into itself. In general, $\mu \circ T^{-1}$ is not a Hida measure. It is an interesting problem to find conditions on T such that $\mu \circ T^{-1}$ is also a Hida measure. More generally, we can take a Hida measure ν and ask the same question, i.e., find the conditions on a nonlinear transformation T such that $\nu \circ T^{-1}$ is also a Hida measure. This type of question can be regarded as a generalized Girsanov theorem.

Chapter 9

§9.1 An original motivation for Hida to introduce the theory of white noise is to express Brownian functionals as functions of white noise by regarding $\{\dot{B}(t); t \in \mathbb{R}\}$ as a coordinate system. This leads to white noise differentiation $\partial/\partial\dot{B}(t)$ (denoted by ∂_t for simplicity). The white noise derivative $\partial_t \varphi$ of a white noise functional φ was originally defined in Hida [45] by the U-functional as $U[\partial_t \varphi](\xi) = \partial U(\xi)/\partial\delta\xi(t)$ (the right hand side being a functional derivative). The U-functional is what we call S-transform nowadays. In Kuo [106] $\partial_t \varphi$ was defined in terms of Fréchet derivative of φ by $\partial_t \varphi(x) = \langle\delta_t, \varphi'(x)\rangle$. In Potthoff and Yan [152] the Gâteaux derivative was used to define $D_y\varphi$ for $y \in \mathcal{S}'(\mathbb{R})$ and $\varphi \in (\mathcal{S})$. In particular, $\partial_t \varphi = D_{\delta_t}\varphi$.

A good way to understand $D_y\varphi$ is to regard it as a bilinear mapping of y and φ. Then we can easily discuss restriction and extension of $D_y\varphi$, i.e., $D_{(\cdot)}(\cdot)$ is continuous (1) from $\mathcal{E}' \times (\mathcal{E})_\beta$ into $(\mathcal{E})_\beta$ and (2) from $\mathcal{E} \times (\mathcal{E})_\beta^*$ into $(\mathcal{E})_\beta^*$.

§9.2 In finite dimensional distribution theory, we have $(\partial/\partial x_k)^* = -\partial/\partial x_k$ (due to the fact that the Lebesgue measure is used). But the multiplication by x_k can not be expressed in terms of $\partial/\partial x_k$ and $(\partial/\partial x_k)^*$. In white noise distribution theory, the situation is different due to the fact that the Gaussian measure μ is used. In fact the sum $D_y + D_y^*$ equals to multiplication by $\langle \cdot, y \rangle$.

The renormalization $:\dot{B}(t_1)\cdots\dot{B}(t_n):$ can be expressed in terms of the adjoint operator ∂_t^* by $:\dot{B}(t_1)\cdots\dot{B}(t_n): = \partial_{t_1}^*\cdots\partial_{t_n}^* 1$. This implies that $:e^{\dot{B}(t)}: = e^{\partial_t^*} 1$.

§9.3 Similar to $D_y\varphi$, the multiplication $Q_y\varphi$ is better understood as a bilinear mapping, i.e., $Q_{(\cdot)}(\cdot)$ is continuous (1) from $\mathcal{E} \times (\mathcal{E})_\beta$ into $(\mathcal{E})_\beta$, (2) from $\mathcal{E} \times (\mathcal{E})_\beta^*$ into $(\mathcal{E})_\beta^*$, and (3) from $\mathcal{E}' \times (\mathcal{E})_\beta$ into $(\mathcal{E})_\beta^*$. For any $\xi \in \mathcal{E}$, we have $Q_\xi = D_\xi + D_\xi^*$ as continuous operators from $(\mathcal{E})_\beta$ into itself and $\widetilde{Q}_\xi = \widetilde{D}_\xi + D_\xi^*$ as continuous operators from $(\mathcal{E})_\beta^*$ into itself. On the other hand, for $y \in \mathcal{E}'$, we have $Q_y = D_y + D_y^*$ as continuous operators from $(\mathcal{E})_\beta$ into $(\mathcal{E})_\beta^*$. In particular, $\dot{B}(t) = \partial_t + \partial_t^*$ as continuous operators from $(\mathcal{S})_\beta$ into $(\mathcal{S})_\beta^*$.

§9.4   The Gross differentiation D_h is used in abstract Wiener space theory. For instance, let f be a bounded Lip-1 function on an abstract Wiener space (H, B). Then the convolution $\mu * f$ of f with the standard Gaussian measure μ on B is smooth in H-directions (see Kuo [102]), i.e., infinitely Gross differentiable.

The integral $\int_T \partial_t^* X(t)\, d\nu(t)$ is a Hitsuda-Skorokhod integral (see §13.4). In an abstract Wiener space, $\nabla^* X$ is the divergence of X (see Goodman [39] and Kuo [98]).

Chapter 10

§10.2   The main results in this section are Theorems 10.2, 10.5, and 10.6. They are due to Hida, Obata, and Saitô (see Hida et al. [60]) for the case $\beta = 0$.

§10.3   The formulas $\Delta_G = \int_{\mathbb{R}} \partial_t^2\, dt$ (Theorem 10.13) and $N = \int_{\mathbb{R}} \partial_t^* \partial_t\, dt$ (Theorem 10.16) have been known for some time; e.g., in Kuo [107]. But for a rigorous definition of the integrals, we need to use the concept of integral kernel operators.

§10.4   It is well-known that $N\varphi(x) = -\Delta_G \varphi(x) + \langle x, \nabla \varphi(x) \rangle$. Thus by Theorem 10.18 we have $\Lambda\varphi(x) = \langle x, \nabla \varphi(x) \rangle$. However, if this were to be taken as the definition of the lambda operator Λ, then we need to check several technical details to show that Λ is well-defined. Thus it is more direct to define Λ as in Definition 10.17. The lambda operator Λ is used in the proof of Theorem 12.8 and in a formula relating the number operator and the Fourier transform (see the paragraph after the proof of Theorem 12.39).

§10.5   The formula in Theorem 10.23 can be written in another form $T_y = \sum_{n=0}^{\infty} \frac{1}{n!} D_y^n$. This formula was proved in abstract Wiener space theory in Y.-J. Lee [120] and in white noise distribution theory in Potthoff and Yan [152].

§10.6   The symbol of a continuous linear operator from $(\mathcal{S})_\beta$ into $(\mathcal{S})_\beta^*$ was introduced by Obata in [140]. It is a generalization of the S-transform. Just as the S-transform uniquely determines a generalized function, the symbol uniquely determines a continuous linear operator Ξ from $(\mathcal{S})_\beta$ into

$(\mathcal{S})^*_\beta$. For Theorem 10.28, we explain only the idea how to obtain the series representation of Ξ from its symbol. See Obata [142] for the details of proof.

- **Fundamental solutions:** Consider the space $\mathcal{D}$ of test functions (C^∞-functions on $\mathbb{R}^n$ with compact support) and the space $\mathcal{D}'$ of distributions. Let $p(D)$ be a partial differential operator with constant coefficients. The Malgrange-Ehrenpreis theorem states that $p(D)$ has a fundamental solution $F \in \mathcal{D}'$, i.e., F satisfies the equation $p(D)F = \delta_0$. The proof of this theorem uses the Fourier transform. Since the white noise Fourier transform is available (see Chapter 11), it is natural to ask whether it is possible to generalize this theorem to white noise distribution theory. In order to do so, we need to find the white noise analogue of the operator $p(D)$. In the finite dimensional case we have $(\partial/\partial x_j)^* = -\partial/\partial x_j$. The operator $\partial/\partial x_j$ in the expression of $p(D)$ should be replaced by $-(\partial/\partial x_j)^*$. Thus the white noise analogue of $p(D)$ is the operator $\sum_{j=0}^n \Xi_{j,0}(\theta_j)$. A fundamental solution of this operator is a generalized function Φ in $(\mathcal{E})^*_\beta$ such that $\sum_{j=0}^n \Xi_{j,0}(\theta_j)\Phi = \widetilde{\delta}_0$. Here $\widetilde{\delta}_0$ is the Kubo-Yokoi delta function at 0. In fact, it is interesting to generalize not just the Malgrange-Ehrenpreis theorem, but the whole theory of partial differential operators with constant coefficients to white noise distribution theory. See further notes and comments for Chapters 11 and 12 below.

Chapter 11

§11.1   Lemma 11.5 with $\beta = 0$ is due to Obata [142]. But our proof is different. The continuity of the Fourier transform from $(\mathcal{E})^*$ into itself (Theorem 11.6 with $\beta = 0$) has been proved in several different ways (see Hida et al. [57], the book by Hida et al. [58], Y.-J. Lee [122], and Yan [178]). In view of the definition of the Fourier transform (Definition 11.1) and the characterization theorem for generalized functions (Theorem 8.2), it is clear that the Fourier transform is a mapping from $(\mathcal{E})^*_\beta$ into (in fact, onto) itself for any $\beta \in [0,1)$.

Consider the finite dimensional Gel'fand triple $\mathcal{H}_\beta(\mathbb{R}^n) \subset L^2(\mathbb{R}^n, \mu_n) \subset \mathcal{H}^*_\beta(\mathbb{R}^n)$ (see notes and comments for §7.1). We know that $\mathcal{S}'(\mathbb{R}^n) \subset \mathcal{H}^*(\mathbb{R}^n)$. The Fourier transform on $\mathcal{S}'(\mathbb{R}^n)$ has been extended to $\mathcal{H}^*(\mathbb{R}^n)$ in Kubo and Kuo [90] and further extended to $\mathcal{H}^*_\beta(\mathbb{R}^n)$ in K. Sim Lee [119]. This can be regarded as the finite dimensional version of the white noise Fourier transform.

§11.4 The restriction of the Fourier transform to $(\mathcal{E})_\beta$ is a continuous linear operator from $(\mathcal{E})_\beta$ into $(\mathcal{E})^*_\beta$. Hence by Theorem 10.28 it can be represented by a series of integral kernel operators. Theorem 11.23 describes the kernel functions $\theta_{j,k}$'s in this representation.

§**11.5**   The Fourier-Gauss transform $\mathcal{G}_{a,b}$ acting on L^2-spaces was defined
in Y.-J. Lee [120] [121]. It has been shown in Y.-J. Lee [122] that $\mathcal{G}_{a,b}$ is
continuous from the space $\mathcal{A}$ into itself ($\mathcal{A} \equiv \mathcal{A}_0$, see §15.2 for the definition
of $\mathcal{A}_\beta$). Since $\mathcal{A} = (\mathcal{E})$ (Theorem 15.14 for the case $\beta = 0$), the operator $\mathcal{G}_{a,b}$
is continuous from $(\mathcal{E})$ into itself. We prove the continuity of $\mathcal{G}_{a,b}$ directly
as an operator of $(\mathcal{E})_\beta$ in Theorem 11.29.

The Fourier-Wiener transform as introduced in Cameron and Martin [19]
is defined by

$$\widehat{\varphi}(y) = \int_{\mathcal{E}'} \varphi(\sqrt{2}\, x + iy)\, d\mu(x).$$

Obviously, this transform is the Fourier-Gauss transform $\mathcal{G}_{\sqrt{2},i}$. Note that
$\mathcal{G}_{\sqrt{2},i}^{-1} = \mathcal{G}_{\sqrt{2},-i}$ (by Theorem 11.30). Moreover, $\mathcal{G}_{\sqrt{2},i}$ is a unitary operator
of $(\mathcal{E}_p)_\beta$ for any $p \geq 0$ and any $0 \leq \beta < 1$ (by Theorem 11.34).

In finite dimensional distribution theory the Fourier transform $\widehat{F}$ of $F \in$
$\mathcal{S}'(\mathbb{R}^n)$ is defined by $\langle \widehat{F}, f \rangle = \langle F, \widehat{f} \rangle$, $f \in \mathcal{S}(\mathbb{R}^n)$. Thus the extension of the
Fourier transform to $\mathcal{S}'(\mathbb{R}^n)$ is actually the adjoint of the Fourier transform
acting on $\mathcal{S}(\mathbb{R}^n)$. This is not the case for the white noise Fourier transform.
However, the Fourier transform is in fact the adjoint of *some* continuous
linear operator of $(\mathcal{E})_\beta$. This fact is proved in Hida et al. [57] and Y.-J. Lee
[122] independently. We use the notion of symbol to give an easy proof of
this fact in Theorem 11.32.

§**11.6**   Lemma 11.35 and Theorem 11.38 for the case $\beta = 0$ are due to Hida
et al. [57]. The proofs for general β follow more or less the same arguments
as in Hida et al. [57] (see also the book by Hida et al. [58]) except for Lemma
11.37.

§**11.7**   Theorem 11.40 was first proved in Kuo [109]. Most results in this
section for the case $\beta = 0$ have appeared in Hida et al. [57], the book by
Hida et al. [58], and Kuo [109]. The Fourier-Mehler transform has a close
connection with the Lévy Laplacian (§12.3); see Saitô [160].

§**11.8**   Initial value problems for powers of the Gross Laplacian Δ_G and
the number operator N have been studied in recent papers (Chung and Ji
[22] [23]).

● **Generalizations of finite dimensional results:**   A very natural task
is to generalize finite dimensional results concerning the Fourier transform
to the white noise Fourier transform; e.g., the Hausdorff-Young inequality,
Heisenberg uncertainty principle, Lax duality theorem, and Paley-Wiener
theorem, just to name a few. See the book by Rauch [153] or the book by
Reed and Simon [156].

Chapter 12

§12.1 The fact that $\frac{1}{2}\Delta_G$ is the infinitesimal generator of the semigroup $\{P_t; t \geq 0\}$ was first proved in Gross [42] (see also Kuo [98]). The white noise version of this fact is given in Theorem 12.1.

§12.2 The heat equation associated with an operator of the form $a\Delta_G + bN$ has been studied in a recent paper (Chung and Ji [22]).

§12.3 Theorem 12.11 has been known for some time (see Hida [47] and Kubo and Takenaka [95]). However, the domain of Δ_L has to be chosen carefully to include the space (L^2).

§12.4 In order to study deeper properties of the Lévy Laplacian Δ_L, we need to define Δ_L in terms of the S-transform. There have been several different conditions being imposed on the domain of Δ_L; e.g., see the book by Hida et al. [58], Kuo [110], and Kuo et al. [113]. The domain of the Lévy Laplacian as given in Definition 12.16 is quite natural and contains the domain of the Volterra Laplacian (§12.7).

The Lévy Laplacian has been studied in function spaces in Polishchuk [149]. In a recent paper [160], Saitô has constructed a semigroup whose infinitesimal generator is given by the Lévy Laplacian Δ_L.

§12.5 For the original idea of the mean value property of the Lévy Laplacian, see Lévy [127]. Theorem 12.27 has appeared in Obata [137] and Kuo et al. [113]. A result similar to Theorem 12.28 has been obtained in Zhang [180].

§12.6 A relationship between the Gross Laplacian and the Lévy Laplacian was first obtained in Kuo et al. [113] (see also the book by Hida et al. [58]). Theorem 12.29 is a more general form of this relationship.

§12.7 Similar to the Lévy Laplacian, there have been several different conditions being imposed on the domain of Δ_V; e.g., see the book by Hida et al. [58], Kuo [110], and Kuo et al. [113]. The domain of Δ_V as given in Definition 12.30 ensures that $\Delta_V \subset \Delta_L$. Thus from remark (a) of Theorem 12.36 we have the inclusions

$$(\mathcal{S})_\beta \subset (L^2) \cap \Delta_V \subset (L^2) \cup \Delta_V \subset \Delta_L \subset (\mathcal{S})_\beta^*.$$

§12.8 The identities in equations (12.36), (12.37), and (12.39) are obtained in Kuo [108]. The identity in equation (12.38) is due to Hida and Saitô [62].

§12.9 Theorems 12.40, 12.41, and 12.42 are due to Hida, Obata, and Saitô [60]. Our proofs, being different from theirs, are quite elementary.

For further notes and comments, see infinite dimensional rotation groups below.

- **Commutation relations:** The operators Δ_G, Δ_G^*, N, Δ_V, and Δ_L satisfy the following commutation relations:

$$[N, \Delta_G] = -2\Delta_G \text{ on } (\mathcal{E})_\beta, \qquad [N, \Delta_G^*] = -2\Delta_G^* \text{ on } (\mathcal{E})_\beta^*,$$
$$[N, \Delta_L] = -2\Delta_L \text{ on } \mathcal{D}(\Delta_L), \qquad [N, \Delta_V] = -2\Delta_V \text{ on } \mathcal{D}(\Delta_V),$$
$$[\Delta_G, \Delta_L] = 0 \text{ on } (\mathcal{S}), \qquad [\Delta_G, \Delta_V] = 0 \text{ on } (\mathcal{S}),$$
$$[\Delta_L, \Delta_G^*] = 2I \text{ on } \mathcal{D}(\Delta_L), \qquad [\Delta_V, \Delta_G^*] = 4N \text{ on } \mathcal{D}(\Delta_V).$$

These identities can be easily verified by direct calculations (see the book by Hida et al. [58]). Actually, we have $\Delta_G \Delta_L = \Delta_L \Delta_G = 0$ (so that $[\Delta_G, \Delta_L] = 0$) and $\Delta_G \Delta_V = \Delta_V \Delta_G = \Delta_G^2$ (so that $[\Delta_G, \Delta_V] = 0$). However, we do not know the commutation relation between the Lévy Laplacian and the Volterra Laplacian.

- **Quantum probability and Yang-Mills equations:** The Lévy Laplacian has a very important application in quantum probability. In particular, it has been shown in Accardi et al. [1] [2] that a parallel transport is associated to a connection satisfying the Yang-Mills equations if and only if it satisfies the Laplace equation for the Lévy Laplacian on the path space. For the white noise setup of quantum probability, see Obata [141] [143].

- **Infinite dimensional rotation groups:** Consider the Gel'fand triple $\mathcal{E} \subset E \subset \mathcal{E}'$. Let $\mathcal{O}(\mathcal{E}; E)$ denote the set of all linear homeomorphisms g from $\mathcal{E}$ onto itself such that $|g\xi|_0 = |\xi|_0$ for all $\xi \in \mathcal{E}$ ($|\cdot|_0$ is the norm on E). Obviously $\mathcal{O}(\mathcal{E}; E)$ is a group. Endow $\mathcal{O}(\mathcal{E}; E)$ with the compact-open topology. It becomes a topological group and is often referred to as an *infinite dimensional rotation group*. For each $g \in \mathcal{O}(\mathcal{E}; E)$, its adjoint g^* is a homeomorphism from $\mathcal{E}'$ onto itself. Let $\Gamma(g)$ be the second quantization of $g \in \mathcal{O}(\mathcal{E}; E)$. It is easy to check that for any $\varphi \in (\mathcal{E})$,

$$\Gamma(g)\varphi(x) = \varphi(g^*x), \quad x \in \mathcal{E}'.$$

Moreover, $\Gamma(g)$ is a linear homeomorphism from $(\mathcal{E})$ onto itself such that $\|\Gamma(g)\varphi\|_0 = \|\varphi\|_0$ for all $\varphi \in (\mathcal{E})$ ($\|\cdot\|_0$ is the norm on (L^2)). Its adjoint $\Gamma(g)^*$ is a linear homeomorphism from $(\mathcal{E})^*$ onto itself. A continuous linear operator Ξ from $(\mathcal{E})$ into $(\mathcal{E})^*$ is said to be *rotation invariant* if

$$\Gamma(g)^* \Xi \Gamma(g) = \Xi, \quad \forall g \in \mathcal{O}(\mathcal{E}; E).$$

The first two facts below have been proved in Obata [142, Theorems 5.5.3 and 5.5.4], while the third fact can be proved by the same argument:

(1) Let $\Xi \colon (\mathcal{E}) \to (\mathcal{E})^*$ be a rotation invariant continuous integral kernel operator. Then Ξ is a linear combination of operators $(\Delta_G^*)^a N^b \Delta_G^c$, ($a, b, c$: nonnegative integers).

(2) Let $\Xi : (\mathcal{E}) \to (\mathcal{E})$ be a rotation invariant continuous integral kernel operator. Then Ξ is a linear combination of operators $N^b \Delta_G^c$, $(b, c :$ nonnegative integers).

(3) Let $\Xi : (\mathcal{E})^* \to (\mathcal{E})^*$ be a rotation invariant continuous integral kernel operator. Then Ξ is a linear combination of operators $(\Delta_G^*)^a N^b$, $(a, b :$ nonnegative integers).

A very important aspect of the infinite dimensional rotation group $\mathcal{O}(\mathcal{E}; E)$ is its structure. Since $\mathcal{O}(\mathcal{E}; E)$ is a rather big group (e.g., it is not locally compact), it is desirable to find interesting subgroups. Obviously $\mathcal{O}(\mathcal{E}; E)$ contains all finite dimensional rotations. An important subgroup is the Lévy group (see Lévy [127]) which can not be approximated by finite dimensional rotations. The Lévy group has been studied in Hida [55] and Obata [136]. It is closely related to the Lévy Laplacian. For the infinite dimensional rotation group $\mathcal{O}(\mathcal{E}; L^2(\mathbb{R}))$, there are subgroups called whiskers in Hida [45]. A *whisker* is a one-parameter subgroup $\{g_t; t \in \mathbb{R}\}$ of $\mathcal{O}(\mathcal{E}; L^2(\mathbb{R}))$ satisfying the following conditions:

(a) g_t is continuous in t.

(b) $g_s g_t = g_{s+t}$.

(c) $(g_t \xi)(u) = \xi(\psi_t(u)) \sqrt{|\psi_t'(u)|}$ for some ψ such that ψ_t is a diffeomorphism of $\overline{\mathbb{R}}$ for each t.

Examples of whiskers are shift, dilation, and conformal transformation. For the infinite dimensional rotation group $\mathcal{O}(\mathcal{E}; L^2(\mathbb{R}^d))$, whiskers can be defined in a similar way. In this case, we have whiskers such as shifts $\{s_t^j; 1 \leq j \leq d\}$, isotropic dilation τ_t, special conformal transformations $\{\kappa_t^j; 1 \leq j \leq d\}$, and special orthogonal group $SO(d)$. The group generated by $\{s_t^j, \tau_t, \kappa_t^j, SO(d); 1 \leq j \leq d\}$ forms a $(d+2)(d+1)/2$ dimensional Lie group. For the details about these infinite dimensional rotation groups, see Hida [45] [47] [49] [50] [55].

Chapter 13

§13.2 The main purpose of this section is to describe the ideas of defining Pettis and Bochner integrals for Banach space valued functions. Even though the space $(\mathcal{E})_\beta^*$ in the Gel'fand triple $(\mathcal{E})_\beta \subset (L^2) \subset (\mathcal{E})_\beta^*$ is not a Banach space, we can still use the same ideas to define these integrals for $(\mathcal{E})_\beta^*$-valued functions in the next section.

§13.3 Defining a white noise integral in terms of the S-transform has been known for some time. It turns out that this is equivalent to the Pettis integral. Theorem 13.4 is quite useful for white noise integrals in the Pettis sense, while Theorem 13.5 is for those in the Bochner sense.

Since conditions for Bochner integrability are quite strong, the white noise integral in the Pettis sense is more appropriate (e.g., see Example 13.8). The representation of Donsker's delta function in Example 13.9 plays a crucial role for Feynman integrals in Chapter 14. White noise integrals for the multidimensional case have been studied in Redfern [154].

§13.4 In general the integral $\int_a^b \partial_t^* \Phi(t)\, dt$ is just a white noise integral and defines a generalized function in $(\mathcal{E})_\beta^*$. Only when $\int_a^b \partial_t^* \Phi(t)\, dt$ is an element in (L^2), it is called a Hitsuda-Skorokhod integral. Thus a Hitsuda-Skorokhod integral defines a random variable. The usage of this term is different from that in the book by Hida et al. [58]. Theorem 13.12 is due to Kubo and Takenaka [94]. Theorem 13.16 has been proved in Kuo and Russek [116] (see also the book by Hida et al. [58]).

§13.5 Theorem 13.19 is due to Hitsuda [64]. The generalized Itô's formula in Theorem 13.23 is due to Kubo [88]. The well-known Tanaka formula is an important application of Theorem 13.23. Observe that Theorem 13.21 is an anticipatory Itô's formula for a Wiener integral. It is natural to ask whether Theorem 13.21 is still true when $X(t)$ is a Hitsuda-Skorokhod integral. In fact, it is an interesting problem to obtain Itô's formula for Hitsuda-Skorokhod integrals.

§13.6 The forward and backward integrals were introduced in Kuo and Russek [116]. The forward integral $\int_a^b \varphi(t)\, dB(t^+)$ can be regarded as an anticipatory extension of the Itô integral. The generalization of Itô's formula to forward integrals has been proved in Asch and Potthoff [8] (see also the book by Hida et al. [58]).

§13.7 This section is about stochastic integral equations, i.e., the solution is a stochastic process $X(t)$ such that $X(t) \in (L^2)$. When a given equation is anticipatory, the solution can be rather surprising. For instance, Example 13.30 (due to Buckdahn [17]) shows that an anticipating initial condition can make a big difference to the sample path property of the solution. Theorems 13.33 and 13.34 give solutions to two linear equations which are anticipatory in different ways. Examples 13.35 and 13.36 show that the solution of a very simple anticipatory equation can be rather complicated. The equation in Example 13.36 has also been studied in Kuo and Potthoff [114]. But the solution given in that paper is wrong.

An anticipative Girsanov type theorem has been obtained in Buckdahn [18] and León and Protter [124]. Hitsuda-Skorokhod integral is not used in these papers. It would be interesting to derive their results from the viewpoint of the Hitsuda-Skorokhod integral.

For a comparison of the Hitsuda-Skorokhod integral to other anticipatory stochastic integrals, see Gawarecki and Mandrekar [37].

§13.8 This section is about white noise integral equations, i.e., the solution $X(t)$ is a generalized function in $(\mathcal{E})_\beta^*$ for each t. Equation (13.74) has been studied in Kuo and Potthoff [114]. But the conditions assumed on f in Kuo and Potthoff [114] are too strong to be useful. The conditions in Theorem 13.43 are quite natural and weak enough to cover a large class of white noise integral equations in equation (13.74). More significantly, the integral in equation (13.74) is a white noise integral in the Pettis sense (see Definition 13.42). The solution $X(t)$ in Example 13.46 is a generalized function in $(\mathcal{S})_\beta^*$ for some $0 < \beta < 1$. This shows the need to use the space $(\mathcal{S})_\beta^*$ for nonzero β.

• **Realizations of solutions of white noise integral equations:** There are two types of realization of solutions. The first one is related to Hida complex measures. In several examples in §13.8, the solutions are induced by Hida measures. In general it would be interesting to find conditions on $X(a)$ and f such that the solution of equation (13.74) is induced by a Hida complex measure and, in particular, by a Hida measure. On the other hand, from the probabilistic viewpoint, it is desirable to realize $X(t)$ as a stochastic process, i.e., $X(t)$ is a random variable for each t. In order to do so, note that by Theorem 8.23 we have $\cup_{p>1} L^p(\mu) \subset (\mathcal{S})^*$. Thus we may ask whether the solution $X(t)$ belongs to $L^p(\mu)$ for each t. If so, then $X(t)$ is a random variable. This kind of realization will be useful to study probabilistic problems related to the solution.

• **Anticipating stochastic integrals and stochastic integral equations:** There are other approaches to define anticipating stochastic integrals. For instance, see Berger and Mizel [12], Nualart and Pardoux [135], Ocone [145], Ogawa [147], and Üstünel [170]. Some anticipating stochastic integral equations have been studied in Ocone and Pardoux [146] and Pardoux and Protter [148]. It would be quite useful to establish a theorem for the existence and uniqueness of a solution of a Hitsuda-Skorokhod type equation.

• **Infinite dimensional stochastic integral equations:** Consider a stochastic integral equation of Itô type taking values in $\mathbb{R}^n$

$$X(t) = x + \int_0^t f(X(s)) \, dW(t) + \int_0^t g(X(s)) \, ds,$$

where f and g satisfy the Lipschitz and growth conditions so that this equation has a unique solution. Let $\nu_{t,x}$ be the law of $X(t)$ starting at x, i.e.,

$$\nu_{t,x}(C) = \text{Prob}\{X(t) \in C \,|\, X(0) = x\}, \quad C \in \mathcal{B}(\mathbb{R}^n).$$

Under certain conditions the Malliavin calculus asserts that $\nu_{t,x}$ is absolutely continuous with respect to the Lebesgue measure and the Radon-Nikodym derivative can be expressed in terms of Donsker's delta function.

Consider an infinite dimensional stochastic integral equation taking values in a white noise space $(\mathcal{E}', \mu)$. For example, a measure-valued stochastic integral equation can be regarded as such an equation taking values in the white noise space $(S'(\mathbb{R}), \mu)$. Let $\nu_{t,x}$ be the law of the solution $X(t)$ starting at $x \in \mathcal{E}'$. Since there is no Lebesgue measure on $\mathcal{E}'$ $(\dim \mathcal{E}' = \infty)$, we may try to use the Gaussian measure μ. But in general $\nu_{t,x}$ is not absolutely continuous with respect to μ. Thus we need to use another concept to study the law of $X(t)$. From the white noise viewpoint, it is natural to ask whether $\nu_{t,x}$ is a Hida measure. If so, then we can study the induced generalized function. More precisely, the problem is to find conditions on f and g such that $\nu_{t,x}$ is a Hida measure. In that case, the induced generalized function should satisfy some white noise integral equation. It would be interesting to study the relationship between the given $\mathcal{E}'$-valued stochastic integral equation and the induced $(\mathcal{E})^*_\beta$-valued white noise integral equation. On the other hand, we can study the law $\nu_{t,x}$ of $X(t)$ within the scope of differentiable measures. Roughly speaking, the derivative of a measure m in the direction h is defined by

$$(m'(\cdot), h) = \lim_{\epsilon \to 0} \frac{m(\cdot + \epsilon h) - m(\cdot)}{\epsilon}.$$

For the precise definition, see Averbuh et al. [9], Kuo [100] [101] or the survey paper Bogachëv and Smolyanov [14]. Then we may ask whether the law $\nu_{t,x}$ of $X(t)$ is a differentiable measure and, if so, whether $\nu_{t,x}$ satisfies some differential equation. Furthermore, it is worthwhile to explore the connection between differentiable measures and Hida measures.

Stochastic integral equations taking values in the space $S'(\mathbb{R}^d)$ have been studied in Bojdecki and Gorostiza [15] and Gorostiza and Nualart [40]. See also the recent monograph Kallianpur and Xiong [77] for infinite dimensional stochastic integral equations.

• **Generalized conditional expectation:** Let $\mathcal{A}$ be a sub-σ-algebra of $\mathcal{B}(\mathcal{E}')$. Is it possible to define the generalized conditional expectation $E(\Phi|\mathcal{A})$ for $\Phi \in (\mathcal{E})^*_\beta$? For the σ-algebra $\mathcal{B}_t = \sigma\{\langle \cdot, 1_{[0,s)} \rangle; s \leq t\}$, this has been defined in Hida [48]. If $E(\Phi|\mathcal{A})$ can be defined, it will be quite useful for the study of white noise integral equations. For instance, we can discuss the martingale property of the solution $X(t)$ of a white noise integral equation and consider prediction problems related to $X(t)$. For conditional expectations related to quantum probability, see the recent paper by Obata [144].

• **Stochastic partial differential equations:** White noise distribution theory has been used to study a certain kind of problems in stochastic partial differential equations. For instance, see Benth and Streit [11], Chow [20] [21], and Lindstrøm et al. [129] [130].

• **Stochastic variational equations:** To study problems involving a multidimensional time parameter, we can take the basic Gel'fand triple to be $\mathcal{E} \subset L^2(\mathbb{R}^d) \subset \mathcal{E}'$. This allows us to apply white noise distribution theory to random fields. In particular, we can study stochastic variational equations such as the Tomonaga-Schwinger equation. See Hida [51], Hida et al. [59], Si Si [166], and the recent papers by Hida [52] [53] [54] [55].

Chapter 14

There are several methods to define Feynman integrals. For instance, see Albeverio and Høegh-Krohn [6], DeWitt-Morette [31] [32], Itô [69] [70], Johnson and Kallianpur [74], Johnson and Skoug [75], Kallianpur et al. [76], and Nelson [133].

§14.1 In the informal derivation of the Feynman integral in Albeverio and Høegh-Krohn [6], $f(x_0)$ is used in the integral over $\mathbb{R}^n$. We have relabelled the variables to get $f(x_n)$ in the integrand. This is more convenient for deriving $f(y(t))$ in equation (14.1).

§14.2 The renormalization in Lemma 14.1 can be done in a more general way. Consider the Gel'fand triple $\mathcal{E} \subset E \subset \mathcal{E}'$. Let K be a positive trace class operator of E such that $I + K$ is invertible. Take an abstract Wiener space $(E, \mathcal{E}'_p)$ for some $p \geq \alpha/2$ (see §4.2). Then for any $h \in E$ we have the equality from Kuo [97]

$$\int_{\mathcal{E}'} \exp\left[-\frac{1}{2}\langle Kx, x \rangle + i\langle x, h \rangle \right] d\mu(x)$$

$$= \left(\det(I + K) \right)^{-1/2} \exp\left[-\frac{1}{2}\langle (I + K)^{-1}h, h \rangle \right].$$

Let $\varphi(x) = \exp\left[-\frac{1}{2}\langle Kx, x \rangle \right]$. Then $E\varphi = \left(\det(I + K) \right)^{-1/2}$. Define the renormalization $\mathcal{N}\varphi$ of φ to be $\mathcal{N}\varphi \equiv \varphi/E\varphi$. Then by the above equality the $\mathcal{T}$-transform of $\mathcal{N}\varphi$ is given by

$$\mathcal{T}(\mathcal{N}\varphi)(\xi) = \exp\left[-\frac{1}{2}\langle (I + K)^{-1}\xi, \xi \rangle \right], \quad \xi \in \mathcal{E}_c.$$

But $S\Phi(\xi) = \mathcal{T}\Phi(-i\xi)\exp\left[-\frac{1}{2}\langle \xi, \xi \rangle \right]$. Hence we have

$$S(\mathcal{N}\varphi)(\xi) = \exp\left[-\frac{1}{2}\langle (I + K)^{-1}K\xi, \xi \rangle \right], \quad \xi \in \mathcal{E}_c.$$

Now, observe that the right hand side of the last equality does not require that K be a trace class operator. In fact, suppose K is a symmetric operator on E such that $I + K$ is invertible and $(I + K)^{-1}K$ is a continuous linear operator from $\mathcal{E}_{p,\beta}$ into $\mathcal{E}_{q,\beta}^*$ for some $p, q \geq 0$. Then the function on the right hand side of the last equality satisfies the conditions in Theorem 8.2. Hence there exists a generalized function, denoted by $\mathcal{N} \exp\left[-\frac{1}{2}\langle K\cdot, \cdot\rangle \right]$, with S-transform

$$S\left(\mathcal{N} \exp\left[-\frac{1}{2}\langle K\cdot, \cdot\rangle \right]\right)(\xi) = \exp\left[-\frac{1}{2}\langle (I+K)^{-1}K\xi, \xi\rangle \right], \quad \xi \in \mathcal{E}_c.$$

This renormalization defines a Gaussian white noise function with operator K. When $K = c^{-1}I$, we get the Gaussian white noise function g_c in Example 8.7. When K is the multiplication operator by $c^{-1}1_{[0,t)}$, we get Lemma 14.1.

Chapter 15

§15.1   For positive generalized functions in the Malliavin calculus, see Itô [72] and Sugita [169]. The positivity of Donsker's delta function in Example 15.9 was proved by a different method in Shieh and Yokoi [165].

§15.2   The construction of Lee does not use the Wiener-Itô decomposition. A test function in $\mathcal{A}_\beta$ satisfies analyticity and growth conditions. This is in the same spirit as in the finite dimensional case. On the other hand, a test function $\varphi \in (\mathcal{E})_\beta$ can be represented by $\varphi(x) = \langle\!\langle :e^{\langle \cdot, x\rangle}:, \Theta\varphi\rangle\!\rangle$. This implies the same analyticity and growth conditions as in Lee's construction. Thus it is not surprising that $\mathcal{A}_\beta = (\mathcal{E})_\beta$. Notice that by Lemma 15.11 $(E, \mathcal{A}_{p,\beta})$ is an abstract Wiener space for some $p > 0$.

§15.3   The assertion $(\mathcal{E})_\beta \subset \cap_{1 \leq p < \infty} L^2(\nu)$ in Theorem 15.15 for any Hida measure ν should be compared with the inclusion $\cup_{p>1} L^p(\mu) \subset (\mathcal{E})_\beta^*$ in Theorem 8.23 for the Gaussian measure μ.

● **Dirichlet forms and quantum field theory:** A positive generalized function Φ is called strictly positive if $\nu(C) > 0$ for any nonempty open subset C of $\mathcal{E}'$, where ν is the Hida measure inducing Φ. The strictly positive generalized functions are very useful for the study of Dirichlet forms and the constructive quantum field theory. See Albeverio et al. [4] [5] and Hida et al. [61].

● **Generalized functions induced by Hida complex measures:** The set of generalized functions induced by Hida measures is characterized in Theorem 15.3 in terms of positivity. However, the positivity of a generalized function is not easy to check. Thus it is desirable to find sufficient conditions for a generalized function to be induced by a Hida measure. On the other hand, it is worthwhile to generalize Theorems 15.3 and 15.17 to Hida complex measures.

Appendix B: Miscellaneous Formulas

1 Hermite polynomials

Definition

$$H_n(x) = (-1)^n e^{x^2} D_x^n e^{-x^2}, \quad n \geq 0$$

Generating function

$$e^{2tx - t^2} = \sum_{n=0}^{\infty} \frac{t^n}{n!} H_n(x)$$

Examples

$$H_0(x) = 1, \quad H_1(x) = 2x, \quad H_2(x) = 4x^2 - 2, \quad H_3(x) = 8x^3 - 12x$$
$$H_4(x) = 16x^4 - 48x^2 + 12, \quad H_5(x) = 32x^5 - 160x^3 + 120x$$

Identities

$$H_{n+1}(x) - 2xH_n(x) + 2nH_{n-1}(x) = 0$$
$$H_n'(x) = 2nH_{n-1}(x)$$
$$H_n''(x) - 2xH_n'(x) + 2nH_n(x) = 0$$

Values at zero

$$H_n(0) = \begin{cases} 0, & n : \text{odd} \\ \dfrac{n!(-1)^{n/2}}{(n/2)!}, & n : \text{even} \end{cases}$$

Orthonormal basis for $L^2(\mathbb{R}, dx)$: Hermite functions

$$e_n(x) = \frac{1}{\sqrt{\sqrt{\pi}\, 2^n n!}} H_n(x) e^{-x^2/2}, \quad n \geq 0$$

Properties of Hermite functions

$$\left(-\frac{d^2}{dx^2} + x^2 + 1\right)e_n = (2n+2)e_n$$

$$\sup_{t\in\mathbb{R}} |e_n(t)| = O(n^{-1/12})$$

$$e_n'(x) = \sqrt{\frac{n}{2}}\, e_{n-1}(x) - \sqrt{\frac{n+1}{2}}\, e_{n+1}(x)$$

$$x e_n(x) = \sqrt{\frac{n}{2}}\, e_{n-1}(x) + \sqrt{\frac{n+1}{2}}\, e_{n+1}(x)$$

---

2 Hermite polynomials with parameter σ^2

Definition

$$:x^n:_{\sigma^2} = (-\sigma^2)^n\, e^{x^2/2\sigma^2} D_x^n\, e^{-x^2/2\sigma^2}$$

Relationship with H_n

$$:x^n:_{\sigma^2} = \frac{1}{(\sqrt{2}\,)^n}\, \sigma^n H_n\big(x/\sqrt{2}\,\sigma\big), \quad H_n(x) = 2^n :x^n:_{1/2}$$

Generating function

$$\exp\left[tx - \tfrac{1}{2}\sigma^2 t^2\right] = \sum_{n=0}^{\infty} \frac{t^n}{n!}\, :x^n:_{\sigma^2}$$

Examples

$$:x^0:_{\sigma^2} = 1, \quad :x^1:_{\sigma^2} = x, \quad :x^2:_{\sigma^2} = x^2 - \sigma^2, \quad :x^3:_{\sigma^2} = x^3 - 3\sigma^2 x$$

$$:x^4:_{\sigma^2} = x^4 - 6\sigma^2 x^2 + 3\sigma^4, \quad :x^5:_{\sigma^2} = x^5 - 10\sigma^2 x^3 + 15\sigma^4 x$$

Relationship with ordinary powers

$$:x^n:_{\sigma^2} = \sum_{k=0}^{[n/2]} \binom{n}{2k}(2k-1)!!\,(-\sigma^2)^k x^{n-2k}$$

$$x^n = \sum_{k=0}^{[n/2]} \binom{n}{2k}(2k-1)!!\,\sigma^{2k}\, :x^{n-2k}:_{\sigma^2}$$

(Convention: $(2k-1)!! = (2k-1)(2k-3)\cdots 3\cdot 1$, $(-1)!! = 1$)

Identities

$$:x^{n+1}:_{\sigma^2} - x :x^n:_{\sigma^2} + \sigma^2 n :x^{n-1}:_{\sigma^2} = 0$$

$$\frac{d}{dx} :x^n:_{\sigma^2} = n :x^{n-1}:_{\sigma^2}$$

$$\left(\sigma^2 \frac{d^2}{dx^2} - x\frac{d}{dx} + n\right) :x^n:_{\sigma^2} = 0$$

Product

$$:x^n:_{\sigma^2} :x^m:_{\sigma^2} = \sum_{k=0}^{n \wedge m} k!\binom{n}{k}\binom{m}{k}\sigma^{2k} :x^{n+m-2k}:_{\sigma^2}$$

Orthonormal basis for $L^2\left(\mathbb{R}, \frac{1}{\sqrt{2\pi}\,\sigma}e^{-x^2/2\sigma^2}\,dx\right)$

$$\xi_n(x;\sigma^2) = \frac{1}{\sqrt{n!}\,\sigma^n} :x^n:_{\sigma^2}, \quad n \geq 0$$

$$\sup_{\sigma>0,x\in\mathbb{R}} \left|\xi_n(x;\sigma^2)e^{-x^2/4\sigma^2}\right| = O(n^{-1/12})$$

Translation and scaling

$$:(x+y)^n:_{\sigma^2} = \sum_{k=0}^{n}\binom{n}{k} :x^{n-k}:_{\sigma^2}\, y^k$$

$$:(ax)^n:_{\sigma^2} = \sum_{k=0}^{[n/2]}\binom{n}{2k}(2k-1)!!(-\sigma^2)^k a^{n-2k}(1-a^2)^k :x^{n-2k}:_{\sigma^2}$$

Fourier-Gauss transforms (cf. §11.5)

$$\frac{1}{\sqrt{2\pi}\,\sigma}\int_{-\infty}^{\infty} :(x+y)^n:_{\sigma^2}\, e^{-x^2/2\sigma^2}\, dx = y^n$$

$$\frac{1}{\sqrt{2\pi}\,\sigma}\int_{-\infty}^{\infty} (ix+y)^n\, e^{-x^2/2\sigma^2}\, dx = :y^n:_{\sigma^2}$$

Change of parameters

$$:(\tau^{-1}\sigma x)^n:_{\sigma^2} = (\tau^{-1}\sigma)^n :x^n:_{\tau^2}$$

Sum of a series

$$\sum_{n=0}^{\infty}\frac{t^n}{n!} :x^n:_{1}^{2} = (1-t^2)^{-1/2}\exp\left[\frac{t}{1+t}x^2\right], \quad |t| < 1$$

3   Wick tensors of generalized functions

Definition (§5.1)

$$:x^{\otimes n}: = \sum_{k=0}^{[n/2]} \binom{n}{2k} (2k-1)!! \, (-1)^k x^{\otimes(n-2k)} \widehat{\otimes} \tau^{\otimes k}$$

$$(\tau : \text{the trace operator})$$

Ordinary tensor (§5.1)

$$x^{\otimes n} = \sum_{k=0}^{[n/2]} \binom{n}{2k} (2k-1)!! \, :x^{\otimes(n-2k)}: \, \widehat{\otimes} \tau^{\otimes k}$$

Exponential function (§5.2)

$$\exp\left[\langle x, h \rangle - \tfrac{1}{2}|h|_0^2\right] = \sum_{n=0}^{\infty} \frac{1}{n!} \langle :x^{\otimes n}:, h^{\otimes n} \rangle$$

Examples

$$:x^{\otimes 0}: = 1, \quad :x^{\otimes 1}: = x, \quad :x^{\otimes 2}: = x^{\otimes 2} - \tau$$

$$:x^{\otimes 3}: = x^{\otimes 3}: -3x \widehat{\otimes} \tau, \quad :x^{\otimes 4}: = x^{\otimes 4} - 6x^{\otimes 2} \widehat{\otimes} \tau + 3\tau^{\otimes 2}$$

$$:x^{\otimes 5}: = x^{\otimes 5} - 10x^{\otimes 3} \widehat{\otimes} \tau + 15x \widehat{\otimes} \tau^{\otimes 2}$$

Recursion formula

$$:x^{\otimes(n+1)}: = :x^{\otimes n}: \widehat{\otimes} x - n \, :x^{\otimes(n-1)}: \widehat{\otimes} \tau$$

Translation (Lemma 7.16)

$$:(x+y)^{\otimes n}: = \sum_{k=0}^{n} \binom{n}{k} :x^{\otimes(n-k)}: \widehat{\otimes} y^{\otimes k}$$

Scaling (Lemma 11.17)

$$:(ax)^{\otimes n}: = \sum_{k=0}^{[n/2]} \binom{n}{2k} (2k-1)!! a^{n-2k} (a^2-1)^k :x^{\otimes(n-2k)}: \widehat{\otimes} \tau^{\otimes k}$$

Representation (§6.1)

$$\sum_{n=0}^{\infty} \langle :x^{\otimes n}:, f_n \rangle = \sum_{n=0}^{\infty} \langle x^{\otimes n}, g_n \rangle$$

$$g_n = \sum_{k=0}^{\infty} \binom{n+2k}{2k} (2k-1)!!(-1)^k \langle \tau^{\otimes k}, f_{n+2k} \rangle$$

4 Delta functions

Dirac delta function

$$\delta_a(x) = \frac{1}{2\pi} \int_{\mathbb{R}} e^{iu(x-a)} \, du$$

$$= \frac{1}{\sqrt{2\pi}\,\sigma} e^{-a^2/2\sigma^2} \sum_{n=0}^{\infty} \frac{1}{n!\sigma^{2n}} :a^n:_{\sigma^2} :x^n:_{\sigma^2}$$

Donsker's delta function (Example 7.4)

$$\delta_a(B(t)) = \frac{1}{\sqrt{2\pi t}} e^{-\frac{a^2}{2t}} \sum_{n=0}^{\infty} \frac{1}{n!t^n} :a^n:_t \langle :\cdot^{\otimes n}:, 1_{[0,t]}^{\otimes n} \rangle$$

Kubo-Yokoi delta function (§7.2)

$$\tilde{\delta}_x = \sum_{n=0}^{\infty} \frac{1}{n!} \langle :\cdot^{\otimes n}:, :x^{\otimes n}: \rangle$$

5 Stirling formula and bounds

Stirling formula

$$n! \approx \sqrt{2n\pi} \left(\frac{n}{e}\right)^n$$

Lower and upper bounds

$$\sqrt{2n\pi} \left(\frac{n}{e}\right)^n \le n! \le e\sqrt{n} \left(\frac{n}{e}\right)^n$$

6 Inequalities $(0 \leq \beta < 1)$

$$\sum_{n=0}^{\infty} \frac{1}{(n!)^{1-\beta}} |a|^n \leq 2^\beta \exp\left[(1-\beta)2^{\frac{\beta}{1-\beta}}|a|^{\frac{1}{1-\beta}}\right]$$

$$\sum_{n=0}^{\infty} \frac{1}{(n!)^{1+\beta}} |a|^n \leq \exp\left[(1+\beta)|a|^{\frac{1}{1+\beta}}\right]$$

7 Polarization identity

$F(v_1,\ldots,v_n)$: a symmetric n-linear functional

$A(v) = F(v,\ldots,v)$

$$F(v_1,\ldots,v_n) = \frac{1}{n!}\sum_{k=1}^{n}(-1)^{n-k}\sum_{j_1<\cdots<j_k} A(v_{j_1}+\cdots+v_{j_k})$$

$$F(v_1,\ldots,v_n) = \frac{1}{2^n n!}\sum_{\epsilon_1=\pm 1,\ldots,\epsilon_n=\pm 1}\epsilon_1\cdots\epsilon_n\, A(\epsilon_1 v_1+\cdots+\epsilon_n v_n)$$

Bibliography

[1] Accardi, L., Gibilisco, P., and Volovich, I. V.: The Lévy Laplacian and the Yang-Mills equations; *Preprint* (1992)

[2] Accardi, L., Gibilisco, P., and Volovich, I. V.: Yang-Mills gauge fields as harmonic functions for the Lévy Laplacian; *Preprint* (1993)

[3] Adams, R. A.: *Sobolev Spaces.* Academic Press, 1975

[4] Albeverio, S., Hida, T., Potthoff, J., Röckner, M., and Streit, L.: Dirichlet forms in terms of white noise analysis I–Construction and QFT examples; *Rev. Math. Phys.* **1** (1990) 291–312

[5] Albeverio, S., Hida, T., Potthoff, J., Röckner, M., and Streit, L.: Dirichlet forms in terms of white noise analysis II–Closability and diffusion processes; *Rev. Math. Phys.* **1** (1990) 313–323

[6] Albeverio, S. and Høegh-Krohn, R.: *Mathematical Theory of Feynman Path Integrals.* Lecture Notes in Math. **523**, Springer-Verlag, 1976

[7] Arnold, L.: *Stochastic Differential Equations.* John Wiley & Sons, 1974

[8] Asch, J. and Potthoff, J.: Itô's lemma without non-anticipatory conditions; *Probab. Th. Rel. Fields* **88** (1991) 17–46

[9] Averbuh, V. I., Smoljanov, O. G., and Fomin, S. V.: Generalized functions and differential equations in linear spaces I, differentiable measures; *Trans. Moscow Math. Soc.* **24** (1971) 140–184

[10] Bargmann, V.: On a Hilbert space of analytic functions and an associated integral transform, I; *Comm. Pure Appl. Math.* **14** (1961) 187–214

[11] Benth, F. E. and Streit, L.: The Burgers equation with a non-Gaussian random force; *Preprint* (1995)

[12] Berger, M. A. and Mizel, V. J.: An extension of the stochastic integral; *Annals of Probability* **10** (1980) 435–450

[13] Bochner, S. and Martin, W. T.: *Several Complex Variables.* Princeton University Press, 1948

[14] Bogachëv, V. I. and Smolyanov, O. G.: Analytic properties of infinite-dimensional distributions; *Russian Math. Surveys* **45**:3 (1990) 1–104

[15] Bojdecki, T. and Gorostiza, L. G.: A nuclear space of distributions on Wiener space and application to weak convergence; *Pitman Research Notes in Math. Series* **310** (1994) 60–74, Longman Scientific & Technical

[16] Buckdahn, R.: Skorohod's integral and linear stochastic differential equations; *Preprint* (1987)

[17] Buckdahn, R.: Anticipating linear stochastic differential equations; *Lecture Notes in Control and Information Sciences* **136** (1989) 18–23, Springer-Verlag

[18] Buckdahn, R.: Anticipative Girsanov transformations; *Probab. Th. Rel. Fields* **89** (1991) 211–238

[19] Cameron, R. H. and Martin, W. T.: Fourier-Wiener transforms of functionals belonging to L_2 over the space C; *Duke Math. J.* **14** (1947) 99–107

[20] Chow, P. L.: Generalized solution of some parabolic equations with a random drift; *J. Appl. Math. Optim.* **20** (1989) 81–96

[21] Chow, P. L.: Stationary solutions of some parabolic Itô equations; *Pitman Research Notes in Math. Series* **310** (1994) 42–51, Longman Scientific & Technical

[22] Chung, D. M. and Ji, U. C.: Transformation groups on white noise functionals and their applications; *Preprint* (1995)

[23] Chung, D. M. and Ji, U. C.: Some Cauchy problems in white noise analysis and associated semigroups of operators; *Preprint* (1995)

[24] Chung, K. L. and Williams, R. J.: *Introduction to Stochastic Integration.* Birkhäuser, 1983

[25] Cochran, W. G., Kuo, H.-H., and Sengupta, A.: A new class of white noise generalized functions; *Preprint* (1995)

[26] Corwin, L. and Greenleaf, F.: *Representations of Nilpotent Lie Groups and Their Applications.* Cambridge University Press, 1990

[27] de Falco, D. and Khandekar, D. C.: Applications of white noise calculus to the computation of Feynman integrals; *Stochastic Processes and Their Applications* **29** (1988) 257-266

[28] de Faria, M., Hida, T., Streit, L., and Watanabe, H.: Intersection local times as generalized white noise functionals; *BiBoS preprint* (1994)

[29] de Faria, M., Potthoff, J., and Streit, L.: The Feynman integrand as a Hida distribution; *J. Math. Phys.* **32** (1991) 2123–2127

[30] de Faria, M. and Streit, L.: Some recent advances in white noise analysis; *Pitman Research Notes in Math. Series* **310** (1994) 52–59, Longman Scientific & Technical

[31] DeWitt-Morette, C.: Feynman's path integral definition without limiting procedure; *Comm. Math. Phys.* **28** (1972) 47–67

[32] DeWitt-Morette, C.: Feynman path integrals, from the prodistribution definition to the calculation of glory scattering; *Acta Physics Austriaca, Suppl.* **26** (1984) 101–170

[33] Dôku, I., Kuo, H.-H., and Lee, Y.-J.: Fourier transform and heat equation in white noise analysis; *Pitman Research Notes in Math. Series* **310** (1994) 60–74, Longman Scientific & Technical

[34] Feller, W.: *An Introduction to Probability Theory and Its Applications,* Vol. II. John Wiley & Sons, 1971

[35] Fernique, M. X.: Intégrabilité des vecteurs Gaussiens; *Academie des Sciences, Paris, Comptes Rendus* **270** Séries A (1970) 1698–1699

[36] Fleming, W. H. and Rishel, R. W.: *Deterministic and Stochastic Optimal Control.* Springer-Verlag, 1975

[37] Gawarecki, L. and Mandrekar, V.: Itô-Ramer, Skorokhod and Ogawa integrals with respect to Gaussian processes and their interrelationship; *Chaos Expansions, Multiple Wiener-Itô Integrals and Their Appls.*, C. Houdré and V. Pérez-Ableu (eds.) (1994) 349–373, CRC Press

[38] Gel'fand, I. M. and Vilenkin, N. Y.: *Generalized Functions*, Vol. 4. Academic Press, 1964

[39] Goodman, V.: A divergence theorem for Hilbert space; *Trans. Amer. Math. Soc.* **164** (1972) 411-426

[40] Gorostiza, L. G. and Nualart, D.: Nuclear Gelfand triples on Wiener space and applications to trajectorial fluctuations of particle systems; *J. Funct. Anal.* **125** (1994) 37–66

[41] Gross, L.: Abstract Wiener spaces; *Proc. 5th Berkeley Symp. Math. Stat. and Probab.* **2**, part 1 (1965) 31-42, University of California Press, Berkeley

[42] Gross, L.: Potential theory on Hilbert space; *J. Funct. Anal.* **1** (1967) 123–181

[43] Gross, L. and Malliavin, P.: Hall's transform and the Segal-Bargmann map; *Preprint* (1995)

[44] Hida, T.: *Stationary Stochastic Processes.* Princeton University Press, 1970

[45] Hida, T.: *Analysis of Brownian Functionals.* Carleton Mathematical Lecture Notes **13**, 1975

[46] Hida, T.: Generalized multiple Wiener integrals; *Proc. Japan Acad.* **54A** (1978) 55–58

[47] Hida, T.: *Brownian Motion.* Springer-Verlag, 1980

[48] Hida, T.: Causal calculus and an application to prediction theory; *Prediction Theory and Harmonic Analysis*, V. Mandrekar and H. Salehi (eds.) (1983) 123–130, North-Holland

[49] Hida, T.: Brownian motion and its functionals; *Ricerche di Matematica* **34** (1985) 183–222

[50] Hida, T.: Infinite-dimensional rotation group and unitary group; *Lecture Notes in Math* **1379** (1989) 125–134, Springer-Verlag

[51] Hida, T.: Stochastic variational calculus; *Lecture Notes in Control and Information Sciences* **176** (1992) 123–134, Springer-Verlag

[52] Hida, T.: White noise analysis and applications; *Stochastic Analysis and Applications in Physics*, A. I. Cardoso et al. (eds.) (1994) 119–131, Kluwer Academic Publishers

[53] Hida, T.: Some recent results in white noise analysis; *Pitman Research Notes in Math. Series* **310** (1994) 111–116, Longman Scientific & Technical

[54] Hida, T.: Random fields as generalized white noise functionals; *Acta Appl. Math.* **35** (1994) 49–61

[55] Hida, T.: White noise analysis: An overview and some future directions; *IIAS Reports* 1995-001 (1995)

[56] Hida, T. and Ikeda, N.: Analysis on Hilbert space with reproducing kernel arising from multiple Wiener integral; *Proc. 5th Berkeley Symp. on Math. Stat. and Probab.* **2**, part 1 (1967) 117–143, University of California Press, Berkeley

[57] Hida, T., Kuo, H.-H., and Obata, N.: Transformations for white noise functionals; *J. Funct. Anal.* **111** (1993) 259–277

[58] Hida, T., Kuo, H.-H., Potthoff, J., and Streit, L.: *White Noise: An Infinite Dimensional Calculus.* Kluwer Academic Publishers, 1993

[59] Hida, T., Lee, K.-S., and Si Si: Multidimensional parameter white noise and Gaussian random fields; *Balakrishnan Volume*, A. B. Aries (ed.) (1987) 177–183, Optimization Software

[60] Hida, T., Obata, N., and Saitô: Infinite dimensional rotations and Laplacians in terms of white noise calculus; *Nagoya Math. J.* **128** (1992) 65–93

[61] Hida, T., Potthoff, J., and Streit, L.: Dirichlet forms and white noise analysis; *Commun. Math. Phys.* **116** (1988) 235–245

[62] Hida, T. and Saitô, K.: White noise analysis and the Lévy Laplacian; *Stochastic Processes in Physics and Engineering*, S. Albeverio et al. (eds.) (1988) 177–184, Reidel Publishing Company

[63] Hille, E. and Phillips, R. S.: *Functional Analysis and Semigroups.* Amer. Math. Soc. Colloquium Publ. **31**, 1957

[64] Hitsuda, M.: Formula for Brownian partial derivatives; *Second Japan-USSR Symp. Probab. Th.* **2** (1972) 111-114

[65] Hitsuda, M.: Formula for Brownian partial derivatives; *Publ. Fac. of Integrated Arts and Sciences, Hiroshima University*, Series III, Vol. 4 (1978) 1–15

[66] Huang, Z.-Y.: Quantum white noises–White noise approach to quantum stochastic calculus; *Nagoya Math. J.* **129** (1993) 23–42

[67] Ikeda, N. and Watanabe, S.: *Stochastic Differential Equations and Diffusion processes.* North-Holland/Kodansha, 1981

[68] Itô, K.: Multiple Wiener integral; *J. Math. Soc. Japan* **3** (1951) 157–169

[69] Itô, K.: Wiener integral and Feynman integral; *Proc. Fourth Berkeley Symp. Math. Statist. Prob.* **II** (1960) 227–238

[70] Itô, K.: Generalized uniform complex measures in the Hilbertian metric space with their application to the Feynman integral; *Proc. Fifth Berkeley Symp. Math. Statist. Prob.* **II** (1965) 145–161

[71] Itô, K.: Extension of stochastic integrals; *Proc. Intern. Symp. Stochastic Differential Equations*, K. Itô (ed.) (1978) 95–109, Kinokuniya

[72] Itô, K.: Positive generalized functionals on $(\mathbb{R}^{\infty}, \mathcal{B}^{\infty}, N^{\infty})$; *White Noise Analysis-Mathematics and Applications*, T. Hida et al. (eds.) (1990) 166–179, World Scientific

[73] Ito, Y., Kubo, I. and Takenaka, S.: Calculus on Gaussian white noise and Kuo's Fourier transformation; *White Noise Analysis-Mathematics and Applications*, T. Hida et al. (eds.) (1990) 180-207, World Scientific

[74] Johnson, G. W. and Kallianpur, G.: Homogeneous chaos, p-forms, scaling and the Feynman integral; *Trans. Amer. Math. Soc.* **340** (1993) 503–548

[75] Johnson, G. W. and Skoug, D. L.: Feynman integrals of nonfactorable finite-dimensional functionals; *Pacific J. Math.* **45** (1973) 257–267

[76] Kallianpur, G., Kannan, D., and Karandikar, R. L.: Analytic and sequential Feynman integrals on abstract Wiener and Hilbert spaces, and a Cameron-Martin formula; *Ann. Inst. Henri Poincaré* **21** (1985) 323–361

[77] Kallianpur, G. and Xiong, J.: *Stochastic Differential Equations in Infinite Dimensional Spaces.* IMS Lecture Notes – Monograph Series, Vol. 26, 1995

[78] Kang, S.-J.: Heat and Poisson equations associated with number operator in white noise analysis; *Soochow J. Math.* **20** (1994) 45–55

[79] Khandekar, D. C. and Streit, L.: Constructing the Feynman integrand; *Annalen der Physik* **1** (1992) 49–55

[80] Kondratiev, Yu. G.: Nuclear spaces of entire functions in problems of infinite-dimensional analysis; *Soviet Math. Dokl.* **22** (1980) 588–592

[81] Kondratiev, Yu. G., Leukert, P., Potthoff, J., Streit, L., and Westerkamp, W.: Generalized functionals in Gaussian spaces–the characterization theorem revisited; *Preprint* (1995), to appear in J. Funct. Anal.

[82] Kondratiev, Yu. G. and Samoilenko, Yu. S.: Integral representation of generalized positive definite kernels of an infinite number of variables; *Soviet Math. Dokl.* **17** (1976) 517–521

[83] Kondratiev, Yu. G. and Streit, L.: A remark about a norm estimate for white noise distributions; *Ukrainean Math. J.* **44** (1992) 832–835

[84] Kondratiev, Yu. G. and Streit, L.: Spaces of white noise distributions: Constructions, Descriptions, Applications. I; *Reports on Math. Phys.* **33** (1993) 341–366

[85] Krée, P.: Solutions faibles d'équations aux dérivées fonctionnelles, I; *Lecture Notes in Math.* **410** (1974) 142–181, Springer-Verlag

[86] Krée, P.: Solutions faibles d'équations aux dérivées fonctionnelles, II; *Lecture Notes in Math.* **474** (1975) 16–47, Springer-Verlag

[87] Krée, P.: Calcul d'intégrales et de dérivées en dimension infinie; *J. Funct. Anal.* **31** (1979) 150–186

[88] Kubo, I.: Itô formula for generalized Brownian functionals; *Lecture Notes in Control and Information Scis.* **49** (1983) 156–166, Springer-Verlag

[89] Kubo, I.: A direct setting of white noise calculus; *Pitman Research Notes in Math. Series* **310** (1994) 152–166, Longman Scientific & Technical

[90] Kubo, I. and Kuo, H.-H.: Finite dimensional Hida distributions; *J. Funct. Anal.* **128** (1995) 1–47

[91] Kubo, I. and Kuo, H.-H.: A simple proof of Hida distribution characterization theorem; *Exploring Stochastic Laws, Festschrift in Honor of the 70th Birthday of V. S. Korolyuk*, A. V. Skorokhod & Yu. V. Borovskikh (eds.) (1995) 243–250, International Science Publishers

[92] Kubo, I. and Takenaka, S.: Calculus on Gaussian white noise I; *Proc. Japan Acad.* **56A** (1980) 376–380

[93] Kubo, I. and Takenaka, S.: Calculus on Gaussian white noise II; *Proc. Japan Acad.* **56A** (1980) 411–416

[94] Kubo, I. and Takenaka, S.: Calculus on Gaussian white noise III; *Proc. Japan Acad.* **57A** (1981) 433–437

[95] Kubo, I. and Takenaka, S.: Calculus on Gaussian white noise IV; *Proc. Japan Acad.* **58A** (1982) 186–189

[96] Kubo, I. and Yokoi, Y.: A remark on the space of testing random variables in the white noise calculus; *Nagoya Math. J.* **115** (1989) 139–149

[97] Kuo, H.-H.: Integration by parts for abstract Wiener measures; *Duke Math. J.* **41** (1974) 373–379

[98] Kuo, H.-H.: *Gaussian Measures in Banach Spaces.* Lecture Notes in Math. **463**, Springer-Verlag, 1975

[99] Kuo, H.-H.: Potential theory associated with Uhlenbeck-Ornstein process; *J. Funct. Anal.* **21** (1976) 63–75

[100] Kuo, H.-H.: The chain rule for differentiable measures; *Studia Math.* **63** (1978) 145–155

[101] Kuo, H.-H.: Differential calculus for measures on Banach spaces; *Lecture Notes in Math.* **644** (1978) 270–285, Springer-Verlag

[102] Kuo, H.-H.: Integration in Banach spaces; *Notes in Banach Spaces*, H. Elton Lacey (ed.) (1980) 1-38, University of Texas Press, Austin

[103] Kuo, H.-H.: On Fourier transform of generalized Brownian functionals; *J. Multivariate Analysis* **12** (1982) 415–431

[104] Kuo, H.-H.: Donsker's delta function as a generalized Brownian functional and its application; *Lecture Notes in Control and Information Sciences* **49** (1983) 167–178, Springer-Verlag

[105] Kuo, H.-H.: Fourier-Mehler transforms of generalized Brownian functionals; *Proc. Japan Acad.* **59A** (1983) 312–314

[106] Kuo, H.-H.: Brownian functionals and applications; *Acta Appl. Math.* **1** (1983) 175–188

[107] Kuo, H.-H.: On Laplacian operators of generalized Brownian functionals; *Lecture Notes in Math.* **1203** (1986) 119–128, Springer-Verlag

[108] Kuo, H.-H.: The Fourier transform in white noise calculus; *J. Multivariate Analysis* **31** (1989) 311-327

[109] Kuo, H.-H.: Fourier-Mehler transforms in white noise analysis; *Gaussian Random Fields*, K. Itô and T. Hida (eds.) (1991) 257–271, World Scientific

[110] Kuo, H.-H.: Lectures on white noise analysis; *Soochow J. Math.* **18** (1992) 229–300

[111] Kuo, H.-H.: Convolution and Fourier transform of Hida distributions; *Lecture Notes in Control and Information Sciences* **176** (1992) 165–176, Springer-Verlag

[112] Kuo, H.-H.: Analysis of white noise functionals; *Soochow J. Math.* **20** (1994) 419–464

[113] Kuo, H.-H., Obata, N., and Saitô, K.: Lévy Laplacian of generalized functionals on a nuclear space; *J. Funct. Anal.* **94** (1990) 74–92

[114] Kuo, H.-H. and Potthoff, J.: Anticipating stochastic integrals and stochastic differential equations; *White Noise Analysis–Mathematics and Applications*, T. Hida et al. (eds.) (1990) 256–273, World Scientific

[115] Kuo, H.-H., Potthoff, J., and Streit, L.: A characterization of white noise test functionals; *Nagoya Math. J.* **121** (1991) 185–194

[116] Kuo, H.-H. and Russek, A.: White noise approach to stochastic integration; *J. Multivariate Analysis* **24** (1988) 218–236

[117] Kuo, H.-H. and Shieh, N. R.: A generalized Itô's formula for multidimensional Brownian motions and its applications; *Chinese J. Math.* **15** (1987) 163–174

[118] Lascheck, A., Leukert, P., Streit, L., and Westerkamp, W.: More about Donsker's delta function; *Soochow J. Math.* **20** (1994) 401–418

[119] Lee, K. Sim: On the characterization of finite dimensional Hida distributions; Ph. D. dissertation, Louisiana State University, 1993

[120] Lee, Y.-J.: Integral transforms of analytic functions on abstract Wiener spaces; *J. Funct. Anal.* **47** (1983) 153–164

[121] Lee, Y.-J.: Unitary operators on the space of L^2-functions over abstract Wiener spaces; *Soochow J. Math.* **13** (1987) 165–174

[122] Lee, Y.-J.: Analytic version of test functionals, Fourier transform and a characterization of measures in white noise calculus; *J. Funct. Anal.* **100** (1991) 359–380

[123] Lee, Y.-J.: A characterization of generalized functions on infinite-dimensional spaces and Bargman-Segal analytic functions; *Gaussian Random Fields*, K. Itô and T. Hida (eds.) (1991) 272–284, World Scientific

[124] León, J. A. and Protter, P.: Some formulas for anticipative Girsanov transformations; *Chaos Expansions, Multiple Wiener-Itô Integrals and Their Applications*, C. Houdré and V. Pérez-Ableu (eds.) (1994) 267–291, CRC Press

[125] Lévy, P.: *Leçons d'analyse fonctionnelle.* Gauthier-Villars, 1922

[126] Lévy, P.: *Processus Stochastiques et Mouvement Brownien.* Gauthier-Villars, 1948

[127] Lévy, P.: *Problèmes Concrets d'Analyse Fonctionnelle*. Gauthier-Villars, 1951

[128] Lévy, P.: Random functions: a Laplacian random function depending on a point of Hilbert space; *Univ. Calif. Publ. Statistics* **2** (1956) 195–206

[129] Lindstrøm, T., Øksendal, B., and Ubøe, J.: Stochastic differential equations involving positive noise; *Stochastic Analysis*, M. Barlow and N. Bingham (eds.) (1991) 261–303, Cambridge University Press

[130] Lindstrøm, T., Øksendal, B., and Ubøe, J.: Stochastic modelling of fluid flow in porous media; *Preprint* (1991)

[131] McKean, H. P.: *Stochastic Integrals*. Academic Press, 1969

[132] Meyer, P. A. and Yan, J. A.: A propos des distributions sur l'espace de Wiener; *Lecture Notes in Math.* **1247** (1987) 8–26, Springer-Verlag

[133] Nelson, E.: Feynman integrals and the Schrödinger equation; *J. Math. Phys.* **5** (1964) 332–343

[134] Nualart, D.: *The Malliavin Calculus and Related Topics*. Springer-Verlag, 1995

[135] Nualart, D. and Pardoux, E.: Stochastic calculus with anticipating integrands; *Probability Theory and Related Fields* **78** (1988) 535–581

[136] Obata, N.: Analysis of the Lévy Laplacian; *Soochow J. Math.* **14** (1988) 105–109

[137] Obata, N.: The Lévy Laplacian and the mean value theorem; *Lecture Notes in Math.* **1379** (1989) 242–253, Springer-Verlag

[138] Obata, N.: A characterization of the Lévy Laplacian in terms of infinite dimensional rotation groups; *Nagoya Math. J.* **118** (1990) 111–132

[139] Obata, N.: Rotation-invariant operators on white noise functionals; *Math. Z.* **210** (1992) 69–89

[140] Obata, N.: An analytic characterization of symbols of operators on white noise functionals; *J. Math. Soc. Japan* **45** (1993) 421–445

[141] Obata, N.: Operator calculus on vector-valued white noise functionals; *J. Funct. Anal.* **121** (1994) 185–232

[142] Obata, N.: *White Noise Calculus and Fock Space*. Lecture Notes in Math. **1577**, Springer-Verlag, 1994

[143] Obata, N.: Integral kernel operators on Fock space and quantum Hitsuda-Skorokhod integrals; *Preprint* (1994)

[144] Obata, N.: Conditional expectation in classical and quantum white noise calculi; *Analysis of Operators on Gaussian Space and Quantum Probability Theory*, RIMS Kokyuroku **923** (1995) 154–190

[145] Ocone D.: Anticipating stochastic calculus and applications; *White Noise Analysis-Mathematics and Applications*, T. Hida et al. (eds.) (1990) 298–314, World Scientific

[146] Ocone D. and Pardoux, E.: A generalized Itô-Wentsel formula: Application to a class of anticipating stochastic differential equations; *Ann. Instit. Henri Poincaré* **25** (1989) 39–71

[147] Ogawa, S.: Quelques propriétés de l'integrale stochastique du type noncausal; *Japan J. Appl. Math.* **1** (1984) 405–416

[148] Pardoux, E. and Protter, P.: Stochastic Volterra equations with anticipating coefficients; *Annals of Probability* **18** (1990) 1635–1655

[149] Polishchuk, E. M.: *Continual Means and Boundary Value Problems in Function Spaces*. Birkhäuser, 1988

[150] Potthoff, J. and Streit, L.: A characterization of Hida distributions; *J. Funct. Anal.* **101** (1991) 212–229

[151] Potthoff, J. and Streit, L.: Generalized Radon-Nikodym derivatives and Cameron-Martin theory; *Gaussian Random Fields*, K. Itô and T. Hida (eds.) (1991) 320–331, World Scientific

[152] Potthoff, J. and Yan, J. A.: Some results about test and generalized functionals of white noise; *Probability Theory*, L. H. Y. Chen et al. (eds.) (1992) 121–145, Walter de Gruyter & Co.

[153] Rauch, J.: *Partial Differential Equations.* Springer-Verlag, 1991

[154] Redfern, M.: White noise approach to multiparameter stochastic integration; *J. Multivariate Analysis* **37** (1991) 1–23

[155] Reed M. and Simon, B.: *Methods of Modern Mathematical Physics I: Functional Analysis.* Academic Press, 1972

[156] Reed M. and Simon, B.: *Methods of Modern Mathematical Physics II: Fourier Analysis, Self-adjointness.* Academic Press, 1975

[157] Saitô, K.: Itô's formula and Lévy's Laplacian; *Nagoya Math. J.* **108** (1987) 67–76

[158] Saitô, K.: Lévy's Laplacian in the infinitesimal generator; *Research Report, Meijo Univ.* **28** (1988) 1–5

[159] Saitô, K.: Itô's formula and Lévy's Laplacian II; *Nagoya Math. J.* **123** (1991) 153–169

[160] Saitô, K.: A group generated by the Lévy Laplacian and the Fourier-Mehler transform; *Pitman Research Notes in Math. Series* **310** (1994) 274–288, Longman Scientific & Technical

[161] Schneider, W. R.: Grey noise; *Stochastic Processes, Physics and Geometry*, S. Albeverio et al. (eds.) (1990) 676–681, World Scientific

[162] Segal, I. E.: Tensor algebras over Hilbert spaces; *Trans. Amer. Math. Soc.* **81** (1956) 106–134

[163] Segal, I. E.: Mathematical characterization of the physical vacuum for a linear Bose-Einstein field; *Illinois J. Math.* **6** (1962) 500–523

[164] Segal, I. E.: The complex wave representation of the free Boson field; *Topics in Functional Analysis: Essays Dedicated to M. G. Krein on the Occasion of His 70th Birthday*, Advances in Math.: Supplementary Studies, I. Gohberg and M. Kac (eds.), Vol. 3 (1978) 321–344, Academic Press

[165] Shieh, N. R. and Yokoi, Y.: Positivity of Donsker's delta function; *White Noise Analysis-Mathematics and Applications*, T. Hida et al. (eds.) (1990) 374–382, World Scientific

[166] Si Si: Variational calculus for Lévy's Brownian motion; *Gaussian Random Fields*, K. Itô and T. Hida (eds.) (1991) 364–373, World Scientific

[167] Skorokhod, A. V.: On a generalization of a stochastic integral; *Theory Probab. Appl.* **20** (1975) 219–233

[168] Streit, L. and Hida T.: Generalized Brownian functionals and the Feynman integral; *Stochastic Processes and Their Applications* **16** (1983) 55–69

[169] Sugita, H.: Positive generalized Wiener functions and potential theory over abstract Wiener spaces; *Osaka J. Math.* **25** (1988) 665–696

[170] Üstünel, A. S.: La formule de changement de variables pour l'integrale anticipante de Skorokhod; *C. R. Acad. Sci. Paris* **303** (1986) 329–331

[171] Watanabe, H.: The local time of self-intersections of Brownian motions as generalized Brownian functionals; *Lett. Math. Phys.* **23** (1991) 1–9

[172] Watanabe, H.: Donsker's δ-function and its applications in the theory of white noise analysis; *Stochastic Processes*, a Festschrift in Honor of G. Kallianpur, S. Cambanis et al. (eds.) (1993) 337–339, Springer-Verlag

[173] Watanabe, S.: Malliavin calculus in terms of generalized Wiener functionals; *Lecture Notes in Control and Information Sciences* **49** (1983) 284–290, Springer-Verlag

[174] Watanabe, S.: Donsker's δ-functions in the Malliavin calculus; *Stochastic Analysis, Liber Amicorum for Moshe Zakai*, E. Mayer-Wolf et al. (eds.) (1991) 495–502, Academic Press

[175] Watanabe, S.: Some refinements of Donsker's delta functions; *Pitman Research Notes in Math. Series* **310** (1994) 308–324, Longman Scientific & Technical

[176] Wiener, N.: Differential spaces; *J. Math. Physics* **58** (1923) 131–174

[177] Wiener, N.: The homogeneous chaos; *Amer. J. Math.* **60** (1938) 897–936

[178] Yan, J. A.: Sur la transformée de Fourier de H. H. Kuo; *Lecture Notes in Math.* **1372** (1989) 393–394, Springer-Verlag

[179] Yokoi, Y.: Positive generalized white noise functionals; *Hiroshima Math. J.* **20** (1990) 137–157

[180] Zhang, Y.: The Lévy Laplacian and Brownian particles in Hilbert spaces; *J. Funct. Anal.* **133** (1995) 425–441

List of Symbols

$\mathcal{A}_{p,\beta}^*$	Dual space of $\mathcal{A}_{p,\beta}$	327
$B(t)$	Brownian motion	1
$\dot{B}(t)$	White noise	1, 21
$\widetilde{D}_\eta$	Extension of D_η for $\eta \in \mathcal{E}$	109
$d\mathbf{t}$	$d\nu(t_1) \cdots d\nu(t_n)$	121
$\mathcal{D}_L$	The set of L-functionals	207
$\mathcal{D}_V$	The set of V-functionals	224
$\mathrm{Dom}(\Delta_L)$	Domain of Δ_L	201
D_y	Gâteaux derivative in the y-direction	103
D_y^*	Adjoint operator of D_y	111
E	A basic Hilbert space	25
$\mathcal{E}$	projective limit of $\{\mathcal{E}_p\}$	25
$\mathcal{E}'$	Dual space of $\mathcal{E}$	25
$(\mathcal{E})$	Space of test functions with $\beta = 0$	27
$(\mathcal{E})^*$	Space of generalized functions with $\beta = 0$	27
$(\mathcal{E})_\beta$	Space of test functions	30
$(\mathcal{E})_\beta^*$	Space of generalized functions	30
$\mathcal{E}_p$	$\{\xi; |\xi|_p < \infty\}$	25
$\mathcal{E}_p'$	Dual space of $\mathcal{E}_p$	26
$(\mathcal{E}_p)$	$\{\varphi; \|\varphi\|_p < \infty\}$	27
$(\mathcal{E}_p)^*$	Dual space of $(\mathcal{E}_p)$	27
$(\mathcal{E}_p)_\beta$	$\{\varphi; \|\varphi\|_{p,\beta} < \infty\}$	30
$(\mathcal{E}_p)_\beta^*$	Dual space of $(\mathcal{E}_p)_\beta$	30
$\mathcal{F}$	Fourier transform	149
$\mathcal{F}_\theta$	Fourier-Mehler transform	178
F_L''	Lévy part of F	207
F_r''	Regular part of F	207
$\mathbb{F}_{V,f}$	Feynman integrand	310
$\mathcal{G}_\theta$	$\mathcal{G}_\theta^* = \mathcal{F}_\theta$	181
$\mathcal{G}_{a,b}$	Fourier-Gauss transform	164
$g_{x,c}$	Gaussian white noise function	83
$H_n(x)$	Hermite polynomials of degree n	17
$\mathbb{I}_{V,f}$	Feynman integral	310

Index